T0340193

Safe Robot Navigation Among Moving and Steady Obstacles

Safe Robot Navigation Among Moving and Steady Obstacles

Alexey S. Matveev
Andrey V. Savkin
Michael Hoy
Chao Wang

AMSTERDAM • BOSTON • HEIDELBERG • LONDON
NEW YORK • OXFORD • PARIS • SAN DIEGO
SAN FRANCISCO • SINGAPORE • SYDNEY • TOKYO
Butterworth-Heinemann is an imprint of Elsevier

Butterworth Heinemann is an imprint of Elsevier
The Boulevard, Langford Lane, Kidlington, Oxford OX5 1GB, UK
225 Wyman Street, Waltham, MA 02451, USA

Library of Congress Cataloging-in-Publication Data
A catalog record for this book is available from the Library of Congress

British Library Cataloguing in Publication Data
A catalogue record for this book is available from the British Library

For information on all Butterworth Heinemann publications
visit our website at http://store.elsevier.com/

ISBN: 978-0-12-803730-0

Publisher: Joe Hayton
Acquisition Editor: Sonnini Yura
Editorial Project Manager: Mariana Kuhl Leme
Editorial Project Manager Intern: Ana Claudia A. Garcia
Production Project Manager: Sruthi Satheesh
Marketing Manager: Louise Springthorpe
Cover Designer: Victoria Pearson

Contents

Preface

Collision-free navigation of mobile robots in environments cluttered with both static and dynamic obstacles is a fundamental problem of robotics. In order to safely operate in a priori unknown complex environments, autonomous unmanned vehicles should be equipped with an automatic navigation system that allows them to avoid collisions with obstacles. Moreover, the capacity of safe navigation among moving and steady obstacles is a key issue for most-real life applications of mobile robotics. Despite extensive research, this fundamental problem still represents a real challenge. This book is primarily a research monograph that presents, in detail and in a unified manner, recent advancements in this area. In doing so, the book extends its scope to obstacles undergoing general motions, including rotations and deformations, i.e., changes of the shape and size, and gives considerable attention to reactive algorithms and rigorous mathematical studies of the proposed navigation solutions, showing the interplay between mathematics and robotics. At the same time, the theoretical results are systematically confirmed via computer simulation studies and real-world experimental validation of the developed algorithms with various mobile robots.

The main intended audience for this monograph encompasses students at postgraduate and graduate levels, as well as professional researchers and industry practitioners working in a variety of areas, notably robotics, control engineering, computer sciences, and applied mathematics, who have an interest in the fascinating and cutting-edge discipline of mobile robotics.

This book is essentially self-contained; the only prerequisite is familiarity with basic undergraduate level mathematics. The theories and algorithms presented are discussed fully and illustrated by examples; the proposed navigation solutions are computationally efficient and can be easily instrumented in practice. Therefore, some basic experience in robotics and programming is enough to implement these algorithms directly from the book. We hope readers find this volume useful, clear, and interesting, gain a deeper insight into the challenging field of collision-free robot navigation, and are encouraged to explore it further by following a similar road.

The material in this book is the result of the joint research of the authors from 2009 to 2015. Some of its parts have separately appeared in journal and conference papers. The manuscript integrates them into a unified whole, highlights their coherence and connections among them, supplements them with new original findings of the authors, and presents the entire material in a systematic way.

In preparation of this work, the authors wish to acknowledge the financial support they received from the Australian Research Council; Alexey Matveev gratefully acknowledges the support from RSF 14-21-00041 in preparation of the book except for Chapter 7, and the support by grant 6.38.230.2015 from the Saint Petersburg State University in preparation of Chapter 7. The authors are also grateful for the support they received throughout the production of this book from the School of Electrical Engineering and Telecommunications at the University of New South Wales, Sydney, Australia and the Faculty of Mathematics and Mechanics at the Saint Petersburg State University, Saint Petersburg, Russia.

Furthermore, Alexey Matveev is grateful for the enormous support he has received from his wife Elena and daughter Julia. Andrey Savkin is indebted to the endless love and support he has received from his wife Natalia, son Mikhail, and daughter Katerina. Michael Hoy is thankful for the continuous support of his loving fiancée Reneé. Chao Wang is thankful for the support he has received from his family and friends.

Alexey S. Matveev
Andrey V. Savkin
Michael Hoy
Chao Wang

Abbreviations

AM avoidance maneuver

APF artificial potential field

DMPC decentralized model predictive control

DRCA distributed reactive collision avoidance

GPT generator of planned trajectories

GPPT generator of presumable planned trajectories

GRPPT generator of refined presumable planned trajectories

IT initial turn

LiDAR light detection and ranging

LQG linear quadratic Gaussian

LTM longitudinal tracking module

MPC model predictive control

NLVO nonlinear velocity obstacle

PD proportional derivative

PFM potential field method

PGL proposed guidance law

PTM path-tracking module

RCA reciprocal collision avoidance

RHC receding horizon control

RPI robustly positively invariant

SMEC sliding motion along the equidistant curve

SMT straight motion to the target

TPM trajectory planning module

TTM trajectory tracking module

UAV unmanned aerial vehicle

VO velocity obstacle

VOA velocity obstacle approach

VSS variable structure system

i.e. that is (*Latin: id est*)

e.g. for example (*Latin: exempli gratia*)

etc. and so forth, and so on, continuing in the same way (*Latin: etcetera*)

Frequently used notations

$:=$	"is equal by definition, is set to be"
$\Rightarrow$	entailment: "$A \Rightarrow B$" means A implies B
$\Leftrightarrow$	equivalence: "$A \Leftrightarrow B$" means A equivalent to B
$\equiv$	identity: "$A \equiv B$" means A and B are identical, i.e., are the same
$\wedge$	logical "and"
$\vee$	logical "or"
$\forall$	"for all"
$\exists$	"exists, there exists"
$\overset{\mathbf{A}}{=}$	"the relation is true by virtue of A"; here $=$ can be replaced by $\leq, \Rightarrow$ and any other relation
$\{a_1, \ldots, a_n\}$	the set composed by the elements $a_1, \ldots, a_n$
$\emptyset$	the empty set
$\{a : C(a)\}$	the set of all a for which the claim $C(a)$ is true; e.g., $\{a : a \leq 1\}$ is the set of all a not exceeding 1
$A \cup B$	the union of the sets A and B
$A \cap B$	the intersection of the sets A and B
$A \setminus B$	the difference of the sets A and B: the set of all points from A that do not belong to B
$f(\cdot)$	dot in parentheses signals that the preceding symbol denotes a function
$\mathbb{R}$	the field of real numbers
$\mathbf{T}$	transposition
$\mathbf{col}\{a_1, \ldots, a_n\}$	the column composed of the listed elements: $\mathbf{col}\{a_1, \ldots, a_n\} = (a_1, \ldots, a_n)^{\mathbf{T}}$
$\mathbb{R}^n$	the Euclidean space of all column-vectors $\boldsymbol{a} = \mathbf{col}\{a_1, \ldots, a_n\}$ with real entries
$\langle \cdot ; \cdot \rangle$	the standard inner product in a Euclidean space: $\langle \boldsymbol{a}; \boldsymbol{b} \rangle := \sum_{i=1} a_i b_i, \quad \boldsymbol{a} = \mathbf{col}\{a_1, \ldots, a_n\}, \boldsymbol{b} = \mathbf{col}\{b_1, \ldots, b_n\}$
$\Vert \cdot \Vert$	the standard Euclidean norm: $\Vert \boldsymbol{a} \Vert := \sqrt{\langle \boldsymbol{a}; \boldsymbol{a} \rangle}$
$\max\{a_1, \ldots, a_n\}$	the maximum among the listed numbers
$\min\{a_1, \ldots, a_n\}$	the minimum among the listed numbers
sgn	the signum function: $\mathbf{sgn}\, a = 1$ if $a > 0$, $\mathbf{sgn} a = -1$ if $a < 0$, and $\mathbf{sgn}\, 0 = 0$

$[a]_+$	the positive portion of a real number a: $[a]_+ := a$ if $a \geq 0$, otherwise $[a]_+ := 0$
$[a]_-$	the negative portion of a real number a: $[a]_- := -a$ if $a \leq 0$, otherwise $[a]_- := 0$
$\lceil a \rceil$	the integer ceiling of a real number a: the least integer that is no less than a
$\lfloor a \rfloor$	the integer floor of a real number a: the largest integer that is no greater than a
$a = b \mod c$	this means that $a - b$ is divisible by c
$a_i \uparrow$	"the sequence is nondecreasing": $a_{i+1} \geq a_i \quad \forall i$;
$a_i \uparrow \infty$	"the sequence is nondecreasing and grows without limits"
$t \approx \tau$	"t is sufficiently close to τ"
$t \approx \infty$	"t is sufficiently large"
$t \to \tau-$	"t converges to τ from the left"
$t \to \tau+$	"t converges to τ from the right"
$[\boldsymbol{a}, \boldsymbol{b}] \subset \mathbb{R}^n$	the closed segment with end-points $\boldsymbol{a}$ and $\boldsymbol{b}$: $[\boldsymbol{a}, \boldsymbol{b}] = \{\boldsymbol{x} : \boldsymbol{x} = (1-\theta)\boldsymbol{a} + \theta\boldsymbol{b} \quad \text{for some} \quad 0 \leq \theta \leq 1\}$
$(\boldsymbol{a}, \boldsymbol{b}) \subset \mathbb{R}^n$	the open segment with end-points $\boldsymbol{a}$ and $\boldsymbol{b}$: $(\boldsymbol{a}, \boldsymbol{b}) = \{\boldsymbol{x} : \boldsymbol{x} = (1-\theta)\boldsymbol{a} + \theta\boldsymbol{b} \quad \text{for some} \quad 0 < \theta < 1\}$
$(\boldsymbol{a}, \boldsymbol{b}] \subset \mathbb{R}^n$	the semi-open segment with end-points $\boldsymbol{a}$ and $\boldsymbol{b}$: $(\boldsymbol{a}, \boldsymbol{b}] = \{\boldsymbol{x} : \boldsymbol{x} = (1-\theta)\boldsymbol{a} + \theta\boldsymbol{b} \quad \text{for some} \quad 0 < \theta \leq 1\}$
$\mathbf{dist}\,[\boldsymbol{r}; D]$	the distance from the point $\boldsymbol{r}$ to the set D: $\mathbf{dist}\,[\boldsymbol{r}; D] := \inf_{\boldsymbol{r}_* \in D} \|\boldsymbol{r} - \boldsymbol{r}_*\|$
$\mathbf{dist}\,[D', D'']$	the distance between two sets D', D'': $\mathbf{dist}\,[D', D''] := \inf_{\boldsymbol{r}' \in D', \boldsymbol{r}'' \in D''} \|\boldsymbol{r}' - \boldsymbol{r}''\|$
∂D	the boundary of the set D
C^k-smooth	k times continuously differentiable
smooth	C^k-smooth with large enough k

Introduction

1

1.1 Collision-free navigation of wheeled robots among moving and steady obstacles

Autonomous navigation of unmanned vehicles is a classic research area in robotics, which gives rise to a whole variety of approaches well documented in literature. Both single and multiple coordinated mobile robots offer great perspectives in many applications, such as industrial, office, and agricultural automation; search and rescue; and surveillance and inspection (we refer the reader to [1, 2] for a more comprehensive list of potential applications). For example, because of lightweights, inexpensive components, and low power consumptions, unmanned aerial and ground vehicles have been used greatly for plenary surveillance and for a variety of applications in hazardous and complex environments to mitigate the risk for humans; see, e.g., [3–6] and references therein. Such applications often involve limitations on communications that require the robotic vehicle to operate autonomously for extended periods of time and distances. In such situations, unmanned vehicles should be equipped with an automatic navigation system by which they can move autonomously and safely operate in populated environments. In all these cases, navigation involves a series of common problems, with collision avoidance in some form being almost universally required.

Moreover, the capacity of safely operating in dynamic and a priori unknown environments is a key issue for most real-life applications of mobile robotics. Despite extensive research, this fundamental problem still represents a real challenge, mostly because of the numerous uncertainties inherent in typical scenarios. This challenge can be much enhanced by deficiency in perception abilities and computational power of the robot, as well as by restrictions on its mobility due to nonholonomic kinematic constraints, limited control range, and under-actuation.

In order to operate in a cluttered environment, an autonomous unmanned vehicle should be able to detect and avoid the enroute obstacles. Typical objectives include, but are not limited to, overall movement in a given direction or reaching a target point through the obstacle-free part of the environment [7]. This may involve bypassing an obstacle, especially a long one, in close range within a safety margin [8]. This maneuver is similar to border patrolling, which mission is of self-interest for many applications. Substantial effort has been made in robotics research for solution of these problems. The rich variety of available relevant algorithms will be specifically surveyed in a special Chapter 3. With a focus on the planning horizon, they can be very broadly classified into global and local planners [9].

Safe Robot Navigation Among Moving and Steady Obstacles. http://dx.doi.org/10.1016/B978-0-12-803730-0.00001-9

Global sensor-based planners use both a priori and sensory information to build a comprehensive model of the environment and then try to find a completed and best possible navigation solution on the basis of this model [7, 10–16]. In effect, this means the environment is assumed to be known to a substantial extent prior to the solution of a navigation task. Within this framework, a variety of techniques has been developed. Their general survey is postponed until Chapter 3 but can also be found in, e.g., [17, 18]; some samples intended to handle dynamic scenes are given by velocity obstacles [17, 19], nonholonomic planners [20], and state-time space [21–23] approaches. Global planners can often be accompanied with guarantees of not only collision avoidance but also achieving a global navigation objective provided that certain general assumptions about the scene are satisfied. On the negative side, global planners are, by and large, computationally expensive and hardly suit real-time implementation. NP-hardness, the mathematical seal for intractability, was established for even the simplest problems of dynamic motion planning [24]. A partial remedy was offered in the form of randomized architectures [25, 26]. At the same time, all global planners are hardly troubled, up to failure in path generation, by unpredictability of the scene, as well as by data incompleteness and erroneousness typical for onboard perception. These disadvantages are shared by hybrid approaches that use global planning as a backbone of navigation [15, 20, 27–30].

Conversely, local planners use onboard sensory data about only a nearest fraction of an unknown environment for iterative re-computation of a short-horizon path [9, 31–33]. This reduces the computational burden toward implementability in real time but makes the ultimate result of iterations an open issue. A detailed overview of the respective obstacle avoidance techniques will be given in Chapter 3. Many of them, e.g., the dynamic window [34, 35], curvature velocity [36], lane curvature [37], boundary following [32, 33], and tangent graph based [16] approaches, treat the obstacles as static. This is a particular case of predictably moving obstacles, which are assumed by methods like velocity obstacles [17, 19], collision cones [38], or inevitable collision states [39, 40]. These methods tend to be computationally expensive, are based on access to the obstacles' full velocities, and assume a modest or minor rate of their change. However, velocity estimation remains a challenging task in practice, and predictability of the scene ranges from full to none in the real world [41]. A medium level of predictability is that with uncertainty, where non-conservative estimates of future obstacle positions can be put in place of exact prognosis [17, 42]. However, these and other approaches [19, 38–40] take in effect excessive precautions against collisions with obstacles. As a result, they may be stuck in cluttered scenes and tend toward bypassing dense clusters of obstacles as a whole, even if a better and sometimes the only option is a permeating route. In hardly predictable complex environments, safety typically concerns only a nearest future, and its propagation until the end of the experiment is not guaranteed [42]. Some local planners, such as Virtual Force Field [43], Potential Field [44, 45], Vector Field Histogram [46], Certainty Grid [47], Nearness Diagram [48] methods, use elements of global modeling by assuming awareness about the scene above the level given by the current sensory data. A common problem with local methods is that many of them are heuristic and not based on mathematical models such as kinematic equations of the vehicles and nonholonomic constraints on their motion, which is a severe limitation in practice. For dynamic environments, fully actuated robots were mostly studied up to now.

Because of inevitable failure scenarios, a common deficiency of the previous research on local planners is the lack of global convergence results that guarantee achieving the primary objective in dynamic environments [49]. At best, rigorous analysis examined an isolated bypass of an obstacle during which the other obstacles were neglected until the bypass end, with an idea that thereafter, the robot focuses on the main goal. However, in cluttered dynamic scenes, bypasses may be systematically intervened by companion obstacles so that no bypass is completed, whereas the robot almost constantly performs obstacle avoidance. Ultimate goal was left, by and large, beyond the scope of theoretical analysis, especially for cluttered unpredictable environments, like a dense crowd of people. However, it is in these cases that rigorous quantitative delineation between failure and success scenarios is highly important since by its own right, any experimentation is not convincing enough due to horizonless diversity of feasible scenarios. Another deficiency is that moving obstacles were viewed as rigid bodies undergoing only translational motions and often of the simplest shapes (e.g., discs [45, 50–52] or polygons [53, 54]). Finally, assumed awareness of the obstacles often meant access to their possibly "invisible" parts (in order to determine, e.g., the disc center [45, 50, 51] or angularly most distant polygon vertex [54]) or full velocity [45, 51, 54].

The basic strong and weak points of local planners attain apotheosis at reactive controllers. For them, the planning horizon collapses into a point so that the controller directly converts the current observation into the current control. Purely reactive approaches are exemplified by [51, 55–59], as well as by biologically inspired methods [32, 33, 52, 60, 61]. Examples concerning nonholonomic robots include artificial potential approach, combined with sliding mode control for gradient climbing [50, 53], and kinematic control based on polar coordinates and Lyapunov-like analysis [51]. Some local planners, such as Virtual Force Field [43], Potential Field [44, 45], Vector Field Histogram [46], Certainty Grid [47], Nearness Diagram [62] methods, in fact combine reactive control with elements of global modeling by assuming awareness about the scene above the level given by the snapshot of the sensory data. Another example of such local path planners is the biologically inspired Bugs family of algorithms [11, 12, 63, 64] motivated by bugs crawling along a wall.

This monograph is aimed at overcoming some of the afore-mentioned deficiencies in the previous research on navigation of mobile robots among obstacles. It deals with both steady and moving obstacles; the scope of the book extends to not only arbitrarily shaped obstacles but also obstacles undergoing arbitrary motions, including translations, rotations, and deformations, i.e., changes of sizes and shapes. Except for Chapter 4, where elements of the global planning approach are employed to handle global optimization issues, this research text develops local techniques, with a main focus on reactive algorithms. Unlike many other publications in robotics, the proposed navigation solutions are based on rigorous mathematical theory, and systematic theoretical justification of their convergence and performance can be viewed as a novel and distinguishing feature of this book. Mostly, its chapters deliver a consistent message that even purely reactive navigation laws have a capacity of being supplied with mathematically rigorous guarantees of achieving a global navigation objective. Though we sometimes refer the reader to textbooks for technical mathematical facts, important theoretical developments are given with care.

The theoretical part of the work presented in this monograph is confined to models conventionally used to describe wheeled robots. However, the results of this part extend in effect on any controlled system obeying the same governing equations; some typical but not exhaustive examples include, under certain circumstances, fixed wing aircrafts, torpedo-like underwater vehicles, and spacecrafts. Principal solutions are discussed with respect to these basic models, which imparts the feature of "hardware independency" to the major messages from this book. At the same time, experimental validation of the proposed navigation solutions is a crucial part in robotics research. In this book, the proposed navigation algorithms were tested with real wheeled unmanned ground vehicles such as laboratorial wheeled robots Pioneer P3-DX, an intelligent autonomous wheelchair, and an autonomous wheeled mobile hospital bed. A detailed description of the employed experimental equipment will be offered in Section 1.4.

Another essential trait of this monograph is systematic account for motion constraints inherent in real-world wheeled robots. Their mathematical models employed in most of the technical chapters explicitly incorporate nonholonomic kinematics, limited control range, and under-actuation; which substantially intensifies the challenge of safe navigation among obstacles. The defiance from this triplet of oppressive properties is somewhat reduced only in three of nine chapters: Chapters 8 and 12 deal with fully actuated robots instead of under-actuated ones, whereas only limitations on the control range are taken into account in Chapter 11.

The organization of this book is problem-oriented, not technique-oriented. Each chapter is relatively self-contained and is devoted to detailed discussion of a certain problem that arises in the rapidly developing area of mobile robotics. On the other hand common ideas and methods thread through chapters so that the entire book can be viewed as an "advocate" for their usefulness as an effective apparatus of wide applicability. In particular, the monograph is featured by systematic use of switched controllers so that in closed loop, the examined systems fall into the class of hybrid dynamical systems [65–69].

The goal of this monograph is to report on some recent progress in the area of navigation of wheeled robots among steady and moving obstacles, and to present, in an unified form, a computationally efficient, reliable, and easily implementable algorithmic framework for control of mobile robots, so that the ultimate goal of their applications can be achieved with a high degree of certainty. Such a framework is very important in the face of ever-increasing real-world applications of wheeled robots since it carries a potential to transcend technological evolution in mobile robotics.

1.2 Overview and organization of the book

In Chapter 2, we briefly recall a few basic concepts and facts of the classic sliding mode control theory, focusing attention on those that are used in this book.

Chapter 3 provides a review of techniques related to navigation of unmanned vehicles through unknown environments cluttered with obstacles, especially those that rigorously ensure collision avoidance (given certain assumptions about the system). This topic continues to be an active area of research, and we highlight some directions in which available approaches may be improved. The chapter discusses models of the sensors and vehicle kinematics, as well as assumptions about the environment and performance criteria. Methods applicable to stationary obstacles, moving obstacles, and scenarios with multiple vehicles are covered. In preference to global approaches based on full knowledge of the environment, particular attention is given to reactive methods based on only local sensory data, with a special focus on recently proposed navigation laws based on the sliding mode and model predictive control.

Chapter 4 deals with the problem of optimal navigation in the presence of obstacles. Specifically, a Dubins-car-like mobile robot travels among smooth and possibly non-convex obstacles, with a constraint on the curvature of their boundaries. The minimal-length path to a given steady target is explicitly found in the case where the scene is known in advance. Based on this solution, a novel reactive randomized algorithm of robot navigation in an unknown environment is proposed. It is proven that under this navigation law, the robot avoids collisions and reaches the steady target with probability 1. The performance of the algorithm is confirmed by computer simulations and outdoor experiments with a Pioneer P3-DX mobile wheeled robot.

Chapter 5 discusses the problem of reactively navigating a Dubins-car-like robot along an equidistant curve of an environmental object or, otherwise stated, following the boundary of this object with a predefined range margin. Two sensing scenarios are considered. In one of them, navigation is based on access to the distance to the nearest point of the boundary. The other scenario assumes that the distance to the boundary is measured perpendicularly to the robot's centerline, and the angle of incidence of this perpendicular to the boundary is also accessible. This situation holds if, e.g., the measurements are supplied by range sensors rigidly mounted to the robot's body at nearly right angles, or by a single sensor scanning a nearly perpendicular narrow sector. For both scenarios, reactive sliding mode control laws are proposed that drive the robot at a pre-specified distance from the boundary and thereafter maintain this distance. This is achieved without estimation of the boundary curvature and holds for boundaries with both convexities and concavities. Mathematically rigorous analysis of the proposed control laws is provided, including explicit account of the global geometry of the boundary. Computer simulations and experiments with real wheeled robots confirm the applicability and performance of the proposed guidance approach.

In Chapter 6, border-patrolling algorithms are applied to a more general problem of reactively navigating a nonholonomic Dubins-car like robot to a target through an environment cluttered with static obstacles. The proposed navigation strategy combines motions straight to the target, when possible, with bypassing obstacles in a close range with the aid of border patrolling algorithms, and also includes a set of rules regulating switches between these two regimes. Mathematically rigorous analysis of the navigation strategy is provided. Its efficiency is also illustrated by computer simulations.

In Chapter 7, we present a method for guidance of a Dubins-car-like mobile robot with saturated control toward a target in a steady simply connected maze-like environment. The robot always has access to the target relative bearing angle and the distance to the nearest point of the maze if it is within the given sensor range. The proposed control law is composed by biologically inspired reflex-level rules. Mathematically rigorous analysis of this law is provided, and its convergence and performance are confirmed by computer simulations and experiments with real robots.

Chapter 8 introduces and studies a simple biologically-inspired strategy for navigation a Dubins-car-like robot toward a target while avoiding collisions with moving obstacles. A mathematically rigorous analysis of the proposed approach is provided. The performance of the algorithm is demonstrated via experiments with real robots and extensive computer simulations.

Chapter 9 deals with a hard scenario where the environment of operation is cluttered with obstacles that may undergo arbitrary motions, including rotations and deformations, i.e., changes of shapes and sizes. A sliding mode based strategy for a unicycle-like robot navigation and guidance is presented. It is then applied to the problems of patrolling the border of a moving and deforming domain and reaching a target through a dynamic environment cluttered with moving obstacles. Mathematically rigorous analysis of the proposed approach is provided. The convergence and performance of the algorithm are demonstrated via experiments with real robots and extensive computer simulation.

Chapter 10 presents a novel reactive algorithm for collision free navigation of a nonholonomic robot in unknown complex dynamic environments with moving obstacles. The novelty is related to employment of an integrated representation of the information about the environment, which does not require separating obstacles and approximating their shapes by discs or polygons, and is very easy to obtain in practice. Moreover, the presented algorithm does not require any information on the obstacles' velocities. Under the examined navigation algorithm, the robot efficiently seeks a short path through a crowd of moving or steady obstacles, rather than avoiding the crowd as a whole, like many other navigation algorithms do. A mathematically rigorous analysis of the proposed approach is provided. The performance of the algorithm is demonstrated via experiments with a real robot and extensive computer simulations.

Chapter 11 introduces and examines a novel purely reactive algorithm to navigate a planar mobile robot in densely cluttered environments with unpredictably moving and deforming obstacles. The algorithm is supplied with mathematically firm guarantees of not only avoidance of collisions with the obstacles but also achieving the global objective of perpetual drift in the desired direction. The efficiency of the method is illustrated by computer simulations.

While Chapters 4–11 deal with navigation of a single robot, Chapter 12 considers a multiple robot scenario. The objective is to simultaneously navigate several mobile robots with limited sensing and communication capabilities through an unknown static environment. In this chapter, we propose a decentralized, cooperative, reactive, model predictive control based collision avoidance scheme that plans short-range paths in the currently sensed part of the environment, and show that it is able to prevent collisions from occurring. The proposed scheme combines a main path-planning system with an auxiliary controller, which is employed to follow previously planned paths whenever the main system fails to update the path. Simulations and real-world testing in various scenarios confirm the method's validity.

For the convenience of the reader, cross-references between chapters are kept to a minimum. Therefore, numbering of formulas and theorem-like objects (theorems, lemmas, assumptions, etc.) starts over with each new section and uses the section number followed by dot "." and the object's sequential number within the section. For example, Definition 2.1 refers to the first definition from the second section in the current chapter. In rare occasions when a reference to not the current chapter is, however, made, the chapter number is added to the beginning; for example, Assumption 3.4.2 refers to the second assumption in the fourth section from Chapter 3. The section number consists of two component parts separated by dot ".", the figure before the dot refers to the chapter number and the figure after the dot refers to the position of the section within the chapter. Thus, Section 5.2 refers to the second section in Chapter 5. The following template is used for numbering subsections: "the section number"."the position of the subsection within the section". For example, Section 7.3.4 refers to the fourth subsection in Section 7.3. Since most of figures and tables are concentrated in the concluding sections of the chapters, the figures and tables are numbered per chapter, such as Fig. 4.3 referring to the third figure in Chapter 4. Unlike numbering of other items, equation numbers appear within parentheses, e.g., (3.8).

1.3 Sliding mode control

Due to the well-known benefits such as high insensitivity to noises and disturbance rejection feature, robustness against uncertainties and unmodeled dynamics, good dynamic response, good transient performance, reduced complexity of design and tuning, capacity to decouple a highly dimensional problem into a set of lower dimensional ones, ease of implementation, etc. [70], the sliding mode approach attracts a steady stream of interest in the area of motion control (we refer the reader to [71, 72] for detailed surveys). In particular, the origins of sliding mode based motion control with obstacle avoidance can be traced back to early 1990s [73, 74]. The listed benefits of the sliding mode approach is a particular concern for wheeled mobile robots, which are intrinsically subjected to various uncertainties, including, but not limited to, parametric disturbances, such as changes in mass, inertia, and friction, uncertainties of an inhomogeneous terrain, such as unexpected slippery conditions, upward and downward ramps, rough surfaces, tire sliding on ice and rolling on sand, uncertainties of aerodynamic forces, uncertainties in actuators and sensors, e.g., in magnifications factors, focal length, and orientation of a video-camera, etc. Furthermore highly complex, coupled, and nonlinear dynamics of typical wheeled robots entail that the control system design is commonly based on account of not the entire dynamics, leaving noticeable discrepancies between the model and the real dynamics of the robot. Further, the need for precise control and high performance is ever increasing in real-world applications.

The technical part of this book systematically applies the sliding mode approach to various ever open problems in the area of navigation, guidance, and control of wheeled mobile robots, though the status of the concerned sliding mode controller may vary from "primary" down to "secondary" between chapters. General issues on design and analysis of sliding mode controllers are well documented in numerous monographs and textbooks; see, e.g., [70, 75–80] and literature therein. For convenience of the reader, basic technicalities of sliding mode control will be recalled in Chapter 2 to the extent needed in this book.

The major obstacle to implementation of sliding mode controllers is a harmful phenomenon called "chattering," i.e., undesirable finite frequency oscillations around the ideal trajectory due to unmodeled system dynamics and constraints. The problem of chattering elimination and reduction has extensive literature; see, e.g., [78, 81–90] and literature therein. It offers a variety of effective approaches, including continuous approximation of the discontinuity, inserting low-pass filters/observers into the control loop, combining sliding mode and adaptive control or fuzzy logic techniques, higher order sliding modes, etc.

In the research presented in this book, we judged the occurrence of chattering mostly at the stage of real-world experiments. Continuous approximation of the discontinuous control law in a boundary layer [70, 86] was used to eliminate chattering in some cases and as a prophylactic measure against chattering in other cases, whereas there were cases where anti-chattering treatment was not needed. These outcomes are specific to experimental platforms employed in our research. Whether chattering be encountered when implementing the proposed control laws on different platforms, it can be handled in accordance with the treatment recipes offered in the afore-mentioned literature.

1.4 Experimental equipment

In experimental verification of the algorithms developed in this book, the following three types of wheeled mobile robots were used:

- laboratorial wheeled robots Pioneer P3-DX;
- an intelligent autonomous wheelchair system;
- an autonomous hospital bed system.

1.4.1 Laboratorial wheeled robot Pioneer P3-DX

Most of the experiments presented in the book were performed with laboratorial wheeled robots Pioneer P3-DX (abbreviated as P3-DX) (Fig. 1.1). This differential drive robot from "ActivMedia Robotics" is intended for classroom and laboratory use. Its key components that are important for the experiments described in this book are as follows:

Figure 1.1 Pioneer P3-DX mobile robot used for experiments.

- P3-DX has a tough aluminum body, two driving wheels with diameter of 195 mm at the sides of the body, and a small caster wheel at the centerline. The robot is 381 mm wide and 455 mm long; its base platform is 237 mm high and has a weight of 9 kg. P3-DX may carry a payload of up to 23 kg.
- Two motors allow P3-DX to travel as fast as 1.2 m/s in both forward and backward direction, and to rotate at an angular velocity up to 300 deg/s.
- The power to run the motors is supplied by multiple (up to 3) hot-swappable 12 V batteries, which allow P3-DX to be functional up to 10 h, provided the batteries are fully charged.
- P3-DX uses a high-performance microcontroller with advanced embedded robot control software based on the 32-bit Renesas SH2-7144 RISC microprocessor with 32K RAM and 128K FLASH. The microcontroller handles low-level details of mobile robotics, including maintaining the robot's drive speed and heading, as well as acquiring and pre-processing sensor readings.
- P3-DX is equipped with 500-tick encoders to maintain accurate dead reckoning data. Based on the signals from the encoders, the on-board micro-controller provides the relative position, direction, and velocity information.
- P3-DX has four RS-232 serial ports configurable from 9.6 to 115.2 kilobaud, two 8-bit bumpers/digital input connectors, user I/O port with 8-bits digital I/O, analog input, and 5/12 VDC power heading correction gyro port, tilt/roll sensor port, joystick port, and microcontroller HOST serial connector. Via the last serial link, the microprocessor of the robot can be connected with a computer responsible for providing higher-level robotics controls. In our experiments, WIFI connection was employed.
- P3-DX is equipped with 16 ultrasonic sonars (8 at the front of P3-DX and 8 at the rear) that provide the distance and angle information to the obstacle if it enters the scan range.
- P3-DX is also equipped with a SICK LMS-200 Lidar range finder. It is mounted at the top of the P3-DX platform and provides the distance from the P3-DX to the obstacles within a

scan angle of 190°. This range finder reaches distances of 80 m with a systematic error of ±25 mm. Its range resolution is 1 mm and its angular resolution is 0.25°.

- An off-board laptop computer (Windows 7 operating system, 4 GB RAM, dual core CPU running at 2.66 GHZ) is connected to the on-board microcontroller of the P3-DX using WIFI. Via this link, the computer acquires data from the microcontroller and in return supplies it with higher-level control commands.
- Advanced Robotics Interface for Applications (ARIA 2.7.0) software, including ARNetworking, is employed. ARIA is an object-oriented Linux platform C++-based open-source library that provides a robust client-side interface to the robot's microcontroller, thus giving dynamic control of the robot's velocity, heading, relative heading, etc.

More detailed information about the P3-DX is available at http://www.mobilerobots.com/researchrobots/pioneerp3dx.aspx.

1.4.2 Intelligent autonomous wheelchair system

The system is build of the basis of a standard wheelchair featured by two front caster wheels (see Fig. 1.2) and two rear driving wheels (see Fig. 1.3). The wheelchair is powered by a 24 V battery. The potential of autonomy and intelligence of the wheelchair is underlaid by its equipment with many accessories, including an on-board computer, DC/AC converter, laser range finder, LCD monitor, encoders, data acquisition devices, etc. The key components that are important to the experiments described in this book are as follows:

- A notebook with Windows XP, 2 GB RAM, and dual-core CUP running at 1.66 GHz. The necessary data are sent to the notebook for computation of the control signals, which is implemented in LabWindows. These signals are then sent to a drive system to control the movement of the wheelchair.

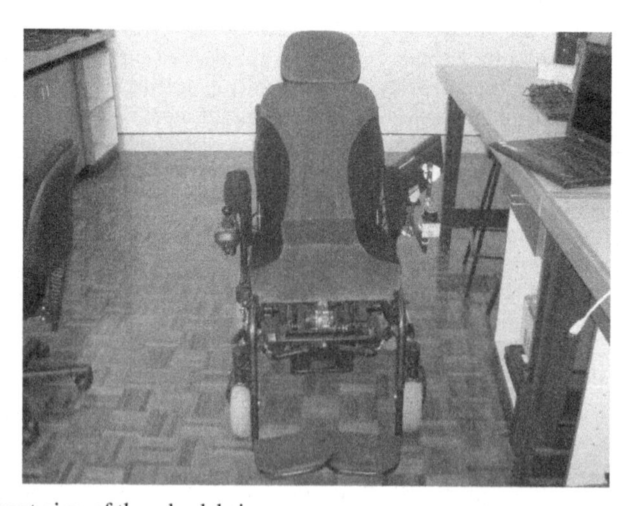

Figure 1.2 Front view of the wheelchair.

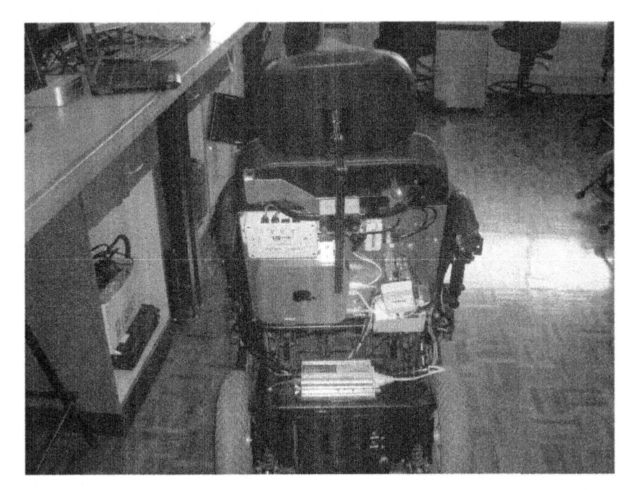

Figure 1.3 Rear view of the wheelchair.

- The URG-04LX laser is mounted on a rack at the front of the wheelchair, as is shown in Fig. 1.2. It provides the distance from the wheelchair to the obstacles in the environment of operation. The device has a maximum scan angle of 240°, maximum scan radius of 4 m, pitch angle of 0.36°, and accuracy of 1% of the measurement.
- USB1 adapter receives data from the encoders, which count the revolution of both wheels, and sends it to the notebook for evaluation of the position and orientation of the wheelchair.
- The control signals computed by the notebook are sent to the motor of the wheelchair using the National Instrument DAQ device NI USB-6008.

1.4.3 Autonomous hospital bed system

Some of the experiments covered in this book were carried out with an intelligent hospital bed system. This subsection presents its detailed description.

The hospital bed system, called Flexbed, is a motorized electric-powered hospital bed that allows for semi-autonomous and autonomous control of its movements. This is the major distinction from conventional hospital beds whose movements rely on human labor. The appearance of Flexbed and its key components are shown in Fig. 1.4. The key components of Flexbed are as follows:

- **Driving wheels**
  Flexbed features two driving wheels with diameter of 320 mm located at the sides of the base sections.
- **Encoders**
  The incremental quadrature encoders (model 775, see [91] for the specifications) have been attached to both driving wheels. Each of these encoders generates two square waves in incremental quadrature mode that provide the position, direction, or velocity information of the wheels.

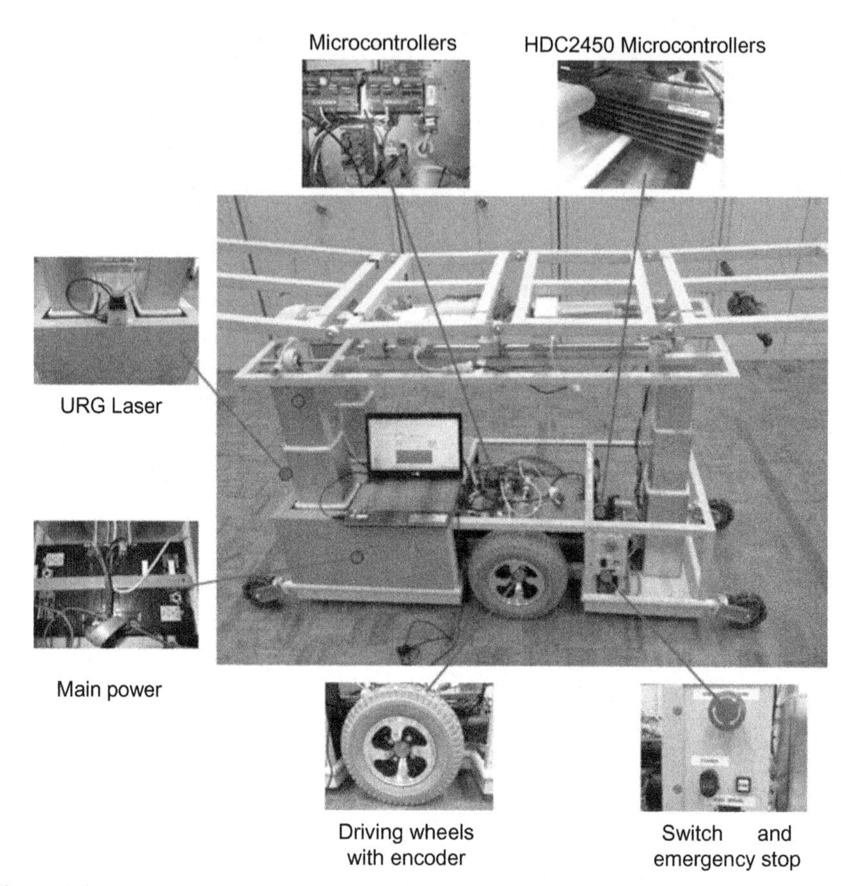

Figure 1.4 Key components in Flexbed.

- **Support lifts**
  The bed is supported by three support lifts. There is one DC actuator inside each of the support lifts for independent extension or contraction movement of each lift. These movements allow personnel to adjust the bed to the Trendelenburg position, anti-Trendelenburg position, tilt to the right or to left position from motor controllers implemented in iPad. The available connection between iPad and Flexbed is RS232 serial cable or WiFi.
- **Power supply**
  The motors for driving Flexbed and the actuators in the support lifts are powered by two 12 V batteries in series. The power for the other electronics is supplied from a separate 12 V battery. The power switch is located on one side of Flexbed together with the emergency stop button.
- **Range sensors**
  Flexbed allows various range sensors to be connected, such as the Hokuyo URG-04LX laser range finder and Microsoft Kinect. In the experiments described in this book, the Hokuyo URG-04LX laser range finder was used as the range sensor. It is mounted in the front of Flexbed and provides the distance data from Flexbed to the obstacles. It has a maximum

range of 240° scan angle with a maximum scan radius of 4 m. The pitch angle is 0.36° with accuracy of 1% of the measurement.

- **Motor controller**
  RoboteQ HDC2450 is used as the motor controller of Flexbed. The motor controller is responsible for receiving information from the encoders and sending control signals to the motor system via a serial or USB interface. Furthermore, this motor controller can also be manually driven via a utility provided by RoboteQ.
- **Embedded control board**
  The embedded control board used in Flexbed is the Keil MCB2300 micro-controller board. For implementation and error correction, we utilize the Labview Embedded Module for ARM micro-controllers to interact with the relevant devices. The Virtual Instrument program generated by Labview can be easily converted into C/C++ language by a built-in program of Labview and downloaded to MCB2300. Labview runs on a notebook (Windows 7 operating system with dual-core CPU running at 2.66 GHz with 4 GB RAM).

Fundamentals of sliding mode control

2

2.1 Introduction

In this chapter, we briefly cover a few basic concepts and facts of classic sliding mode control theory, focusing on those that will be used in this book later on.

2.2 Sliding motion

The phenomenon of sliding motion may appear in Variable Structure Systems (VSS) under certain circumstances, whereas variance of structure may be intentionally injected into a system by feedback with the capacity of instantly switching between distinct controls. A somewhat typical birth of a VSS by an "invariable" controlled system can be illustrated by a plant of the form

$$\dot{x} = a(\boldsymbol{x}) + b(\boldsymbol{x})u, \tag{2.1}$$

where $\boldsymbol{x} = \boldsymbol{x}(t) \in \mathbb{R}^n$ is the state, $u = u(t) \in \mathbb{R}$ is the control, and $a(\boldsymbol{x}), b(\boldsymbol{x}) \in \mathbb{R}^n$ are smooth functions of the state. Let this system be fed back by a controller assuming only two values $u_+ \neq u_-$. Also, let they be assumed in respective half-spaces separated by a smooth hyper-surface $S \subset \mathbb{R}^n$ given by the equation $g(\boldsymbol{x}) = 0$:

$$u = u(\boldsymbol{x}) := \begin{cases} u_+ & \text{if } g(\boldsymbol{x}) \geq 0, \\ u_- & \text{if } g(\boldsymbol{x}) < 0. \end{cases} \tag{2.2}$$

Then the closed-loop system is described by the differential equation

$$\dot{\boldsymbol{x}} = f(\boldsymbol{x}) := \begin{cases} f_+(\boldsymbol{x}) & \text{if } g(\boldsymbol{x}) \geq 0, \\ f_-(\boldsymbol{x}) & \text{if } g(\boldsymbol{x}) < 0 \end{cases} \tag{2.3}$$

with smooth functions $f_\pm(\boldsymbol{x}) := a(\boldsymbol{x}) + b(\boldsymbol{x})u_\pm$. Meanwhile, typically the right-hand side of (2.3) instantly jumps as the state $\boldsymbol{x}$ crosses the surface S. This means discontinuity of the right-hand side and may be interpreted as switch from one continuous structure to another.

In the case of *transversal intersection* where the vector fields given by both $f_-(\cdot)$ and $f_+(\cdot)$ dictate to cross the surface S in a common direction, as in Fig. 2.1(a), one of these fields drives the solution of (2.3) to the surface, whereas the other

Safe Robot Navigation Among Moving and Steady Obstacles. http://dx.doi.org/10.1016/B978-0-12-803730-0.00002-0

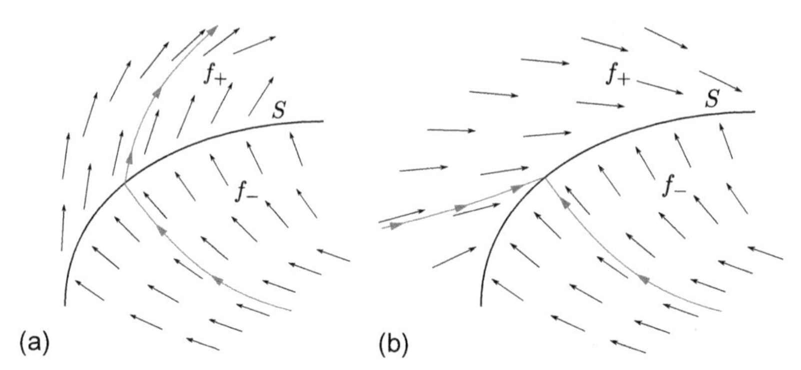

Figure 2.1 (a) Solution pierces the discontinuity surface; (b) Sliding surface.

field immediately withdraws it from the surface so that the solution instantly pierces the surface and goes away from it. The sliding mode phenomenon is associated with another situation where the crossing directions given by the fields not only are opposite but also every field tries to forward the solution to the "zone of responsibility" of the companion field: the field $f_-(\cdot)$ into $\{x : g(x) > 0\}$, and the field $f_+(\cdot)$ into $\{x : g(x) < 0\}$, as in Fig. 2.1(b). For the regular surface (the gradient $\nabla g(\boldsymbol{x})$ is nonzero everywhere on S), this situation at point $\boldsymbol{x} \in S$ can be more precisely characterized as follows:

$$\langle f_-(\boldsymbol{x}); \nabla g(\boldsymbol{x})\rangle = \lim_{\boldsymbol{x}_* \to \boldsymbol{x}: g(\boldsymbol{x}_*)<0} \frac{\mathrm{d}g}{\mathrm{d}t}(\boldsymbol{x}_*) > 0 \quad \text{and}$$

$$\langle f_+(\boldsymbol{x}); \nabla g(\boldsymbol{x})\rangle = \lim_{\boldsymbol{x}_* \to \boldsymbol{x}: g(\boldsymbol{x}_*)>0} \frac{\mathrm{d}g}{\mathrm{d}t}(\boldsymbol{x}_*) < 0, \tag{2.4}$$

where $\langle \cdot ; \cdot \rangle$ is the standard inner product in $\mathbb{R}^n$, the derivative $\frac{\mathrm{d}}{\mathrm{d}t}$ is along the solution of (2.3), and $\boldsymbol{x}_*$ converges to $\boldsymbol{x}$ so that it always remains on a definite side from S. A surface or its part every point $\boldsymbol{x}$ of which satisfies (2.4) is said to be *sliding*.

While (2.3) implies that the sliding surface attracts and absorbs solutions from both its sides (see Fig. 2.1(b)), the existence and behavior of the solution after arrival at this surface is not so straightforward from (2.3), so a somewhat intricate analysis is required. Typically it practices approximation of the system (2.3) by a system with a smooth right-hand side, with the idea that the limits of the solutions of the latter should be treated as the solutions of the former, where the limits are as the mismatches between the models decay to zero. These limits lie on the sliding surface and are called *sliding solutions* or *motions* (see Fig. 2.2(a)); the regime of the system's operation that gives rise to a sliding motion is called *sliding mode*. Some particulars may depend on approximation option, whose underlying basis may vary from rather abstract, "mathematical" concepts of closeness between functions to a deeper look at the structure and physics of the process, including better account for micro-dynamics of the real switching process that were neglected in the simplified model of the form (2.3) by assuming the capacity to switch instantly [92]. One of the widely accepted outcomes of such studies, which combines strong mathematical backing with practical

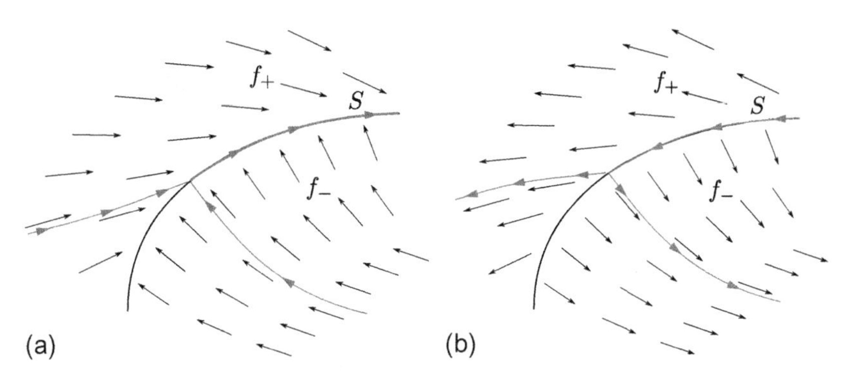

Figure 2.2 (a) Sliding motion over a sliding surface; (b) Repulsive surface.

relevance, is the *solution in Filippov's sense* [93], and this book deals exclusively with these solutions.

This solution $\boldsymbol{x}(t)$ is uniquely determined by the following two properties:

- the solution goes over the sliding surface $\boldsymbol{x}(t) \in S\ \forall t$;
- its derivative belongs to the straight segment $[f_-(\boldsymbol{x}(t)), f_+(\boldsymbol{x}(t))]$ whose end-points are given by the vector-fields $f_-(\cdot), f_+(\cdot)$ associated with (2.3).

The second requirement means that

$$\dot{\boldsymbol{x}}(t) = [1-\theta(t)]f_-[\boldsymbol{x}(t)] + \theta(t)f_+[\boldsymbol{x}(t)], \quad \text{with some} \quad \theta(t) \in [0,1]. \tag{2.5}$$

Given that the solution starts from S, the first requirement means the derivative $\dot{\boldsymbol{x}}(t)$ is always tangential to the surface:

$$\frac{\mathrm{d}}{\mathrm{d}t}g[\boldsymbol{x}(t)] = \langle \nabla g[\boldsymbol{x}(t)]; \dot{\boldsymbol{x}}(t)\rangle = 0. \tag{2.6}$$

As a result, we see that the above two properties uniquely determine the parameter $\theta(t)$ and the derivative $\dot{\boldsymbol{x}}(t)$ as functions of the state $\boldsymbol{x} = \boldsymbol{x}(t)$:

$$\begin{aligned} \theta(t) = \theta(\boldsymbol{x}) &:= \frac{\langle \nabla g[\boldsymbol{x}]; f_-[\boldsymbol{x}]\rangle}{\langle \nabla g[\boldsymbol{x}]; f_-[\boldsymbol{x}]\rangle - \langle \nabla g[\boldsymbol{x}]; f_+[\boldsymbol{x}]\rangle}, \\ \dot{\boldsymbol{x}} &= [1-\theta(\boldsymbol{x})]f_-[\boldsymbol{x}] + \theta(\boldsymbol{x})f_+[\boldsymbol{x}]. \end{aligned} \tag{2.7}$$

The last differential equation uniquely determines the sliding motion (given an initial state on the surface S). It is called the *equation of sliding motion* and describes the *sliding mode equivalent dynamics* of the system. This equation results from closing the loop of the original system (2.1) with the smooth feedback control:

$$u = u(\boldsymbol{x}) := [1-\theta(\boldsymbol{x})]u_- + \theta(\boldsymbol{x})u_+,$$

which is called the *equivalent control*; it may be treated as governing the evolution of the system in sliding mode.

Apart from the two already considered cases the discontinuity surface may also be *repulsive* (*two side repelling*). In this case, the crossing directions given by the fields are still opposite, as for the case of sliding surface, but contrary to this case, every of them tries to pull the solution in its own "zone of responsibility": the field $f_-(\cdot)$ into $\{x : g(x) < 0\}$, and the field $f_+(\cdot)$ into $\{x : g(x) > 0\}$, as in Fig. 2.2(b). As a result, these fields block solutions from arrival at the surface S by driving them out of it. The notion of the solution in the Filippov sense is still applicable on such a surface; however, it becomes ill-posed: such a solution may leave S and go to any side of S at any instant of time; thus uniqueness and stability of solution are lost in this case.

The considered cases do not exhaust all possibilities; for example, one or both of the fields $f_-(\cdot), f_+(\cdot)$ may be tangential to the surface at the examined point, the surface may be only piecewise smooth and have singularities, the controls u_+ and u_- may be not constant but smoothly vary with state in (2.2), the controller (2.2) may generate more than two values and be associated with more than one discontinuity surface, etc. The notion of solution in Filippov's sense remains viable in all these and many other cases. Now we briefly recall it with respect to a rather general situation.

2.3 Filippov solutions

We consider a general time-varying differential equation in $\mathbb{R}^n$:

$$\dot{\boldsymbol{x}} = f[\boldsymbol{x}, t]. \tag{3.1}$$

Here, the right-hand side is defined for almost all $(\boldsymbol{x}, t)$ from an open region $\mathcal{R} \subset \mathbb{R}^{n+1}$, i.e., for all but a set of the zero Lebesgue measure.[1] We assume that the function $f(\cdot)$ is measurable and for any compact subset $\mathcal{K} \subset \mathcal{R}$, there exists an integrable function $\alpha(t)$ such that $\|f(\boldsymbol{x}, t)\| \leq \alpha(t)$ for almost all $(\boldsymbol{x}, t) \in \mathcal{K}$. Let

$$B(\boldsymbol{x}, \delta) := \{\boldsymbol{x}' : \|\boldsymbol{x}' - \boldsymbol{x}\| \leq \delta\}$$

stand for the ball of the radius δ centered at $\boldsymbol{x}$, and let $f(Q, t)$ be the set of all values taken by the function $\boldsymbol{x} \mapsto f(\boldsymbol{x}, t)$ on arguments $\boldsymbol{x}$ that simultaneously belong to its domain of definition and the set Q.

Filippov solution [93] is an absolutely continuous function $\boldsymbol{x}(t) \in \mathbb{R}^n, t \in [t_0, t_1]$ such that $[x(t), t] \in \mathcal{R}\ \forall t \in [t_0, t_1]$ and for almost all $t \in [t_0, t_1]$,

$$\dot{\boldsymbol{x}}(t) \in F[\boldsymbol{x}(t), t], \tag{3.2}$$

where

$$F[\boldsymbol{x}, t] := \bigcap_{\delta > 0} \bigcap_{\mathcal{N}} \overline{\mathrm{co}} f\, [B(\boldsymbol{x}, \delta) \setminus \mathcal{N}, t] .$$

Here, the second intersection is over all sets $\mathcal{N} \subset \mathbb{R}^n$ of the zero Lebesgue measure, and the following notations are used:

[1] We recall that the Lebesgue measure rigorously extends the standard measure of length, area, or volume for $n = 1, 2, 3$, respectively, on an arbitrary dimension; it is also known under the name *n-volume*.

- $\overline{\mathbf{co}}\,A$—the closed convex hull of the set $A \subset \mathbb{R}^n$, i.e., the smallest closed convex set containing A;
- $B \setminus A$—the difference of the sets B and A, i.e., the set of all elements of B that do not belong to A.

The closed convex hull is equal to the closure of the convex hull $\mathbf{co}\,A$, i.e., of the smallest convex set containing A. In turn, $\mathbf{co}\,A$ is the set of all convex combinations of points in A. We recall that the convex combination of points $a_1, \dots, a_k$ is their weighted average:

$$a = \sum_{i=1}^{k} \theta_i a_i \tag{3.3}$$

with non-negative weights $\theta_i \geq 0$ whose sum equals 1:

$$\sum_{i=1}^{k} \theta_i = 1.$$

By the Carathéodory theorem [94, Th. 17.1], only convex combinations (3.3) with $k = n + 1$ addends can be considered when forming $\mathbf{co}\,A$ for $A \subset \mathbb{R}^n$.

The differential inclusion (3.2) is well defined: for almost all t concerned by this inclusion (i.e., such that $(x, t) \in \mathcal{R}$ for some x), the set $F(x, t)$ on the right-hand side is non-empty for all x such that $(x, t) \in \mathcal{R}$; moreover, this set is compact and convex.

If the right-hand side of the differential equation (3.1) is continuous, then the set $F(x, t)$ contains only one point $f(x, t)$ and so Filippov solution reduces to the ordinary solution. For the differential equation (2.3) considered in the previous section,

$$F(x,t) = \begin{cases} f_-(x) & \text{if } g(x) < 0, \\ f_+(x) & \text{if } g(x) > 0, \\ \mathbf{co}\,\{f_-(x), f_+(x)\} & \text{if } g(x) = 0. \end{cases}$$

Here $\mathbf{co}\,\{f_-(x), f_+(x)\}$ is the convex hull of the set $\{f_-(x), f_+(x)\}$ consisting of two points $f_-(x)$ and $f_+(x)$. This is nothing but the straight segment with the end-points $f_-(x)$ and $f_+(x)$. Hence, the definitions of solutions of (3.1) given in this and the previous sections, respectively, are in fact the same.

Now we revert to a general equation (3.1). It can be shown that given initial data $x(t_0) = x_0$, Filippov's solution necessarily exists at least locally [93, 95]. Any such solution can be extended to a maximal solution, which is defined as a solution that cannot be extended to the right. The property of local existence implies that maximal solution is defined on a semi-open interval $[t_0, \tau)$, where $\tau > t_0$ may be infinite. The finiteness of τ is necessarily accompanied by the fact that as $t \to \tau-$, the solution $[x(t), t]$ leaves any compact subset $K \subset \mathcal{R}$ of the domain $\mathcal{R}$ where (3.1) is defined: there exists $\tau_K \in [t_0, \tau)$ such that $[x(t), t] \notin K$ for all $t \in [\tau_K, \tau)$ [93, 95]. This property can be interpreted as convergence to the boundary of $\mathcal{R}$. If $\mathcal{R}$ is the entire space $\mathcal{R} = \mathbb{R}^{n+1}$, the boundary is empty and the discussed property means that the solution "blows up" for a finite time: $\|x(t)\| \to \infty$ as $t \to \tau-$. In this case, the derivative $\dot{x}(t)$

cannot remain bounded or even summable as $t \to \tau-$, i.e., $\int_{t_0}^{\tau} \|\dot{x}(t)\| \, dt = \infty$ [93, 95]. Most differential equations that will be encountered in this book have bounded right-hand sides. For them, the last observation implies that any solution can be infinitely extended.

As for the uniqueness of Filippov's solution, the situation with repulsive surface shows that this is a more intricate matter in general. In this book, we typically do not come into respective details and when dealing with a system governed by a discontinuous control law, address all possible Filippov's solutions.

Survey of algorithms for safe navigation of mobile robots in complex environments

3

3.1 Introduction

This chapter broadly documents methods relating to navigation of unmanned vehicles as applicable to collision avoidance, which still remains a subject of intensive ongoing research. Among various proposals, this chapter mainly focuses on recent ones and those capable of rigorous collision avoidance, i.e., for which it is rigorously guaranteed that collisions never occur, provided that some technical assumptions are fulfilled. Provable collision avoidance is a highly desirable trait in autonomous navigation systems since it shows that the system safely operates under a broad range of circumstances, rather than just those examined during testing.

Overall, navigation systems underlaid by more general assumptions about the problem would be considered superior. A list of some typical samples of assumptions, which is not intended to be exhaustive, is as follows:

- Vehicle models vary in complexity from velocity controlled linear models to realistic car-like models (see Section 3.2 for details). In this scope, analysis of the collision avoidance problem is simpler for velocity controlled models, partly since the vehicle is assumed to can stop instantly if required. However, this is physically unrealistic and therefore, a more complex model that better characterizes the mobile robot is desirable for use during analysis.
- Different levels of knowledge about the environment, including the obstacles and other robots, are required by different navigation strategies. This ranges from comprehensive abstracted a priori knowledge about the set of obstacles down to realistic and incomplete noisy sensory data obtained on-the-fly from range-finding sensors. In addition to the realism of the sensor model, the requirements to the scope of "perception" significantly vary between approaches; in some cases, methods that utilize only very limited information (such as the minimum distance to the nearest obstacle) are required.
- A whole variety of assumptions about the shapes of obstacles have been set forth, which ranges from limiting obstacles to some simplest geometric shapes to imposing only slight requirements on their basic geometrical properties. To ensure correct behavior when operating near an obstacle with limited information available, it is often necessary to presume smoothness of the obstacle boundary. Further, approaches with more flexible assumptions about obstacles would be more widely applicable to real-world scenarios.
- Uncertainty is inevitable in real robotic systems, and proving collision avoidance for a nominal model of the mobile robot does not always imply that the so justified correct

Safe Robot Navigation Among Moving and Steady Obstacles. http://dx.doi.org/10.1016/B978-0-12-803730-0.00003-2

behavior will be exhibited in practice. To reflect this, disturbances from the nominal model can be explicitly taken into account, which involves assumptions about them. An overview of the basic types of uncertainty encountered in mobile robots is available in, e.g., [96].

While being able to prove collision avoidance under broad circumstances is one of the most desirable features, some examples of other features that determine effectiveness of navigation laws are listed as follows:

- Provable satisfaction of the robot's goals is highly desirable. When possible, this can be generally shown by providing an upper bound on the time in which the mobile robot completes a finite task, however conservative this may be. Proving goal satisfaction is possibly less critical than proving collision avoidance, so long as it can be experimentally demonstrated that non-convergence is virtually non-existent. Typically, implications of non-convergence are less disastrous than those of collisions.
- In many applications, the computational capacity of the robot control unit is limited, and approaches with lower computational cost are favored. With ever-increasing computer power, this concern is mainly focused on small and fast robots, such as miniature UAVs (unmanned air vehicles), that require high update rates, challenging the limited computational faculties available. In practice, the computational burden is highly dependent on many factors, e.g., programming efficiency, which impedes direct comparison of the computational performance of different approaches. However, in general, it can be said that controllers based on closed-form expressions are faster than those based on local planning, which in turn are faster than those based on global planning.
- Many navigation laws were constructed for continuous-time models. Virtually all digital control systems are updated in discrete time. Thus, navigation laws constructed in discrete time are more suitable for direct implementation.

A key distinction between different approaches is concerned with the amount and content of information about the workspace that is assumed to be available to the robotic vehicle. When full information about the environment is present, *global path planning* methods may be used to find the optimal path. When only local information is available, *sensor-based* methods are used. A subset of sensor-based methods is constituted by *reactive* methods, which are characterized by direct mapping of the sensor reading into control, with no memory involved.

Although global methods are not directly applicable to the problem of collision avoidance in unknown environments, we discuss some of these approaches in the current chapter. This is because they could be adapted for unknown environments, sometimes with relatively little modification. Therefore, a review of collision avoidance methods would be incomplete without including them.

Recently, *Model Predictive Control* (MPC) architectures have been widely applied to collision avoidance problems. This approach seems to show great potential in providing efficient navigation, and easily extends to robust and non-linear problems; see Section 3.3 for details. It also has many favorable properties compared to, e.g., the commonly used *Artificial Potential Field* (APF) and *Velocity Obstacles* (VO) based methods, which could be generally more conservative when extended to higher-order vehicle models. MPC demonstrates many features desirable for sensor-based navigation, including applicability to problems of boundary following (see Section

3.4.1.2), avoidance of moving obstacles (see Section 3.5), and coordination of multiple mobile robots (see Section 3.6).

In this chapter, we review both local and some global approaches, together with methods applicable to multiple robots and moving obstacles. In doing so, we survey various types of mobile robots and sensor models, together with assumptions about static and moving obstacles.

The main material of this chapter was originally published in [97].[1]

3.1.1 Exclusions

Because of the breadth of the subject matter, we sharpen the focus of this survey by basically ignoring the following areas, which will be concerned only briefly and only when this is needed to better highlight the main topics:

- *Mapping algorithms.* Mapping is becoming fairly popular in real-world robotics concerned with exploration of unknown environments; see, e.g., [98–100]. Correspondingly there are collision avoidance techniques that require some kind of map to operate. However, a vast number of approaches to mapping have been proposed in recent years. Therefore, this review would be too broad if we were to include them. Additionally, the focus of this chapter is on local collision avoidance, which can be achieved without a map in many cases. At the same time, we will briefly discuss the *DistBug* class of algorithms (see, e.g., [101, 102]), which are possibly among the simplest examples of non-reactive navigation methods: a single point is stored to detect circumnavigation of a particular obstacle. Some examples of more advanced techniques based on mapping include *Combinatorial Mapping* (see, e.g., [103, 104]), *Occupancy Grids* (see, e.g., [100]), and *Simultaneous Localization and Mapping* (see, e.g., [98, 99]).
- *Path-tracking controllers.* Their design continues to be an important and non-trivial problem, partly due to various non-idealities and intricacies of the real world. Several types of collision avoidance approaches assume the presence of an accompanying path following navigation law, which plays an auxiliary role in the control architecture. From a broader perspective, path tracking is constituted as a separate, self-sufficient, and long-standing problem, well documented in specialized literature. Samples of path-tracking methods developed for mobile robots are available in, e.g., [105, 106].
- *High-level decision making.* Hierarchical control seems to be common and classic in complex real-world robotic vehicles. This structure means that a high-level planner provides general navigational directions, while a low layer attempts to follow the commands given by the higher layer and may bear the responsibility for local navigation tasks, such as collision avoidance. While the high layer is important for successful convergence in some situations, by and large, it is too abstracted from our basic focus on collision avoidance. In some cases, it is, however, troublesome to separate with a certainty the navigation system into layers; this may be viewed as if a lower level collision avoidance system acquires the "higher-level" feature of provably convergent navigation (e.g., to a target point); see, e.g., [30]. These cases are within the scope of interest in this chapter. However, we systematically exclude from consideration the situation where the provable convergence is basically due to an apparently separate higher level control module, with its own individual algorithmic logics.

[1] Hoy et al. [97]. Copyright ©2015 Cambridge University Press. Reprinted with permission.

- *Motor schemas, subsumption, and behavior based architectures.* In general, these basically refer to methods that attempt to replicate the assumed behavior of animals by building a library of various behaviors and selecting an appropriate behavior according to the situation; see, e.g., [107–110]. These approaches have a clear historic importance, but mainly focus on tasks more general than just collision avoidance. At the same time, it is fairly typical for these methods to achieve collision avoidance in the vein of the artificial potential field paradigm, which is separately covered in this review. Also, they do not tend to be associated with rigorously provable collision avoidance, which is among the main concerns of this review.
- *Implementation of path planning algorithms.* This issue is relatively decoupled from the algorithm itself, so that in many cases, a common approach may be used with several types of algorithms. Therefore, their discussion may be separated. This review effectively focuses on the parameters and constraints given to path planning systems, and the subsequent effect of the output. Implementation issues are well concerned in many other surveys; see, e.g., [111, 112].
- *Specific missions, including robotic swarming, formation control, target searching, area patrolling, and target visibility maintenance.* For them, the primary objective is not collision avoidance. Therefore, the respective developments will be concerned in this chapter only if the underlying collision avoidance method is not documented elsewhere.
- *Iterative learning, fuzzy logic, and neural networks.* These fertile approaches are able to generate promising experimental results and moreover, are well suited to some applications; see, e.g., [113]. However, when applied directly to a robot's motion, it is generally rather difficult to supply them with firm guarantees of motion safety; see, e.g., [114]. At the same time, these approaches may be indirectly addressed by this chapter so far as they underlay, more or less, some of the planning algorithms discussed here.

The remainder of the chapter is structured as follows. In Section 3.2, the problem of navigating cluttered environments is described. In Section 3.3, MPC-based navigation systems are discussed, while Section 3.4 is focused on methods of sensor-based navigation. Section 3.5 reviews methods for dealing with moving obstacles, and Section 3.6 is concerned with multiple robots scenarios.

3.2 Problem considerations

In this section, we outline some of the real-world factors that influence designs of navigation systems. It would be considered an advantage of a designed navigation system to take these factors into account since this facilitates its real-world implementation.

3.2.1 Environment

Our definition of a cluttered environment consists of a two- or three-dimensional workspace, which contains a set of simple, closed, untraversable obstacles. The last means that the robot is not allowed to overlap with obstacles. The area outside the obstacle is considered homogeneous and equally easy to navigate in all its parts. Examples of cluttered environments include, but are not limited to, offices, man-made structures, and urban environments. A classification of objects in an urban environment is offered in [115].

The mobile robot is spatially modeled as either a point, circle, or polygon in virtually all approaches. Polygons can be conservatively bounded by a circle, so polygonal robot shapes are generally used for planning tight maneuvers in close proximity to obstacles (e.g., when moving through close obstacles), where an enclosing circle would exclude marginally viable trajectories.

3.2.2 Kinematics of mobile robots

Many types of mobile robots have to operate in cluttered environments; such as ground mobile robots, UAVs, surface vessels and underwater unmanned vehicles. Further, the term *dynamic* is used to refer to models based on resolution of physical forces, whereas *kinematic* refers to models based on more abstracted control inputs. Most popular kinematic models of vehicles can be generally categorized into three types—*holonomic*, *unicycle*, and *bicycle*; the differences between them are characterized by different kinematic constraints. Detailed reviews of various vehicle models are available in, e.g., [116–119].

- *Holonomic kinematics.* Models of this class are most popular in studies of mobile robots that have control capability in any direction. Holonomic kinematics are encountered on, e.g., helicopters and certain types of wheeled robots equipped with omni-directional wheels. These models involve no notion of body orientation for the purposes of path planning, and only the positional coordinates are taken into account. However, orientation may become a consideration when applying the resultant navigation law to real vehicles, though this is decoupled from planning.
- *Unicycle kinematics.* These models describe mobile robots that are associated with a particular angular orientation, which determines the direction of the velocity vector. Changes to the orientation are limited by a turning rate constraint. Unicycle models may be used to describe various types of robotic vehicles, such as differential drive wheeled mobile robots, unmanned aerial and underwater vehicles, missiles, fixed wing aircrafts, etc., which in certain circumstances, have similar governing differential equations; see, e.g., [32, 33, 118, 120–128] and references therein.
- *Bicycle kinematics.* This class of models is popular in studies of car-like mobile robots that have a pair of steerable wheels (classically the front wheels) separated from a pair of fixed wheels (classically the rear wheels). Kinematically this implies that the maximum turning rate is proportional to the robot's speed. This places an absolute bound on the curvature of any path that may be followed by the robot, regardless of speed. This constraint often necessitates high-order planning to navigate in confined environments.

Nonholonomic constraints are generally only a limiting factor at low speed; for example, realistic vehicle models would likely be also subjected to absolute acceleration bounds, which limit their maneuverability at high speed. More complex kinematics are also encountered in literature, but uncommon.

In addition to these basic variants of kinematics, the associated linear and angular variables may be either *velocity controlled* or *acceleration bounded.* Mobile robots with acceleration-bounded control inputs are in general much harder to navigate; velocity controlled robots may stop instantly at any time if required.

When predicting robots' actual motion, these nominal models are invariably subject to disturbance. Models of disturbance correlate with kinematic models.

- *Holonomic models.* Disturbance models commonly consist of bounded additions to the translational control inputs; see, e.g., [129].
- *Unicycle models.* Bounded addends to the control inputs can be combined with a bounded difference between the robot's orientation and actual velocity vector; see, e.g., [130]. More realistic models of differential drive mobile robots are also available, which are based on modeling wheel slip rates; see, e.g., [131, 132].
- *Bicycle models.* Disturbances can be modeled as slide slip angles on the front and rear wheels; see, e.g., [133]. Alternatively, more realistic disturbance models of car-like mobile robots are available, which include factors such as suspension and type adhesion; see, e.g., [134, 135].

Mobile robots with bicycle kinematics and realistic robots with minimum speed constraints are subject to absolute bounds on their path curvature. This places a global limit on the types of environments they can successfully navigate through; see, e.g., [136, 137]. Lower bounds on allowable speeds further complicate the planning system. For example, instead of halting, the robot must follow some "holding pattern" at the termination of a trajectory.

3.2.3 Sensor data

Most autonomous vehicles base their navigation decisions on data reported by on-board sensors, which provide some information about the vehicle's immediate environment. The main types of sensor models can be broadly categorized as follows:

- *Abstract sensor models.* This label is applied to any sensor-based method where the navigation law is assumed to know, with certainty, whether a given point lies within a given obstacle. Usually any occluded region (without a line-of-sight to the robot) is considered to be a part of an obstacle. Though it is impossible to perfectly identify the boundary of an obstacle using a physical sensor, LiDAR (Light Detection and Ranging) sensing technology is currently capable of achieving accuracies high enough for any sampling effects to be of minor concern. This may be achieved via, e.g., even the simplest, piecewise linear interpolation of the detected points. However, with lower resolutions, this model may be unsuitable for navigation law design.
- *Ray-based sensor models.* These models inform the navigation law of the distance to the obstacles in a finite number of directions around the robot; see, e.g., [9, 138, 139]. This is a physically more realistic model of laser-based sensors compared to the abstract sensor model, and may be suitable to account for effects from low-resolution sensors. A reduced version of this model underlies some boundary following applications, where only a single detection ray in a fixed direction (relative to the robot's body) is used.
- *Minimum distance measurements.* This model reports the distance to the nearest obstacle point. This may be realized by certain types of wide aperture sensors, e.g., acoustic or optic flow sensors. Using this type of measurement leads to less efficient movement patterns during obstacle avoidance since, e.g., it may be not immediately clear which side of the robot the obstacle is on; see, e.g., [33].

- *Tangent sensors.* This sensor model reports the angles to the visible edges of an obstacle as seen by the robot; see, e.g., [104, 140]. This can be realized by a camera sensor, provided that a method of detecting obstacle edges from a video stream is available; see, e.g., [141].
- *Optic flow sensors.* This model reports the average rate of the pixel flow across a camera sensor; see, e.g., [142, 143]. It was shown that based on only this rate, an effective navigation command may be generated and good practical results may be achieved. However, rigorously provable collision avoidance is very far from being common in research based on this model.

There are many ways in which noise and distortion may be compensated when using these models. By and large, these ways are quite specific to individual approaches. Some examples include linear quadratic Gaussian optimization [144], sliding mode based tracking control [145], and H_∞ tracking control [106].

3.2.4 Optimality criteria

There are several methods of preferentially choosing one possible trajectory over another. Many path optimization algorithms may be implemented with various such measures or combinations of them. Common options are listed below:

- *Minimum distance traveled.* This metric is used in the majority of path-planning schemes. Its operational convenience is partly due to the fact that it can be decoupled from the achievable profile of the robot's speed. The classic result by Dubins [146] displays the optimal bounded-curvature path of a mobile robot between two configurations in a workspace without obstacles: the optimal path consists of a sequence of no more than three motion primitives. One of them is going straight ahead and the two other are turning as sharply as possible to the left and right, respectively. Similar results are available for vehicles with actuated speed [147] and for velocity controlled, omni-directional vehicles [148]. However, these results were not much employed in practical path planning mainly due to their abstracting from obstacles, which typically have a complex and inneglectable effect on any optimal path. The minimum distance path may also be constructed from the *Tangent Graph* of the obstacle set, as was shown in [13] for the case when acceleration constraints are absent, and in [16] for the general case.
- *Minimum wheel rotation.* This basically applies to differential drive wheeled mobile robots and in most cases, is only subtlety different from the above minimum distance metric. However, this criterion may entail better performance in some situations, especially when fine movements are required; see, e.g., [149].
- *Minimum time.* The traversal time of a path depends on the speed profile of the mobile robot and thus on kinematic (and possibly dynamic) constraints. In most situations, this metric would be more appropriate than the minimum distance for selecting the most efficient maneuvers. This criterion is often used in MPC-based approaches; see, e.g., [129].
- *Minimum control effort.* This criterion is well suited to mobile robots with limited energy supply, e.g., spacecraft or passive robotic vehicles. However, it is invariably combined with another measure to avoid the "do nothing" solution.
- *Optimal surveillance rate.* In unknown environments, it may be better to select trajectories that minimize the occluded part of the scene; see, e.g., [150]. In cases where the occluded parts of the environment must be treated as unknown dynamic obstacles, this carries potential to more efficiently traverse through the scene. This method commonly relies on stochastic inferences about the unknown portion of the workspace.

Other requirements that can be applied to paths include higher-order curvature rate limits, which may be useful to produce smoother trajectories; see, e.g., [151].

3.2.5 Biological inspiration

Researchers in the area of robot navigation in complex environments find much inspiration from biology, where the problem of controlled animal motion is a central one. This is prudent since biological systems are highly efficient and refined, while equivalent robotic systems are in relative infancy. Animals, such as insects, birds, or mammals, are believed to use simple, local motion control rules that result in remarkable and complex intelligent behaviors. Therefore, it does not come as a surprise that biologically inspired or biomimetic algorithms of collision-free navigation play an important part in this research field.

For example, the idea of navigation along an equiangular spiral and the rules for local obstacle avoidance proposed in [32, 52, 126] are inspired by biological examples. That idea stems from the observation that peregrine falcons, which are among the fastest birds on earth, plummet toward their targets at speeds of up to 200 miles an hour along an equiangular spiral [152]. Meanwhile the above rules resemble a similar obstacle avoidance strategy known in biology as "negotiating obstacles with constant curvatures" and exemplified by a squirrel running around a tree; see, e.g., [153]. These ideas and rules in reactive collision avoidance robotic systems are discussed in more details in Section 3.4.1. The sliding mode control based methods of obstacle avoidance discussed in Section 3.4.1.2 are also inspired by biological examples such as the near-wall behavior of a cockroach [154]. Another example is the Bug family algorithms, which are also inspired by bugs crawling along a wall; see Section 3.4.1.3. Motor schemas and subsumption architectures may also be considered as biologically inspired; see, e.g., [107–110]. However, these topics are among the exclusions from this review.

Optical flow navigation is another important class of biologically inspired navigation methods. The remarkable ability of honeybees and other insects like them to navigate effectively using very little information is a source of inspiration for control strategies design. In particular, the work of Srinivasan et al. [155] explains the use of optical-flow in honeybee navigation, where a honeybee makes a smooth landing on a surface without any knowledge of its vertical height above the surface. As it is commonly observed in insects flights, the navigation command is derived from the average rate of pixel flow across an optical sensor; see, e.g., [142, 143].

Other approaches that use a camera to locate targets may also be classed as biologically inspired. For example, the navigation approach in [122] allows a robot to converge to a target solely using measurements of the target's location in the camera frame.

Many ideas in multi-robot navigation are also inspired by biology, where the problem of animal aggregation is central in both ecological and evolutionary theory. Animal aggregations, such as schools of fish, flocks of birds, groups of bees, or swarms of social bacteria, are believed to use simple, local motion coordination rules at the individual level that result in remarkable and complex intelligent behavior at the group level; see, e.g., [156, 157]. Such intelligent cooperative behavior is expected from very-large-scale robotic systems, the interest in which is rapidly growing because of the decreasing costs of robots. In more details, this issue is discussed in Section 3.6.

There is also evidence that higher animals use approaches resembling Model Predictive Control to avoid obstacles via some type of planning into the future [158]. MPC-based navigation laws are discussed in Section 3.3.

3.2.6 Implementation examples

There are many excellent reviews of current applications and implementations of real-world autonomous vehicles; see, e.g., [159]. With no intention to be exhaustive, we limit ourselves to only one particular application, mainly on the ground, that it is directly concerned in the main body of this book.

Semi-autonomous wheelchairs and mobile robotic hospital beds are among recent applications in which a navigation law must be designed to prevent collisions while executing high-level direction commands; see, e.g., [160–162]. The methods described in this review are highly relevant for these applications since safety is a fundamental concern for them. Several original collision avoidance approaches proposed for wheelchairs and mobile hospital beds (see, e.g., [161]) will be discussed in Section 3.5.3.2.

3.2.7 Summary of the methods reviewed

A broad summary of these methods is given in Fig. 3.1, where the availability of certain traits is shown. This summary gives only subjective, generalizing judgements,

		Single vehicle	Moving obstacles	Multi-vehicle	Boundary following
MPC	Standard	U*,A,M,F,T	U*,A,M,F,T*		
	Robust	U*,A,M,F,R,T			
	Local planning	U,A,M,F,R	A,M,F		U,A,M,F,T
DMPC	Distributed optimization			U*,A,M,R	
	Synchronous			U*,A,M,F,R	
Boundary following	Minimal information		U*,A*,M,F,T*		U,A*,M,F,T
	Full information		U*,A*,M,F,T*		U,A*,M,F,T
	Bug algorithms				U,F,T
Velocity obstacle	VO/NLVO	U*,A*,F,T*	U*,A*,M,F,T*		
	DRCA/RCA		U,A*,M*,F,R*	U,A*,M,F,R,T*	
Other	APF	U,A*,M,F,T*	U,A*,M,F	A*,M*,F,T	U,M,F,T
	Tangent following	U,A*,M,F,T	U*,A*,M,F,T		U,A*,M,F,T
	Other reactive	U,A*,M,F,T	U,A*,M,F,T*	U,F	U,M,F,T
	Hybrid logic	A,M,F,T		A,M,F,T	

Key:	
U	Unknown environment
A	Acceleration bounded
M	Unicycle or bicycle model
F	Fast computation
R	Robust to disturbance
T	Provably convergent
*	Some disadvantages

Figure 3.1 Summary of the traits characterizing the methods discussed in this review. The symbol * signals that the method would not be completely characterized as exhibiting the corresponding trait (and thus further work could possibly be done).

and should not be considered a definitive comparison. A more detailed discussion of these methods will be given in the remainder of this chapter.

3.3 Model predictive control

There are many examples of path-following systems that plan an obstacle-avoiding path off-line and are able to robustly follow it, even if subjected to bounded disturbances. However, they typically rely on a priori access to an extensive pool of perfect knowledge about the environment, which is not conducive to on-line collision avoidance.

Model Predictive Control (MPC) (also known as *Receding Horizon Control*, RHC) is increasingly being applied to on-line navigation problems. This method is capable of combining path planning with on-line stability and convergence guarantees; see, e.g., [163, 164]. This is basically due to performing the path-planning process at every time instant, although only the initial control related to the optimal trajectory is applied to the mobile robot; such a process is illustrated in Fig. 3.2. In most cases, the optimum is sought among paths that terminate with an invariant state (the robot is either stationary or in a loiter circle), whereas optimization means minimization of some cost-to-go function. When repeating this process at every time-step, the controller can be shown to exhibit stability in many cases.

In this section, we mainly discuss approaches concerned with known environments. At the same time, many of these methods can be extended to cases where information is restricted. The main discussion of MPC approaches applicable to unknown environments is reserved until Section 3.4.2.

3.3.1 Robust MPC

The key advantage of MPC lies with its robust variants, which are able to account for bounded disturbances. These variants can be categorized into three main classes:

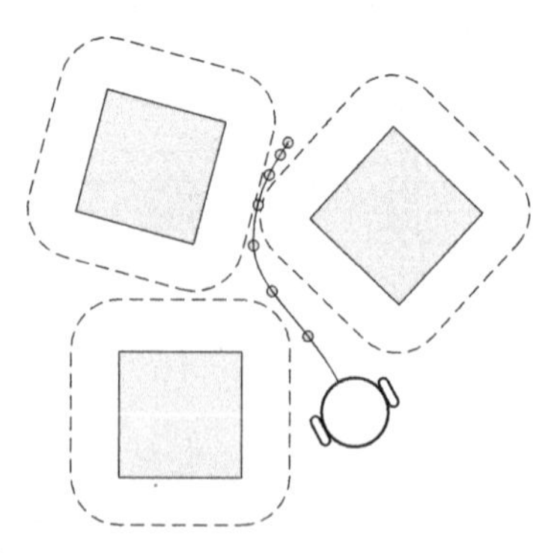

Figure 3.2 Illustration of the MPC approach to robot navigation. At each control update, an optimal path is planned into the future; only the initial control input corresponding to the optimal solution is implemented on the robot.

- *Min-max MPC.* In this formulation, optimization is performed with respect to all possible evolutions of the disturbance; see, e.g., [165]. In many cases, this can be reduced to a linear robust control problem. However, relatively high computational cost of solution generally precludes it from being used for real navigation.
- *Constraint tightening MPC.* Here the state constraints are dilated by a pre-specified margin so that the found trajectory can be guaranteed from collisions with the obstacles even when disturbance causes the state to drift toward the constraints imposed by obstacles; see, e.g., [129, 166, 167]. The typical basic argument shows existence of a future viable trajectory by using a feedback, though the feedback input is not directly applied to the mobile robot.

  This approach is commonly used for robot navigation problems—for example, a system has been described where an obstacle avoiding trajectory is found based on minimization of a cost functional compromising the control effort and maneuver time [129]. In this case, convergence to the target and the ability to overcome bounded disturbances can be shown.
- *Tube MPC.* This method uses an independent reference model of the system, and employs feedback to ensure that the actual state converges to the nominal state; see, e.g., [168]. In contrast, the constraint tightening system would essentially treat the nominal state as actual at each time step. Tube MPC is more conservative than constraint tightening since it wouldn't take advantage of favorable disturbance. Thus, it doesn't offer significant benefits for robot navigation problems when a linear model is used. However, it is useful for robust nonlinear MPC (see, e.g., [169]), and problems where only partial state information is available; see, e.g., [170]. Also, any approach that includes path following with bounded deviation (see, e.g., [171]) is somewhat equivalent to tube MPC.

For robust MPC, the amount of separation required from the state constraints on an infinite horizon is determined by the *Robustly Positively Invariant* (RPI) set. This is the set of all possible state deviations that may be introduced by disturbance while a particular disturbance rejection law is operating. Techniques have been developed to efficiently calculate the smallest possible RPI set [172].

If disturbance is Gaussian rather than bounded, the MPC problem may be reformulated stochastically so that the probability of collision is required to be below an arbitrarily small level; see, e.g., [173, 174].

3.3.2 Nonlinear MPC

Current approaches to MPC-based robot navigation generally rely on linear kinematic models, usually with double integrator dynamics. While many path-planning approaches exist for robots with nonholonomic kinematics, it is generally harder to show their stability and robustness [175]. Approaches to robust nonlinear MPC are generally of the tube MPC type [169]. In these cases, a nonlinear trajectory tracking system can be used to ensure that the actual state converges to the nominal state. A proposition has been made to also use sliding mode control laws for the axillary system (for which disturbance rejection is typically easier to show), though such systems typically require continuous time analysis; see, e.g., [176].

In terms of robot navigation problems, examples of the MPC approach that deal with unicycle kinematics while having disturbance present have been proposed; see, e.g., [145, 171, 177, 178]. Similar methods guarantee convergence using passivity-based arguments [179]. However, it seems that more general applications of nonlinear MPC to robot navigation problems should be possible.

There are methods in which MPC is applied to robot navigation problems other than performing rigorously safe path planning. In some cases, the focus is shifted toward controlling a robot's dynamics; see, e.g., [135, 180, 181]. The respective studies typically use a realistic robot model during planning, and are able to give good practical results, though guarantees of safety are currently easier with more kinematic models. In other cases, MPC may be used to regulate the distance to obstacles (see, e.g., [182]); a more detailed discussion of this issue is given in Section 3.4.1.4.

3.3.3 Planning algorithms

Global path planning is a relatively well-studied research area supplied with many thorough reviews; see, e.g., [111, 112]. MPC may be implemented with a number of different path-planning algorithms. The main relevant measure of algorithm quality is *completeness*, which indicates whether calculation of a valid path can be guaranteed whenever one exists. Some common global path-planning algorithms are summarized as follows:

- *Rapidly-exploring random trees.* This method is based on building a tree of possible actions to connect initial and goal configurations; see, e.g., [183, 184]. Some variants are provably asymptotically optimal [184].
- *Graph search algorithms.* Examples include A* and D* algorithms (see, e.g., [185] and [186], respectively), and Fast Marching; see, e.g., [187]. Most methods hybridize the environment into either a square graph, an irregular graph [188], or a Voronoi diagram [187], where the latter is the skeleton of points that separates all obstacles. A search can then be performed to calculate the optimal sequence of node transitions. In addition, this may be used as the first step to find a bounded area within which further path-planning operations can take place [189].
- *Optimization of predefined paths.* Examples include Bezier curves [190], splines [191], and polynomial basis functions [20]. While these are inherently smoother, showing completeness when using them may be more difficult in some situations.
- *Artificial potential field methods.* These methods will be introduced in Section 3.4.3.1, as they are also ideally suited to online reactive navigation of robots (without path planning). These can also be used as path planning approaches, essentially by using more information about the environment; see, e.g., [192, 193]. However, the resultant trajectories would not be optimal in general.
- *Mathematical programming and optimization.* This usually is achieved using *Mixed Integer Linear Programming* constraints to model obstacles as multiple convex polygons [194]. Currently, this is commonly used for MPC approaches.
- *Tangent graph based planning.* This limits the set of trajectories to cotangents between obstacles and obstacle boundary segments, from which the minimum distance path being found in general [16, 195]. The problem of shortest path planning in a known environment for unicycle-like mobile robots with a hard constraint on the robot's angular speed was solved in [16]. It was assumed that the environment consists of a number of possibly non-convex obstacles with a constraint on the curvatures of their boundaries, along with a steady target that should be reached by the robot. It has been proved that the shortest (minimal in length) path

consists of edges of the so-called tangent graph. Therefore, the problem of the shortest path planning is reduced to a finite search problem.

- *Evolutionary algorithms, simulated annealing, particle swarm optimization.* These are based on a population of possible trajectories, which follow some update rules until the optimal path is reached; see, e.g., [196, 197]. However, these approaches seem to be suited to complex constraints, and may have slower convergence for normal path planning problems.
- *Partially observable Markov decision processes.* This approach is based on calculating a type of decision tree for different realizations of uncertainty. It also employs probabilistic sampling to generate plans that may be used for navigation over long time frames; see, e.g., [198]. However, this may not be necessary for all MPC-based navigation problems.

3.4 Sensor-based techniques

Sensor-based navigation techniques typically employ only limited local knowledge about the environment, exemplified by data obtained from range sensors, video cameras, or optic flow sensors. Global sensor-based planners use the sensory information (possibly combined with a priori knowledge, if applicable) to build a comprehensive model of the observed chunk of the environment and to find the best complete trajectory through it [7, 14, 15]. Within this framework, several techniques (surveyed in, e.g., [17, 18]) have been developed even for dynamic scenes, including nonholonomic planners [20], velocity obstacles [17, 19], and state-time space [22, 23] approaches. While the computational capabilities of unmanned vehicles are continually increasing, global planning problems will always feature NP-hardness: this mathematical seal for intractability was established for even the simplest problems of dynamic motion planning [24]. This is an important consideration especially for, e.g., micro UAVs, which must react quickly to new information despite having limited computational faculties available.

On the other hand, local path planners use onboard sensors to locally observe a nearest fraction of an unknown environment for iterative re-computation of a short-horizon trajectory [9, 31, 32]. This reduces the calculation time and creates a potential for employment in certain real-time guidance systems. Many of the related techniques, such as the dynamic window [34, 35], curvature velocity [36], and lane curvature [37] approaches treat the obstacles as static. On the other hand, approaches like velocity obstacles [19], collision cones [38], or inevitable collision states [39, 40] assume deterministic knowledge about the obstacle velocity and a moderate rate of its change.

In the marginal case where the planning horizon collapses into an infinitesimally short time interval, the local planner acts as a reactive feedback controller: it directly maps the current observation into the current control. Examples can be found where a general technique for design of reactive controllers is offered; see, e.g., [199]. Other examples include artificial potential approach, combined with sliding mode control for gradient climbing [50, 53], and kinematic control based on polar coordinates and Lyapunov-like analysis [51]. Until now, fully actuated, velocity controlled robots were mostly studied in this area, and the obstacles were interpreted as rigid bodies of the simplest shapes (e.g., discs [50, 51] or polygons [53, 54]), the sensory data were assumed to be enough to determine the location of obstacle characteristic points concerned with its global geometry (e.g., the disc center [50, 51] or angularly most distant polygon vertex [54]) and to provide access to its full velocity [50, 51, 54]. Furthermore, rigorous justification of the global convergence of the proposed algorithm is rarely encountered.

In this section, we mainly focus on local sensor-based planners, with particular attention given to reactive algorithms.

3.4.1 Obstacle avoidance via boundary following

Temporarily following the boundary of an obstacle is a standard method employed by many obstacle avoidance algorithms. According to this method, the robot directly pursues the main control objective until a threat of collision with an obstacle is detected. After this, the robot bypasses the obstacle by following its boundary by temporarily putting aside the main objective. The converse switch from the boundary following holds as soon as a leaving condition is satisfied. This condition should guarantee that firstly, resuming direct pursuit of the objective does not cause a collision threat at least initially and secondly, the control objective will be ultimately achieved. In this context, the distance to the followed boundary is relatively insignificant; it may be time-varying but should not excessively decay and increase to ensure safety and to exclude collisions with companion obstacles, respectively. It should be also noted that boundary following is a self-contained navigation task of immediate interest for border patrolling and structure inspection [5, 200], bottom following by autonomous underwater vehicles [201], lane following by autonomous road vehicles [202], and other missions, where traveling along the boundary at a pre-specified distance from it is an essential requirement.

3.4.1.1 Distance based

In many approaches, boundary following can be rigorously achieved by only measuring the minimum distance to the obstacle; see, e.g., [33, 61, 203]. For example, in [33] the navigation strategy was based on a sliding mode navigation law. This strategy uses the minimum distance to the obstacle as input and is suitable for guiding nonholonomic mobile robots traveling at constant speeds. In [61, 203], a similar sliding mode argument is used, with the only required input being the rate of change of the distance to the obstacle.

Other approaches have been proposed that use a single distance measurement at a specific angle relative to the robot's body; see, e.g., [204, 205]. In [32, 126], the controller is fed by purely the length of the detection ray, whereas in [204, 205], additional information is also collected to estimate the tangential angle of the obstacle at the intersection point, while [206] also requires the curvature of the obstacle's boundary. Additional information would presumably result in improved behavior, but comparisons of closed loop performance are difficult.

Other work using similar assumptions is focused on following straight walls; see, e.g., [207–210]. However, it seems that, at least theoretically, navigation laws capable of tracking contours are more general and therefore superior.

In most of these examples, the desired behavior can be rigorously shown. However, the common limitation is that the robot must travel at a constant speed, and this speed must be set conservatively in accordance with the smallest feature of the obstacle. In some cases, simple heuristics can partially solve this problem: by instructing the robot to instantly stop and turn in place if the obstacle distance becomes too small, collision may be averted [204].

3.4.1.2 Sliding mode control

Due to the well-known benefits such as stability under large disturbances, robustness against system uncertainties, good dynamic response, simple implementation, etc. [70], the sliding mode approach has attracted an increasing interest in the area of mobile robotics. Examples include but are not limited to target following [125, 128], environmental extremum seeking [127], and trajectory tracking [211, 212].

Sliding mode based boundary following with a pre-specified margin was addressed in [33, 205] for a planar under-actuated mobile robot modeled as unicycle. It travels forward with a constant speed and is controlled by the angular velocity upper-limited by a given constant. In the literature, this model is also known under the name *Dubins car*; this model captures the capability of the mobile robot to travel forward with the given speed along planar curves whose curvature radius exceeds a certain threshold. In [33], a discontinuous controller was proposed that drives the distance between the robot and a given static obstacle to a predefined value and thereafter makes the robot circulating at this predefined distance around the obstacle. This controller originates from the equiangular navigation and guidance law [126] and assumes the robot has access to the current distance to the obstacle and the rate at which this reading evolves over time (computed via, e.g., numerical differentiation), but no further sensing capabilities are needed.

The control law from [33] is fed by input variables whose computation may require wide-aperture sensing and intensive pre-processing, like many other controllers proposed in the extensive literature on reactive path and boundary following. For example, reactive (i.e., mapless) vision navigation (for which excellent surveys are available in [142, 213]), employs the capability of the visual sensor to capture and memorize a whole chunk of the environment. It is typically based on extraction of certain image features and estimation of their motion within a sequence of images, which requires intensive image processing. Some other examples of perceptually and computationally demanding input variables not confined to the area of visual navigation include the closest point on the obstacle boundary, the distance to this point, or the value of another function determined by the entire boundary; see, e.g., [61, 119, 203, 208–210, 214–220] for representative samples. Another such variable employed by many proposed controllers (see, e.g., [206, 218, 219]) is the boundary curvature, which is particularly sensitive to corruption by measurement noises since this is a second derivative property.

Sliding mode controllers fed by non-demanding input variables immediately provided by certain perceptually deficient sensory systems are offered in [205, 221]. The paper [205] deals with the problem of boundary following with a pre-specified margin by a Dubins-car-like mobile robot based on only the distance along and the reflection angle of the ray perpendicular to the robot's centerline. This situation holds if, e.g., the measurements are supplied by several range sensors rigidly mounted to the robot's body at nearly right angles from its centerline or by a single sensor scanning a nearly perpendicular narrow sector. This perception scheme is used in some applications to reduce the complexity, cost, weight, and energy consumption of the sensor system and to minimize the detrimental effects of mechanical external disturbances on the measurements. However, the related deficit of sensor data makes most known navigation solutions inapplicable in this case especially if this deficit is enhanced by low computation power of the digital processor, which is characteristic for, e.g., miniature robots. Moreover, this gives rise to special challenges, such as an inability to detect a threat of head-on collision in certain situations (e.g., motion over a circle centered at the obstacle boundary) or strong sensitivity of the overall output of the sensor system to the robot's posture.

In [205], a solution to the boundary following problem was given in the form of a switching control law; sliding mode effects play an essential role in achieving the control objective. Some other reactive controllers fed by data from perpendicularly mounted sensors have been proposed in [32, 206]. The control law from [32] is aimed at pure obstacle avoidance, with no objective to follow the boundary with a pre-specified margin. Boundary following with a given margin by a Dubins-car like robot was addressed in [206] via a hybrid (non-sliding mode) strategy of switching between Lyapunov-based highly nonlinear control laws in order to overcome singularities caused by concavities of the tracked curve. In doing so, restrictions on the steering control were neglected by assuming that the robot is capable of making arbitrarily sharp turns, and noise-sensitive estimates of the boundary curvature were essentially employed.

Another situation concerned with perceptually deficient sensory systems may occur in cases where the obstacle is constituted by an area which the robot is forbidden or unwelcome, rather than incapable, to penetrate. This may be a radioactively or chemically contaminated area, a region of hazardous weather conditions such as hurricanes or that filled by fire, vapor, poisonous gases, or contaminant clouds. Some such areas can be sensed as a whole by means of, e.g., remote capabilities of satellites or radars, which opens the door to computation or estimation of the distance to the obstacle, the curvature of its boundary, or other variables used as regular inputs by many available controllers. However, there are many scenarios where such observation is troublesome, e.g., because of obstructions in urban, forest, or indoor environments, or is not precise or frequent enough, like, e.g., in fire spreads [222], or is infeasible since observation is physically based on immediate contact with the sensed entity, like a transparent fluid, gas, or radiation. For many such cases, the "forbidden" area is that where the value of an unknown scalar environmental field exceeds a given threshold, and the sensors observe this value in a point-wise fashion, i.e., at the location of the sensor. In this case, boundary following takes the form of tracking an environmental level set, and the control objective shapes into driving the field value, treated as the system's output, to a predefined value.

The problem of tracking environmental level sets has gained much interest in the control literature. Most of the related publications fall into two categories [223]: one of them assumes access to the field gradient or derivative-dependent[2] information (see, e.g., [224–227]) and the other is the gradient-free approach; see, e.g., [222, 228–230]. Gradient-based contour estimation by multiple sensor platforms was studied in [224, 225, 231]. The centralized methods developed in [224, 231] originate from the "snake" algorithms in image segmentation; the cooperative algorithm from [225] considers the optimal spread of the sensors over the estimated contour to minimize latency. Artificial potential approach based on direct access to the gradient was applied in [227] to decentralized cooperative boundary tracking by a team of velocity-controlled points. By assuming the ability to move over the isoline for granted, as well as access to its curvature and tangent, [232] proposes an algorithm for uniform cooperative distribution of mobile sensors along the estimated boundary in order to ensure its optimal polygonal approximation. Collaborative estimation of the gradient and Hessian of a scalar field corrupted by noise was used in [226] to develop a control law driving the center of a rigid formation of multiple mobile sensors along level curves.

In practice, derivative information is in fact unavailable in many cases, whereas its estimation requires access to the field values at several locations. Teams of mobile sensors have extended capacity of the latter thanks to collaborative sensing and data exchange. However, even in this case, limitations on communication may require a mobile sensor to operate individually for considerable time and distance. The practical scenario of a single mobile sensor with access to only point-wise measurements of only the field value is the main motivation for gradient-free approaches.

Gradient-free bang-bang type steering controllers were reported in [223, 233]. The control is via switches between alternative steering angles depending on whether the current field value is above or below the threshold of interest. In [228], similar in spirit approach with a larger set of alternatives was applied to an underwater vehicle equipped with a profile sonar. These methods typically result in zigzagging behavior. This is in line with the approach to the gradient climbing, which arranges for acquisition of extra information via extra maneuvers by "dithering" the sensor position [234–236]. However, systematic superfluous maneuvers may be required for this, whereas the multiple sensor scenario means more complicated and costly hardware. A method to control an unmanned aerial vehicle based on segmentation of the infrared local images of a forest fire was proposed in [237]. These works rely, more or less, on heuristics and in fact offer no rigorous and completed justification of the proposed control laws. A linear PD controller fed by the current field value was proposed in [238] for steering a unicycle-like mobile robot along a level curve of a field given by a radial harmonic function, and a local convergence result was established for a robot with unlimited control range.

[2] For example, the curvature of the isoline.

The sliding mode approach was employed in [221] to design a controller that ensures tracking the desired environmental level set, does not employ gradient estimation, and is non-demanding with respect to both computation and motion. It is assumed that the on-board control system of the robot has access only to the field value at the robot's current location and is capable of accessing the rate at which this reading evolves over time. The paper [221] discloses requirements of the isoline necessary for the mobile robot with limited turning capacity to be capable of tracking the isoline. It is rigorously proved that whenever these minimal requirements are met in a slightly enhanced form and are extended on the transient, the propose controller does ensure global convergence to and subsequent following along the required isoline provided that the controller parameters are properly tuned, with explicit recommendations on their choice being offered.

The major obstacle to implementation of sliding mode controllers is a harmful phenomenon called "chattering" [70], i.e., undesirable finite frequency oscillations around the ideal trajectory due to unmodeled system dynamics and constraints. The problem of chattering elimination and reduction has an extensive literature (see, e.g., [87] for a survey), which offers a variety of effective approaches, including continuous approximation of the discontinuity, inserting low-pass filters/observers into the control loop, combining sliding mode and adaptive control techniques, higher-order sliding modes, etc. The issue of chattering was addressed in this book via computer simulations and real-world experiments.

A sample of a completed reactive control strategy of target reaching with obstacles avoidance that employs sliding mode-based boundary following as a hint to bypass en-route obstacles can be found in [33]. The considered scenario assumes that the workspace contains a steady point target and several disjoint steady obstacles. Under some technical assumptions, it is shown that the proposed control strategy does bring the robot to the target in a finite time while always respecting a given safety margin. The proofs are basically indifferent to the particulars of the boundary-following control law so that the convergence has solid potential to remain true when using another convergent boundary following algorithm respecting the same margins during transients. The aforementioned assumptions stipulate in particular that the obstacles are convex and substantially spaced. Sliding mode boundary following was in effect concerned in [61], though the proposed control strategy does not explicitly offer to follow the obstacle boundary.

3.4.1.3 Bug algorithms

The Bug family of algorithms [11, 12, 63, 239–247] originated in [248, 249] are among the first-target reaching methods for which global convergence in complex scenes has been rigorously established; for a systematic survey of this family, we refer the reader to [216, 250]. Most Bug-family algorithms directly employ boundary following in close range as a hint to bypass an encountered obstacle. Motion toward the target is resumed as soon as a leaving condition is satisfied, which is designed to ensure global convergence and for which verification often requires memorizing some prior sensory information. At the same time, these algorithms typically do not include any specific method of boundary following but instead take the capability of such following for granted. In this sense, they can be viewed as higher-level control strategies rather than completed methods of reactive navigation. What is more, practical implementation of this and other assumed capabilities (e.g., for instantly turning) not only may constitute a separate engineering problem but also may be impossible due to kinematic or dynamic constraints that cannot be ignored in practice. Implications of this impossibility basically lie in an uncharted territory.

A completed Bug-type reactive navigation strategy for target reaching with obstacle avoidance for a nonholonomic under-actuated and control-saturated mobile robot of the Dubins-car type was proposed in [61]. Based on access to the target-bearing angle where the sum of full turns of the target line-of-sight is reckoned, the reactive control law was shown to be capable of achieving the global robotics task: target reaching in a simply connected maze-like environment. This holds with the fixed turn direction when reaching the maze boundary provided that the robot's initial location and the target are not deep inside the maze. Otherwise, the claim is true with probability 1 if this direction is sometimes randomly updated. All these are valid if the robot is maneuverable enough to cope with the narrows and contortions of the maze, with constructive conditions for this being provided. Limited randomization permits resolving navigational limit cycles without violation of the reactive nature of the overall algorithm. A number of deterministic methods to prevent these cycles is known (see, e.g., [251] and the above literature on the Bug-type algorithms), but they hardly can be classified as purely reactive.

3.4.1.4 Full information based

In situations where more information about the obstacle is available, a clearer view of the immediate environment can be recreated. This means that more informed navigation decisions may be made. This can lead to desirable behaviors, which include, e.g., variable speed and offset distance from the obstacle. This also carries the potential to loosen assumptions on the obstacle shape and curvature. An example of such behavior may be slowing down at concavities of a boundary and speeding up otherwise, or completely skipping concavities of sufficiently small size, which may introduce singularities into the motion [33, 206].

One such approach using abstract obstacle information is the VisBug class of algorithms, which navigates toward a visible edge of an obstacle inside the detection range; see, e.g., [63, 101, 247]. However, these algorithms are concerned with the overall strategy, and are not concerned with details relating to the robot's kinematics or the sensor model. Several approaches have been able to account for the robot's dynamics, but still have inadequate models of the sensors. For example, in [252] the *joggers problem* was proposed, where the idea was to guarantee safe navigation by ensuring that the robot can stop in the currently sensed obstacle-free set. However, an abstract sensor model was used, which presumes that the robot has continuous knowledge about the obstacle set. In [253], boundary following was achieved by picking *instant goals* based on observable obstacles. A ray-based sensor model was used, though a velocity controlled holonomic model of the robot was assumed. In [254], instant goals are also employed, and allowance is made for the robot's kinematics; however, a ray-based obstacle sensor model was not in use.

An MPC-based approach to boundary following has been proposed, which generates avoidance constraints and suitable target points to achieve boundary following [255]. This was found to give better performance than existing methods when applied to acceleration-constrained mobile robots, and may be a first step to applying MPC to boundary following problems.

3.4.2 Sensor-based path planning

Trajectory planning using only sensor information was originally termed the *joggers problem*, since the mobile robot must always maintain a path that brings it to a halt within the currently sensor area; see, e.g., [252, 256].

The classic *Dynamic Window* (see, e.g., [35, 257, 258]) and *Curvature Velocity Method* (see, e.g., [140, 259]) can be interpreted as a planning algorithm with a prediction horizon of a single time step [257]. To this end, the range of considered control inputs is limited to those bringing the robot to a halt within the sensor visibility area, using only circular paths. This can also be easily extended to other robot's shapes and models [260]; further, measures are available that reduce oscillatory behavior [261]. A wider range of admissible trajectory shapes has also been considered [262]. The Lane-Curvature Method (see, e.g., [263]) and the Beam-Curvature Method (see, e.g., [140, 259]) are both variants based on a different trajectory selection process from a similar class of possible trajectories.

In all these cases, justification for collision avoidance is based on essentially the same argument: the robot can stop while moving along the chosen trajectory. The differences in performance are mainly heuristic.

Approaches similar to the dynamic window have also been extended to cases where disturbance is present, using an approach similar to the tube MPC [264]. In addition, navigation systems that generate obstacle constraints by processing information from a ray-based sensor model have been proposed [255, 265].

MPC type approaches have previously been used to navigate mobile robots in unknown environments; see, e.g., [266–268]. Here, a MPC algorithm is combined with some type of mapping algorithm; however, in this case, some of the rigorous guarantees normally provided in MPC approaches are harder to show. Robust MPC may also be used in unknown environments; see, e.g., [264]. However, there appears to be room for more research in this particular area.

An interesting approach to collision avoidance using these types of methods is to estimate obstacle positions based on bearing measurements combined with some state estimation method [269]. In this case, observability constraints can be taken into account during planning.

When compared to potential field methods, MPC methods generally perform better as they consider a more optimal path that plans ahead as obstacles are approached. They are also less conservative, bringing the robot closer to the edge of its control capability.

3.4.3 Other reactive methods

In this section, we describe methods that do not explicitly generate a path around obstacles and do not explicitly perform boundary following.

Many approaches to this problem assume holonomic velocity controlled mobile robots. In this case, this is not a severe limitation, as techniques for extending such methods to arbitrary dynamics—including acceleration constrained vehicles—are available; see, e.g., [262, 270, 271]. These techniques are based on transformations that provide a zone around the vehicle that essentially accounts for perturbations introduced by the dynamics, and may be applied to a range of navigation approaches. This means that the use of the simple model is of a lesser concern. Alternatively, a method has been proposed that guarantees collision avoidance by ensuring that the distance to obstacles is always greater than the stopping distance [272]. This approach may be useful if little is known about the robot's model.

3.4.3.1 Artificial potential field methods

A classic approach to reactive collision avoidance is to construct a virtual potential field that causes repelling from obstacles and attraction to the target. These are termed *Artificial Potential Field* (APF) methods. Note that they are different from APF-based path planning methods: in this section, methods are considered that compute the control input on-line based on the available information.

Several improvements are listed as follows:

- *Unicycle kinematics.* Performance can be improved on mobile robots with unicycle type kinematics. Specifically, this can be achieved by moving the robot's reference point slightly away from the center of the robot; see, e.g., [273, 274].
- *Local minima avoidance.* The shape of the potential field can be designed to flow around obstacle concavities. Some respective findings are based on harmonic potential fields and provide better performance with local minima, though in general, they cannot be fully solved by using deterministic reactive algorithms; see, e.g., [275, 276].
- *Closed-loop performance.* Alteration to the shape of the potential field may lead to an improvement to the closed loop performance; see, e.g., [277, 278]. Additionally, reduction of oscillations in narrow corridors may be achieved; see, e.g., [274, 279]. However, in general, the closed loop trajectories of APF-based methods would not be optimal.
- *Limited obstacle information.* There are examples where only the nearest obstacle point is available [280]. Approaches that assume global knowledge about the workspace would not be suitable for sensor-based navigation.
- *Actuator constraints.* Examples that focus on satisfying actuator constraints are also available; see, e.g., [281, 282]. However, these methods do not generally directly achieve acceleration bounds.

Artificial potential field methods have lower computational requirements than local planning approaches, but this is becoming less of a concern with ever-increasing computational powers of mobile robots.

3.4.3.2 Uncategorized approaches

There are many other approaches that achieve collision avoidance but do not fit into the above categories:

- The *Safe Maneuvering Zone* is suited for kinematic unicycle model with saturation constraints, when the nearest obstacle point is known [283]. This is somewhat similar to the *Deformable Virtual Zone*, where the navigation is based on a function of obstacle detection ray length [9], though collision avoidance is not explicitly proven in [9].
- The *Vector Field Histogram* directs the robot toward sufficiently large gaps between detection rays [284]. The *Nearness Diagram* is an improved version of this method; it employs a number of behaviors for a number of different situations, providing good performance even in particularly cluttered environments; see, e.g., [28, 62].
- The method of probabilistically convergent on-line navigation involves randomly choosing tangents to travel down (see, e.g., [16]), or employs the deterministic *TangentBug* algorithm; see, e.g., [12]. Tangent events can be detected from a ray-based sensor model (see, e.g., [140]) or by processing data from a camera sensor; see, e.g., [141]. This results in an abstract tangent sensor, which reports the angles to tangents around the robot. A common method of achieving obstacle avoidance is to maintain a fixed angle between the tangent and the robot's motion; see, e.g., [141, 285].
- A collision avoidance system based on MPC has been proposed and shown to successfully navigate real-world helicopters in unknown environments based on the nearest obstacle point within the visibility radius [182]. However, this is less concerned with planning safe trajectories, than with controlling vehicle dynamics.
- A different class of navigation laws is based on the *Voronoi Diagram*, which essentially describes the set of points equidistant from adjacent obstacles. In general, it leads to longer paths than the tangent graph, though it represents the smallest set of trajectories that span the free space in an environment. Navigation laws have been developed to equalize the distance to obstacles, when a velocity controlled unicycle kinematic model is assumed; see, e.g., [286].
- A method has been proposed that considers the set of velocities that avoid obstacles without performing explicit path planning [144]. This method uses concepts borrowed from LQG feedback control to account for uncertainty, and is applicable to various types of linear models of mobile robots.

3.5 Moving obstacles

Certain types of mobile robots unavoidably encounter moving obstacles, which are generally more challenging to avoid than static counterparts. The main factors that underlie the hardness of this problem are as follows: the need to characterize possible actions of another object; the increased complexity of the search space and terminal constraints in the case of path planning; and additional conservativeness in the case of sensor-based systems.

At one extreme, an obstacle translating with a constant velocity vector may be accounted for by merely calculating the future position of the obstacle. The other extreme is an obstacle pursuing the robot, for which the set of potential locations grows polynomially along the planning horizon. Several offerings also describe integrated approaches, including estimation of obstacle motion from LiDAR sensors [287]. However, discussion is focused on the avoidance behavior in this section.

General planning algorithms suited for dynamic environments are also available; but without assumptions about the obstacles, it is impossible to guarantee existence of a viable path; see, e.g., [288]. When planning in known environments, states that necessarily lead to collision—the *Inevitable Collision States*—may also be abstracted and used to assist planning [289]. If the motion of the robot is known stochastically, the overall probability of collision for a probational trajectory may also be computed based on the expected behavior of the obstacles; see, e.g., [290].

3.5.1 Human-like obstacles

Several works attempt to characterize the motion of moving obstacles. For avoiding humans, several models of a socially acceptable pedestrian behavior are available; see, e.g., [291–294]. An approach that avoids obstacles based on the concept of personal space has been proposed and works well in practice [292]. Other approaches that can avoid human-like obstacles while also considering the reciprocal effect of the robot's motion have also been proposed [293, 294].

3.5.2 Known obstacles

Obstacles translating at a constant speed and in a constant direction may be avoided using the concept of a *velocity obstacle*; see, e.g., [19, 295, 296]. This is essentially the set of robot's velocities that will result in collision with the obstacle; so by avoiding these velocities, collisions may be prevented. This result may be extended to arbitrary (but known) obstacle paths and more complex robot's kinematics by using the concept of *nonlinear velocity obstacle*; see, e.g., [17]. The velocity obstacle method also extends to 3D spaces; see, e.g., [297, 298].

3.5.3 Kinematically constrained obstacles

When obstacles are only known to satisfy nominal kinematic constraints, the set of possible obstacle positions grows drastically over time.

3.5.3.1 Path-based methods

There are three basic methods of planning trajectories that avoid such obstacles:

- Ensuring that whenever a collision could possibly occur, the robot is stationary—this is referred to as *passive motion safety*; see, e.g., [299, 300]. In some situations, it is impossible to show any higher form of collision avoidance; in any case, this approach partly relies on the behavior of obstacles to avoid collisions.

- Ensuring that the robot can move arbitrarily far away from the obstacle over a infinite horizon. This is discussed in [301], where the time minimal paths were calculated. Similar examples of approaches include more allowance for other uncertainties; see, e.g., [174].
- Ensuring the mobile robot lies in a set of points that cannot be easily reached by the obstacle. This is proposed in [42], where a non-empty set of points is found under certain assumptions that lies just behind the obstacles movement direction. This allows avoidance over a infinite horizon, while being possibly less conservative than the previous option.

When performing path planning in a sensor-based paradigm, the main additional assumption is that any occluded part of the workspace must be considered as a potential dynamic obstacle [299, 302]. Naturally this makes the motion of a mobile robot even more conservative.

3.5.3.2 Reactive methods

When moving obstacles are present in the workspace, it is still possible to design reactive navigation strategies that can provably prevent collisions, at least with some more restrictive assumptions about the obstacle motion. These methods are outlined as follows:

- When obstacles are sufficiently spaced (so that multiple obstacles must not be simultaneously avoided), an extension of the velocity obstacle approach has been designed to prevent collisions [161]. When using this method, the robot is effectively steered toward the projected edge of the obstacle.
- Some artificial potential field methods have been extended to moving obstacles, though without rigorous justification; see, e.g., [274, 303].
- Some sliding mode boundary following techniques reported in Section 3.4.1.2 have been successfully extended to handle moving obstacles with provable collision avoidance, assuming that the obstacles are sufficiently spaced and their motion and/or deformation satisfy some technical smoothness constraints; see, e.g., [52, 60].

In particular, Matveev et al. [60] examines the navigation strategy proposed in [33]. It consists of properly switching between moves to the target in straight lines, when possible, and sliding mode based bypassing enroute obstacles at a pre-specified distance by applying a sliding mode control law. While the convergence and performance of this method were shown in [33] only for static scenes with convex obstacles, Matveev et al. [60] demonstrated the viability of this strategy for robots traveling in dynamic environments cluttered with arbitrarily shaped obstacles. They are not assumed to be rigid or even solid: they are continuums of arbitrary and time-varying shapes undergoing general motions, including rotations and deformations. Since the strategy includes a controller that ensures patrolling of the boundary of a moving domain at a pre-specified distance, the proposed solution can be used for pure border surveillance, which is of self-interest. It should be noted that the theoretical guarantees given in [60] concern only boundary following problem by proving non-local collision-free convergence of the robot to the pre-specified distance to the moving and deforming boundary. Global target reaching in cluttered environments was demonstrated by computer simulations and experiments with a real wheeled robot.

In [304], the results of [60] on target reaching with obstacle avoidance were extended on a planar differential drive robot, exemplified by an automated intelligent wheelchair. This robot has two independently actuated driving wheels mounted on the same axle and may be castor wheels, which do not affect its mobility. The driving wheels roll without sliding. The robot is controlled by the angular velocities of the left and right driving wheels, which are limited by a common and given constant. As in [60], obstacles are continuums of arbitrary and time-varying shapes undergoing general motions, including rotations and deformations. The main theoretical result guarantees non-local collision-free convergence of the robot to the desired distance to the moving and deforming boundary while always respecting a given safety margin provided that some technical and partly unavoidable assumptions are satisfied. Global convergence to the target is illustrated via computer simulations and experiments with a real wheelchair. Furthermore, the algorithm of [60] was successfully applied to the problem of collision-free navigation of intelligent robotic hospital beds for critical neurosurgery patients [162].

A quite simple and computationally efficient biologically inspired algorithm of collision-free navigation among moving obstacles was proposed in [52]. The proposed navigation strategy is based on switching between moving to the target along straight lines, when possible, and a sliding mode obstacle avoidance navigation law. Mathematically rigorous analysis of the algorithm was given for the case of round obstacles moving with constant velocities. Computer simulations and experiments with a real robot show that the algorithm outperforms some well-known other methods such as Artificial Potential Field and velocity obstacle based navigation laws.

3.6 Multiple robot navigation

Navigation of multiple robots has gained much interest in recent years. As autonomous vehicles are used in greater concentrations, the probability of multiple vehicle encounters correspondingly increases, and new methods are required to avoid collisions. A typical scenario is shown in Fig. 3.3.

The study of decentralized control laws for groups of mobile robots has emerged as a challenging new research area in recent years; see, e.g., [305, 306] and references therein. Broadly speaking, this problem falls within the domain of decentralized control, but the unique aspect of it is that groups of mobile robots are dynamically decoupled, meaning the motion of one robot does not directly affect that of the others. This scenario is viewed as that of a networked control system, which is an active field of recent research. For examples of more generalized work in this area, see, e.g., [307–310]. One of the important applications of navigation of multi-robot systems is sensing coverage [311–313]. To improve coverage and reduce the cost of deployment in a geographically-vast area, employing a network of mobile sensors for the coverage is an attractive option. Three types of coverage problems for robotics were studied in recent years: blanket coverage [314, 315], barrier coverage [316, 317], and sweep coverage [316, 318]. Combining existing coverage algorithms with effective collision avoidance methods is an open practically important problem.

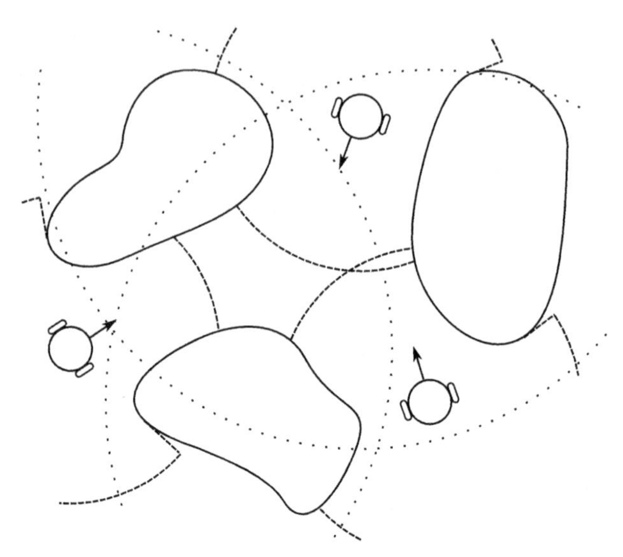

Figure 3.3 Diagram of the multi-vehicle collision avoidance problem. This includes communication, sensing, and dynamic constraints.

While there is extensive literature on centralized navigation of multiple robots, it is only briefly mentioned here, since it is generally not applicable to arbitrarily scalable on-line collision avoidance systems. Examples of off-line path planning systems that can find near optimal trajectories for a set of vehicles are available; see, e.g., [190, 319]. Another variation of this problem involves a precomputed prescription of the paths to be followed, where the navigation law must only find an appropriate velocity profile that avoids collisions; see, e.g., [320, 321].

One of the issues facing navigation systems for multi-robot problems is deadlock and livelock. In the broadest definition, deadlock occurs when robots do not converge to their destinations, and instead converge to an arrangement that cannot be resolved using the nominal controller. Similarly, the term "livelock" means the robots continue to move indefinitely, but without converging to their respective destinations. These are situations that are very non-trivial to solve, but certain methods are currently able to provably achieve target convergence.

3.6.1 Communication types

There are three common modes of communication in multiple robot collision avoidance systems:

- *Direct state acquisition.* This can be achieved without communication by using only sensor information to measure the states of the surrounding mobile robots, and is employed in many non-path based reactive approaches.
- *Single direction broadcasting.* In addition to the physical state of the robot, additional variables are also transmitted; they are usually related to the current trajectory of the robot. This is often used in building path constraints based on mutual exclusion, where the projected

states of the other robots are avoided during planning, and is occasionally referred to as *sign board* communication.

- *Two-way communication.* This can range from simple acknowledgement signals to fully decentralized optimization algorithms. These are commonly used for decentralized MPC, though some MPC variants have been proposed where sign boards are sufficient.

A variety of models of communication delay and error are considered in networked navigation problems. The ability to cope with communication delays, packet dropouts, and finite communication ranges is definitely desirable in any navigation system.

3.6.2 Reactive methods

The most basic form of this problem considers only a small number of mobile robots with at most a single point obstacle. For example, navigation laws have been proposed to avoid collisions between two vehicles traveling at constant speeds with turning rate constraints; see, e.g., [322, 323]. A common example of this type of systems is an *Air Traffic Controller*. However, such navigation systems do not directly relate to avoiding collisions in cluttered environments [324].

3.6.2.1 Potential field methods

Potential field methods may be constructed to repel companion mobile robots. In some ways, this approach is more satisfactory than the equivalent methods applied to static obstacles: for example, local minima are less of an issue in the absence of contorted obstacle shapes. Methods have been proposed that avoid collision between an unlimited number of velocity controlled unicycles or velocity controlled linear vehicles [325–327]. One variant, called the *multi-vehicle navigation function*, is able to show convergence to targets in the absence of obstacles. However, these still use similar types of repulsive and attractive fields; see, e.g., [328–330].

Other variants also include measures to provably maintainable cohesion between groups (see, e.g., [331]), while others have also been applied to robots with limited sensing capabilities [332]. Navigation function methods have also been combined with MPC; this supplies some type of planning into the future to reject disturbances [333]. Navigation function methods have also been applied to 3D scenarios [334].

In cases where finite acceleration bounds are present (but still without any nonholonomic constraints), a mutual repulsion based navigation system with a more sophisticated avoidance function has been proven to avoid collisions for up to three mobile robots [335]. When more robots are present, it is possible to back-step the additional dynamics into a velocity controlled model [282, 336]. Furthermore, recently proposed control methods can achieve collision avoidance for Lagrangian systems with bounded inputs [336].

Some other methods provide good practical results, though without focusing on mathematical analysis of collision avoidance; see, e.g., [280, 337, 338]. Many of these methods can be extended to static obstacles, and these combined systems are achieved by the same avoidance functions as in the single robot case; see, e.g., [282].

3.6.2.2 Reciprocal collision avoidance methods

Approaches termed *Reciprocal Collision Avoidance* (RCA) achieve collision avoidance by assuming each robot takes some of the responsibility for each pairwise conflict, with the resulting constraints forming a set of viable velocities from which a selection can be made using linear programming; see, e.g., [339, 340]. Some interesting extensions have been proposed to the RCA concept; for example, it has been applied to both nonholonomic vehicles and linear vehicles with acceleration constraints, while maintaining collision avoidance [340–342]. The method may be extended to arbitrary models of mobile robots, as rigorous avoidance is achieved via addition of a generic bounded-deviation path-tracking system [340, 343, 344]. These methods are also able to include collision avoidance of static obstacles, which easily integrates into the navigation framework.

This idea is somewhat similar to a previous method based on collision cones and called *Implicit Cooperation* [345]. Another method has also been proposed that is based on collision cones; this method is called *Distributed Reactive Collision Avoidance*. It has the benefit of showing achievement of the robot's objective in limited situations, ensuring that minimum speed constraints are met when global information is available, and showing robustness to disturbances [346, 347].

3.6.2.3 Hybrid logic approaches

For these approaches, discrete logic rules are used to coordinate robots. In most cases, this is achieved through segregation of the workspace into cells such that each cell can hold at most one robot; see, e.g., [348–351]. Then collisions can be prevented by devising a scheme where many robots do not attempt to occupy the same cell simultaneously. Additionally, many methods of integrating such approaches with control of the robot's dynamics have been proposed; see, e.g., [352].

Generation of cells may be on-line and ad-hoc. This is useful when minimum speed constraints are present: the robots may be instructed to maintain a circular holding pattern, and then to appropriately shift the pattern, when safe. In this case, various options for the shifting logic have been proposed; for example, some of them are based on robot priority [353], or traffic rules [354]. These methods are able to show convergence to the desired states, thus avoiding deadlock.

3.6.3 Decentralized MPC

While optimal centralized MPC is theoretically able to coordinate groups of mobile robots, the underlying optimization process appears to be too complex for many scalable real-time applications. Examples of centralized MPC for multiple robot systems are available; see, e.g., [355].

Decentralized variants of MPC in general do not specifically address the problem of deadlock. For example, in [356], a distributed navigation system is proposed that is able to plan near optimal solutions that robustly prevent collisions, allow altruistic behavior between the robots, and monotonically decrease the global cost function.

However, this does not equate to deadlock avoidance, which can currently only be solved in general by using discrete graph based methods. One very simple approach has been proposed to avoid this problem without a graph abstraction, though it has several limitations [357].

A review of general decentralized MPC methods is available [358], along with a review specific to robot navigation [359]. There are currently four main methods of generating de-conflicted trajectories that seem suitable for coordination of multiple robots:

- Decentralized optimization may find a near-optimal solution for a multi-agent system via dual decomposition, thus supplying a set of trajectories for a team of mobile robots; see, e.g., [360–362]. While this is more efficient than centralized optimization, it requires many iterations of communication exchange among the robots in order to converge to a solution. Other decentralized planning algorithms may also be effective; for example, decentralized random tree based methods have been proposed that allow for alteration of neighbors plans [363].
- Other approaches are based on multiplexed MPC (see, e.g., [166, 364]) and sequential decentralization; see, e.g., [365–367]. The robust control input for every robot may be computed by updating the trajectory for each robot sequentially, at least when they are close. While multiplexed MPC is suited to real-time implementation, its possible disadvantage is that path planning cannot be simultaneously performed by two adjacent robots. However, this framework has been extended to provide collision avoidance in vehicle formation problems [368].
- Another possible solution is to request acknowledgement signals before implementing a possible trajectory; it has the benefit of not requiring the robots to be synchronized. This method seems an effective solution [369, 370], but interaction between the robots may cause planning delays under certain conditions.
- Approaches have been proposed that admit a single communication exchange per control update [177, 371, 372]. This is done by including a coherence objective to prevent any robot from changing its planned trajectory significantly after transmitting it to other robots. Another approach uses various types of constraints to avoid coherence objectives, though it only works with a limited class of planning algorithms [145].

MPC may also easily include maintenance of objectives other than collision avoidance. For example, radio propagation models have been included in a path-evaluation function in order that communication between robots be maintained [373, 374].

When compared to potential field based methods, MPC more naturally accounts for obstacles and complex models of the robot. Preliminary results have shown that MPC-based methods have better closed-loop performance than potential field based ones [145].

Shortest path algorithm for navigation of wheeled mobile robots among steady obstacles

4

4.1 Introduction

In this chapter, we first consider the problem of global shortest path planning in a known environment. The robot should reach a given steady target in a scene cluttered with obstacles. The obstacles are possibly non-convex, the curvature of their boundaries does not exceed a certain threshold. In this context, we prove that the shortest (minimal in length) path consists of edges of the so-called tangent graph (T-graph), which will be specified in Remark 3.1. Therefore, shortest path planning is reduced to a finite search problem. Furthermore, we also consider a problem of on-line reactive navigation in an unknown environment. We propose a randomized navigation algorithm and prove that it drives the robot to the target with probability 1, while avoiding collision with the obstacles during the maneuver.

In this chapter, we study a unicycle like mobile robot described by the standard nonholonomic model with a hard constraint on the angular velocity. As was pointed out in Section 3.2, motions of many wheeled robots, missiles, and unmanned aerial vehicles can be described by this model. Unlike many other papers on this area of robotics, which present heuristics-based navigation strategies, we give mathematically rigorous analysis of the proposed navigation algorithms, along with complete proofs of the stated theorems. A Hamilton-Jacobi-Bellman equation based approach to the shortest path-planning problem in environments with obstacles was proposed in [375]. It should be also remarked that many papers on this topic (see, e.g., [10–13]) do not assume nonholonomic constraints on a robot's motion, which are a severe, yet realistic, limitation. The performance of the proposed real-time navigation strategy is confirmed with extensive computer simulations and outdoor experiments with a Pioneer P3-DX mobile wheeled robot.

The main results of this chapter were originally published in [16].[1] A preliminary version of some of these results was also reported in the conference paper [376].

The reminder of the chapter is organized as follows. Section 4.2 presents the system description and the assumptions. Section 4.3 considers the problem of shortest path planning. The problem of reactive navigation in unknown environments is introduced

[1] Savkin and Hoy [16]. Copyright ©2013 Cambridge University Press. Reprinted with permission.

Safe Robot Navigation Among Moving and Steady Obstacles. http://dx.doi.org/10.1016/B978-0-12-803730-0.00004-4

and solved in Section 4.4. Computer simulations for the proposed reactive navigation algorithm are given in Section 4.5, while Section 4.6 presents experiments with a Pioneer P3-DX mobile wheeled robot.

4.2 System description and main assumptions

We consider a planar vehicle or wheeled mobile robot modeled as a unicycle. It travels with a constant speed and is controlled by the angular velocity limited by a given constant. The model of the robot is as follows (see Fig. 4.1):

$$\begin{aligned} \dot{x} &= v\cos\theta, & x(0) &= x_0, \\ \dot{y} &= v\sin\theta, & y(0) &= y_0, \\ \dot{\theta} &= u \in [-\overline{u}, \overline{u}], & \theta(0) &= \theta_0. \end{aligned} \tag{2.1}$$

Here **col** (x, y) is the vector of the robot's Cartesian coordinates, θ gives its orientation, and v and u are the speed and angular velocity, respectively. The maximal angular velocity $\overline{u}$ is given. The minimum turning radius of the robot amounts to

$$R_{\min} = \frac{v}{\overline{u}}. \tag{2.2}$$

Therefore, any path of the robot (2.1) is a plane curve satisfying the following constraint on its so-called *average curvature* (see, e.g., [146]): let $P(s)$ be the parametrization of this path by the arc length, then

$$\|P'(s_1) - P'(s_2)\| \leq \frac{|s_1 - s_2|}{R_{\min}} \quad \forall s_1 \neq s_2. \tag{2.3}$$

Here $\|\cdot\|$ denotes the standard Euclidean vector norm and $'$ stands for differentiation with respect to s. We employ the average curvature because it needs only C^1-smoothness of the path, whereas the "standard" curvature introduced in textbooks on differential geometry (see, e.g., [377]) requires C^2-smoothness and so may not exist at some points of the robot's path.

There is a steady point-wise target $\mathcal{T}$ and disjoint obstacles $D_1, \ldots, D_k$ in the plane. The objective is to drive the robot to the target through the obstacle-free part of the plane, while respecting a given safety margin $d_0 > 0$.

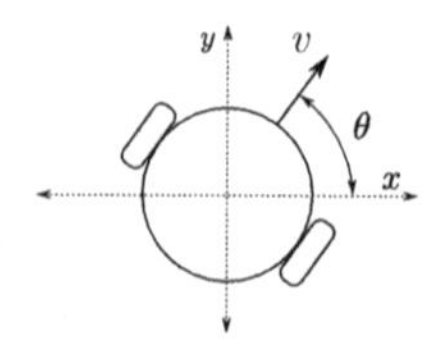

Figure 4.1 Unicycle model of a wheeled robot.

We recall that the distance **dist** $[\boldsymbol{p}; D]$ from a point $\boldsymbol{p}$ to a set D is defined as

$$\mathbf{dist}\,[\boldsymbol{p}; D] := \min_{\boldsymbol{q} \in D} \|\boldsymbol{p} - \boldsymbol{q}\|.$$

Here min is achieved since D is closed, and **dist** $[\boldsymbol{p}; D] = 0$ if $\boldsymbol{p} \in D$.

Definition 2.1. The d_0*-neighborhood* of the domain $D \subset \mathbb{R}^2$ is the set formed by all points at the distance no greater than d_0 from D, i.e.,

$$\mathcal{N}[D, d_0] := \{\boldsymbol{p} \in \mathbb{R}^2 : \mathbf{dist}\,[\boldsymbol{p}; D] \le d_0\}.$$

Definition 2.2. Let $\boldsymbol{r}(t) = \mathbf{col}\,[x(t), y(t)]$ be a path of the robot, which starts at $t = 0$ and terminates at some time $t = t_f \ge 0$. This path is said to be *target reaching with obstacle avoidance* if $\boldsymbol{r}(t_f) = \mathcal{T}$ and $\boldsymbol{r}(t) \notin \mathcal{N}[D_i, d_0]$ for all $i, t \in [0, t_f]$.

Assumption 2.1. *For any i, the set $\mathcal{N}[D_i, d_0]$ is closed, bounded, and linearly connected.*

Assumption 2.2. *The sets $\mathcal{N}[D_i, d_0]$ and $\mathcal{N}[D_j, d_0]$ do not overlap for any $i \neq j$.*

If Assumption 2.2 does not hold, target reaching with obstacle avoidance may be impossible; see Fig. 4.2.

Assumption 2.3. *For any i, the boundary ∂D_i of the obstacle D_i is a closed, regular, and analytic curve.*

Assumption 2.4. *For any i, the boundary $\partial D_i(d_0)$ of the neighborhood $\mathcal{N}[D_i, d_0]$ is a closed, regular, and analytic curve. The curvature $\kappa_i(\boldsymbol{p})$ of this curve at any its point $\boldsymbol{p}$ does not exceed $\frac{1}{R_{\min}}$.*

Here we use the standard definition of the signed curvature from differential geometry; see, e.g., [377]. This curvature is well-defined at any point of the boundary $\partial D_i(d_0)$ since $\partial D_i(d_0)$ is regular and analytic; the curvature assumes nonnegative and negative values on convexities and concavities of the boundary, respectively. Thus, Assumption 2.4 limits only the curvature of convex pieces of the boundaries.

We also assume that the initial position $\boldsymbol{r}(0) = \mathbf{col}\,[x(0), y(0)]$ of the robot is far enough from the obstacles and the target.

Assumption 2.5. *For any i, the following inequalities hold: $\|\mathcal{T} - \boldsymbol{r}(0)\| \ge 8R_{\min}$ and* **dist** $[\boldsymbol{r}(0); \mathcal{N}(D_i, d_0)] \ge 8R_{\min}$.

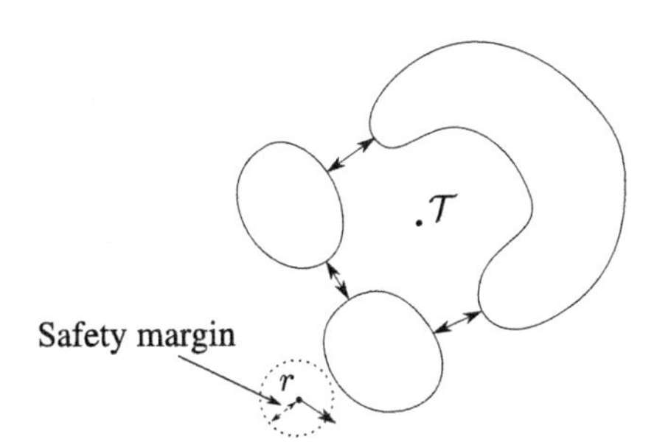

Figure 4.2 Target $\mathcal{T}$ cannot be safely reached: the spacing between obstacles is too small.

Definition 2.3. An *initial circle* is any of two circles with the radius $R_{\min}$ that crosses the initial robot's position $\boldsymbol{r}(0)$ tangentially to its initial heading $\theta(0)$.

Assumption 2.6. *The initial heading $\theta(0)$ of the robot is not tangent to any boundary $\partial D_i(d_0)$.*

4.3 Off-line shortest path planning

In this section, we describe the shortest target reaching paths with obstacle avoidance.

Definition 3.1. A straight line L is called a *tangent line* if it satisfies at least one of the following conditions:

1. The line L is simultaneously tangent to two boundaries $\partial D_i(d_0)$ and $\partial D_j(d_0)$, where $i \neq j$;
2. The line L is tangent to a boundary $\partial D_i(d_0)$ at two different points;
3. The line L is simultaneously tangent to a boundary $\partial D_i(d_0)$ and an initial circle;
4. The line L is tangent to a boundary $\partial D_i(d_0)$ and crosses the target $\mathcal{T}$;
5. The line L is tangent to an initial circle and crosses the target $\mathcal{T}$.

Points where a tangent line touches either $\partial D_i(d_0)$ or an initial circle are called *tangent points*. We also consider segments of tangent lines limited by a couple of tangent points such that the interior of the segment is disjoint with all boundaries $\partial D_i(d_0)$ and the disks bounded by initial circles. Such a segment is called

(OO)-segment if it is of types 1 or 2;
(CO)-segment if it is of type 3;
(OT)-segment if it is of type 4;
(CT)-segments if it is of type 5.

Definition 3.2. An arc of a boundary $\partial D_i(d_0)$ between two tangent points is called a *(B)-segment* if the curvature of the boundary is non-negative everywhere on this segment; see Fig. 4.3(a). An arc of an initial circle between the initial robot position $\boldsymbol{r}(0)$ and a tangent point is called a *(C)-segment*; see Fig. 4.3(b).

Now we are in a position to present the main result of this section.

Theorem 3.1. *Suppose that Assumptions 2.1–2.6 hold. Then there exists a shortest (minimum length) target reaching path with obstacle avoidance. Furthermore, a shortest target reaching path consists of $n \geq 2$ segments $S_1, S_2, \ldots, S_n$ such that*

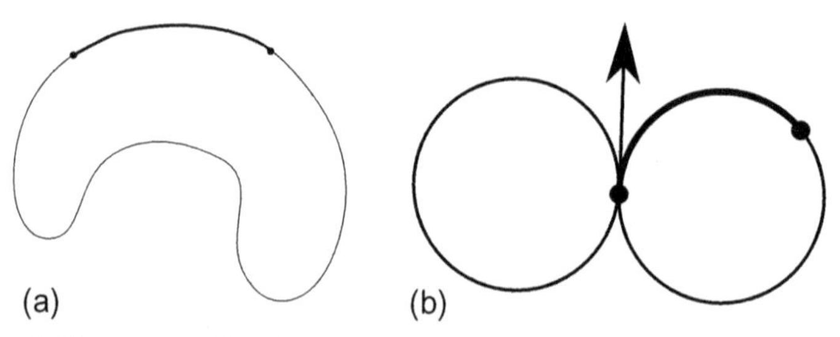

Figure 4.3 (a) (B)-segment; (b) (C)-segment.

if $n = 2$ then S_1 is a (C)-segment and S_2 is a (CT)-segment. If $n \geq 3$ then S_1 is a (C)-segment, S_2 is a (CO)-segment, and S_n is a (OT)-segment. If $n > 3$ and $3 \leq k \leq n-1$, then any S_k is either (OO)-segment or (B)-segment.

Proof. Existence is proved by merely retracing the respective arguments from [146].

Based on existence, we examine a shortest target reaching path P. Our first step is to show that P does not go inside the initial circles. To this end, we consider the circle C of radius $6R_{\min}$ centered at $\boldsymbol{r}(0)$. By Assumption 2.5, the target and all obstacles are outside C. Let $\boldsymbol{p}_1$ be the point of C at which the path P leaves C for the last time, and let θ_1 be the trajectory heading at $\boldsymbol{p}_1$. Thus, P consists of two segments $(\boldsymbol{r}(0), \boldsymbol{p}_1)$ and $(\boldsymbol{p}_1, \mathcal{T})$, where $(\boldsymbol{p}_1, \mathcal{T})$ is outside C. Among all paths that connect $\boldsymbol{r}(0)$ and $\boldsymbol{p}_1$, have the respective headings $\theta(0)$ and θ_1 at these points, and satisfy the average curvature condition (2.3), we pick a path P_1 of the minimum length. It follows from the main result of [146] that such a path exists and belongs to the disk of radius $8R_{\min}$ centered at $\boldsymbol{r}(0)$. Assumption 2.5 implies that P_1 does not intersect any $\partial D_i(d_0)$. Since P_1 is a minimum length path, its length L_1 does not exceed the length L of $(\boldsymbol{r}(0), \boldsymbol{p}_1)$. Were $L_1 < L$, the path composed of P_1 and $(\boldsymbol{p}_1, \mathcal{T})$ would be a shorter target reaching path than P, in violation of the choice of P. Hence, $L_1 = L$ and $(\boldsymbol{r}(0), \boldsymbol{p}_1)$ is also a shortest path satisfying (2.3). Then it follows from the main result of [146] that $(\boldsymbol{r}(0), \boldsymbol{p}_1)$ consists of segments of two or less minimum radius circles and a straight line segment, and $(\boldsymbol{r}(0), \boldsymbol{p}_1)$ does not cross the interiors of the both initial circles. Hence, P does not go inside the initial circles indeed.

Neglecting the average curvature constraint (2.3), we now consider all paths that connect $\boldsymbol{r}(0)$ and $\mathcal{T}$ and do not cross the interiors of the sets $\mathcal{N}[D_i, d_0]$ and initial circles. Let P_2 be the path that has the minimum length among them. Since the path P does not go inside any of two initial circles by the foregoing, $L(P_2) \leq L(P)$, where $L(\cdot)$ stands for the path's length. We are going to prove that P_2 is constituted by pieces of the boundaries of $\mathcal{N}[D_i, d_0]$ with non-negative curvature, arcs of initial circles, and straight line segments tangentially connecting points on these curves.

Suppose to the contrary that the path P_2 contains a point $\boldsymbol{p}_2$ that can be attributed to none of the above curves. As follows from Theorem 4 of [13], $\boldsymbol{p}_2$ is outside the sets $\mathcal{N}[D_i, d_0]$ and initial circles, and is not an interior point of a straight line segment of P_2. Therefore, in a vicinity of $\boldsymbol{p}_2$, a sufficiently small segment of P_2 can be replaced by a straight line segment, which does not intersect the sets $\mathcal{N}[D_i, d_0]$ and initial circles; see Fig. 4.4. This results in a path with a shorter length than P_2, in violation of the choice of P_2. Thus, we see that P_2 is composed of the aforementioned pieces, arcs, and segments. It remains to note that due to Theorem 5 in [13], any of these straight line segments connects either two tangent points or a tangent point and the target, with tangentially touching the concerned tangent points.

Finally Assumption 2.6 implies that the first part of P_2 is an arc of one of the initial circles. Since all involved pieces of the boundaries have curvature less or equal to $\frac{1}{R_{\min}}$, and all straight line segments of P_2 tangentially connect tangent points (or a tangent point and the target), the path P_2 satisfies (2.3). This completes the proof of Theorem 3.1.

□

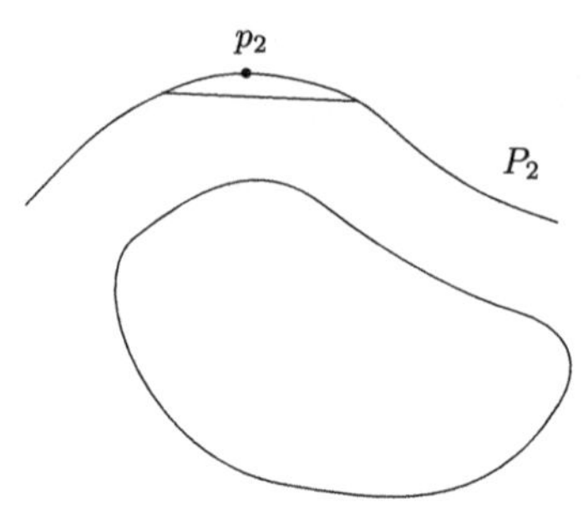

Figure 4.4 Reducing the path's length via replacement of a curvy segment by a straight line segment.

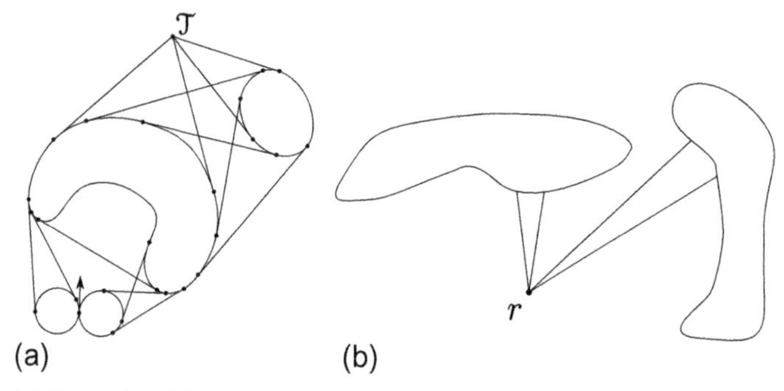

Figure 4.5 (a) Example of the tangent graph, with the target point $\mathcal{T}$; (b) The robot r is equipped with some type of vision sensor that is able to detect obstacles, tangents, and the target, so long as they are in line-of-sight.

Remark 3.1. Theorem 3.1 reduces the problem of finding the shortest target reaching path to a search among all paths in a special geometric graph in the plane, called the *tangent graph*. The set of its vertices consists of the robot's initial position, the target $\mathcal{T}$, and all tangent points, whereas the edges are the segments introduced in Definitions 3.1 and 3.2. This graph is finite since all the boundaries $\partial D_i(d_0)$ are analytic by Assumption 2.4. The problem of finding the shortest path in the graph is a finite optimization problem that can be solved by the standard dynamic programming.

Figure 4.5(a) shows an example of a tangent graph.

4.4 On-line navigation

Now we consider the case where the robot does not know the obstacles and location of the target a priori. The robot is equipped with a vision type sensor, which provides access to the coordinates of the point-wise target and points on the boundaries $\partial D_i(d_0)$ wherever the view from the sensor to the respective point is not obstructed by obstacles; see Fig. 4.5(b).

To simplify matters, we impose the following.

Assumption 4.1. *Any tangent point belongs to only one tangent line.*

To proceed, we need some definitions.

Let the robot move over the boundary of an obstacle or initial circle and let it reach a tangent point from which a straight line edge of the tangent graph outgoes. If at this moment the robot's heading is in the direction of this edge, like in Fig. 4.6(a), we say that the robot reaches an *exit tangent point*. Then the robot can pass from motion over the boundary to motion over this edge, whereas in the "otherwise" case this is impossible; see Fig. 4.6(b). If that straight line edge is a (OT) or (CT) segment, then the concerned exit tangent point is said to be the *T-type*. Otherwise, this edge is a (OO) or (CO) segment, and the point is said to the of *O-type*.

To navigate the robot, we propose the following probabilistic algorithm, which employs a parameter $0 < p < 1$:

A1: The robot starts to move along any of two initial circles.
A2: If it reaches an exit tangent point of T-type when moving along an obstacle boundary or initial circle, it then moves over the concerned (OT) or (CT) edge.
A3: If the robot reaches an exit tangent point of O-type when moving along an obstacle boundary or initial circle, it starts to move along the concerned (CO) or (OO) edge with probability p, and continues motion over the boundary or initial circle with probability $(1 - p)$.
A4: Whenever the robot reaches a tangent point on an obstacle boundary when moving over a (CO) or (OO) segment, it starts to move over this boundary.
A5: Whenever the robot reaches the target, the maneuver is terminated.

Now we are in a position to present the main result of this section.

Theorem 4.1. *Suppose that Assumptions 2.1–2.6 and 4.1 hold. Then with probability* 1, *the algorithm* ***A1–A5*** *builds a target reaching path with obstacle avoidance.*

Proof. The algorithm **A1–A5** defines a time-homogeneous Markov chain in which the states are the vertices of the tangent graph. With probability 1, the system cannot leave any state associated with the target. In other words, such a state is absorbing [378, Ch. 11]. Furthermore, there is no other absorbing state. It is also obvious that this absorbing state can be reached with non-zero probability from any other state of the chain, not necessarily in one step. This means the Markov chain itself is absorbing

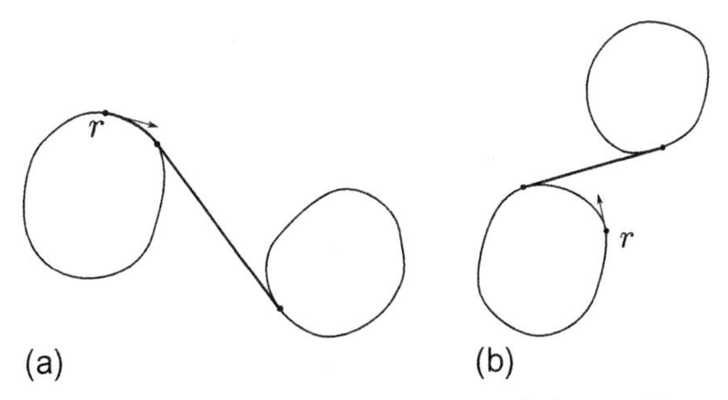

Figure 4.6 Robot may proceed over the tangent segment only in case (a).

[378, Ch. 11], and therefore implies that with probability 1, one of the absorbing states is inevitably reached [378, Ch. 11]. This completes the proof. □

In conclusion, we note that Theorem 4.1 does not guarantee that the path built by the algorithm **A1–A5** has the minimum length with probability 1. At the same time, it shares many basic features (highlighted in Theorem 3.1) with the shortest path.

4.5 Computer simulations

In this section, we present the results of simulation tests that deal with the robot (2.1) driven in accordance with the reactive navigation strategy **A1–A5**. To implement this strategy, we employ a lower-level guidance controller based on the approach proposed in [33] and reported in Chapters 5 and 6. This is a discontinuous controller that switches between three regimes of operation R_1, R_2, R_3. The robot turns as sharply as possible in mode R_1, performs a pure pursuit of a current destination point in mode R_2, and follows the boundary of the nearest obstacle in mode R_3. The respective control inputs are

$$u(t) = \begin{cases} \pm\overline{u} & \text{in } R1, \\ \Gamma \mathbf{sgn}\,[\beta(t)]\,\overline{u} & \text{in } R2, \\ \Gamma \mathbf{sgn}\,\left[\dot{d}(t) + \chi(d(t) - d_0)\right]\overline{u} & \text{in } R3. \end{cases} \tag{5.1}$$

Here

- $\beta(t)$ is the signed angle between the robot's heading and the ray from the robot to the current destination point, which is selected on-line (see (5.4) further);
- $d(t)$ is the minimum distance to the nearest obstacle;
- Γ is set to $+1$ in R_3 if the nearest obstacle is on the left, and also in R_3; $\Gamma = -1$ in R_3 if the nearest obstacle is on the right;
- $\chi(\cdot)$ is a linear function with saturation

$$\chi(z) = \begin{cases} \gamma r & \text{if } |z| < \delta, \\ \gamma\delta\mathbf{sgn}\,(z) & \text{otherwise}, \end{cases} \tag{5.2}$$

where $\gamma, \delta > 0$ are tunable parameters of the controller.

Switching between the regimes is regulated by the following rules, which employ one more parameter $d_{\text{trig}} > d_0$:

$$\begin{array}{llll} R1 \xrightarrow{\hspace{2cm}} R2 & \text{whenever} & CO \text{ or } CT \text{ is detected}, \\ R2 \xrightarrow{\hspace{2cm}} R3 & \text{whenever} & d(t) < d_{\text{trig}}, \dot{d}(t) < 0, \\ R3 \xrightarrow{\text{with probability } p} R2 & \text{whenever} & \begin{cases} OT \text{ is detected} \\ OO \text{ is detected} \end{cases}. \end{array} \tag{5.3}$$

In fact, switching to pursuit of a current destination point was artificially disabled for a pre-specified short time whenever the decision to ignore this point is made.

In the simulation tests and real-world experiments described in Section 4.6, the simplest two-point difference approximation of $\dot{d}$ from (5.1) was used. By assuming a LIDAR-type sensor, the robot was given access to the distance d_i to obstacles in finitely many directions, evenly distributed with spacing $\Delta\theta$; here the index $i = 0$ is associated with the direction in front of the robot. To detect a tangent point straight ahead, it was checked whether $|d_0 - d_{\pm 1}|$ exceeds a pre-specified threshold d_{thresh}. To decide whether the nearest obstacle is to the left or to the right, two immediately adjacent obstacle detections were compared. To calculate β, an intermediate target $\mathcal{T}_{\text{int}}$ was calculated via iteration of the following steps:

(1) Initially when $R2$ is engaged, the $\mathcal{T}_{\text{int}}$ is set to be a constant offset from the detected tangent point:

$$\mathcal{T}_{\text{int}} = \begin{bmatrix} x(t) + \cos(\theta(t) + i\Delta\theta + \Gamma \tan^{-1}(d_0/d_{\text{cen}}(t))) \\ y(t) + \sin(\theta(t) + i\Delta\theta + \Gamma \tan^{-1}(d_0/d_{\text{cen}}(t))) \end{bmatrix} \tag{5.4}$$

with $i := 0$, where $d_{\text{cen}}(t)$ is the estimated distance to the tangent point.

(2) β is calculated by finding the angle between the robot's heading and a line connecting the robot and $\mathcal{T}_{\text{int}}$.

(3) In subsequent time steps, a successive tangent point is computed: out of all couples of adjacent detection rays in some given "detection region" such that $\eta_i - \eta_{i+\Gamma} > d_{\text{thresh}}$ and the transversal direction is congruent, the associated target point (according to (5.4)) with the smallest Euclidean distance to $\mathcal{T}_{\text{int}}$ is selected.

Remark 5.1. While this calculation calls for an estimate of position to be available to the robot, this estimate only needs to be accurate for a relatively short time between control updates. Thus, in the studied case, robot odometry is sufficient since it gives an accurate estimate over short time intervals.

The parameters used for simulations are shown in Table 4.1; the navigation law was updated at 10 Hz. In Fig. 4.7, it can be seen that the robot converges to the target

Table 4.1 Parameters used for simulations

$\overline{u}$	1.3 rad/s	v	1.5 ms^{-1}	d_{tar}	5 m	d_{trig}	10 m
γ	0.33	δ	15 m	p	0.7	d_{thresh}	10 m

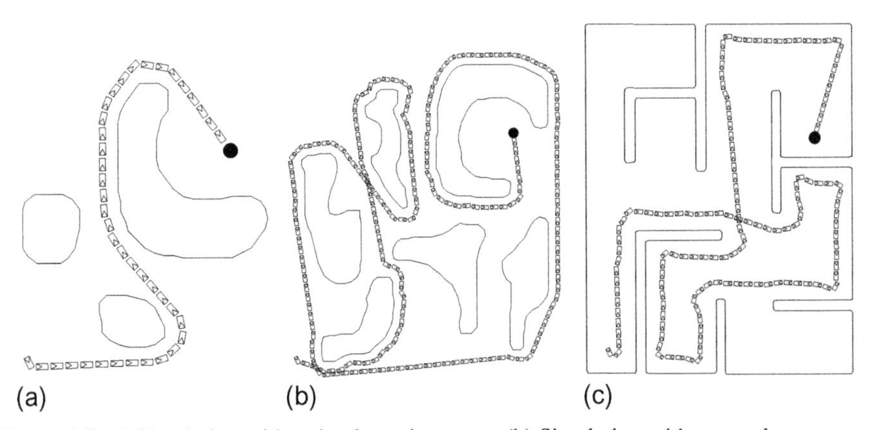

Figure 4.7 (a) Simulation with a simple environment; (b) Simulation with a complex environment; (c) Simulation with a more challenging environment.

without any problem, as is expected. Different sequences of random numbers would of course lead to different paths around the obstacles.

4.6 Experiments with a real robot

They were carried out with a Pioneer P3-DX mobile robot described in Section 1.4.1. A LIDAR device with an angular resolution of 0.5^o was used to detect tangent points straight ahead the robot, as well as obstacles in its vicinity. In the scenario tested, the robot was not provided with a target, rather it was allowed to patrol the area indefinitely. Odometry information available to the robot was used over a single time step to compensate for the movement of the previously calculated tangent point relative to the robot, as was explained in the previous section. Range readings over a pre-specified maximum threshold $R_{\max}$ were truncated, in order to prevent any object outside the test area from influencing the results.

Preventive measures were also taken against control chattering, which may cause detrimental effects when using real robots [32, 87]. In (5.1), the signum function was replaced by a saturation function given by (5.2) with $\gamma = \delta = 1$, and high-frequency components of the controller output were filtered away. The parameters employed in the experiments are given in Table 4.2, where the maximum turning rate is not shown since it appeared to be much larger than the rates requested in the experiments; therefore, its effect was void.

The robot successfully navigates around the obstacles without collision, as indicated in Fig. 4.8.

Table 4.2 Parameters used for experiments with a real robot

v	$0.25\,\text{ms}^{-1}$	d_{tar}	$1\,\text{m}$	d_{trig}	$1.5\,\text{m}$	γ	0.1
δ	$1.0\,\text{m}$	p	0.7	$R_{\max}$	$6\,\text{m}$	d_{thresh}	$1\,\text{m}$

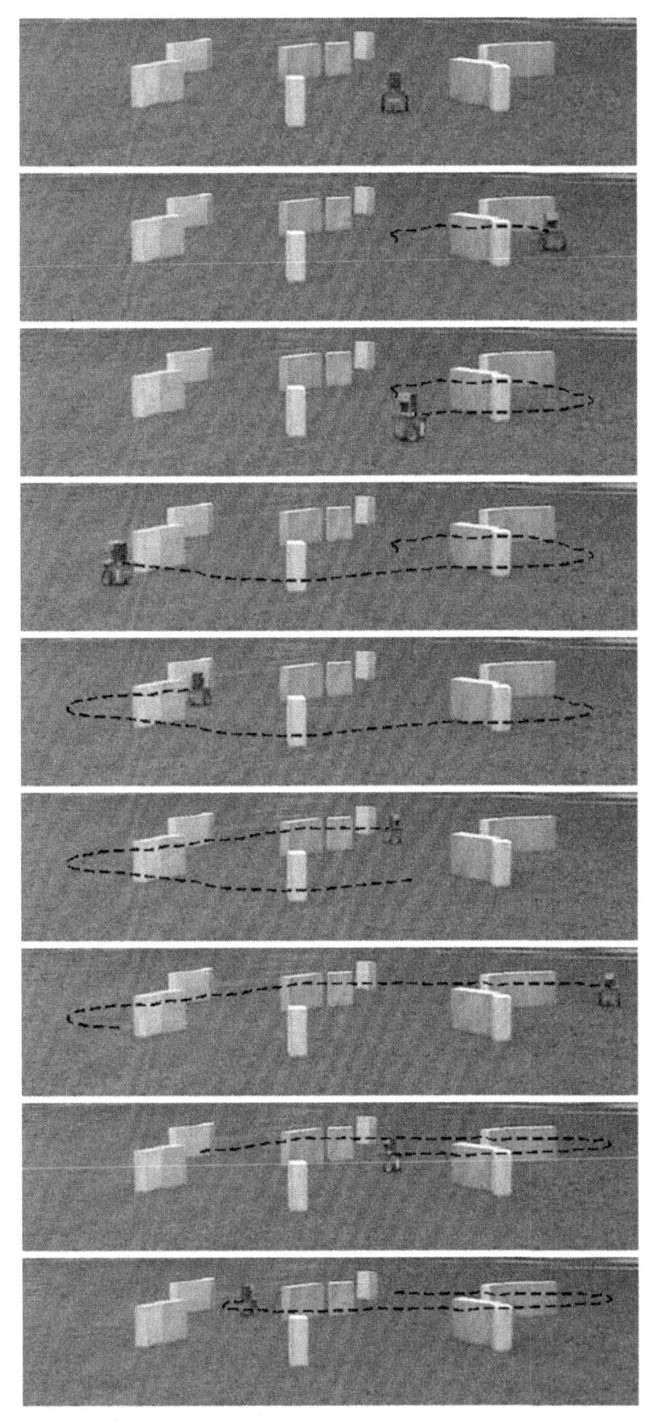

Figure 4.8 Sequence of images showing the tangent following navigation law in action. This was done without a target, so that the robot patrols the area indefinitely.

Reactive navigation of wheeled robots for border patrolling

5.1 Introduction

Safe navigation of a mobile robot along the boundary of an environmental object is a fundamental problem in mobile robotics research. For example, temporarily following the boundary of an obstacle is a standard method employed by many obstacle avoidance algorithms; see, e.g., [8]. Moreover, the navigation task of boundary following is of interest in its own right for border patrolling and structure inspection [200], bottom following by autonomous underwater vehicles [201], lane following by autonomous road vehicles [202], border surveillance with UAVs, and other missions, where traveling along the boundary at a pre-specified distance from it is an essential requirement. In particular, the use of UAVs for border-security missions has gained a lot of attention over the past few years [5, 379, 380] since UAVs carry a potential to greatly improve the border surveillance over remote areas. They have quicker responses and can provide real-time information to a ground station. Some kinds of UAVs can fly over 30 h without refueling, which is considerable compared with just over 2 h average flight time for the helicopters [379]. The ability of loitering for such a long time enables UAVs to cover greater and further areas, which was not possible by means of the conventional manned aircrafts. The constant visual coverage achieves when UAV flies autonomously over a region of interest, which could be on the top of border or close to it with a predefined margin [381].

Many approaches to solving the problem of boundary following have been proposed in the literature. They can be broadly classified based on the type and amount of the required sensor data. If full information about the surrounding area is available, the general approach is to navigate the robot toward the edge of the observable part of the obstacle, as is exemplified in the TangentBug-like algorithms [12, 101]. There are several modifications of this method supporting different degrees of model and sensor sophistication. For example in [382], the robot picks a probational target point based on a series of obstacle detections. Many path planning and model predictive control algorithms also fall in this category. They generate a probational path from the obstacle representation based on the sensor data at each time step and under appropriate circumstances, do ensure boundary following; see, e.g., [267, 268]. A combination of model predictive control with sampled obstacle information has also been proposed [383]. However many of these methods do not care about maintaining a given distance to the boundary. Furthermore, these methods

Safe Robot Navigation Among Moving and Steady Obstacles. http://dx.doi.org/10.1016/B978-0-12-803730-0.00005-6

typically require relatively bulky multi-directional range sensors and extensive data pre-processing, which may not be suitable for, e.g., miniature robots. The focus on real-time implementation with minimization of consumed sensing and computational resources have motivated intensive interest to reactive controllers for which the current control is a simple reflex-like response to the current observation.

Reactive path following has an extensive literature. However many of the proposed feedback control laws are fed by an input variable whose computation may require wide-aperture sensing and intensive pre-processing. For example, reactive (i.e., mapless) vision navigation, which excellent surveys are available in [142, 213], employs the capability of the visual sensor to capture and memorize a whole chunk of the environment and is typically based on extraction of certain image features and estimation of their motion within a sequence of images, which requires intensive image processing. Some other examples of perceptually and computationally demanding input variables not confined to the area of visual navigation can be found in, e.g., [33, 119, 208–210, 214–220, 384]. Furthermore, many of the proposed controllers (see, e.g., [206, 218, 219]) are fed by the boundary curvature, which is particularly sensitive to corruption by measurement noises since this is a second derivative property.

In this chapter, we consider the problem of reactive boundary following with a pre-specified margin by a planar robotic vehicle. Its kinematics are described by the standard model of the Dubins car type, i.e., a nonholonomic system moving along planar paths of bounded curvature without reversing direction [146]. In the literature, this model is applied to many mechanical systems such as wheeled robots, UAVs, missiles, and autonomous underwater vehicles, as was discussed in Section 3.2.

Two scenarios are considered, depending on the available sensory data:

1. The robot has access only to its distance from the boundary;
2. The robot has access only to the distance along the ray perpendicular to the robot's centerline, and to the reflection angle of this ray.

In case 1, the capacity to evaluate the rate at which the distance reading evolves over time (via, e.g., numerical differentiation) is also assumed.

We recall that the distance from a set to the robot is defined as the minimum among the distances from a point in this set to the robot. Therefore, scenario 1 will be referred to as that with a *minimum distance sensor*. Scenario 2 holds if, e.g., the measurements are supplied by several range sensors rigidly mounted to the robot's body at nearly right angles from its centerline or by a single sensor scanning a nearly perpendicular narrow sector; this perception scheme is used in some applications to reduce the complexity, cost, weight, and energy consumption of the sensor system and to minify the detrimental effects of mechanical external disturbances on the measurements. Scenario 2 will be referred to as that with a *rigidly mounted sensor*.

Scenario 2 involves a series of special challenges, like the inability to detect the threat of head-on collision in certain situations (see Fig. 5.11) or strong sensitivity of the output of the sensor system to the vehicle posture.

In this case, some approaches to the boundary following problem have been proposed in [32, 206]. However, the navigation law from [32] is aimed at pure obstacle avoidance, with no objective to follow the boundary with a pre-specified margin. Boundary following with a given margin was addressed in [206] via a hybrid strategy of switching between Lyapunov-based highly nonlinear navigation laws in order to overcome singularities caused by concavities of the tracked curve. In doing so, restrictions on the steering control were neglected by assuming that the vehicle is capable of making arbitrarily sharp turns, whereas these restrictions are realistic and essentially complicate the controller design by, e.g., giving birth to states from which future collision with the boundary is unavoidable. Furthermore, estimates of the boundary curvature were essentially employed, and theoretical analysis ignored the issue of possible front-end collisions with the boundary during transients due to deficiencies of side-view sensors (see Fig. 5.11), as well as jumps of sensor readings (up to loss of view at the obstacle) possible when the rotating "ray of view" of the sensor first becomes tangential to a convex part of the boundary and then loses touch with it (or vice versa).

For scenario 1, this chapter offers a navigation strategy by which an autonomous vehicle moving with a constant linear velocity can patrol the border of a region in close range. This strategy originates from the equiangular navigation and guidance law [126] and belongs to the class of sliding mode navigation algorithms; see, e.g., [70]. For scenario 2, we propose a novel sliding mode navigation strategy that does not employ curvature estimates and homogeneously handles both concavities and convexities of the followed boundary, as well as transitions between them. Both strategies asymptotically steer the mobile robot to the pre-specified distance to the boundary and afterward ensures stable maintenance of this distance so that the safety requirement is always satisfied. Unlike many papers in the area of robotic guidance, which are based on heuristics, this chapter offers a mathematically rigorous analysis and justification of non-local convergence of the proposed strategy is offered for the both scenarios. For scenario 2, possible abrupt jumps of the sensor readings are taken into account. Furthermore, much attention is given to revealing requirements to the global geometry of the boundary that make it possible to avoid front-end collisions based on only side view sensors, thus making extra front-view sensors superfluous. The convergence and performance of the proposed navigation and guidance laws are confirmed by computer simulations and real-world tests with a Pioneer P3-DX robot, equipped with a SICK LMS-200 Lidar sensor.

The main results of the chapter were originally published in [33, 205]. A preliminary version of some of these results was also reported in the conference paper [385].

The remainder of the chapter is organized as follows. Sections 5.2–5.5 deal with the robot equipped with a minimum distance sensor. Section 5.2 presents the statement of the border patrolling problem and the proposed navigation law. Section 5.3 offers basic definitions and discusses the main assumptions. Mathematically rigorous analysis of the navigation strategy is presented in Section 5.4. Illustrative examples and computer simulations are given in Section 5.5.

Sections 5.6–5.10 consider a scenario with a rigidly mounted sensor. Specifically, Section 5.6 presents the problem setup and navigation law, while Section 5.7 discusses the assumptions of theoretical analysis and tuning of the navigation controller.

The main theoretical results are given in Section 5.8. Sections 5.9 and 5.10 discuss computer simulation tests and experiments with a real wheeled robot, respectively.

5.2 Boundary following using a minimum distance sensor: System description and problem statement

We consider a planar vehicle or wheeled mobile robot modeled as unicycle. It travels with a constant speed and is controlled by the angular velocity limited by a given constant. There is a domain D in the plane. The objective is to patrol the border ∂D of this domain at the given distance d_0 from it and at the given speed v; see Fig. 5.1(a). The robot has access to the current distance $d(t)$ to the border and the rate $\dot{d}(t)$ at which this reading evolves over time.

As in Chapter 4, we employ the classical model of such a vehicle:

$$\begin{aligned} \dot{x} &= v\cos\theta, & x(0) &= x_0, \\ \dot{y} &= v\sin\theta, & y(0) &= y_0, \\ \dot{\theta} &= u \in [-\overline{u}, \overline{u}], & \theta(0) &= \theta_0. \end{aligned} \tag{2.1}$$

Here $\boldsymbol{r} = \mathbf{col}\,(x, y)$ is the pair of the robot's Cartesian coordinates, θ gives its orientation, v and u are the speed and angular velocity, respectively; see Fig. 5.1(b). The maximal angular velocity $\overline{u}$ is given. The minimal turning radius of the robot equals

$$R_{\min} = \frac{v}{\overline{u}}. \tag{2.2}$$

We recall that the distance d from the robot to the domain D is defined as the minimum distance between a point from this domain and the robot: $d(t) := \mathbf{dist}\,[\boldsymbol{r}(t); D]$, where $\mathbf{dist}\,[\boldsymbol{r}; D] = \min_{\boldsymbol{r}' \in D} \|\boldsymbol{r} - \boldsymbol{r}'\|$. This minimum is achieved if D is closed, which is assumed throughout the chapter. Access to $\dot{d}$ may be acquired via, e.g., numerical differentiation.

The border patrolling problem is as follows. Find a navigation law that asymptotically drives the robot to the required distance $d(t) \to d_0 > 0$ to the domain D and ensures the robot traveling along the domain boundary at the given speed v. During the entire maneuver, the distance from the mobile robot to the domain should constantly exceed the given safety margin $d_{\text{safe}} \leq d(t)\ \forall t$, where $0 \leq d_{\text{safe}} < d_0$.

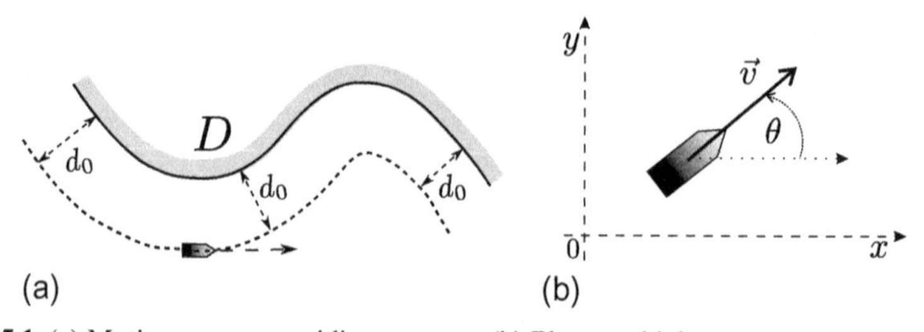

Figure 5.1 (a) Motion over an equidistant curve; (b) Planar vehicle.

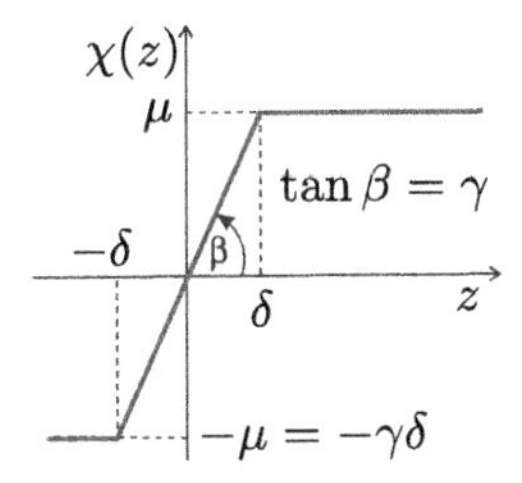

Figure 5.2 Linear function with saturation.

We examine the following navigation algorithm:

$$u(t) = \overline{u} \cdot \mathbf{sgn}\left\{\dot{d}(t) + \chi[d(t) - d_0]\right\}, \quad \text{where} \tag{2.3}$$

$$\chi(z) := \begin{cases} \gamma z & \text{if } |z| \le \delta, \\ \mu \mathbf{sgn}(z) & \text{if } |z| > \delta, \end{cases} \quad \mu := \gamma\delta \tag{2.4}$$

is a linear function with saturation illustrated in Fig. 5.2, $\mathbf{sgn}\,\alpha := 1$ for $\alpha > 0$, $\mathbf{sgn}\,0 := 0$, and $\mathbf{sgn}\,\alpha := -1$ for $\alpha < 0$, and $\gamma > 0, \delta > 0$ are controller parameters.

5.3 Main assumptions of theoretical analysis

We start with a natural technical requirement.

Assumption 3.1. *The domain D is closed and has a C^3-smooth boundary ∂D.*

At the same time, we do not assume that D is convex.

Let $\varkappa(\boldsymbol{r})$ and $R_\varkappa(\boldsymbol{r})$ stand for the signed curvature of the boundary ∂D at the point $\boldsymbol{r} \in \partial D$ and the (unsigned) curvature radius $R_\varkappa(\boldsymbol{r}) := |\varkappa(\boldsymbol{r})|^{-1}$. (We refer the reader to [377, Ch. 1] for the formal definition of the curvature.) Here $\varkappa(\boldsymbol{r}) > 0$ for convexity points, $\varkappa(\boldsymbol{r}) < 0$ for concavity points, and $0^{-1} := +\infty$; see Fig. 5.3(a).

Definition 3.1. A point $\boldsymbol{r} \notin D$ is said to be *multiple* if there exist two different *minimum-distance* points $\boldsymbol{r}_*, \boldsymbol{r}_\star \in D$, i.e., such that $\mathbf{dist}[\boldsymbol{r}; D] = \|\boldsymbol{r} - \boldsymbol{r}_*\| = \|\boldsymbol{r} - \boldsymbol{r}_\star\|$, *fracture-multiple* if $\boldsymbol{r}, \boldsymbol{r}_*, \boldsymbol{r}_\star$ are not co-linear, and *focal* if there is a minimum-distance point $\boldsymbol{r}_* \in D$ such that $\mathbf{dist}[\boldsymbol{r}; D] = \|\boldsymbol{r} - \boldsymbol{r}_*\| = R_\varkappa(\boldsymbol{r}_*)$ and $\varkappa(\boldsymbol{r}_*) < 0$.

We note that $\boldsymbol{r}$ is focal if and only if it is the point of intersection of two infinitesimally close normals to ∂D.

We proceed with two lemmas to illustrate assumptions to be imposed. The first lemma concerns only non-convex domains since, otherwise, there are no multiple and focal points.

Lemma 3.1. *Let the robot travel at the required distance d_0 from the boundary at a nonzero speed. Then its path contains neither fracture-multiple nor focal points.*

Proof. Consider a point $\boldsymbol{r}$ on the path $\boldsymbol{r} = \boldsymbol{r}(\tau)$ and any related minimum-distance point $\boldsymbol{r}_* \in \partial D$. Let $\boldsymbol{\rho}(s) \in \mathbb{R}^2$ stand for a regular parametric representation of the boundary ∂D in a vicinity of $\boldsymbol{r}_* = \boldsymbol{\rho}(s_*)$, where s is the arc length, or natural

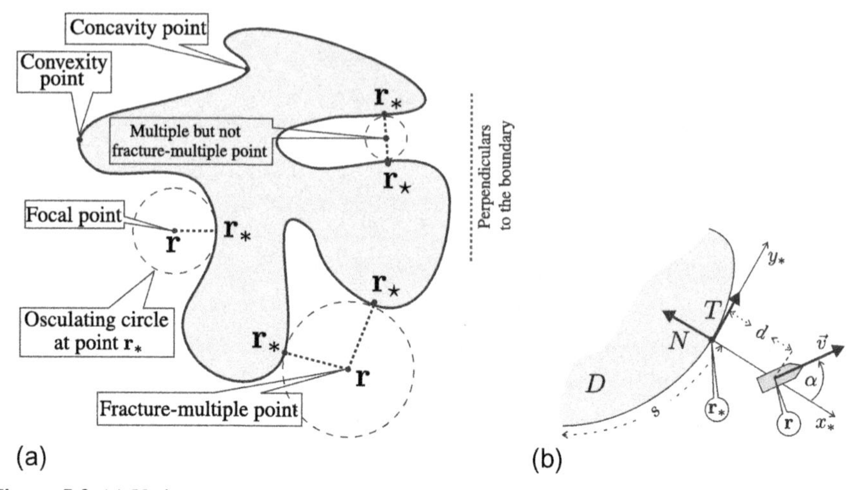

Figure 5.3 (a) Various types of boundary points; (b) Moving coordinate frame.

parameter. We observe that $\Phi(t,s) := 1/2\, \|\boldsymbol{r}(t) - \boldsymbol{\rho}(s)\|^2 \geq 1/2\mathbf{dist}\,[\boldsymbol{r}(t); D]^2 = d_0 = \Phi(\tau, s_*)$. Therefore, the derivatives at the point (τ, s_*) satisfy the relations:

$$\langle \dot{\boldsymbol{r}}(\tau); \boldsymbol{r} - \boldsymbol{r}_* \rangle = \frac{\partial \Phi}{\partial t} = 0 = \frac{\partial \Phi}{\partial s} = -\left\langle \boldsymbol{\rho}'(s_*); \boldsymbol{r} - \boldsymbol{r}_* \right\rangle,$$

$$\Phi'' = \begin{pmatrix} \Phi''_{tt} & \Phi''_{ts} \\ \Phi''_{st} & \Phi''_{ss} \end{pmatrix} \geq 0, \quad (3.1)$$

where $\langle \cdot, \cdot \rangle$ is the standard inner product and the inequality means the matrix is nonnegative definite. It follows from the first equation in (3.1) that for any two minimum-distance points $\boldsymbol{r}_*, \boldsymbol{r}_\star$, the vectors $\boldsymbol{r} - \boldsymbol{r}_*$ and $\boldsymbol{r} - \boldsymbol{r}_\star$ are perpendicular to a common nonzero vector $\dot{\boldsymbol{r}}(\tau)$. Therefore, the points $\boldsymbol{r}, \boldsymbol{r}_*, \boldsymbol{r}_\star$ are co-linear.

Now suppose that $\boldsymbol{r}$ is a focal point, i.e., there is a minimum-distance point $\boldsymbol{r}_* \in D$ such that $d_0 = \|\boldsymbol{r} - \boldsymbol{r}_*\| = R_\varkappa(\boldsymbol{r}_*)$, $\varkappa(\boldsymbol{r}_*) < 0$. Let $\mathbf{T} = \boldsymbol{\rho}'$ be the unit tangent vector and $\mathbf{N}$ be the unit normal vector directed inwards D; see Fig. 5.3(b). By the second equation in (3.1), $\boldsymbol{r}_* - \boldsymbol{r} = d_0 N(s_*)$, and due to the Frenet-Serret formulas [386],

$$\mathbf{T}' = \varkappa \mathbf{N}, \quad \mathbf{N}' = -\varkappa \mathbf{T}. \quad (3.2)$$

Hence, $\Phi''_{ss}(\tau, s_*) = \|\mathbf{T}(s_*)\|^2 + \left\langle \mathbf{T}'(s_*); \boldsymbol{r}_* - \boldsymbol{r} \right\rangle = 1 + \varkappa(\boldsymbol{r}_*) d_0 \left\langle \mathbf{N}(s_*); \mathbf{N}(s_*) \right\rangle = 1 + \varkappa(\boldsymbol{r}_*) R_\varkappa(\boldsymbol{r}_*) = 0$; $\Phi''_{ts}(\tau, s_*) = -\left\langle \dot{\boldsymbol{r}}(\tau); \boldsymbol{\rho}'(s_*) \right\rangle = \pm v \neq 0$, where the last equation results from the equations in (3.1) since $\|\dot{\boldsymbol{r}}(\tau)\| = v$. It follows that the determinant $\det \Phi''(\tau, s_*) = -v^2 < 0$, whereas the inequality from (3.1) implies that this determinant is non-negative. The contradiction obtained proves the lemma. □

By Lemma 3.1, the navigation objective is realistic only if the ideal trajectory of the robot does not contain multiple-fracture and focal points. This trajectory is not known in advance and moreover, the robot may circulate around the domain D. Therefore, it is reasonable to assume that there are no such points on the entire d_0-equidistant curve of the boundary ∂D (i.e., the curve formed by all points at the required distance d_0 from

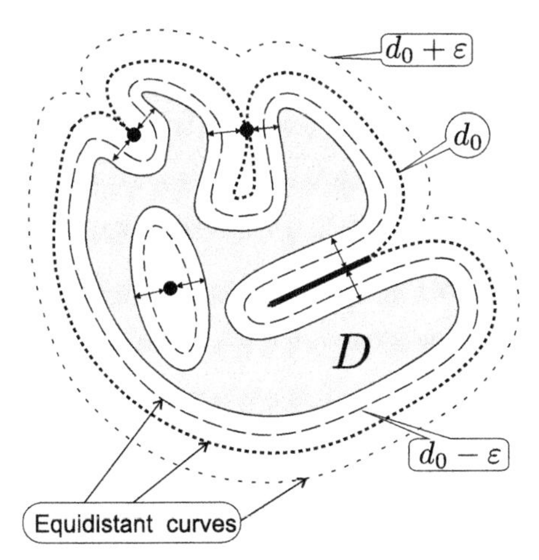

Figure 5.4 Singular multiple points.

∂D outside the domain D). This taboo does not exclude "singular" multiple points $\boldsymbol{r}$ accompanied by exactly two minimum distance ones $\boldsymbol{r}_* \neq \boldsymbol{r}_\star$ for which the points $\boldsymbol{r}, \boldsymbol{r}_*, \boldsymbol{r}_\star$ are co-linear. Such $\boldsymbol{r}$'s either are self-intersection points the d_0-equidistant curve where this curve is tangential to itself (see Fig. 5.4) or are produced by collapse of some part of this curve into a point (see the center of the ellipsoid in Fig. 5.4). Such points may exist only if the domain is non-convex and only for d_0's, which in total, form a set of zero Lebesgue measure. The chance to pick exactly such a d_0 from the continuum of reals is negligible. Therefore, we do not consider this exceptional case. Summarizing, we arrive at the following.

Assumption 3.2. *There are neither multiple nor focal points at the distance d_0 from the domain D.*

This necessarily holds if either the domain D is convex or it is non-convex, the boundary ∂D is compact, and d_0 is small enough.

Lemma 3.2. *Let the robot travel at the distance d_0 from the boundary at the speed $v > 0$. Then $|d_0 + R_\varkappa \mathbf{sgn}\, \varkappa| \geq R_{\min}$ for any point $\boldsymbol{r}$ on its path. Here $R_{\min}$ is the minimal turning radius* (2.2) *of the robot, and the curvature $\varkappa$ and the curvature radius $R_\varkappa$ are computed at the minimum distance point.*

Proof. We employ the notations from the proof of Lemma 3.1. The path of the robot is parametrically represented by $\boldsymbol{r}(s) := \boldsymbol{\rho}(s) - d_0\mathbf{N}(s)$. Invoking (3.2) yields that $\boldsymbol{r}' = (1 + \varkappa d_0)\mathbf{T}, \quad \boldsymbol{r}'' = \varkappa' d_0 \mathbf{T} + \varkappa(1 + \varkappa d_0)\mathbf{N}$. Let $J := \left(\begin{smallmatrix} 0 & 1 \\ -1 & 0 \end{smallmatrix}\right)$, and let $^{\mathbf{T}}$ stand for the transposition. Then the curvature radius $\mathcal{R}$ of the path is given by

$$\mathcal{R} = \frac{\|\boldsymbol{r}'\|^3}{|(\boldsymbol{r}')^{\mathbf{T}} J \boldsymbol{r}''|} = \frac{|1 + \varkappa d_0|^3}{|\varkappa||1 + \varkappa d_0|^2} = |d_0 + R_\varkappa \mathbf{sgn}\, \varkappa|.$$

It remains to note that $\mathcal{R}$ should be no less than the minimal turning radius of the robot. □

As is shown in this proof, $|d_0 + R_\varkappa \mathbf{sgn}\, \varkappa|$ is the curvature radius of the d_0-equidistant curve along which the robot should move to maintain the distance d_0 to the boundary. The conclusion of the lemma comes to the evident condition under which the robot is capable of tracking this curve. These observations also give rise to the following.

Remark 3.1. Let $>$ be put in place of $\geq$ in the necessary condition from Lemma 3.2, and let the resultant condition $|d_0 + R_\varkappa \mathbf{sgn}\, \varkappa| > R_{\min}$ hold everywhere on the d_0-equidistant curve. Then the output d is locally controllable in proximity of this curve: the robot can not only move over the curve but also is able to turn off this curve to any side.

Now we enhance the necessary condition from Lemma 3.2 in the vein of this remark. The next assumption means the maneuverability of the robot slightly exceeds the minimally required level.

Assumption 3.3. *There exists* $\lambda \in (0, 1)$ *such that either*

$$d_0 + \inf_{\boldsymbol{r}_* \in \partial D} R_\varkappa(\boldsymbol{r}_*) \mathbf{sgn}\, \varkappa(\boldsymbol{r}_*) > \frac{R_{\min}}{\lambda} \quad \textit{or}$$

$$\varkappa(\boldsymbol{r}_*) < 0 \; \forall \boldsymbol{r}_* \in \partial D \quad \text{and} \quad \inf_{\boldsymbol{r}_* \in \partial D} R_\varkappa(\boldsymbol{r}_*) > d_0 + \frac{R_{\min}}{\lambda}. \tag{3.3}$$

The second case may hold only for domains with no convexity points at the boundary. Furthermore, (3.3) encompasses Assumption 3.2 in the part concerning focal points.

To establish the convergence result, we need to extend the above assumptions on the initial state of the robot and also ensure that they hold and the safety requirement is met during the transient. To this end, we introduce two definitions.

Definition 3.2. An interval $[d_-, d_+], 0 < d_- < d_+$ of distances d to the boundary ∂D is said to be *regular* if there exists $\lambda \in (0, 1)$ such that Assumptions 3.2 and 3.3 hold with d_0 replaced by any point from this interval, and $d_- > d_{\text{safe}}$.

Therefore, the first inequality from (3.3) should hold with $d_0 := d_-$, whereas the second one should be true with $d_0 := d_+$.

Definition 3.3. The *initial circle* is any of two circles of the radius $R_{\min}$ that pass the initial position $\mathbf{col}\,(x_0, y_0)$ of the robot with the tangential angle θ_0 from (2.1).

In other words, this is the circle along which the robot would move under the constant control $u \equiv \overline{u}$ or $u \equiv -\overline{u}$.

Now we summarize the above discussion by imposing the cumulative assumption, which covers Assumptions 3.2 and 3.3.

Assumption 3.4. *There exists a regular interval* $[d_-, d_+]$ *that contains the desired distance* d_0 *and the distances from the points on the initial circles to* ∂D.

If the domain D is convex, this assumption is implied by the following inequality:

$$\min \{d_0; d(0) - 2R_{\min}\} > \max \{R_{\min} - R_-; d_{\text{safe}}\}, \qquad \text{where} \quad R_- := \inf_{\boldsymbol{r}_* \in \partial D} R_\varkappa(\boldsymbol{r}_*). \tag{3.4}$$

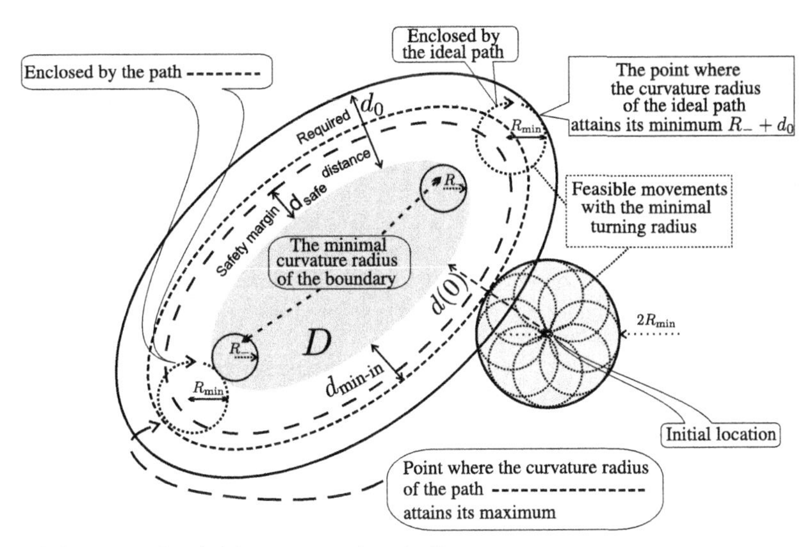

Figure 5.5 Assumption 3.4 for a convex domain D.

Here $d(0) - 2R_{\min}$ is the tight lower bound of the distances $d = d_*$ encountered when moving along feasible initial circles[1]; see Fig. 5.5. Inequality (3.4) means the robot is capable of maintaining both the given speed v and any of these distances d_*, as well as the required one $d_* := d_0$, since the minimal curvature radius $d_* + R_-$ of the corresponding trajectory exceeds the minimal turning radius $R_{\min}$ of the robot. Furthermore, (3.4) means this trajectory meets the safety requirement $d_* > d_{\text{safe}}$.

5.4 Navigation for border patrolling based on minimum distance measurements

Now we are in a position to present the main theoretical result concerning the border patrolling problem in the scenario described in Section 5.2.

Theorem 4.1. *Let Assumptions 3.1 and 3.4 hold and the parameters* $\gamma > 0$ *and* $\delta > 0$ *of the navigation rule* (2.3), (2.4) *be chosen so that the following inequality is true, where* λ *is the constant related to the interval from Assumption 3.4 by Definition 3.2,*

$$\sqrt{1 + \frac{\gamma^2}{(1-\lambda)^2 \overline{u}^2}} < v_* := \frac{v}{\delta\gamma} \quad \left(\overset{(2.4)}{==} \frac{v}{\mu}\right). \tag{4.1}$$

[1] That is, circles of the radius $R_{\min}$ that pass the robot's initial position $\mathbf{col}\,(x_0, y_0)$. Neglecting the actual tangential angle θ_0 simplifies computation at the expense of injecting more conservatism.

Then the navigation law (2.3) *asymptotically drives the robot at the desired distance to the domain D, i.e.,* $d(t) \xrightarrow{t\to\infty} d_0$. *During this motion, the safety requirement is always satisfied:* $d(t) \geq d_{safe}\ \forall t$.

Remark 4.1.

(i) Theorem 4.1 takes into account that the discontinuous controller (2.3) may exhibit a sliding motion [70].

(ii) Inequality (4.1) means that chosen λ and γ, the parameter δ should be small enough. Furthermore, (4.1) $\Rightarrow \mu < v$.

(iii) As $\lambda \uparrow 1$, the distance constraints (3.3) are relaxed, whereas the parameter constraints (4.1) are enhanced. In other words, larger distances give more freedom in the choice of the parameters in the navigation law.

(iv) Theorem 4.1 remains true for domains D with piece-wise smooth boundary, provided that the tangential vector turns toward D when passing any corner point of ∂D. At such points, the curvature radius is assumed to be zero. This theorem also remains true if D is not a domain but is a line segment or a point.

(v) In the case where D degenerates into a point, a comprehensive analysis of the behavior exhibited by the navigation law (2.3) for all feasible parameter values can be carried out. It shows that this law cannot ensure global convergence $d \rightarrow d_0$. For any parameters, there is a non-empty domain in the phase space such that whenever the robot enters this domain, afterward it moves with the constant control $u \equiv \pm\overline{u}$ along a circle of the radius $R_{\min}$. Then the distance d oscillates and does not converge to d_0. Thus, requirements to the robot's initial state and the required distance (like (3.3)) are unavoidable.

The following remark will be justified in the proof of Theorem 4.1.

Remark 4.2. The navigation controller (2.3) arranges border patrolling into two phases. The first of them holds while the argument $S := \dot{d} + \chi(d - d_0)$ of the **sgn** function from (2.3) is nonzero. During this phase, **sgn** S is not altered due to continuity of S and so the robot is driven by the constant control $u \equiv \pm\overline{u}$ along an initial circle. The first phase necessarily terminates: S arrives at zero. Since this time instant a sliding motion over the surface $S = 0$ is maintained and the robot monotonically approaches the desired distance d_0 to the domain. This approach is exponentially fast.[2]

5.4.1 Proof of Theorem 4.1

We introduce the normally oriented moving Cartesian coordinate frame centered at the point $\boldsymbol{r}_*(t) \in D$ nearest to the robot's current position $\boldsymbol{r}(t)$, with the axis of abscissa directed toward the robot; see Fig. 5.3(b). Let α denote the algebraic angle between the velocity vector of the robot and the axis of abscissa.

[2] Due to access to only d and $\dot{d}$, the robot is unable to distinguish between cases (c1) and (c2) from Fig. 5.6. Due to this deficiency in the sensor data, it has to apply a common control in both cases, which in case (c2) results in a less-effective maneuver than in case (c1). Since neither the initial orientation nor location on the initial $d(0)$-equidistant curve are measured, the expected operational zone of the robot is the domain limited by the $\min\{d(0) - 2R_{\min}, d_0\}$- and $\max\{d(0) + 2R_{\min}, d_0\}$-equidistant curves; see Fig. 5.5.

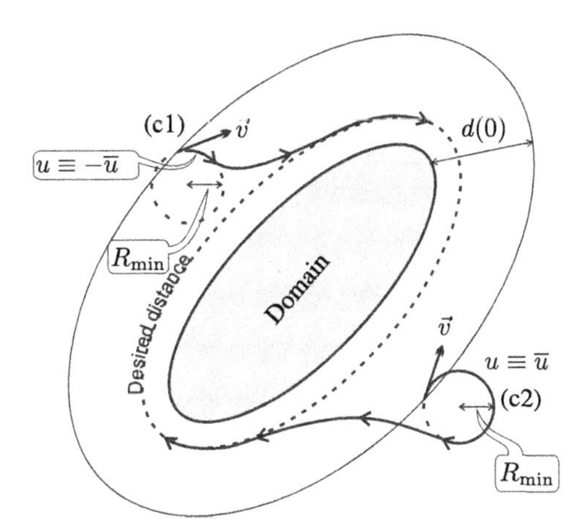

Figure 5.6 Two types of maneuver.

Lemma 4.1. *Let $d \in [d_-, d_+]$, where $[d_-, d_+]$ is the interval from Assumption 3.4. Then the following relations hold:*

$$\dot{d} = v \cos \alpha, \quad \dot{\alpha} = u - \tfrac{\varkappa(\boldsymbol{r}_*)}{1+\varkappa(\boldsymbol{r}_*)d} v \sin \alpha. \tag{4.2}$$

Proof. We employ the parametric representation $\boldsymbol{\rho}(s)$ of ∂D and the vectors $\mathbf{T}$ and $\mathbf{N}$ from the proof of Lemma 3.1. Let $s(t)$ denote the natural parameter corresponding to $\boldsymbol{r}_*(t)$. Then the robot position $\boldsymbol{r}(t) = \boldsymbol{\rho}[s(t)] - d(t)\mathbf{N}[s(t)]$. Therefore,

$$\dot{\boldsymbol{r}} = \mathbf{T}\dot{s} - \dot{d}\mathbf{N} - d\mathbf{N}'\dot{s} \overset{(3.2)}{=\!=} [1 + d\varkappa]\dot{s}\mathbf{T} - \dot{d}\mathbf{N}.$$

On the other hand, $\dot{\boldsymbol{r}} = \mathbf{T}v \sin \alpha - \mathbf{N}v \cos \alpha$. Through equating the matching coefficients in the linear combinations of $\mathbf{T}$ and $\mathbf{N}$ encountered in the previous two equations, we arrive at the first equation from (4.2) and the relation

$$\dot{s} = \frac{v \sin \alpha}{1 + d\varkappa}. \tag{4.3}$$

The angle between $\dot{\boldsymbol{r}}_* = \dot{s}\mathbf{T}$ and the abscissa axis amounts to $\pi \, \mathbf{sgn}\, \dot{s}$. By the well-known formula for relative polar coordinates of two moving points $\boldsymbol{r}(t)$ and $\boldsymbol{r}_*(t)$,

$$\dot{\alpha} = u - d^{-1} [v \sin \alpha - \dot{s}] = u - d^{-1} \left[v \sin \alpha - \frac{v \sin \alpha}{1 + d\varkappa} \right]$$
$$= u - \frac{\varkappa}{1 + \varkappa d} v \sin \alpha.$$

□

The next lemma reveals conditions under which the discontinuous navigation law (2.3) exhibits sliding motion.

Lemma 4.2. *Within the domain $d \in [d_-, d_+]$, the surface $S := \dot{d} + \chi(d - d_0) = 0$ is sliding in the sub-domain*

$$\sin\alpha > 0 \tag{4.4}$$

and two side repelling in the sub-domain

$$\sin\alpha < 0 \tag{4.5}$$

if (4.1) *holds. On this surface within the above domain,*

$$|\sin\alpha| \geq \frac{\sqrt{v^2 - \mu^2}}{v} > 0. \tag{4.6}$$

Proof. Let $\dot{d} + \chi(d - d_0) = 0$ and so $|\dot{d}| = |\eta|$, where $\eta(t) := -\chi(d - d_0)$. By (2.4),

$$|\eta(t)| \leq \mu; \quad |\dot{\eta}| \leq \gamma|\dot{d}| = \gamma|\chi(d - d_0)| \leq \gamma\mu. \tag{4.7}$$

Due to (4.2),

$$v|\cos\alpha| = |\dot{d}| = |\eta| \leq \mu \Rightarrow |\cos\alpha| \leq v^{-1}\mu; |\sin\alpha| = \sqrt{1 - \cos^2\alpha} \Rightarrow (4.6).$$

Furthermore,

$$\dot{S} = \frac{\mathrm{d}}{\mathrm{d}t}\left[\dot{d} + \chi(d - d_0)\right] = \ddot{d} - \dot{\eta} \overset{(4.2)}{=\!=} -v\sin\alpha\left[u - \frac{\varkappa}{1 + \varkappa d}v\sin\alpha\right] - \dot{\eta}$$
$$= \underbrace{\frac{\varkappa}{1 + \varkappa d}v^2\sin^2\alpha - \dot{\eta}}_{b} - vu\sin\alpha; \tag{4.8}$$

$$|b| \overset{|\sin\alpha|\leq 1}{\leq} \frac{v|\varkappa|}{|1 + \varkappa d|}v|\sin\alpha| + |\dot{\eta}|.$$

Here $|\dot{\eta}| \leq \gamma \cdot \mu \cdot 1$ by (4.7) and the last multiplier $1 \leq \frac{v|\sin\alpha|}{\sqrt{v^2 - \mu^2}}$ by (4.6). With regard to the equation $\varkappa^{-1} = R_\varkappa \mathbf{sgn}\,\varkappa$ between the curvature and curvature radius,

$$|b| \leq \left[\frac{v}{|d + R_\varkappa \mathbf{sgn}\,\varkappa|} + \frac{\gamma\mu}{\sqrt{v^2 - \mu^2}}\right]v|\sin\alpha|.$$

Since d belongs to the regular interval $[d_-, d_+]$ from Assumption 3.4, (3.3) holds for $d_0 := d$ by Definition 3.2. Hence,

$$|d + R_\varkappa \mathbf{sgn}\,\varkappa| \geq \lambda^{-1}R_{\min} \overset{(2.2)}{=} (\lambda\overline{u})^{-1}v \tag{4.9}$$

and so

$$|b| \leq \left[\lambda\overline{u} + \frac{\gamma}{\sqrt{v_*^2 - 1}}\right]v|\sin\alpha|, \quad \text{where} \quad v_* := \frac{v}{\mu}.$$

Inequality (4.1) entails that the second addend in the square brackets is less than $(1-\lambda)\overline{u}$. Summarizing and invoking (4.6), we see that $|b| \leq \overline{u}v|\sin\alpha| - \varepsilon$, where $\varepsilon > 0$

is small enough. Therefore, due to (4.8), the signs taken by $\dot{S}$ for $u = \overline{u}$ and $u = -\overline{u}$, respectively, are opposite. If $\sin\alpha > 0$, the sign is opposite to $\mathbf{sgn}\, u \overset{(2.3)}{=} \mathbf{sgn}\, S$; if $\sin\alpha < 0$, the signs are equal. This implies the conclusion of the lemma. □

Lemma 4.3. *If the equation $\dot{d} + \chi(d - d_0) = 0$ becomes true at some time t_0 when $d \in [d_-, d_+]$, then $d \to d_0$ as $t \to \infty$ and $d(t) \in [d_-, d_+]$ $\forall t \geq t_0$. The convergence $d \to d_0$ is exponentially fast.*

Proof. Lemma 4.2 guarantees that first, $\sin\alpha > 0$ at $t = t_0$ and second, this inequality is still valid and sliding motion occurs while $d \in [d_-, d_+]$. During this motion, $\dot{y} = -\chi(y)$ for $y := d - d_0$, where $\chi(y) \cdot y > 0$ $\forall y \neq 0$ and $\chi(0) = 0$. It follows that any solution d of the sliding mode differential equation monotonically converges to d_0. At the same time, $d_0 \in [d_-, d_+]$ by Assumption 3.4. Hence d will never leave the interval $[d_-, d_+]$, the sliding mode will never be terminated, and $d \to d_0$ as $t \to \infty$. Application of the Lyapunov's first method to the equation $\dot{y} = -\chi(y)$ at the equilibrium point $y = 0$ shows that the convergence is exponentially fast. □

Lemma 4.4. *Under the assumptions of Theorem 4.1, both relations $\dot{d} + \chi(d - d_0) = 0, \sin\alpha > 0$ become true at some time t_0, and for the first such a time, $d \in [d_-, d_+]$.*

Proof. If these relations are true initially, the claim is evident. Otherwise, $\dot{d} + \chi(d - d_0) \neq 0$ for $t > 0, t \approx 0$, and until the first time t_0 when the equation becomes true, the robot moves with the constant control $u \equiv \pm\overline{u}$.

Let $u \equiv \overline{u}$ for the definiteness. Now we analyze the motion of the robot driven by the constant control $u \equiv \overline{u}$. By Assumption 3.4, $d \in [d_-, d_+]$ during this motion. Since $d_- > 0$, this means the robot does not hit D. By (4.2),

$$\dot{\alpha} = \overline{u} - \frac{1}{R_\varkappa \mathbf{sgn}\, \varkappa + d} v \sin\alpha,$$

$$\left| \frac{1}{R_\varkappa \mathbf{sgn}\, \varkappa + d} v \sin\alpha \right| \overset{(4.9)}{\leq} \lambda\overline{u} \Rightarrow \dot{\alpha} \geq (1 - \lambda)\overline{u} > 0.$$

Hence, $\dot{d} = v\cos\alpha$ takes the respective values v and $-v$ at some time instants t_+ and t_-. Since $|\chi(d - d_0)| \leq \mu < v$ by (4.1), the expression $\dot{d} + \chi(d - d_0)$ takes the values of opposite signs at t_+ and t_-. It follows that this expression inevitably vanishes during the motion. Lemma 4.2 implies that $\sin\alpha > 0$ at the first such time $t = t_0$ and completes the proof. □

Proof of Theorem 4.1. This theorem is immediate from Lemmas 4.3 and 4.4. □

5.5 Computer simulations of border patrolling with a minimum distance sensor

For computer simulation, we used Matlab and Mobotsim 1.0, which is a configurable 2D simulator of mobile robots with an effective graphical interface [387].

We consider a fixed wing UAV that is supposed to fly at constant speed $v = 50$ m/s along the border of a region with a predefined margin $d_0 = 5$ km and maximum angular velocity $\overline{u} = 1$ rad/s. The controller parameters are $\lambda = 0.5$, $\delta = 10$ m, and $\gamma = 2\,\mathrm{s}^{-1}$. Hence, $\mu := \delta\gamma = 20\,\mathrm{m/s}$ and the condition (4.1) is met. In the controller (2.3), the sign function was replaced by a linear function with saturation, as prophylaxis against chattering; see Section 1.3.

To examine robustness of the proposed navigation law against numerical differentiation errors in conjunction with the measurement noise, we assume the range measurements d are corrupted by a Gaussian white noise with zero mean and standard deviation σ. Although many modern sensors reach distances of several kilometers with accuracies of several tens of centimeters, the focus is on much worse sensing $\sigma = 10\,\mathrm{m}$. This focus on the worst-case scenario motivated us to examine the simplest two-point finite difference derivative estimate, although estimation of derivatives from noise-corrupted data is a well-established discipline offering a variety of more effective approaches. Among them, there is approximation of the transfer function of the ideal differentiator, optimal differentiation based on stochastic models, observers with sliding modes and large gains, difference methods, approximation and Tikhonov regularization, etc.; see, e.g., [388, 389] for recent surveys.

The smooth lines in Figs. 5.7 and 5.8 correspond to the zero noise $\sigma = 0$; in these cases, the UAV successfully flies along the border with the predefined margin. Feeding the controller by noisy measurements of d, along with the exact values of $\dot{d}$, causes

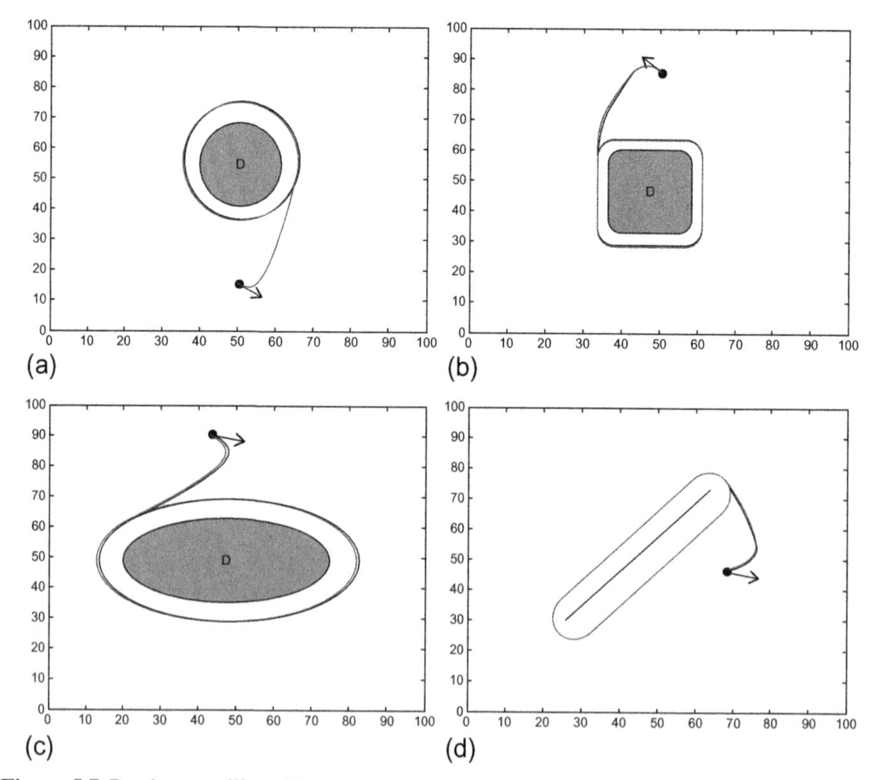

Figure 5.7 Border patrolling: No visible changes when noisy measurements of d and the exact value of $\dot{d}$ are fed into the controller, $\tau = 0.4$ sec.

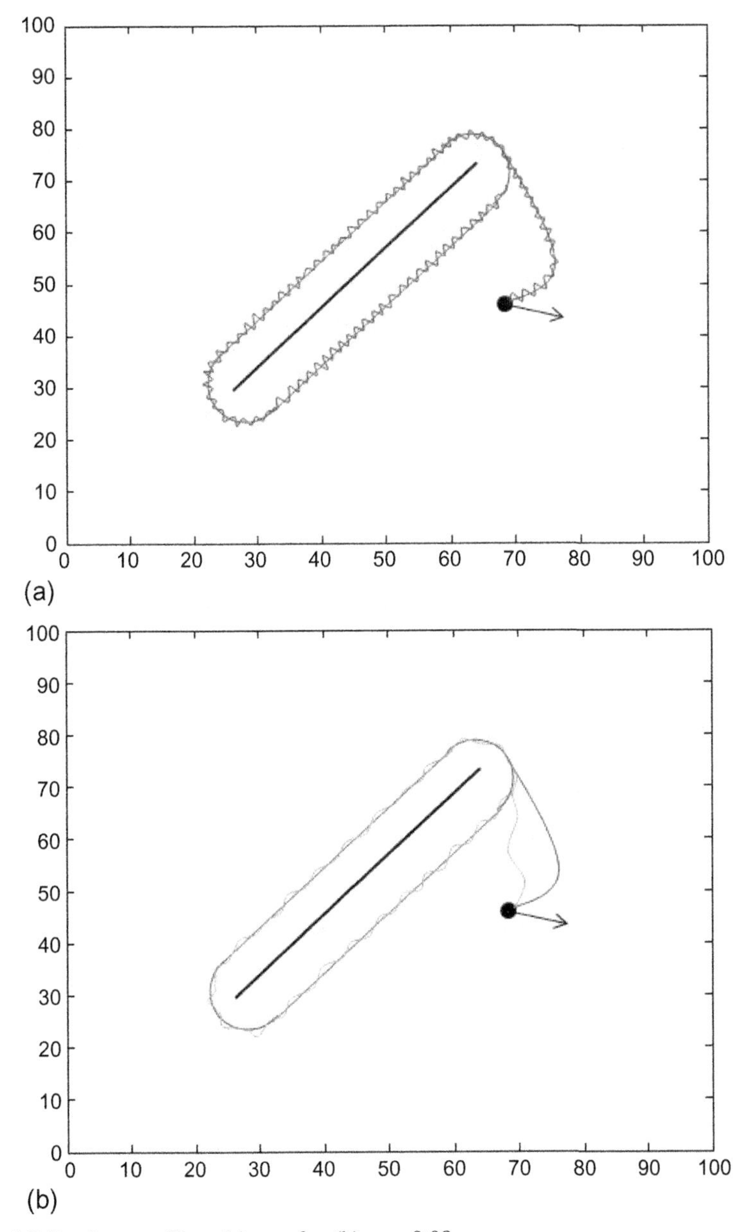

Figure 5.8 Border patrolling: (a) $\tau = 3$ s, (b) $\tau = 0.08$ s.

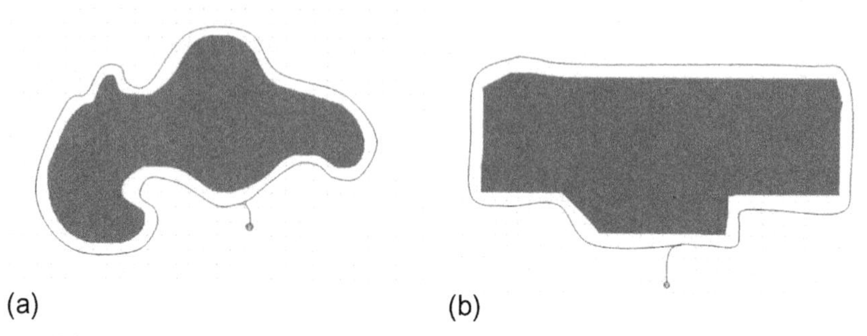

Figure 5.9 Border patrolling by a mobile robot.

no visible changes in the trajectory. This highlights that numerical differentiation is a potential bottleneck in the face of sensor noise.

In the other simulations, the derivative was estimated from the noisy data $\dot{d}(t) \approx \tau^{-1}[d(t) - d(t-\tau)]$. The curvy line in Fig. 5.8(a) illustrates that large τ may cause visible chattering of UAV about the ideal path, whereas the similar line in Fig. 5.8(b) shows that small τ may entail poor stabilization accuracy due to noise amplification. Figure 5.7 deals with a proper choice of the constant τ between these two extremes and displays that the UAV very closely follows the ideal "noise-free" path, thus satisfactorily carrying out the required border patrolling mission. This confirms the reputation of sliding mode controllers as highly tolerable to noises.

Other simulation results are illustrated in Fig. 5.9 in which a mobile robot patrols the border of two random areas at constant speed.

5.6 Boundary following with a rigidly mounted distance sensor: Problem setup

We still consider a Dubins-car type mobile robot traveling in a plane with the constant speed v. It is controlled by the angular velocity u limited by a given constant $\overline{u}$. There also is a domain D with a smooth boundary ∂D in the plane. The objective is to steer the robot to the equidistant curve of the domain D separated from it by the pre-specified distance d_0, and then to drive the robot over this curve; see Fig. 5.1(a).

The key distinction from the previous sections of this chapter is in the perception scenario. Unlike them, now the sensors do not provide access to the distance from the robot to the domain:

$$\mathbf{dist}\,[\boldsymbol{r}; D] = \min_{\boldsymbol{r}' \in D} \|\boldsymbol{r} - \boldsymbol{r}'\|. \tag{6.1}$$

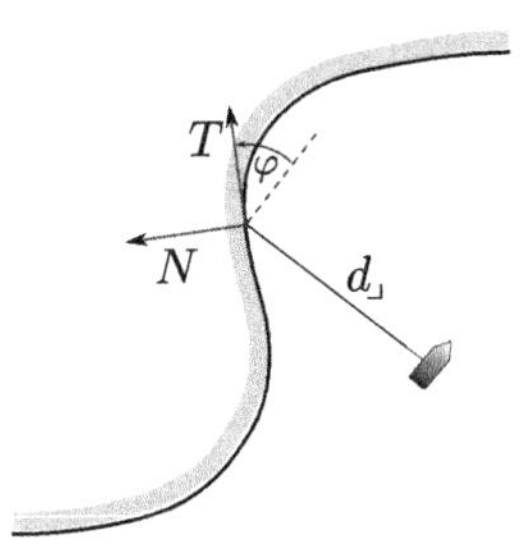

Figure 5.10 Mobile robot with a rigidly mounted range sensor.

Instead the robot is equipped with a narrow-aperture range sensor directed perpendicularly to the robot's centerline and to the left. This sensor provides the distance $d_\lrcorner$ from the robot to the nearest point of D in the sensed direction; see Fig. 5.10. Scan within the aperture and processing of the collected data gives the robot access to the angle φ from its forward centerline to the tangential direction of the boundary at the reflection point. Whenever the sensor does not detect the obstacle, $d_\lrcorner := \infty, \varphi := 0$.

Apart from stable motion over the equidistant curve, it is required to ensure transition to this motion from a given initial state. In doing so, the robot must not collide with the boundary ∂D. Moreover, the distance (6.1) from the robot to the boundary should constantly exceed the given safety margin $d_{\text{safe}} < d_0$.

The kinematics of the robot are still described by (2.1), reproduced here for the convenience of the reader:

$$\begin{array}{lll} \dot{x} = v\cos\theta, & \multirow{2}{*}{$\dot{\theta} = u \in [-\overline{u}, \overline{u}],$} & \boldsymbol{r}(0) = \boldsymbol{r}_0 \notin D, \\ \dot{y} = v\sin\theta, & & \theta(0) = \theta_0. \end{array} \qquad (6.2)$$

The set of initial states $\boldsymbol{r}_0 \notin D$ for which the problem has a solution will be specified later on in Theorem 8.1.

In the remainder of this chapter, we examine the following navigation law:

$$u = \begin{cases} -\overline{u} & \text{if} \quad S := \varphi + \chi[d_\lrcorner - d_0] \leq 0, \\ u_+ := \min\left\{\overline{u}; v d_\lrcorner^{-1}\right\} & \text{if} \quad S > 0. \end{cases} \qquad (6.3)$$

Here $\chi(\cdot)$ is a linear function with saturation (2.4); see Fig. 5.2. The gain coefficient $\gamma > 0$ and the saturation level $\mu \in \left(0, \frac{\pi}{2}\right)$ are design parameters, as before.

5.7 Assumptions of theoretical analysis and tuning of the navigation controller

For the navigation objective to be achievable, the robot should be capable of tracking the d_0-equidistant curve of the boundary ∂D. However, this is impossible if this curve contains cusp singularities, so far as any path of the unicycle (6.2) is everywhere smooth. Such singularities are typically born whenever the boundary contains concavities and the required distance d_0 exceeds the critical value, which is equal to the minimal curvature radius of the concavity parts of the boundary [390]. Moreover, even if there are no singularities, the equidistant curve should not be much contorted since the robot is able to trace only curves whose curvature radius exceeds (2.2). These observations are detailed in the conditions necessary for the d_0-equidistant curve to be trackable by the robot that are given by Lemmas 3.1 and 3.2. Being slightly enhanced by putting the uniformly strict inequality sign in place of the non-strict one, they come to the following nearly unavoidable assumption.

Assumption 7.1. *The following inequalities hold*

$$\begin{aligned} R_{\varkappa}^{+}(D) &:= \inf_{\boldsymbol{r}\in\partial D:\varkappa(\boldsymbol{r})>0} R_{\kappa} > R_{\min} - d_0, \\ R_{\varkappa}^{-}(D) &:= \inf_{\boldsymbol{r}\in\partial D:\varkappa(\boldsymbol{r})<0} R_{\kappa} > d_0 + R_{\min} \end{aligned} \tag{7.1}$$

and the curvature is bounded $K := \sup_{\boldsymbol{r}\in\partial D}|\varkappa(\boldsymbol{r})| < \infty$.

Here $\varkappa = \varkappa(\boldsymbol{r})$ is the signed curvature of ∂D at $\boldsymbol{r}$, $R_{\varkappa} := |\varkappa|^{-1}$ is the curvature radius of the boundary, $R_{\min}$ is the minimal turning radius of the robot (2.2), and inf over the empty set is defined as $+\infty$. The signed curvature is nonnegative on the convexities of the boundary and negative on the concavities.

As is illustrated in Fig. 5.11, the sensor system of the robot is deficient in capability of online detection of head-on collisions with the domain D. In the absence of extra forward-view sensors, a partial remedy may be systematic full turns to accomplish environment mapping within the entire vicinity of the robot given by the sensor range, which however consumes extra resources and may be unacceptable. If no special measures are taken to explore the forward direction, collisions can be excluded only due to special geometric properties of the obstacle D. They should guarantee a certain amount of free forward space on the basis of circumstances sensible in the side direction. These guarantees should cover the entire operational zone including the transient.

Local guarantees of such a kind are given by (7.1). To highlight this, we denote by $\mathcal{N}(\boldsymbol{r})$ the outer normal ray to the boundary ∂D rooted at $\boldsymbol{r} \in \partial D$, and by $\boldsymbol{r}(L)$ the point of $\mathcal{N}(\boldsymbol{r})$ at the distance L from $\boldsymbol{r}$. Then the second inequality from (7.1) implies that some piece $\partial_s D$ of ∂D surrounding $\boldsymbol{r}$ does not intersect the open disk of the radius $d_0 + R_{\min}$ centered at $\boldsymbol{r}(d_0 + R_{\min})$ [390], as is illustrated in Fig. 5.12. In

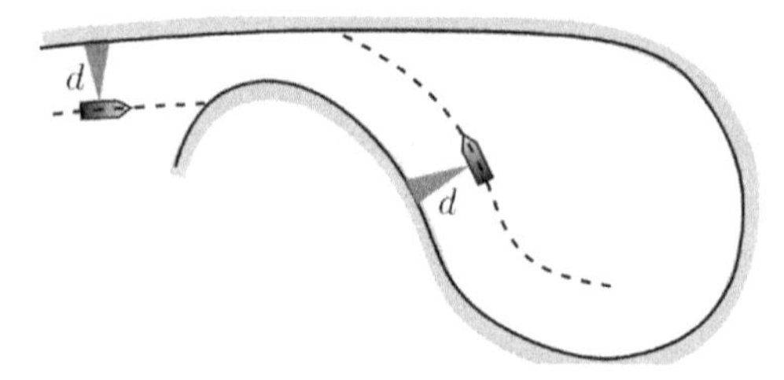

Figure 5.11 Insufficiency of the side sensor to ensure safety.

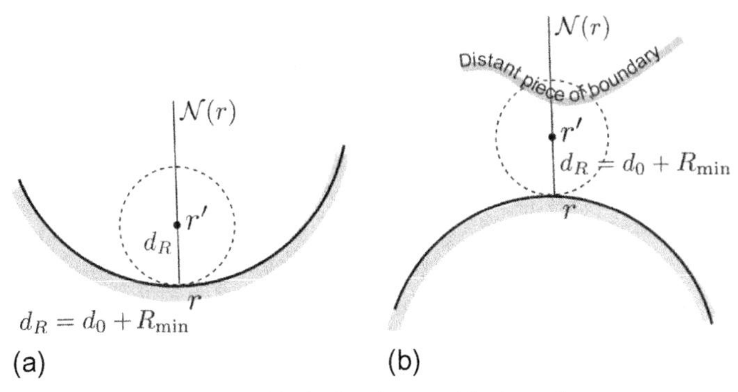

Figure 5.12 Disk free of collision with the local part of the boundary.

turn, this implies that first, the smaller disk of the radius $R_{\min}$ centered at $\boldsymbol{r}(d_0 + R_{\min})$ is separated from $\partial_s D$ by a distance of no less than $d_0 > d_{\text{safe}}$ and second, the smaller open disk $\mathfrak{D}(\boldsymbol{r}, d_0)$ of the radius d_0 centered at $\boldsymbol{r}(d_0)$ does not intersect $\partial_s D$. Now we extend these guarantees on the entire boundary and operational zone, which is assumed to be upper limited $d_\lrcorner \le d_*$ by a constant d_*.

Assumption 7.2. *The following two claims hold:*

(i) *For any $\boldsymbol{r} \in \partial D$, the open disk $\mathfrak{D}(\boldsymbol{r}, d_0)$ is disjoint with the boundary ∂D;*
(ii) *There exist $d_* > d_0$ and $\eta > 0$ such that for any $\boldsymbol{r} \in \partial D$, the set $Q(\boldsymbol{r}, 0)$ introduced in Fig. 5.13(a) is separated from the boundary ∂D by a distance of no less than $d_{safe} + \eta$; see Fig. 5.13(b).*

It follows from Fig. 5.13(a) that $d_{\text{safe}} + \eta < d_0$.

Finally, we assume the distance to D is locally controllable when operating at the safety margin d_{safe}: it can be maintained constant, increased, and decreased by

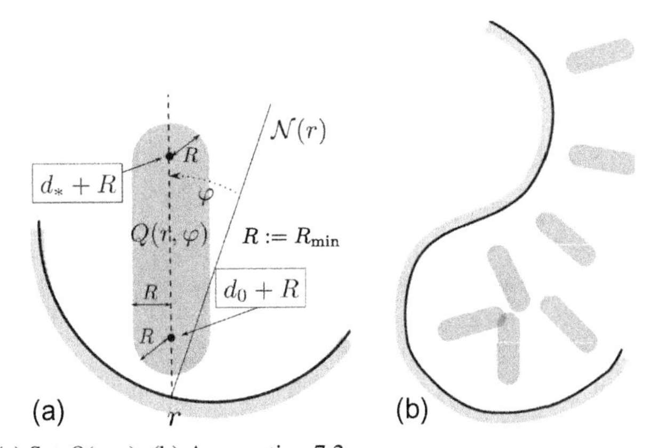

Figure 5.13 (a) Set $Q(\boldsymbol{r}, \varphi)$; (b) Assumption 7.2.

selecting respective controls. By invoking Remark 3.1, this gives rise to the last assumption.

Assumption 7.3. *The following inequality holds:*

$$R_\kappa^+(D) + d_{safe} > R_{\min}, \tag{7.2}$$

where the quantity $R_\kappa^+(D)$ is defined in (7.1).

The similar condition $R_\kappa^-(D) > d_{\text{safe}} + R_{\min}$ for the concavity parts $\varkappa < 0$ of the boundary follows from the second inequality in (7.1) since $d_0 > d_{\text{safe}}$.

5.7.1 Tuning of the navigation controller

Due to (7.1), (7.2), there exists $\mu \in \left(0, \frac{\pi}{2}\right)$ such that

$$R_\varkappa^+(D)\cos\mu + d_{\text{safe}} > R_{\min}, \tag{7.3}$$

$$R_\varkappa^-(D)\cos\mu > R_{\min} + d_0. \tag{7.4}$$

We pick $\eta_* \in (0, \eta)$, where η is taken from (ii) of Assumption 7.2, and by decreasing μ if necessary, ensure the following property, which is possible thanks to Assumption 7.2.

Property 1. For any $\boldsymbol{r} \in \partial D$ and $\varphi \in [-\mu, \mu]$, the following two claims hold:

(i) For any point $\boldsymbol{r}'$ such that $\|\boldsymbol{r}' - \boldsymbol{r}\| \le d_0$ and the angle subtended by $\boldsymbol{r}' - \boldsymbol{r}$ and the normal $\mathcal{N}(\boldsymbol{r})$ equals φ,

(i.1) the straight line segment with the end-points $\boldsymbol{r}$ and $\boldsymbol{r}'$ has only one point $\boldsymbol{r}$ in common with ∂D;

(i.2) the distance from $\boldsymbol{r}'$ to ∂D is no less than d_{safe} if $\|\boldsymbol{r}' - \boldsymbol{r}\| \ge d_{\text{safe}} + \eta_*$;

(ii) the set $Q(\boldsymbol{r}, \varphi)$ is separated from the boundary ∂D by a distance of no less than $d_{\text{safe}} + \eta_*$.

Finally, we put $[s]_+ := \max\{s, 0\}$ and pick γ so that

$$\gamma < \frac{R_\varkappa^+(D)\cos\mu - [R_{\min} - d_{\text{safe}}]_+}{R_\varkappa^+(D)[R_{\min} - d_{\text{safe}}]_+ \sin\mu} \quad \text{if} \quad R_{\min} > d_{\text{safe}}, \tag{7.5}$$

$$\gamma < \frac{R_\varkappa^-(D)\cos\mu - (R_{\min} + d_0)}{(R_{\min} + d_0)R_\varkappa^-(D)\sin\mu}, \tag{7.6}$$

$$\gamma < \frac{\cos\mu}{(R_{\min} + d_0)\sin\mu}. \tag{7.7}$$

This choice is possible, since the right-hand sides of all inequalities are positive due to (7.3) and (7.4).

If the boundary ∂D is compact, inequalities (7.1)–(7.6) can be checked in point-wise fashion. In doing so, any inequality involving $R_\varkappa^+(D)$ should be checked at any point $\boldsymbol{r} \in \partial D$ of convexity $\varkappa(\boldsymbol{r}) > 0$ with substituting $R_\varkappa(\boldsymbol{r})$ in place of $R_\varkappa^+(D)$. Similarly, any inequality involving $R_\varkappa^-(D)$ should be checked at any point $\boldsymbol{r} \in \partial D$ of concavity $\varkappa(\boldsymbol{r}) < 0$ with substituting $R_\varkappa(\boldsymbol{r})$ in place of $R_\varkappa^-(D)$.

5.8 Boundary following with a rigidly mounted sensor: Convergence of the proposed navigation law

Since the controller (6.3) is fed by the side-view observations $d_\lrcorner, \varphi$, prior to putting (6.3) in use, a special maneuver is performed until the domain D becomes visible. If D is initially visible, this is omitted; otherwise, the sharpest clock-wise turn ($u \equiv -\overline{u}$) may be, for example, but not necessarily, performed. Therefore, when examining convergence of the navigation law, we assume the initial state is in the set $\mathcal{V}$ of all states (given by $\boldsymbol{r} \notin D$ and θ) for which the domain is visible. Let $d_\lrcorner(\boldsymbol{r}, \theta)$ denote the corresponding measurement $d_\lrcorner$.

The set $\mathfrak{C}$ of initial states $\boldsymbol{r}, \theta$ from which convergence to the required equidistant curve can be theoretically guaranteed is composed of three parts: $\mathfrak{C}_0, \mathfrak{C}_-$, and $\mathfrak{C}_+$. They contain initial states with $S = 0, S < 0$, and $S > 0$, respectively, where S is defined in (6.3). Here

$$\mathfrak{C}_0 := \{(\boldsymbol{r}, \theta) \in \mathcal{V} : S = 0 \text{ and } d_{\text{safe}} + \eta_* \leq d_\lrcorner \leq d_*\}. \tag{8.1}$$

To introduce $\mathfrak{C}_-$, we first pay attention to the following.

Lemma 8.1. *Under the navigation law* (6.3), *motion with $S < 0$ necessarily terminates with arrival at $S = 0$, provided that the robot does not collide with the domain D.*

The proof of this lemma is united with the proof of a somewhat similar Lemma 8.5, with both being given in Section 5.8.2.

Let $C^-_{\boldsymbol{r},\theta}$ be the circle of radius (2.2) traced clockwise from the initial state $\boldsymbol{r}, \theta$ and $(\boldsymbol{r}_*, \theta_*)$ be the first position on this circle that either belongs to D or is such that $S = 0$. By Lemma 8.1, this position does exist. Let also $\hat{C}^-_{\boldsymbol{r},\theta}$ denote the arc of $C^-_{\boldsymbol{r},\theta}$ between these two positions, and $\mathbf{dist}(A, B) := \inf_{\boldsymbol{r} \in A, \boldsymbol{r}' \in B} \|\boldsymbol{r} - \boldsymbol{r}'\|$ denote the distance between the sets A and B. The second part of the set $\mathfrak{C}$ is given by

$$\mathfrak{C}_- := \Big\{(\boldsymbol{r}, \theta) \in \mathcal{V} : S < 0, \quad \mathbf{dist}[\hat{C}^-_{\boldsymbol{r},\theta}, D] \geq d_{\text{safe}}, \quad \text{and} \quad d_\lrcorner(\boldsymbol{r}_*, \theta_*) \geq d_{\text{safe}} + \eta_*\Big\}. \tag{8.2}$$

Introduction of the last part $\mathfrak{C}_+$ is prefaced by the following.

Lemma 8.2. *Under the navigation law* (6.3) *and for initial states with $S > 0$, there may be the following three scenarios:*

- *(i) With maintaining $S > 0$, the safety margin is violated;*
- *(ii) With respecting the safety margin and maintaining $S > 0$, the robot arrives at a position where the view of D becomes obstructed by another part of D; at this moment, S abruptly jumps down at a negative value;*
- *(iii) With respecting the safety margin and maintaining both the view of D unobstructed and $S > 0$, the robot arrives at $S = 0$.*

The proof of this lemma will be given in Section 5.8.2.

While $S > 0$, the robot moves counter-clockwise over the circle of the radius d centered at the reflection point (which does not move) whenever $d > R_{\min}$. Otherwise, it moves with the maximal turning rate $\overline{u}$ over a circle of radius (2.2).

Finally, we introduce the set $\mathfrak{C}_+$ of all initial states $(\boldsymbol{r}, \theta) \in \mathcal{V}$ for which the following claims hold:

- Scenario (i) from Lemma 8.2 does not hold and $S(0) > 0$;
- In the case (ii) from Lemma 8.2, the robot arrives at a state from the set $\mathfrak{C}_-$ when S becomes negative;
- In the case (iii) from Lemma 8.2, the robot arrives at a state from the set $\mathfrak{C}_0$ when S becomes zero.

Now we are in a position to state the main theoretical result concerned with the navigation law (6.3).

Theorem 8.1. *Let Assumptions 7.1–7.3 be true, and in* (6.3)*, the parameters be chosen so that* (7.3)–(7.7) *and Property 1 hold. If the initial state lies in the set* $\mathfrak{C} := \mathfrak{C}_0 \cup \mathfrak{C}_- \cup \mathfrak{C}_+$, *the robot driven by the navigation and guidance law* (6.3) *does not lose track of the domain D, respects the safety margin, and asymptotically follows the boundary of D at the required distance:* $d_\lrcorner(t) \to d_0, \varphi(t) \to 0$ *as* $t \to \infty$.

The proof of this theorem will be given in Section 5.8.2.

Additional information about the domain D typically allows specifying the assumptions, requirements to the controller parameters, and the estimate of the convergence domain. Now we illustrate this in a simple but instructive case.

5.8.1 Illustrative analysis of the convergence domain

Suppose the region D is convex. It is easy to see that all assumptions (i.e., Assumptions 7.1–7.3) are reduced to the inequality

$$R_{\varkappa}^{+}(D) + d_{\text{safe}} > R_{\min}. \tag{8.3}$$

Therefore, by Remark 3.1, the robot can track any equidistant curve at a distance greater than the safety margin and can pass from it both further from and closer to D.

The requirement (7.3) to the controller parameters shapes into

$$\mu \in \begin{cases} (0, \pi/2) & \text{if} \quad R_{\min} \leq d_{\text{safe}} \quad \text{or} \quad R_{\varkappa}^{+}(D) = \infty, \\ \left(0, \arccos \frac{R_{\min} - d_{\text{safe}}}{R_{\varkappa}^{+}(D)}\right) & \text{otherwise}. \end{cases}$$

This also implies (i.1) in Property 1. In (i.2) and (ii), arbitrary $\eta_* \in (0, d_0 - d_{\text{safe}})$ can be considered. The claims (i.2) and (ii) result from the respective inequalities

$$\cos\mu \geq g_1(\eta_*) := \frac{d_{\text{safe}}}{d_{\text{safe}} + \eta_*}, \quad \cos\mu \geq g_1(\eta_*) := \frac{d_{\text{safe}} + R_{\min} + \eta_*}{d_0 + R}. \tag{8.4}$$

By computing $\min_{\eta_* \in (0, d_0 - d_{\text{safe}})} \max\{g_1(\eta_*); g_2(\eta_*)\}$, we see that modulo the freedom to manipulate with η_*, inequalities (8.4) hold if and only if

$$\mu < \arccos \frac{2d_{\text{safe}}}{\sqrt{R_{\min}^2 + 4d_{\text{safe}}(d_0 + R_{\min})} - R_{\min}}. \tag{8.5}$$

Finally, the controller parameters are chosen as follows:

- $\mu \in (0, \pi/2)$ is subjected
  - to (8.5) if $R_{\min} \le d_{\text{safe}}$ or $R_\varkappa^+(D) = \infty$
  - to (8.5) and the inequality $\mu < \arccos \frac{R_{\min} - d_{\text{safe}}}{R_\varkappa^+(D)}$ otherwise;
- After this, $\gamma > 0$ is chosen subject to (7.5) and (7.7).

The estimate of the convergence domain employs (in (8.1) and (8.2)) the parameter η_* from (i.2) and (ii) of Property 1. Due to (8.4), its choice is limited by

$$d_{\text{safe}} \frac{1 - \cos\mu}{\cos\mu} \le \eta_* \le (d_0 + R_{\min})\cos\mu - (d_{\text{safe}} + R_{\min}).$$

By (8.1) and (8.2), it is beneficial to take the least feasible value $\eta_* = d_{\text{safe}} \frac{1-\cos\mu}{\cos\mu}$, for which

$$d_{\text{safe}} + \eta_* = \frac{d_{\text{safe}}}{\cos\mu}. \tag{8.6}$$

When characterizing the initial states $\boldsymbol{r}$ for which the proposed navigation and guidance law ensures achievement of the navigation objective, we assume the domain D is initially visible by the side-view sensor of the robot, due to the reasons explained at the beginning of Section 5.8. Any such state $\boldsymbol{r}$ is associated with the reflection point $\boldsymbol{r}_*$, the sensed distance d, and the angle ψ from the outer normal to the ray from $\boldsymbol{r}_*$ to $\boldsymbol{r}$; see Fig. 5.14(a). For any possible value $\psi \in (-\pi/2, \pi/2)$, the description of the convergence domain will be given in terms of $d, \boldsymbol{r}_*$. To this end, we denote by $\mathbf{e}(\alpha)$ the unit vector whose polar angle in the Frenet frame of ∂D at $\boldsymbol{r}_*$ equals α, by $\rho_D(\boldsymbol{r}')$ the distance from the point $\boldsymbol{r}'$ to D, and consider separately two cases.

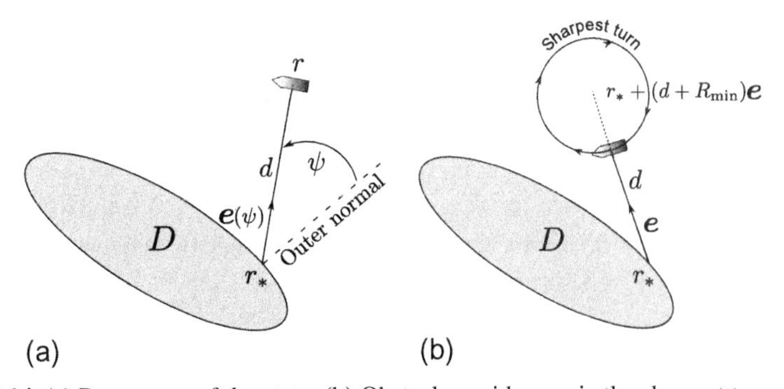

Figure 5.14 (a) Parameters of the state; (b) Obstacle avoidance via the sharpest turn.

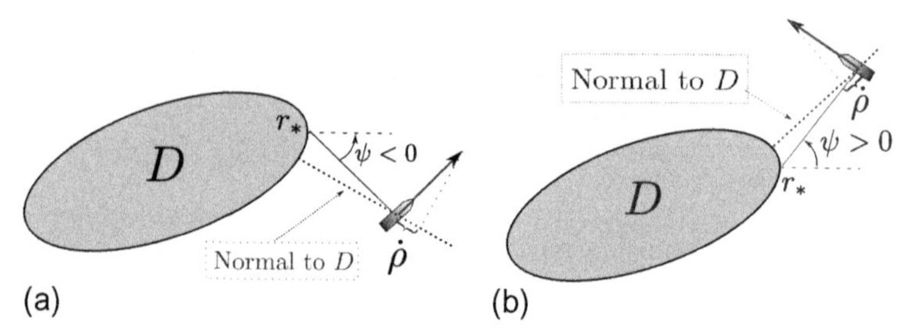

Figure 5.15 Rate at which the distance ρ to D evolves over time.

1. $\boldsymbol{\psi \leq \chi(d_\lrcorner - d_0)}$. To shorten the text and focus on the main ideas, we limit ourselves to the case where $d_{\text{safe}} \geq R$; the "otherwise" case is considered likewise. If $\psi < \chi(d_\lrcorner - d_0)$, (6.3) implies the robot initially moves counter-clockwise over the circle of the radius $d_\lrcorner$ centered at $\boldsymbol{r}_*$ until $S = 0(\Leftrightarrow \psi = \chi(d_\lrcorner - d_0))$. The last event necessarily holds at some time τ since $d_\lrcorner$ is constant and φ decreases. Here $\tau = 0$ if initially $\psi = \chi(d_\lrcorner - d_0)$. The definition of $\mathfrak{C}_+$ requires to arrive at the set (8.1) at $t = \tau$ and to constantly respect the safety margin. By the simple geometric argument illustrated in Fig. 5.15, ρ increases for $\psi < 0$ and decreases for $\psi > 0$. Therefore, it suffices to care of safety only at $t = 0$ and if $d_\lrcorner > d_0$, at $t = \tau$. Thus, the initial state $(\boldsymbol{r}, \theta)$ is in the convergence domain whenever

$$d_\lrcorner \geq d_{\text{safe}} \cos^{-1} \mu, \quad \rho_D[\boldsymbol{r}] \geq d_{\text{safe}} \tag{8.7}$$

$$\text{and} \quad \rho_D[\boldsymbol{r}_* + d_\lrcorner \mathbf{e}(\chi(d_\lrcorner - d_0))] \geq d_{\text{safe}} \quad \text{if} \quad d_\lrcorner > d_0. \tag{8.8}$$

2. $\boldsymbol{\psi > \chi(d_\lrcorner - d_0)}$. Due to (6.3), initially $S < 0$. If $\psi \leq 0$, elementary geometrical argument (see Fig. 5.15(a)) shows that both ρ and d increase until $S = 0$. Hence, it suffices to require that (8.7) holds. For $\psi > 0$, the robot is capable of maintaining the safety margin $\geq d_{\text{safe}}$ only if

$$\mathbf{dist}[C^-_{r,\theta}, D] > d_{\text{safe}}; \tag{8.9}$$

see Fig. 5.14(b). At the same time, this implies the second inequality in (8.2). Therefore, along with (8.9), the last inequality from (8.2) should be imposed:

$$d_\lrcorner(\boldsymbol{r}_*, \theta_*) \geq d_{\text{safe}} \cos^{-1} \mu. \tag{8.10}$$

Further calculations can be carried out and result in a description of the convergence domain in a closed form for specific geometric shapes. We illustrate this in the case where

The domain D is a half-plane. Since $R^+_\varkappa(D) = \infty$, the equivalent (8.3) of our assumptions always holds, and the controller parameters should be chosen so that (7.7) and (8.5) are true. Since for $\psi > 0$, (8.7)–(8.10) follow from $\mathbf{dist}[C^-_{r,\theta}, D] \geq d_{\text{safe}} \cos^{-1} \mu$, the robot driven by the navigation controller (6.3) does not lose track of D, is constantly at a distance of no less than d_{safe} from it, and asymptotically follows the boundary of D at the required distance whenever initially (d, ψ) lies in the shadowed domain from Fig. 5.16.

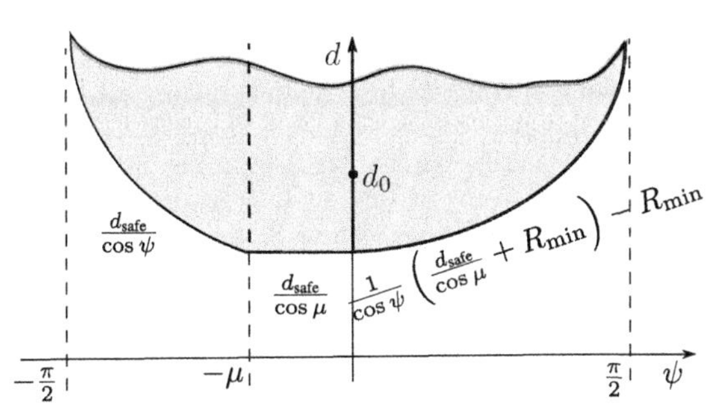

Figure 5.16 Convergence domain.

5.8.2 Proofs of Theorem 8.1 and Lemmas 8.1 and 8.2

We assume the boundary ∂D is oriented so that the domain is to the left when traveling over ∂D. Let $\boldsymbol{\rho}(s)$ be the natural parametric representation of ∂D, where s is the curvilinear abscissa ascending in the positive direction of ∂D. This abscissa is cyclic if ∂D is bounded: s and $s+L$ encode a common point, where L is the perimeter of ∂D. We notationally identify s and $\boldsymbol{\rho}(s)$ and introduce the Frenet frame $\mathbf{T}(s), \mathbf{N}(s)$ of ∂D at the point s ($\mathbf{T}$ is the positively oriented unit tangent vector, $\mathbf{N}$ is the unit normal vector directed inward D). The symbol $s(t)$ stands for the curvilinear abscissa of the sensor reflection point at time t.

Routine kinematic arguments give rise to the following.

Lemma 8.3. *Whenever the sensor sees the obstacle ($d_\lrcorner < \infty$) and $\varphi \neq \pm\frac{\pi}{2}$ mod 2π, the following relations hold:*

$$\dot{\varphi} = \varkappa \dot{s} - u, \quad \dot{d}_\lrcorner = \dot{s}\sin\varphi, \quad \dot{s} = \frac{v - d_\lrcorner u}{\cos\varphi}. \tag{8.11}$$

To proceed, we recall that δ is the saturation threshold in (2.4) and note that the derivative $\gamma_\zeta(d_\lrcorner)$ at the point $d_\lrcorner$ of the function (2.4) is given by

$$\gamma_\zeta = \gamma_\zeta(d_\lrcorner) := \begin{cases} 0 & \text{if } |d_\lrcorner - d_0| > \delta, \\ \gamma & \text{if } |d_\lrcorner - d_0| \leq \delta. \end{cases}$$

From now on, we examine the system (6.2) driven by the navigation controller (6.3). Elementary observations show that (7.5)–(7.7) can be rewritten in the following form

$$\frac{R_\varkappa \cos\mu}{1+\gamma R_\varkappa \sin\mu} + d_{\text{safe}} > R_{\min} \quad \text{if} \quad \varkappa > 0, \tag{8.12}$$

$$R_{\min} + d_0 < \inf_{r\in\partial D:\varkappa(r)<0} \frac{R_\varkappa \cos\mu}{1+\gamma R_\varkappa \sin\mu}, \tag{8.13}$$

$$\gamma < \frac{\cos\mu}{(R_{\min} + d_0)\sin\mu}, \quad \frac{\cos\mu}{\gamma\sin\mu} > R_{\min} + d_0. \tag{8.14}$$

Lemma 8.4. *On the surface*

$$S := \varphi + \chi[d_\lrcorner - d_0] = 0 \tag{8.15}$$

the following relation holds:

$$|\cos\varphi| \geq \cos\mu. \tag{8.16}$$

This surface is a barrier for penetration into the domain $S > 0$ whenever

$$\begin{gathered} \varkappa + \gamma_\zeta^\pm \sin\varphi \leq 0 \quad \text{or} \\ \varkappa + \gamma_\zeta^\pm \sin\varphi > 0 \quad \text{and} \quad R_{\min} < \frac{\cos\varphi}{\varkappa + \gamma_\zeta^\pm \sin\varphi} + d_\lrcorner, \end{gathered} \tag{8.17}$$

where γ_ζ^+, γ_ζ^- are the limits from the right and left, respectively, and in the case of uncertainty $\gamma_\zeta^+ \neq \gamma_\zeta^- \Leftrightarrow |d_\lrcorner - d_0| = \delta$, (8.17) holds for both signs. Whenever

$$\varkappa + \gamma_\zeta^\pm \sin\varphi > 0 \quad \text{and} \quad R_{\min} > \frac{\cos\varphi}{\kappa + \gamma_\zeta^\pm \sin\varphi} + d_\lrcorner, \tag{8.18}$$

the trajectories depart from this surface into the set $S > 0$.

This surface is a barrier for penetration into the domain $S < 0$ whenever

$$\begin{gathered} \varkappa + \gamma_\zeta^\pm \sin\varphi \geq 0 \quad \text{or} \\ \varkappa + \gamma_\zeta^\pm \sin\varphi < 0 \quad \text{and} \quad R_{\min} + \frac{\cos\varphi}{\kappa + \gamma_\zeta^\pm \sin\varphi} + d_\lrcorner < 0. \end{gathered} \tag{8.19}$$

Whenever

$$\varkappa + \gamma_\zeta^\pm \sin\varphi < 0 \quad \text{and} \quad R_{\min} + \frac{\cos\varphi}{\kappa + \gamma_\zeta^\pm \sin\varphi} + d_\lrcorner > 0, \tag{8.20}$$

the trajectories of the closed-loop system depart from this surface into the set $S < 0$.

Proof. Due to (2.4),

$$S = 0 \Rightarrow |\varphi| = |\chi[d_\lrcorner - d_0]| \leq \mu < \frac{\pi}{2} \Rightarrow \text{(8.16)}.$$

Since $\dot{S} = \dot{\varphi} + \gamma_\zeta \dot{d}_\lrcorner$, we have

$$\begin{aligned} \cos\varphi\dot{S} &\overset{(8.11)}{=} (\varkappa + \gamma_\zeta \sin\varphi)(v - d_\lrcorner u) - u\cos\varphi \\ &= v(\varkappa + \gamma_\zeta \sin\varphi) - u\left[\cos\varphi + d_\lrcorner(\varkappa + \gamma_\zeta \sin\varphi)\right]. \end{aligned}$$

Let the examined point converge to a location on the surface $S = 0$ so that $S \neq 0$ and $\mathbf{sgn}\, S \equiv \mathbf{const}$. By (6.3),

$$\overline{L}_+ := \overline{\lim_{S \to 0, S>0}} \cos\varphi \dot{S} \leq \max_{\gamma_* = \gamma_\zeta^+, \gamma_\zeta^-} \Xi_+(\gamma_*),$$

$$\underline{L}_+ := \underline{\lim_{S \to 0, S>0}} \cos\varphi \dot{S} \geq \min_{\gamma_* = \gamma_\zeta^+, \gamma_\zeta^-} \Xi_+(\gamma_*),$$

$$\text{where} \quad \Xi_+(\gamma_*) := v\,(\varkappa + \gamma_* \sin\varphi) - u_+ \left[\cos\varphi + d_\lrcorner\, (\varkappa + \gamma_* \sin\varphi)\right],$$

$$\underline{L}_- := \underline{\lim_{S \to 0, S<0}} \cos\varphi \dot{S} \geq \min_{\gamma_* = \gamma_\zeta^+, \gamma_\zeta^-} \Xi_-(\gamma_*),$$

$$\overline{L}_- := \overline{\lim_{S \to 0, S<0}} \cos\varphi \dot{S} \leq \max_{\gamma_* = \gamma_\zeta^+, \gamma_\zeta^-} \Xi_-(\gamma_*),$$

$$\text{where} \quad \Xi_-(\gamma_*) := v\,(\varkappa + \gamma_* \sin\varphi) + \overline{u}\left[\cos\varphi + d_\lrcorner\, (\varkappa + \gamma_* \sin\varphi)\right].$$

If (8.17) holds, $\overline{L}_+ < 0$, which implies the first claim of the lemma. If (8.18) holds, $\underline{L}_\pm > 0$, which implies the second claim. The other claims are established likewise. □

Corollary 8.1. *The surface* (8.15) *is a barrier for penetration into the domain* $S > 0$ *whenever* $d_\lrcorner \geq d_{safe}$.

Proof. By Lemma 8.4, the claim holds if $\varkappa + \gamma_\zeta^\pm \sin\varphi \leq 0$; otherwise, $\varkappa + \gamma_\zeta^\pm \sin\varphi > 0$, and it suffices to show that

$$\Phi := \frac{\cos\varphi}{\varkappa + \gamma_\zeta^\pm \sin\varphi} + d_\lrcorner > R_{\min}. \tag{8.21}$$

In doing so, we consider separately three cases.

(1) $d_\lrcorner \geq d_0 \Leftrightarrow \varphi = -\chi(d_\lrcorner - d_0) \leq 0$. Then $\varkappa > 0$ and in (8.21),

$$\Phi \geq \frac{\cos\varphi}{\varkappa} + d_{\text{safe}} \geq R_\kappa \cos\mu + d_{\text{safe}} \overset{(7.3)}{>} R_{\min}.$$

(2) $d_\lrcorner < d_0, \varkappa > 0$. For $d_\lrcorner < d_0 - \delta$, the derivative $\gamma_\zeta^\pm = 0$ and so the proof is like above. For $d_\lrcorner \geq d_0 - \delta$,

$$\Phi \geq \frac{R_\varkappa \cos\mu}{1 + \gamma R_\varkappa \sin\mu} + d_{\text{safe}} \overset{(8.12)}{>} R_{\min}.$$

(3) $d_\lrcorner < d_0, \varkappa \leq 0$. Then

$$\gamma_\zeta^\pm \sin\varphi > 0 \Rightarrow d_\lrcorner \geq d_0 - \delta.$$

Hence,

$$\Phi \geq \frac{\cos\mu}{\gamma \sin\mu} + d_{\text{safe}} \overset{(8.25)}{>} R_{\min} \quad \text{if} \quad \varkappa = 0.$$

If $\varkappa < 0$,

$$\Phi \geq \frac{R_\varkappa \cos\varphi}{\gamma R_\varkappa \sin\varphi - 1} + d_\lrcorner \geq \frac{\cos\varphi}{\gamma \sin\varphi} + d_{\text{safe}} \geq \frac{\cos\mu}{\gamma \sin\mu} + d_{\text{safe}} \overset{(8.14)}{>} R_{\min}.$$

□

To proceed, we observe that thanks to the strict inequality signs in (7.3), (7.4), (8.13), and (8.14), there exist $\xi, \xi_+ > 0$ such that

$$\cos \mu R_\varkappa + d_{\text{safe}} > R_{\min} + \xi_+ \quad \text{whenever} \quad \varkappa > 0, \tag{8.22}$$

$$\cos \mu R_\varkappa > R_{\min} + d_0 + \xi \quad \text{whenever} \quad \varkappa < 0, \tag{8.23}$$

$$R_{\min} + d_0 + \min\{\xi, \mu/\gamma\} < \frac{R_\varkappa \cos \varphi}{1 + \gamma R_\varkappa \sin \varphi} \quad \text{if } \varkappa < 0, \tag{8.24}$$

$$\frac{\cos \mu}{\gamma \sin \mu} > R_{\min} + d_0 + \min\{\mu/\gamma, \xi\}. \tag{8.25}$$

Corollary 8.2. *The surface* (8.15) *is a barrier for penetration into the domain* $S < 0$ *whenever* $d_\lrcorner \leq d_0 + \xi$.

Proof. By Lemma 8.4, the claim holds if $\varkappa + \gamma \sin \varphi \geq 0$; otherwise, $\varkappa + \gamma_\zeta^\pm \sin \varphi < 0$, and it suffices to show that

$$\Psi := R_{\min} + \frac{\cos \varphi}{\varkappa + \gamma_\zeta^\pm \sin \varphi} + d_\lrcorner < 0.$$

In doing so, we consider separately three cases.

(1) $d_\lrcorner \leq d_0 \Leftrightarrow \varphi = -\chi(d_\lrcorner - d_0) \geq 0$. Then $\varkappa < 0$ and

$$\begin{aligned} \Psi &\leq R_{\min} - \frac{\cos \varphi}{|\varkappa| - \gamma \sin \varphi} + d_0 + \xi \\ &\leq R_{\min} - \frac{\cos \varphi}{|\varkappa|} + d_0 + \xi \leq R_{\min} - R_\varkappa \cos \mu + d_0 + \xi \overset{(8.23)}{<} 0. \end{aligned}$$

(2) $d_\lrcorner > d_0, \varkappa < 0$. Then for $d_\lrcorner > d_0 + \delta$, the derivative $\gamma_\zeta^\pm = 0$ and so the proof is like above. For $d_\lrcorner \leq d_0 + \delta$,

$$\begin{aligned} \Psi &\leq R_{\min} - \frac{\cos \varphi}{|\varkappa| + \gamma \sin \varphi} + d_0 + \min\{\xi, \delta\} \\ &\leq R_{\min} - \frac{R_\varkappa \cos \varphi}{1 + \gamma R_\varkappa \sin \varphi} + d_0 + \min\{\xi, \mu/\gamma\} \overset{(8.24)}{<} 0. \end{aligned}$$

(3) $d_\lrcorner > d_0, \varkappa \geq 0$. Then $d_\lrcorner \leq d_0 + \delta$ and

$$\Psi = R_{\min} - \frac{\cos \varphi}{\gamma |\sin \varphi| - \varkappa} + d_\lrcorner \leq R_{\min} - \frac{\cos \mu}{\gamma \sin \mu} + d_0 + \min\{\xi, \mu/\gamma\} \overset{(8.25)}{<} 0.$$

□

Lemma 8.5. *During motion with* $S < 0$*, the robot does not lose track of D. If during this motion* $d_\lrcorner$ *arrives at* d_0 *from above,* $d_\lrcorner < d_0$ *from the moment of this arrival and until termination of the motion with* $S < 0$.

Proof of Lemmas 8.1 and 8.5. While $S < 0$, (6.3) yields $u \equiv -\overline{u}$ and the robot moves clockwise along a circle of radius $R_{\min}$. Since

$$S = \varphi + \chi(d_\lrcorner - d_0) < 0 \wedge |\chi(d_\lrcorner - d_0)| \leq \mu \Rightarrow \varphi \leq \mu < \pi/2,$$

the robot does not lose track of D (which is illustrated in Fig. 5.17(a)) and even if $d_\lrcorner$ abruptly changes, $d_\lrcorner$ jumps down; see Fig. 5.17(b).

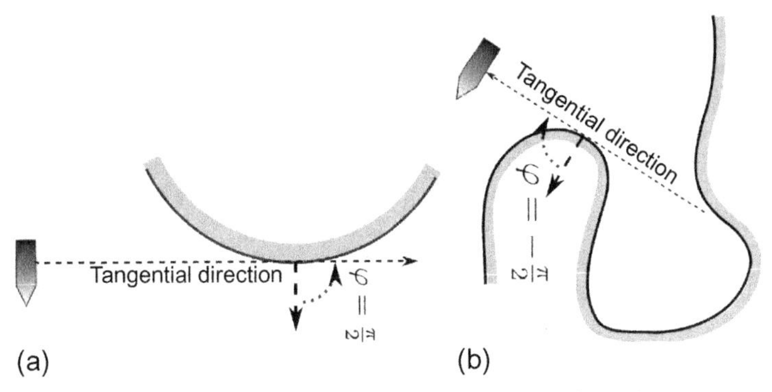

Figure 5.17 (a) Loss of view of the domain; (b) Abrupt change of the distance.

Now suppose that motion with $S < 0$ never terminates and thus the robot repeatedly performs full turns. Since $\dot{s} = \frac{v + d_\lrcorner \overline{u}}{\cos\varphi} > 0$ by (8.11), we see that

$$d_\lrcorner > d_0 \Rightarrow \varphi < 0 \overset{(8.11)}{\Longrightarrow} \dot{d}_\lrcorner < 0. \tag{8.26}$$

Hence, the entire circle cannot lie in the domain $d_\lrcorner \geq d_0$. Moreover, either (a) $\varphi \equiv 0$ during the entire motion over the circle or (b) during the motion in the domain $d_\lrcorner < d_0$, there are time instants τ when $\varphi > 0$.

(a) By (8.11),

$$d_\lrcorner = \text{const}, \dot{s} \equiv v + d_\lrcorner \overline{u}, \dot{\varphi} = 0 \equiv \kappa(v + d_\lrcorner \overline{u}) + \overline{u} \Rightarrow \kappa \equiv -\frac{1}{R_{\min} + d_\lrcorner} < 0$$
$$\Rightarrow R_\kappa = R_{\min} + d_\lrcorner \overset{(7.4)}{\Rightarrow} d_\lrcorner > d_0 \Rightarrow S > 0,$$

in violation of the hypothesis.

(b) For $t > \tau$ and while $d_\lrcorner < d_0, \varphi > 0$, we have $0 < \varphi \leq -\chi(d_\lrcorner - d_0) \leq \mu$. Therefore, due to (i.1) in Property 1, $d_\lrcorner$ and φ smoothly evolve over time and

$$\dot{\varphi} \overset{(8.11)}{=} \varkappa \frac{v + d_\lrcorner \overline{u}}{\cos\varphi} + \overline{u} \begin{cases} > 0 & \text{if } \varkappa \geq 0, \\ \geq \overline{u}|\varkappa| \frac{-(R_{\min}+d_0)+R_\kappa \cos\mu}{\cos\varphi} \overset{(7.4)}{>} 0 & \text{if } \varkappa < 0. \end{cases}$$

Hence, $\varphi > 0$ when $d_\lrcorner$ arrives at d_0 once more, in violation of $S < 0$. Therefore, for $t > \tau$ not only $\dot{\varphi} > 0$ but also $d_\lrcorner < d_0$, which gives rise to a contradiction when the full turn is completed since $\varphi \in [0, \mu]$ is determined by the robot's position.

The contradictions obtained complete the proof. □

Lemma 8.6. *During the sliding motion over the surface $S = 0$, the robot does not lose track of D and the curvilinear abscissa increases $\dot{s} > 0$. Moreover, $\dot{s} \geq \varepsilon > 0$, where ε depends $\varepsilon = \varepsilon(\hat{d})$ only on the upper estimate $\hat{d} \geq d_\lrcorner$ of the distance $d_\lrcorner$.*

Proof. The first claim holds since

$$S = 0 \Rightarrow \varphi = -\chi(d_\lrcorner - d_0) \in [-\mu, \mu] \subset (-\pi/2, \pi/2).$$

Invoking (8.11) yields that

$$0 = \cos\varphi \dot{S} = \cos\varphi \left[\dot{\varphi} + \gamma_\zeta^\pm \dot{d}_\lrcorner\right] = (\varkappa + \gamma_\zeta^\pm \sin\varphi)(v - d_\lrcorner u) - u\cos\varphi \Rightarrow$$

$$u = \frac{v(\varkappa + \gamma_\zeta^\pm \sin\varphi)}{\cos\varphi + (\varkappa + \gamma_\zeta^\pm \sin\varphi)d_\lrcorner} \Rightarrow \dot{s} = \frac{v}{\cos\varphi + (\varkappa + \gamma_\zeta^\pm \sin\varphi)d_\lrcorner}. \tag{8.27}$$

Thus, $\dot{s} \geq \frac{v}{\cos\varphi} \geq \frac{v}{\cos\mu} > 0$ if $(\varkappa + \gamma_\zeta^\pm \sin\varphi) \geq 0$. In the case where $(\varkappa + \gamma_\zeta^\pm \sin\varphi) < 0$, we first note that

$$R_{\min} + \frac{\cos\varphi}{\varkappa + \gamma_\zeta^\pm \sin\varphi} + d_\lrcorner \leq 0,$$

since otherwise the robot immediately enters the domain $S < 0$ due to (8.20), in violation of (2). Therefore,

$$\frac{\cos\varphi}{\varkappa + \gamma_\zeta^\pm \sin\varphi} + d_\lrcorner \leq -R_{\min} \Rightarrow$$

$$\cos\varphi + (\varkappa + \gamma_\zeta^\pm \sin\varphi)d_\lrcorner \geq R_{\min}\left|\varkappa + \gamma_\zeta^\pm \sin\varphi\right| > 0 \tag{8.28}$$

and thus $\dot{s} > 0$ again. To prove the second claim of the lemma, it suffices to note that the modulus of the denominator in (8.27) is upper bounded by $1 + (\gamma + K)\hat{d}$, where K is taken from Assumption 7.1. □

Corollary 8.3. *The distance $d_\lrcorner$ decreases $\dot{d}_\lrcorner < 0$ and the curvilinear abscissa s increases $\dot{s} > 0$ while the robot moves in the domain $S \leq 0$ and $d_\lrcorner > d_0$.*

Proof. Whenever the robot is in the domain $S \leq 0$ and $d_\lrcorner > d_0$, either (a) $S < 0$ or (b) the robot moves over the surface $S = 0$. The claim is justified by (8.26) in case (a) and by Lemma 8.6 in case (b). Since $S \leq 0 \Rightarrow \varphi \leq -\chi(d_\lrcorner - d_0) < 0$, the proof is completed by the second equation in (8.11). □

Lemma 8.7. *During motion in the domain $S > 0$ while respecting the safety margin, the robot does not lose track of D and $\dot{\varphi} \leq -\varepsilon_+ < 0$, where the constant ε_+ does not depend on the robot's state. During motion in the domain $S \geq 0, d_\lrcorner \leq d_0$, the variables $d_\lrcorner, s$ either both increase $\dot{d}_\lrcorner > 0, \dot{s} > 0$ or both stay constant $\dot{d}_\lrcorner = 0, \dot{s} = 0$.*

Proof. We note first that

$$S = \varphi + \chi(d_\lrcorner - d_0) \geq 0 \wedge d_\lrcorner < d_0 \Rightarrow \varphi > 0.$$

If $S > 0$ and $d_\lrcorner \geq R_{\min}$, we have $u = v/d_\lrcorner$ and due to (8.11),

$$\dot{s} = \frac{v - d_\lrcorner u}{\cos\varphi} = 0, \quad \dot{d}_\lrcorner = \dot{s}\sin\varphi = 0, \quad \dot{\varphi} = -\overline{u} < 0.$$

If $S > 0$ and $d_\lrcorner < R_{\min}$, conversely $u = \overline{u}$ and so

$$\dot{s} = \frac{v - d_\lrcorner \overline{u}}{\cos\varphi} > 0, \quad \dot{d}_\lrcorner = \dot{s}\sin\varphi > 0,$$

$$\dot{\varphi} = \varkappa\frac{v - d_\lrcorner \overline{u}}{\cos\varphi} - \overline{u} = \frac{\overline{u}}{\cos\varphi}\left[\varkappa(R_{\min} - d_\lrcorner) - \cos\varphi\right].$$

Therefore, $\dot{\varphi} < -\overline{u}$ if $\varkappa \leq 0$. Otherwise,

$$\dot{\varphi} \leq -\overline{u}/2 \quad \text{if} \quad \varkappa \leq \frac{\overline{u}\cos\varphi}{2(v - d_\lrcorner \overline{u})}; \quad \text{otherwise,}$$

$$\dot{\varphi} \leq \frac{\varkappa\overline{u}}{\cos\varphi}\left[R_{\min} - d_{\text{safe}} - R_\varkappa \cos\mu\right] \overset{(8.22)}{\leq} -\frac{\varkappa\overline{u}}{\cos\varphi}\xi_+ \leq -\frac{\overline{u}^2}{2(v - d_\lrcorner \overline{u})}\xi_+$$

$$\leq -\frac{\overline{u}^2}{2v}\xi_+ < 0.$$

Thus, during motion with $S > 0$, the angle φ decreases and so cannot achieve the value $\pi/2$, whereas

$$S > 0 \Rightarrow \varphi > -\chi(d_\lrcorner - d_0) \geq -\mu > -\pi/2.$$

It follows that the robot does not lose track of D. Under the circumstances, the robot may also undergo sliding motion over the surface $S = 0$. Then $\dot{s} > 0$ again by Lemma 8.6. Hence, by (8.11), $\dot{d}_\lrcorner > 0$. The track of D is not lost by the same lemma. □

Proof of Lemma 8.2. Suppose that scenario (i) does not hold and so the safety margin is respected while $S > 0$. Since $\dot{\varphi} \leq -\varepsilon_+ < 0$ by Lemma 8.7 and $|\chi(d_\lrcorner - d_0)| \leq \mu < \pi/2$, the robot arrives at either $S = 0$ or the position where the view of D becomes obstructed by another part of D, as is illustrated in Fig. 5.17(b). At this moment φ jumps down at the level $\varphi = -\pi/2$ and so S becomes negative. □

Lemma 8.8. *While the robot moves in the set* $\mathfrak{S} := \{S \leq 0, d_\lrcorner \geq d_0\}$ *while respecting the safety margin, it does not lose track of D and the distance d arrives at* d_0 *either for a finite time or asymptotically as* $t \to \infty$. *In the latter case,* $\varphi(t) \to 0$ *as* $t \to \infty$.

Proof. The first claim follows from Lemmas 8.1, 8.5, and 8.6. If initially $d_\lrcorner = d_0$, the second claim is evident. Let initially $d_\lrcorner > d_0$. While $d_\lrcorner > d_0$, the inequality $S \leq 0$ is maintained by Corollary 8.1. Hence, if the domain $\mathfrak{S}$ is left, $d_\lrcorner = d_0$ at the time of departure, which completes the proof. It remains to examine the case where the robot never leaves $\mathfrak{S}$.

By Corollary 8.3, $d_\lrcorner(t) \to d_\infty$ as $t \to \infty$ and $d_\lrcorner(0) \geq d_\lrcorner(t) > d_\infty$ $\forall t$. Suppose that $d_\infty > d_0$, in violation of the claim of the lemma. Then due to Corollary 8.3, we have

$$S \leq 0 \Rightarrow \varphi \leq -\chi(d_\lrcorner - d_0) \leq -\chi(d_\infty - d_0) < 0,$$

$$\dot{d}_\lrcorner \overset{(8.11)}{=} \dot{s}\sin\varphi \leq -\dot{s}\sin\chi(d_\infty - d_0) \Rightarrow s(t) \leq s(0) + \frac{d_\lrcorner(0) - d_\lrcorner(t)}{\sin\chi(d_\infty - d_0)}.$$

Therefore, the increasing function $s(\cdot)$ has a finite limit

$$s_* = \lim_{t\to\infty} s(t) \in \mathbb{R}. \tag{8.29}$$

Whenever the robot undergoes sliding motion over the surface $S = 0$, Lemma 8.6 yields that $\dot{s} \geq \varepsilon[d(0)]$, whereas

$$S < 0 \Rightarrow \dot{s} \overset{(8.11)}{=\!=\!=} \frac{v + d_\lrcorner \overline{u}}{\cos\varphi} \geq v + d_0\overline{u} > 0.$$

Therefore, there is $\varepsilon > 0$ such that $\dot{s}(t) \geq \varepsilon \; \forall t \approx \infty$, in violation of (8.29). The contradiction obtained proves that

$$\lim_{t\to\infty} d_\lrcorner(t) = d_0 \quad \text{and} \quad \overline{\lim_{t\to\infty}} \varphi(t) \overset{S\leq 0}{\leq} \lim_{t\to\infty} -\chi(d_\lrcorner - d_0) = 0.$$

Suppose to the contrary to the last claim of the lemma that $\varphi(t) \not\to 0$ as $t \to \infty$. Then there exist $\alpha > 0$ and a sequence $t_i \uparrow \infty$ such that $\varphi(t_i) \leq -2\alpha \; \forall i$. For $i \approx \infty$, there also exists $\tau_i < t_i$ such that $\varphi(\tau_i) = -\alpha$ and $\varphi(t) \leq -\alpha \; \forall t \in [\tau_i, t_i]$. For these t's, we have $\dot{d} = \dot{s}\sin\varphi \leq -\dot{s}\sin\alpha$ and so

$$s(t_i) - s(\tau_i) \leq \frac{d_\lrcorner(\tau_i) - d_0}{\sin\alpha}.$$

Meanwhile, $S < 0$ and so $u \equiv -\overline{u} \; \forall t \in [\tau_i, t_i]$ and large enough i. Therefore, for these t's,

$$\dot{\varphi}(t) = \varkappa\dot{s} + \overline{u} \geq \overline{u} - K\dot{s},$$

where K is taken from Assumption 7.1. Thus,

$$\begin{aligned} -\alpha \geq \varphi(t_i) - \varphi(\tau_i) &\geq \overline{u}[t_i - \tau_i] - K\frac{d_\lrcorner(\tau_i) - d_0}{\sin\alpha} \\ &\geq -K\frac{d_\lrcorner(\tau_i) - d_0}{\sin\alpha} \to 0 \quad \text{as} \quad i \to \infty. \end{aligned}$$

The contradiction obtained proves that $\varphi(t) \to 0$ as $t \to \infty$ and completes the proof. □

Lemma 8.9. *Let the initial state be in the set* (8.1). *Then the robot respects the safety margin d_{safe}, does not lose track of D, and $d_\lrcorner(t) \to d_0$, $\varphi(t) \to 0$ as $t \to \infty$.*

Proof. If $d_\lrcorner(0) = d_0$, the equation $S = 0$ implies that $\varphi = 0$ and by Corollaries 8.1 and 8.2, the robot traces the d_0-equidistant curve of ∂D. It remains to note that this curve does not collide with ∂D thanks to Assumption 7.2.

Suppose that $d_\lrcorner(0) < d_0$. By Corollaries 8.1 and 8.2, the robot undergoes sliding motion over the surface $S = 0$ while $d_{\text{safe}} \leq d_\lrcorner \leq d_0$. Within this motion, the robot does not lose track of D by Lemma 8.6 and

$$\varphi \overset{S=0}{=\!=} -\chi(d_\lrcorner - d_0), \quad \dot{d} \overset{(8.11)}{=\!=} \dot{s}\sin\varphi = -\dot{s}\sin\chi(d_\lrcorner - d_0),$$

where $\dot{s} > 0$ by Lemma 8.6 and $|\chi(d_\lrcorner - d_0)| \leq \mu < \pi/2$. It follows that $d_\lrcorner$ increases, never reaches the value d_0, and remains in the set $[d_{\text{safe}} + \eta_*, d_0]$. Hence, the robot respects the margin d_{safe} thanks to (i) in Property 1, and sliding motion is never terminated. Moreover, by the second claim of Lemma 8.6, $\dot{s}$ is lower limited by a positive constant, which implies that

$$d_\lrcorner(t) \to d_0 \quad \text{and} \quad \varphi(t) = -\chi[d_\lrcorner(t) - d_0] \to 0 \quad \text{as} \quad t \to \infty.$$

Let $d_\lrcorner(0) > d_0$. If the robot does not trespass the safety margin and $d_\lrcorner$ does not arrive at d_0, the proof is completed by Lemma 8.8. Suppose the converse and denote by τ the first time when either trespass or arrival of $d_\lrcorner$ at d_0 holds.

Suppose that τ is the time of trespass. If $S(\tau) = 0$, then $\varphi(\tau) = -\chi(d_\lrcorner - d_0) \in [-\mu, 0]$ and $d_0 \leq d_\lrcorner \leq d_*$ due to Corollary 8.3. Let $\boldsymbol{r}_*$ be the robot's position at $t = \tau$ and $\boldsymbol{r}$ be the reflection point. Then $\|\boldsymbol{r}_* - \boldsymbol{r}\| = d_\lrcorner \in [d_0, d_*]$ and the angle subtended by the normal $\mathscr{N}(\boldsymbol{r})$ and the vector from $\boldsymbol{r}$ to $\boldsymbol{r}'$ does not exceed μ. Therefore, by (ii) in Property 1, the distance from $\boldsymbol{r}_*$ to ∂D is greater than d_{safe}, in violation of the definition of τ. This contradiction implies that $S(\tau) < 0$.

Let τ_0 be the maximal time $t \in [0, \tau]$ such that $S(t) = 0$. Then $S(\tau_0) = 0$ and for $t \in (\tau_0, \tau]$, we have $S(t) < 0$, $u = -\overline{u}$, and so the robot moves along the circle C of the radius $R_{\min}$ centered at the point $\boldsymbol{r}' = \boldsymbol{r} + \frac{d_\lrcorner + R_{\min}}{d}(\boldsymbol{r}_* - \boldsymbol{r})$. By (ii) in Property 1, C is separated from ∂D by a distance greater than d_{safe}, and so the trespass at $t = \tau$ is impossible. Thus, $d_\lrcorner(\tau) = d_0$.

If $S(\tau) = 0$, the proof is completed by its starting argument. Let $S(\tau) < 0$ and τ_0 still stand for the maximal time $t \in [0, \tau]$ such that $S(t) = 0$. By retracing the arguments from the previous paragraph, we see that while $S(t) < 0$ for $t \geq \tau$, the robot moves over a circle separated from ∂D by a distance greater than $d_{\mathrm{safe}} + \eta_*$. Therefore, the safety margin is respected during this motion. By Lemma 8.1, this circular motion necessarily terminates by arriving at $S(t) = 0$ with $d_\lrcorner < d_0$. Since $d_\lrcorner$ is no less than the distance from the circle to the boundary, we have $d_\lrcorner \geq d_{\mathrm{safe}} + \eta_*$. Thus, we have arrived at the case considered in the second paragraph of the proof. □

Proof of Theorem 8.1. If the initial state lies in the set (8.1), the claim is given by Lemma 8.9. Suppose that it lies in the set (8.2). Then initially $S < 0$ and the robot moves with $u \equiv -\overline{u}$ over a circle of radius $R_{\min}$. During the motion along this circle and until $S = 0$, the safety margin is respected by (8.2), the robot does not lose track of D by Lemma 8.5 and arrives at $S = 0$ by Lemma 8.1. The state of the robot at the moment of arrival belongs to the set (8.1) by (8.2). Therefore, the proof is completed by the foregoing.

If $S(0) > 0$, the proof follows from the foregoing and the definition of the set $\mathfrak{C}_+$, which is given in the paragraph before Theorem 8.1. □

5.9 Computer simulations of border patrolling with a rigidly mounted distance sensor

Simulations were performed using the perfect kinematic model of the robot (6.2). To estimate the angle φ, the tangent at the reflection point was approximated by the secant between this point and another point slightly in front; the angular separation between these points was 9 deg. The navigation law was updated with the sampling period of 0.1 s. Other parameters used for simulation are shown in Table 5.1.

In the first simulation test, the domain D fits the maneuverability of the robot: the minimal turning radius of the robot exceeds the radius required for perfectly tracking the boundary of D with the requested margin d_0. Figure 5.18 shows that after a short transient, the proposed navigation law provides a visibly perfect motion over the desired equidistant curve and successfully copes with both convexities and concavities

Table 5.1 **Parameters used for simulations**

$\overline{u}$	45.8 deg/s	v	0.3 m/s
μ	57.3 deg	γ	171.9 deg/m
d_0	1.0 m		

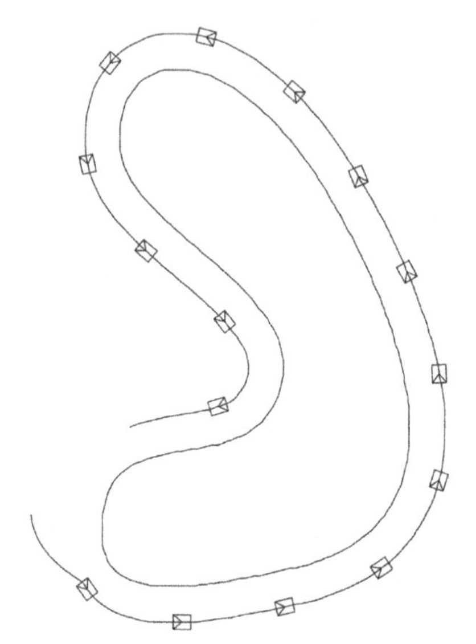

Figure 5.18 Following the boundary that fits the maneuverability of the robot.

of the obstacle, as well as with transitions from convexities to concavities and vice versa. Figures 5.19 and 5.20 provide a closer look at the boundary following errors. After the transient is completed ($t \gtrapprox 18$ s), the error in the true distance to the obstacle (6.1) does not exceed 1 cm, whereas the error in the estimated angular discrepancy φ between the tangent at the reflection point and the robot's centerline does not exceed 12.6°. However, this good exactness proceeds from taking into account only the non-idealities that are due to control sampling and numerical evaluation of the relative tangent angle.

The second group of simulation tests provides deeper insights into the effects of real-live non-idealities on the performance of the closed-loop system. These tests were carried out in the previous scene with additionally taking into account sensor and actuator noises and un-modeled dynamics. To this end, a bounded random and uniformly distributed offset was added to every relevant quantity at each control update. Specifically, the noises added to d and φ were 0.3 m and 17.2°, respectively; the noise added to the control signal was 11.5 deg/s, and the control signal was not allowed to change faster than 4 rad/s^2. These are relatively large noises that would be unlikely met for typical modern sensors and actuators. The test was repeated several times, each with its own realization of the random noises. Ten typical results are

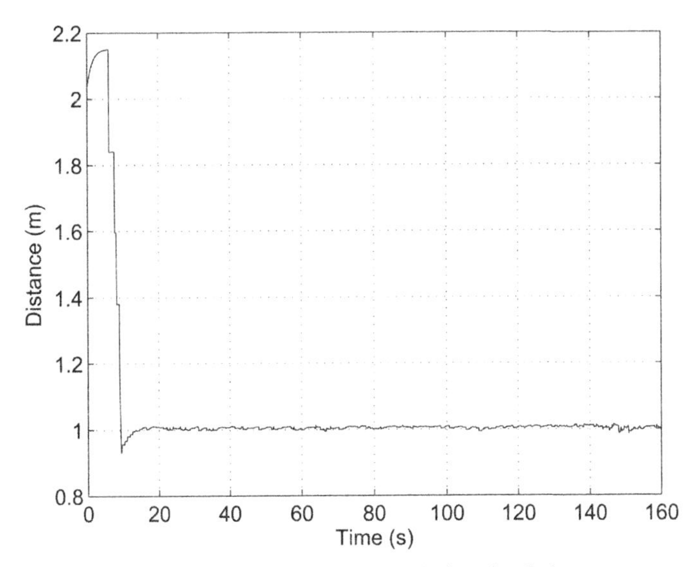

Figure 5.19 Evolution of the distance to the obstacle during simulation.

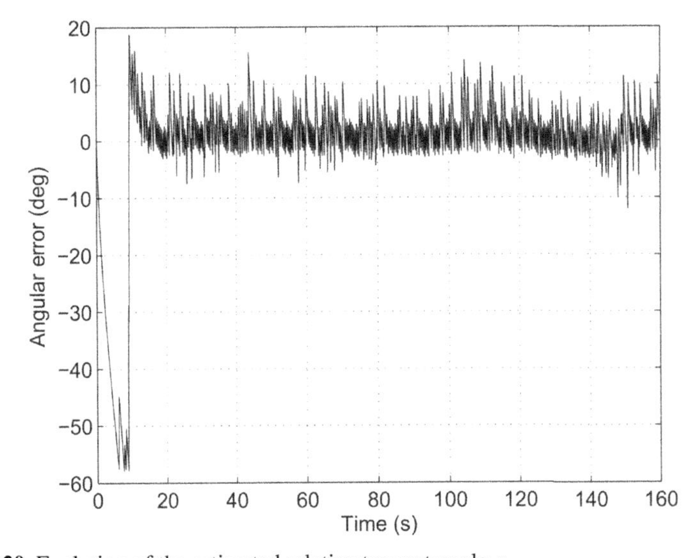

Figure 5.20 Evolution of the estimated relative tangent angle φ.

depicted in Fig. 5.21. They show that the navigation objective is still achieved with the distance error $\leq 0.3\,\text{m}$. Since this is the accuracy of the distance sensor, the result seems to be more than satisfactory.

The purpose of the next test is to examine the performance of the algorithm in the case when the obstacle boundary ∂D contains points where the robot is absolutely

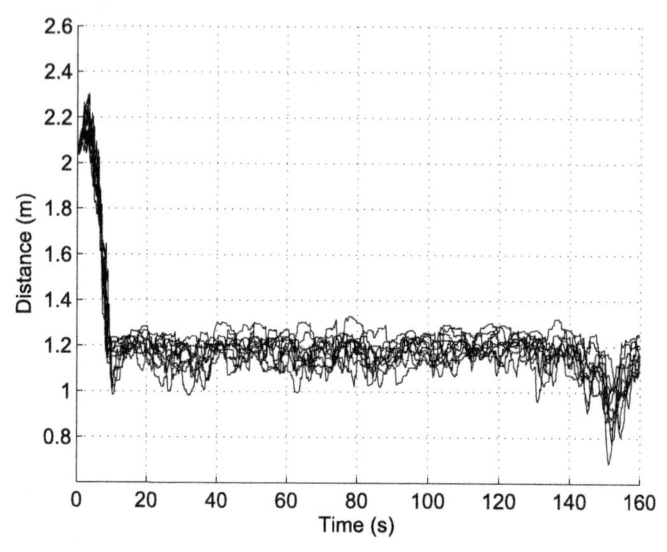

Figure 5.21 Evolution of the distance to the obstacle during simulations with actuator dynamics and sensor noises.

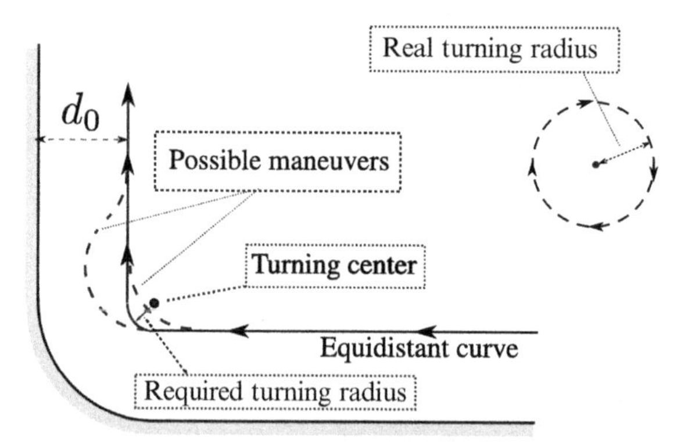

Figure 5.22 Unavoidable disturbance of the distance to D because of the limited turning radius.

incapable of maintaining the required distance d_0 to D because of the limited turning radius (see Fig. 5.22): to move over the equidistant curve, the robot should turn sharper than feasible. Though the main theoretical results of the work are not concerned with this case, it may be hypothesized that if these points constitute only a small piece of the boundary, the overall behavior of the closed-loop system remains satisfactory.[3] To verify this hypothesis and reveal some details, we consider the obstacle depicted in

[3] Since this behavior is expected to be close to that in the absence of such points.

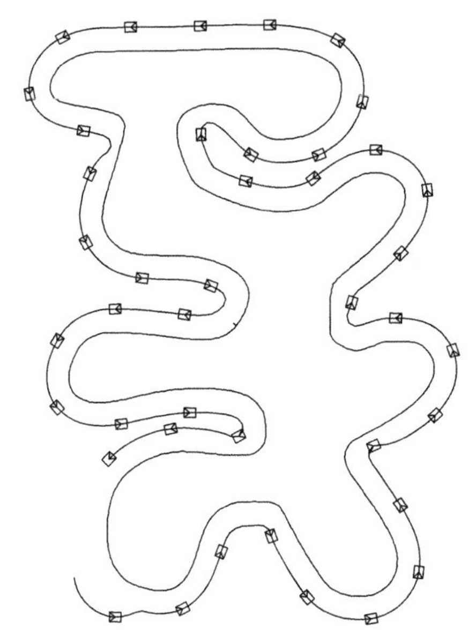

Figure 5.23 Closed-loop trajectory obtained during simulation.

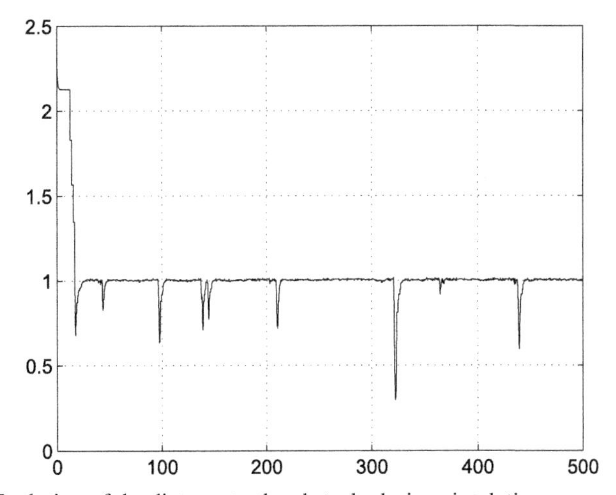

Figure 5.24 Evolution of the distance to the obstacle during simulation.

Fig. 5.23 whose concavities do contain the aforementioned points. The related results are shown in Figs. 5.23, 5.24, and 5.25. All abrupt local both falls of the distance d in Fig. 5.24 and deviations of the angle φ from 0 in Fig. 5.25 hold at $t \approx 90$, 130, 190, 300, 340, 400, and 490 s when the robot passes points where it is absolutely incapable of maintaining the distance to D at level d_0. However, the algorithm demonstrates

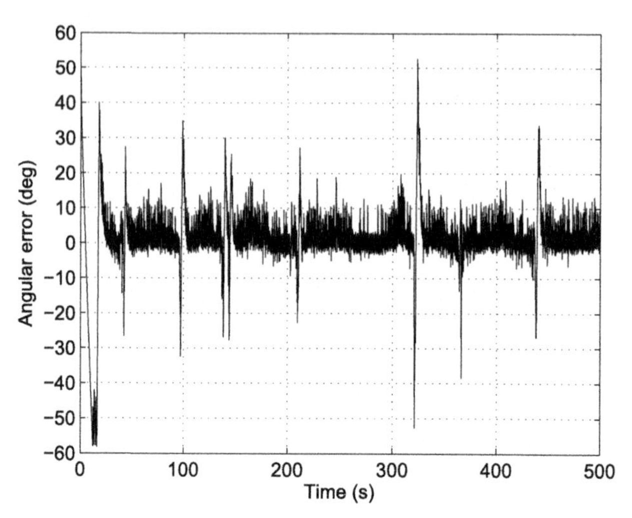

Figure 5.25 Evolution of the estimated relative tangent angle φ during simulation.

a good capability of quickly recovering after these unavoidable distance errors, and except for the related short periods of time and the initial transient, keeps the distance error within a very small bound of 1 cm and always maintains the margin of safety at the level ≥ 0.3 m. This good overall performance is illustrated in Fig. 5.23, where the robot's path looks nearly perfect except for few very local violations of the required distance, with most of them being hardly visible. The worst distance error is observed when passing the left upper concavity in Fig. 5.23. This concavity is similar to that from Fig. 5.22, i.e., it is the right-angle type. Such concavities are challenging for robots equipped with only side-view sensors. For example, when perfectly following the horizontal part of the equidistant curve from Fig. 5.22, the robot with side-view sensor is incapable of detecting the need for turn until the reflection point leaves the flat part of the boundary. Even if the robot performs the sharpest turn after this, its forward advancement toward the vertical part of the boundary may be nearly equal to the minimal turning radius (2.2) (which holds if in Fig. 5.22, the "requested turning radius" is close to 0). For the robot examined in the simulation tests, the minimal turning radius amounts to 0.375 m. Therefore, theoretically the distance to the obstacle may be reduced to $1 - 0.375\,\text{m} = 0.625\,\text{m}$ when passing the above concavity. Practically it reduces to ≈ 0.3 m, where the difference is basically caused by errors in numerical evaluation of the tangent angle φ. Except for this troublesome concavity, the navigation law ensures the safety margin of ≥ 0.6 m according to Fig. 5.24.

In this test as in the first one, only non-idealities related to control sampling and numerical evaluation of the relative tangent angle were taken into account. Simulations were also carried out to test the additional effect of unmodeled dynamics, sensor noise, and actuator noise on the performance of the closed-loop system in the environment from Fig. 5.23. All these phenomena were modeled like in the second group of simulations. Ten typical results corresponding to various realizations of the

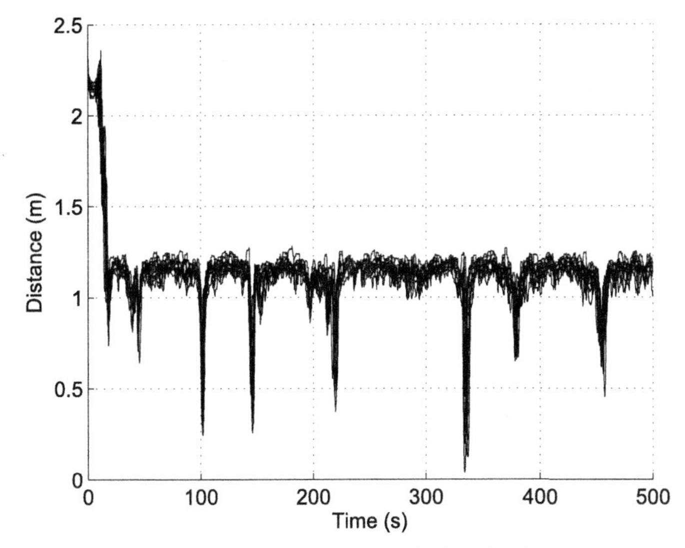

Figure 5.26 Evolution of the distance to the obstacle during simulations with actuator dynamics and sensor noises.

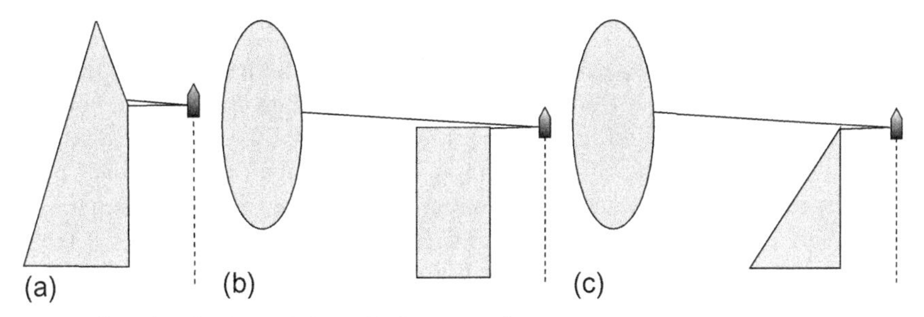

Figure 5.27 Following boundaries with fracture points.

random noises are depicted in Fig. 5.26. Similar to the second group of simulations, the distance accuracy degradation is approximately equal to the error of the distance sensor, which seems to be the fair price for using imperfect sensors. The robot still successfully follows the boundary without collision with the obstacle.

The next series of simulation tests were aimed at illustrating the algorithm performance in environments with many obstacles. The objective of the algorithm is to follow the boundary of the selected obstacle D_0 despite the presence of the others. According to the above discussion, it does follow the boundary provided that firstly, the view of D_0 is not obstructed and secondly, the assumptions of Theorem 8.1 are satisfied, in particular, the boundary of D_0 is smooth everywhere. The first requirement can be typically met by picking the desired distance of boundary following d_0 small enough as compared with spacing between obstacles. Therefore, the focus in the tests was on following boundaries with fractures; see Fig. 5.27.

At the fracture point, the tangent **T** to the boundary and the angle φ between **T** and the robot's centerline, which is used in the navigation law (6.3), are strictly speaking undefined. Therefore, the employed method to access φ may be puzzled at fractures, with the outcome being dependent on the method. In our experiments, the tangent **T** was approximated by the secant between the points $\boldsymbol{r}_*$ and $\boldsymbol{r}_+$ of incidence of the perpendicular ray R_* and a ray R_+ slightly in front, respectively. When arriving at a fracture point, this method implies an abrupt increase of φ. Let for simplicity the robot perfectly follow the boundary $d_\lrcorner \equiv d_0, \varphi \equiv 0$ prior to this event. Then the angle becomes positive $\varphi > 0$. By (6.3), this causes robot rotation about $\boldsymbol{r}_*$ at the distance d_0 from $\boldsymbol{r}_*$; see Fig. 5.28. This rough analysis gives a first evidence that the robot maintains following the boundary of the obstacle at hand, as desired. This is confirmed by the simulations shown in Fig. 5.29. Similar results with slightly worse performance are obtained in the case where the "preceding" ray R_+ is replaced by a ray R_- slightly behind R_*, and the distance $d_\lrcorner$ in the perpendicular direction is computed as $d_\lrcorner := \min\{d_*, d_{\text{safe}}\}$, where d_{safe} and d_* are the distances along the respective rays; see Fig. 5.30. The idea to compute the distance to the nearest obstacle as the minimum of the available distances along two close rays conforms to common sense. If it is illogically set to the maximum of them, the robot correctly passes obtuse fractures but for right and acute ones, starts circular motion about the farthest incidence point in accordance with (6.3), thus switching to a bypass of the competing obstacle; see Fig. 5.31.

Simulations were also carried out to compare the performance of the proposed navigation method with that from [206]. The latter algorithm employs the boundary curvature. By following [206], a geometric estimate of this curvature was used in our simulations. It results from drawing a circle through three boundary points detected by three close rays, i.e., the main perpendicular ray and two auxiliary rays on either side of the main one. The lengths to the boundary along every detection ray were corrupted by iid noises uniformly distributed over the very small interval $[-5\,\text{mm}, 5\,\text{mm}]$. The same noises affected the two-point approximation of the tangent to the boundary. To enhance the difference, both methods were equally challenged by not pre-filtering noisy measurements. The simulation scenario was the obtuse corner from Fig. 5.29. The parameters of the controller from [206] were taken to be $\kappa_M := 0.6; \epsilon := 0.4; \epsilon_2 = 0.2; \mu = 1; \mu_2 = 5$; and $\mu_3 = 5$, which meet the guidelines given in [206]. All simulation tests started at a common position, which corresponds to perfectly following the boundary at the required distance d_0. The tests were repeated 500 times with individual realizations of the noises; for each test, the maximal (over the experiment) deviation from the desired distance to the boundary $\max_t |d_\lrcorner(t) - d_0|$ was recorded.

The overall results are demonstrated by the histogram in Fig. 5.32, which shows the number of experiments with a given maximal deviation. The displayed better performance of the proposed method is presumably due to getting rid of the second derivative property (the boundary curvature) in the navigation law, which is particularly sensitive to both the distance measurement noises and violations of the boundary smoothness.

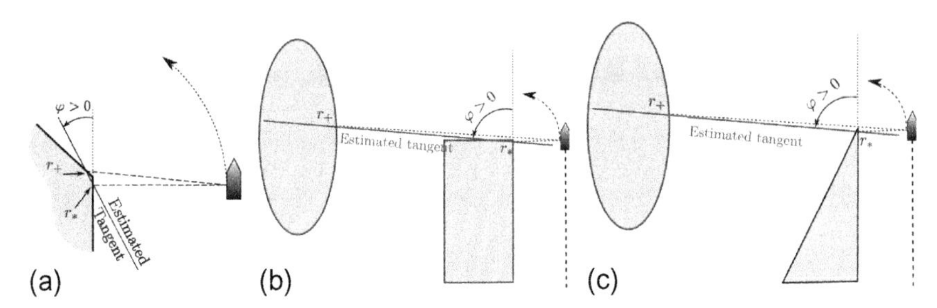

Figure 5.28 Rotation about $\boldsymbol{r}_*$.

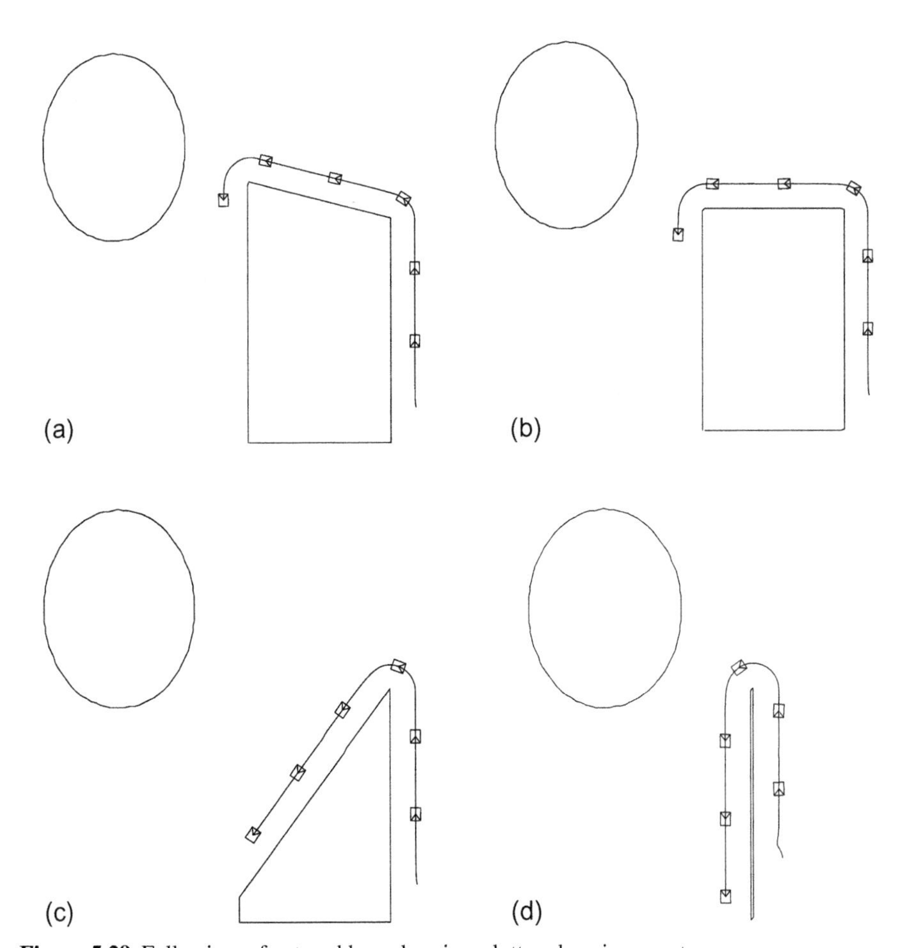

Figure 5.29 Following a fractured boundary in a cluttered environment.

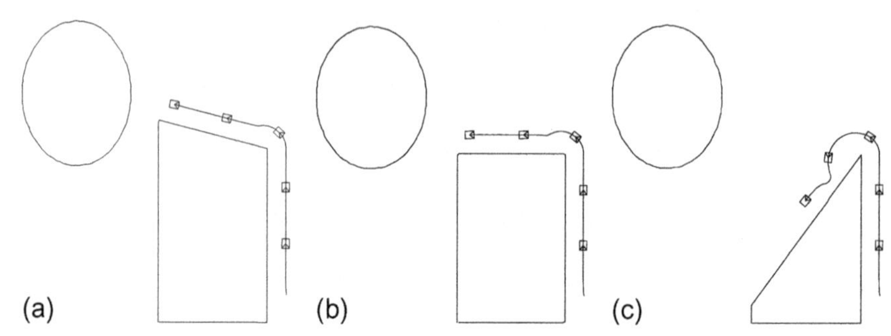

Figure 5.30 Following a fractured boundary with the use of the "following" ray R_{-}.

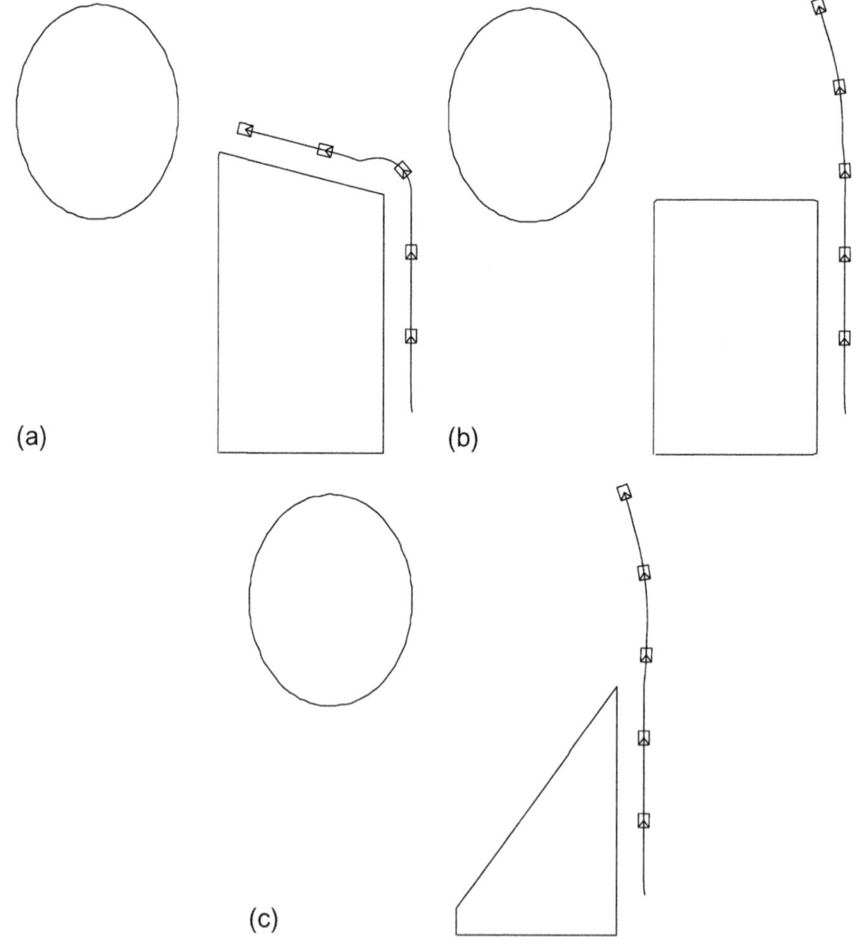

Figure 5.31 Following a fractured boundary in the case where illogically $d := \max\{d_*, d_{\text{safe}}\}$.

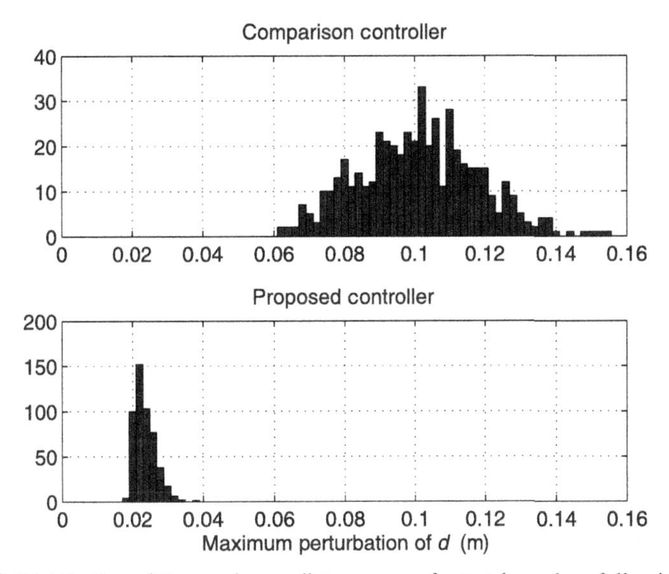

Figure 5.32 Distribution of the maximum distance error for two boundary following methods.

5.10 Experiments with a real robot

Experiments were performed with a Pioneer P3-DX mobile robot described in Section 1.4.1. The scenario with a rigidly mounted sensor was examined. The navigation controller (6.3) was slightly modified by continuous approximation due to the reasons given in Section 1.3.

The control requested by the navigation law was forwarded as the desired steering input to the ARIA library associated with the robot (version 2.7.4). The control was updated at the rate of 0.1 s, and the angle between the two rays used to determine the secant that approximates the tangent to the obstacle boundary was set to be 10 deg. The parameters used in testing are shown in Table 5.2. The path obtained is shown in Fig. 5.33 and is captured by manually marking the position of the robot on the video frames. Figure 5.33 demonstrates that the robot behaves as expected, successfully circling the obstacle with both concavities and convexities without collision and with a safety margin $\geq$0.46 cm. The related tracking measurements are displayed in

Table 5.2 **Parameters employed in experiments with real robot**

$\overline{u}$	28.6 deg/s	v	0.2 ms^{-1}
μ	57.3 deg	γ	171.9 deg/m
d_0	0.5 m		

Figs. 5.34 and 5.35. As seen in Fig. 5.34, the deviation of the distance measurement d from the desired value 0.5 m is always within the interval [−4.0 cm, +6.0 cm] and moreover, is typically in [−1.0 cm, +3.0 cm], except for several very short jumps out of this smaller range. However, after these jumps, the robot quickly recovers. Except for the respective short periods of time, the distance error typically does not exceed 3.0 cm, with the dangerous deviation from the required distance toward the obstacle being no more than 1 cm. As is seen in Fig. 5.35, the estimate of the boundary relative tangential angle φ is subject to a significant amount of noise. This is due to amplifying the noise in the distance measurement during numerical evaluation of φ via approximation of the tangent by the secant between two close points of the boundary. This approximation also appeared to be sensitive to small angular perturbations accompanying the motion of the robot. Even with these detrimental effects, the navigation controller has demonstrated good ability to achieve the desired outcome.

The objective of the second experiment was to test the effect of a thin obstacle for which the obstacle curvature assumptions adopted in the theoretical part of the paper are heavily violated at the end of the obstacle. The results are shown in Figs. 5.36–5.38. In accordance with explanatory remarks in Fig. 5.29, passing the obstacle end is accompanied with detection of a different and farther obstacle, as can be seen in Fig. 5.37, which displays the farthest distance detected within the employed narrow beam of detection rays for illustration purposes. However, due to the reasons disclosed at the end of Section 5.9, the robot successfully copes with the thin obstacle and behaves as expected.

Figures 5.39–5.41 present typical results of experiments in more cluttered environments. As compared with the previous experiments, the speed of the robot was reduced to 0.05 ms^{-1} to make its minimal turning radius (2.2) smaller than spacing between the obstacles. It can be seen that the robot correctly navigates around the selected obstacle despite the presence of others.

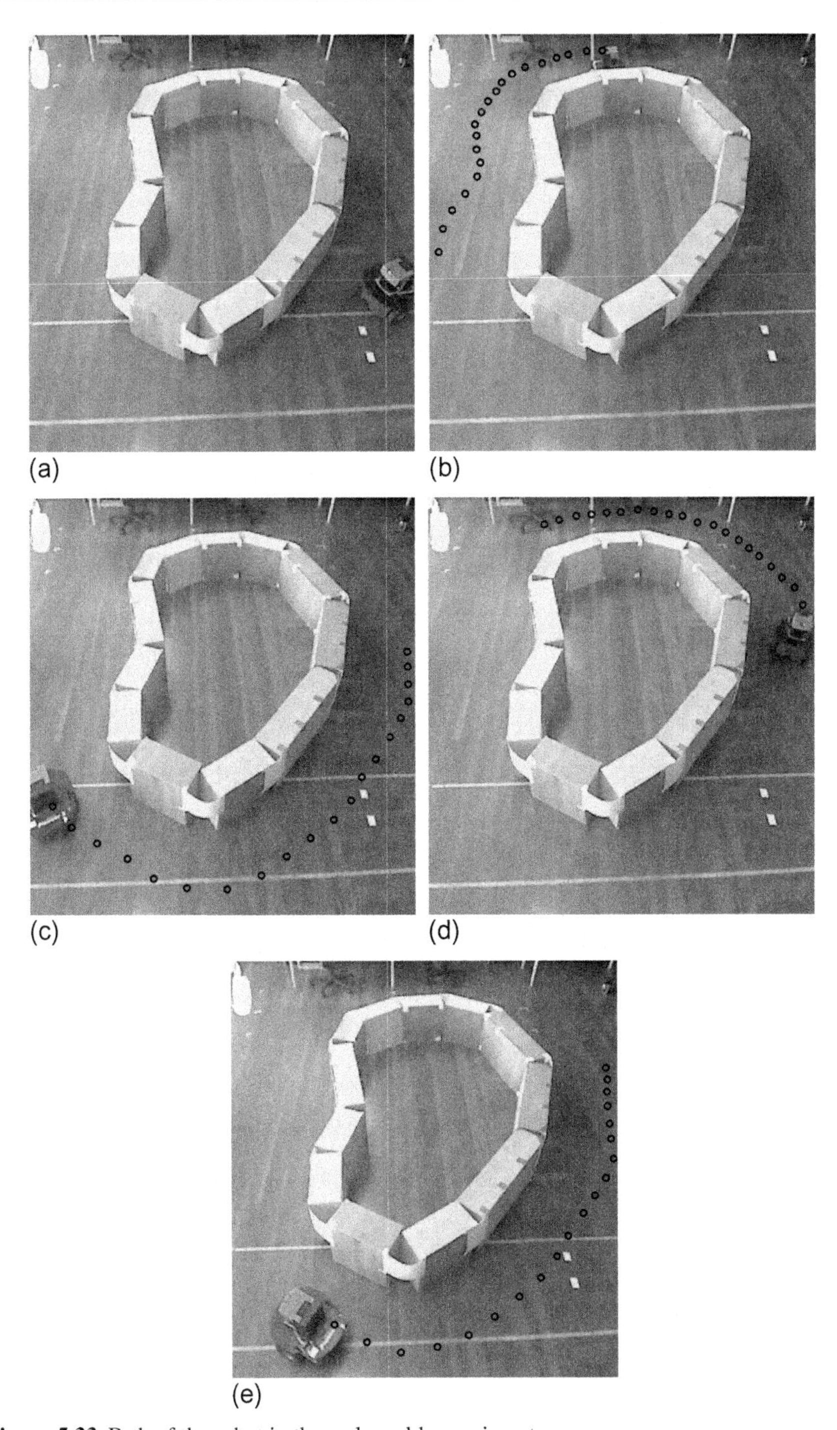

Figure 5.33 Path of the robot in the real-world experiment.

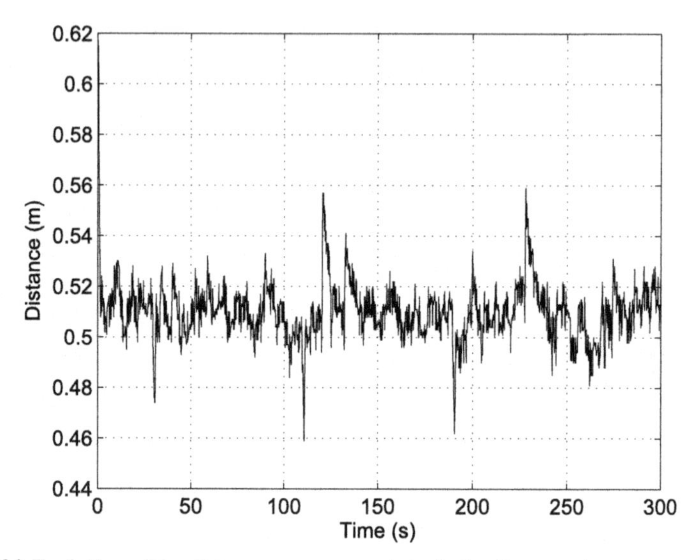

Figure 5.34 Evolution of the distance measurement $d_\lrcorner$ during the experiment.

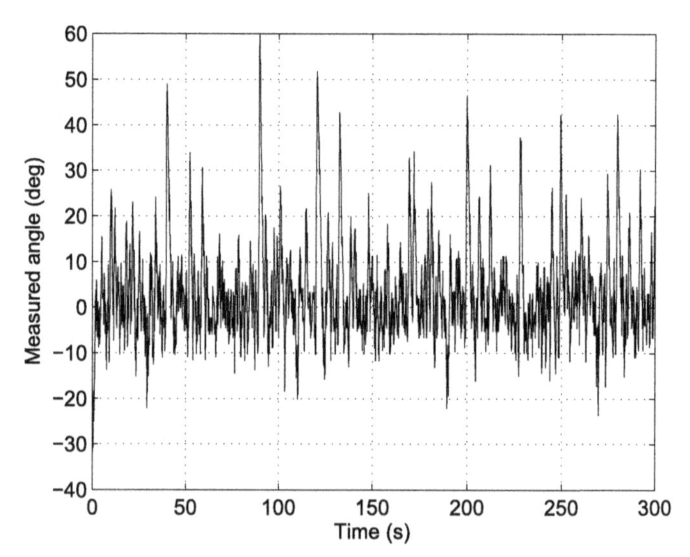

Figure 5.35 Evolution of the estimate of the relative tangent angle φ during the experiment.

Figure 5.36 Path of the robot in the experiment with a thin obstacle.

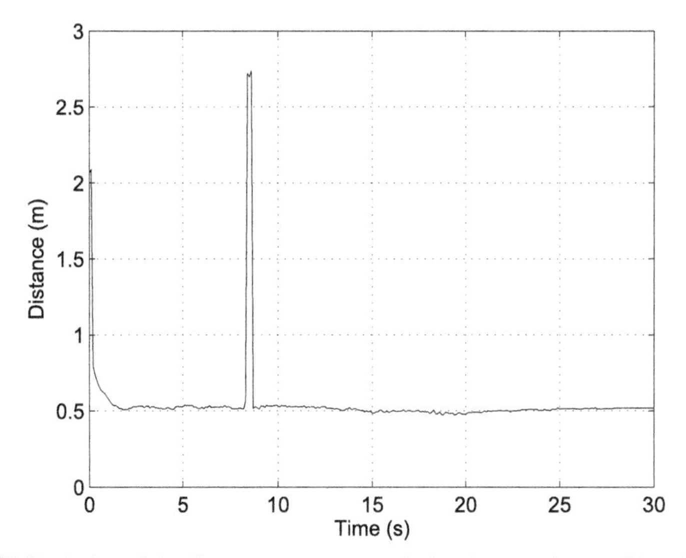

Figure 5.37 Evolution of the distance measurement during the experiment with a thin obstacle.

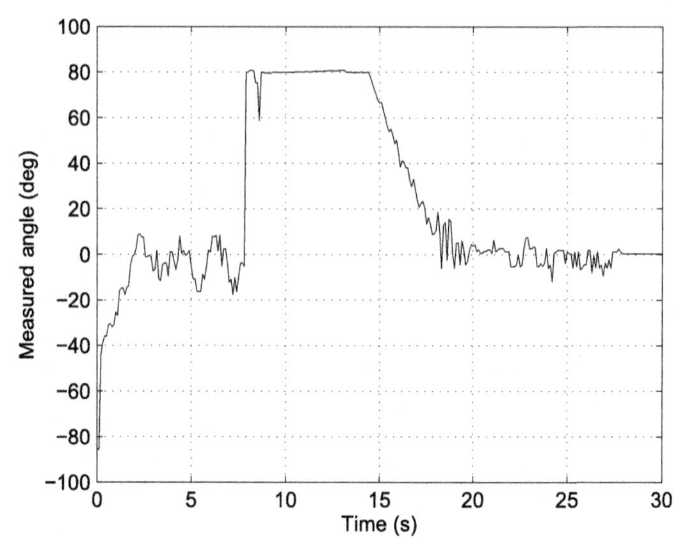

Figure 5.38 Evolution of the estimate of the relative tangent angle φ during the experiment with a thin obstacle.

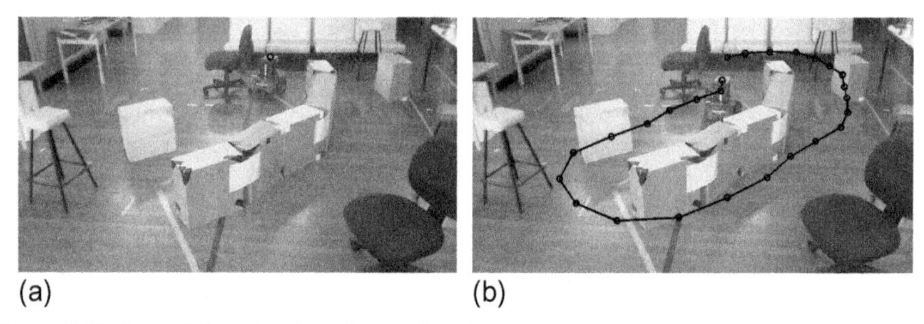

Figure 5.39 Path of the robot in a cluttered environment.

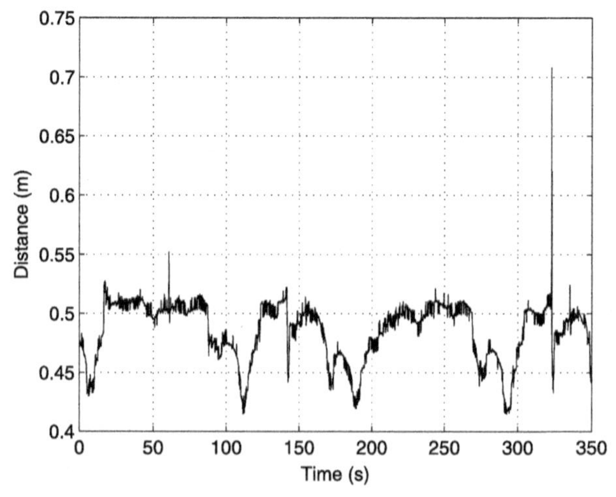

Figure 5.40 Evolution of the distance measurement during the experiment in a cluttered environment.

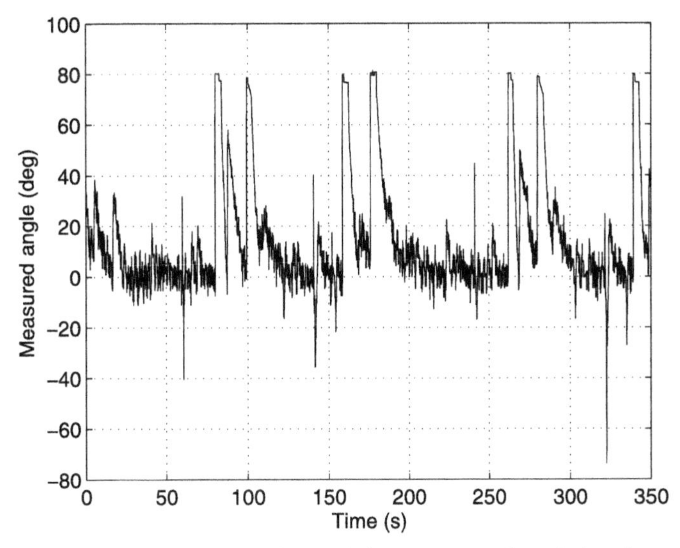

Figure 5.41 Evolution of the estimate of the relative tangent angle φ during the experiment in the cluttered environment.

Safe navigation to a target in unknown cluttered static environments based on border patrolling algorithms

6

In this chapter, we revert to the border patrolling algorithm from Section 5.2, which is based on measurements of the current distance to the obstacle. With a slight modification, this algorithm is applied to the problem of safe guidance of an autonomous wheeled vehicle toward a target in an a priori unknown static cluttered environment. Specifically, this algorithm is used to bypass the enroute obstacles with a predefined safety margin, whereas the robot is directed straight to the target when possible. This is alike in flavor to, e.g., the biologically inspired DistBug algorithm [11], which exhibits a similar behavior. Overall, the presented strategy of target reaching with obstacle avoidance thus includes two regimes of operation, called *modes*: obstacle bypassing and motion without the threat of collision with an obstacle, respectively. An important part of this strategy is the set of rules regulating switches between these modes. Unlike many works in the area of robotic guidance with obstacle avoidance, which are based on heuristics and do not take into account nonholonomic constraints, like [11], this chapter offers a mathematically rigorous analysis and justification of the proposed strategy for a nonholonomic robot. The theoretical results on convergence and performance of the navigation strategy are confirmed by computer simulation tests.

The main results of the chapter were originally published in [33]. The navigation and guidance strategy discussed in this chapter was implemented at robotic wheelchairs and mobile hospital beds for head injury patients and exhibited a promising performance [304, 391].

The body of the chapter is organized as follows. Section 6.1 introduces an algorithm for guidance toward a target while avoiding the enroute obstacles. Section 6.2 presents assumptions of theoretical analysis and the main theoretical results of the chapter. Computer simulations are discussed in Section 6.3.

Safe Robot Navigation Among Moving and Steady Obstacles. http://dx.doi.org/10.1016/B978-0-12-803730-0.00006-8

6.1 Navigation for target reaching with obstacle avoidance: Problem statement and navigation strategy

As in the previous chapter, we consider a planar wheeled mobile robot modeled as unicycle. It travels with a constant speed v and is controlled by the angular velocity u limited by a given constant $\overline{u}$. There is a steady point target $\mathcal{T}$ and several disjoint obstacles $D_1, \dots, D_k$ in the plane. The objective is to drive the robot to the target through the obstacle-free part of the plane $\mathbb{R}^2 \setminus \{D_1 \cup \cdots \cup D_k\}$. Moreover, the distance to any obstacle should not reduce below a given safety margin d_{safe}.

In proximity of any particular obstacle D_i, the robot has access to its distance from this obstacle:

$$d_i := \mathbf{dist}\,[\boldsymbol{r}; D_i] := \min_{\boldsymbol{r}' \in D_i} \|\boldsymbol{r} - \boldsymbol{r}'\|.$$

Here $\boldsymbol{r} = \mathbf{col}\,(x, y)$ is the pair of the robot's Cartesian coordinates, and the minimum is achieved if the obstacle is closed, which is assumed throughout the chapter. The robot also has access to the rate $\dot{d}_i$ at which the distance reading d_i evolves over time via, e.g., numerical differentiation.

The kinematics of the vehicle is described by (2.1), which are reproduced here for the convenience of the reader:

$$\begin{aligned} \dot{x} &= v\cos\theta, & x(0) &= x_0, \\ \dot{y} &= v\sin\theta, & y(0) &= y_0, \\ \dot{\theta} &= u \in [-\overline{u}, \overline{u}], & \theta(0) &= \theta_0, \end{aligned} \tag{1.1}$$

where the angle θ gives the robot's orientation; see Fig. 5.1(b).

To describe the proposed navigation strategy, we introduce the following definition illustrated in Fig. 6.1.

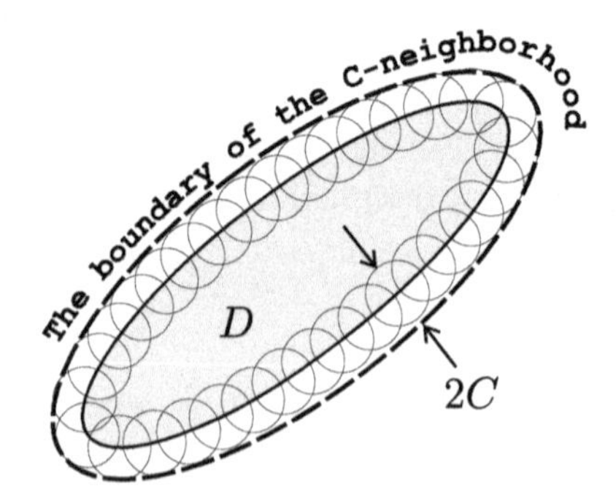

Figure 6.1 C-neighborhood of the set D.

Definition 1.1. For $C > 0$, the *C-neighborhood* of the domain $D \subset \mathbb{R}^2$ is the set formed by all points at the distance $\leq C$ from D, i.e.,

$$\mathcal{N}[C, D] := \left\{ \boldsymbol{r} \in \mathbb{R}^2 : \mathbf{dist}\,[\boldsymbol{r}; D] \leq C \right\}.$$

The proposed navigation strategy consists of switching between two regimes of operation, called the *obstacle avoidance mode* and *free motion mode*, respectively.

- *Obstacle avoidance mode* is used to bypass a particular obstacle D_i. In this mode, the robot is driven by the "border patrolling" navigation law (2.3), where $d(t)$ is replaced by the distance $d_i(t)$ to the obstacle D_i:

$$u(t) = \overline{u} \cdot \mathbf{sgn}\left\{ \dot{d}_i(t) + \chi[d_i(t) - d_0] \right\}, \quad \text{where} \tag{1.2}$$

$$\chi(z) := \begin{cases} \gamma z & \text{if } |z| \leq \delta, \\ \mu \mathbf{sgn}(z) & \text{if } |z| > \delta, \end{cases} \quad \mu := \gamma\delta \tag{1.3}$$

  is a linear function with saturation shown in Fig. 5.2, and d_0 is a parameter of the controller, with the meaning of the "desired distance to the obstacle when bypassing it."
- *Free motion mode* is used when there is no threat of collision with obstacles. In this mode, the robot is driven straight toward the target:

$$u(t) = 0. \tag{1.4}$$

The rule for switching between the modes employs two more parameters:

$$\epsilon > 0 \quad \text{and} \quad C > d_0 + \epsilon. \tag{1.5}$$

Here, C is the distance to an obstacle at which its avoidance is commenced; the maneuver termination is allowed only if the robot is close enough to the obstacle:

$$d_i \leq d_0 + \epsilon.$$

The rule for switching between (1.4) and (1.2) is illustrated in Fig. 6.2, where the initial right turn is caused by the assumed initial dominance of the first addend $\dot{d}_i < 0$ in the sum in the curly brackets from (1.2). Specifically, this rule is as follows:

R1 Switching from (1.4) to (1.2) occurs at any time instant τ when the distance from the mobile robot to the obstacle D_i reduces to the value C, i.e., $d_i(\tau) = C$, whereas $d_i(t) > C$ for $t < \tau, t \approx \tau$;

R2 Switching from (1.2) to (1.4) occurs when $d_i(t) \leq d_0 + \epsilon$ and the robot is oriented toward the target.

6.2 Assumptions of theoretical analysis and convergence of the navigation strategy

We start with a technical assumption.

Assumption 2.1. *Every obstacle D_i is a closed, bounded, and convex set with a smooth boundary ∂D_i.*

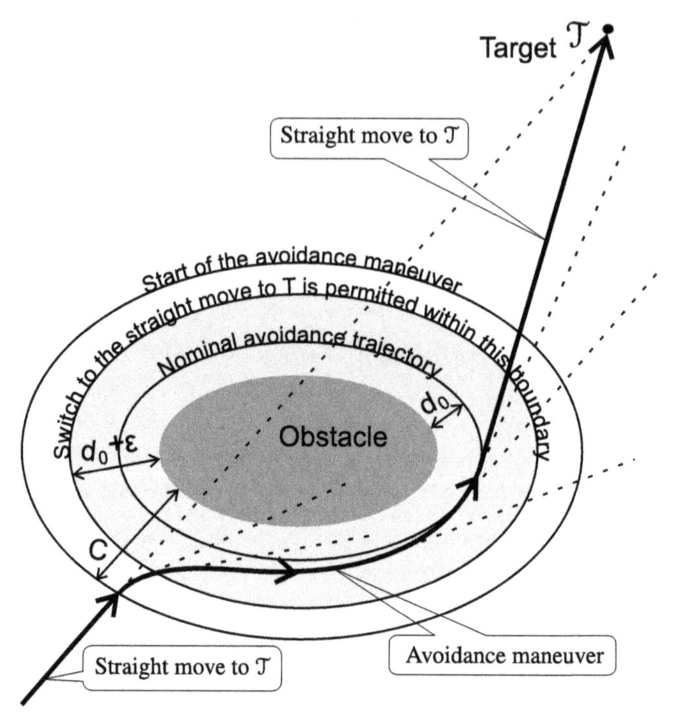

Figure 6.2 Obstacle avoidance.

In the above algorithm, the rule **R1** is also responsible for selection of an obstacle to be bypassed. In doing so, it is tacitly assumed that the situation where the rule **R1** simultaneously selects several obstacles cannot be encountered:

$$\mathcal{N}[C, D_i] \cap \mathcal{N}[C, D_j] = \emptyset \quad \forall i \neq j.$$

To simplify the proof, we require more: during avoidance of any obstacle D_j, the rule **R1** associated with any other obstacle $D_i \neq D_j$ should not be activated. With the footnote from page 72 in mind, all these claims mean that the obstacles should be far enough from each other, as is stated in the following.

Assumption 2.2. *The sets* $\mathcal{N}[C + 2R_{\min}, D_i]$ *and* $\mathcal{N}[C, D_j]$ *are disjoint for* $i \neq j$.

We recall that $R_{\min}$ is the minimal turning radius of the robot and is given by (5.2.2). In terms of the distance **dist** $[D', D''] := \inf_{\boldsymbol{r}' \in D', \boldsymbol{r}'' \in D''} \|\boldsymbol{r}' - \boldsymbol{r}''\|$ between two sets $D', D'' \subset \mathbb{R}^2$, Assumption 2.2 takes the following symmetric form:

$$\textbf{dist}\left[D_i, D_j\right] > 2(C + R_{\min}) \quad \forall i \neq j. \tag{2.1}$$

Since $C > d_{\text{safe}}$, this also guarantees that during avoidance of the obstacle D_i, the safety margins of all other obstacles cannot be violated.

The next assumption ensures the robot is able to recognize the situations from **R1** and **R2** and serve the controller (2.3) during the entire avoidance maneuver.

Assumption 2.3. *The robot measures the angle $\beta(t)$ from its centerline ray to the target $\mathcal{T}$; whenever $d_i \leq C + 2R_{\min}$, it has access to the distance d_i to D_i and its derivative $\dot{d}_i$.*

Assumption 2.4. *Initially the robot is not at the target $\boldsymbol{r}(0) \neq \mathcal{T}$, moves toward it $\beta(0) = 0$, and is outside all sets $\mathcal{N}[C, D_i]$; similarly, $\mathcal{T} \notin \cup_i \mathcal{N}[d_0 + \epsilon, D_i]$.*

Since the controller (5.2.3) is still employed, we need to replicate the condition (5.3.4) that (along with (5.4.1) ensures successful patrolling of the boundary of any obstacle $D := D_i$, putting $d(0) := C$ in (5.3.4) since the avoidance maneuver starts at the distance C from the obstacle. This gives rise to the next assumption.

Assumption 2.5. *The following inequality holds:*

$$\min\{d_0; C - 2R_{\min}\} > \max\{R_{\min} - R_-; d_{\text{safe}}\}. \tag{2.2}$$

Here R_- is defined in (5.3.4) with $D := \cup_{i=1}^{k} D_i$ (which implies that $\partial D = \cup_i \partial D_i$).

The statement of the last assumption is prefaced by two technical definitions.

Definition 2.1. Let $\xi > 0$. The ξ*-face* $\mathcal{F}[\xi, D_i]$ of the obstacle D_i is the set of all points $\boldsymbol{r} \in \partial\mathcal{N}[\xi, D_i]$ such that the straight line segment $(\boldsymbol{r}, \mathcal{T}]$ with the end points $\boldsymbol{r}$ and $\mathcal{T}$ does not intersect $\mathcal{N}[\xi, D_i]$.

Definition 2.2. Let $i \neq j$. The obstacle D_j is said to be *blocking* for D_i if there exists a point $\boldsymbol{r}_0 \in \mathcal{F}[d_0 + \epsilon, D_i]$ such that the segment $[\boldsymbol{r}_0, \mathcal{T}]$ intersects $\mathcal{N}[C, D_j]$.

Assumption 2.6. *If the obstacle D_j is blocking for the obstacle D_i, then for any points $\boldsymbol{r}_i \in \mathcal{F}[d_0 + \epsilon, D_i]$, $\boldsymbol{r}_j \in \mathcal{F}[d_0 + \epsilon, D_j]$, the following inequality holds:*

$$\|\boldsymbol{r}_i - \mathcal{T}\| > \|\boldsymbol{r}_j - \mathcal{T}\|. \tag{2.3}$$

Finally, we observe that by (2.2), there is $\lambda \in \mathbb{R}$ for which

$$\frac{R_{\min}}{\min\{d_0; C - 2R_{\min}\} + R_-} < \lambda < 1, \tag{2.4}$$

and introduce the following.

Definition 2.3. A guidance strategy is said to be *target reaching with obstacle avoidance* if there exists a time $t_f > 0$ such that the target is reached $\boldsymbol{r}(t_f) = \mathcal{T}$ at time t_f and the safety requirement $d_i(t) \geq d_{\text{safe}}\ \forall i, t \in [0, t_f]$ is satisfied.

Theorem 2.1. *Suppose that Assumptions 2.1–2.5 hold and the parameters $\gamma > 0$ and $\delta > 0$ of the navigation law* (1.2) *be chosen so that inequality* (5.4.1) *is true, where λ is a constant satisfying* (2.4)*. Then there exists $\epsilon_* > 0$ such that for any $0 < \epsilon < \epsilon_*$, the navigation rules **R1, R2** provide a target-reaching strategy with obstacle avoidance.*

6.2.1 Proof of Theorem 2.1

In this subsection, we examine the motion of the robot driven by the guidance strategy as described earlier in this section.

Lemma 2.1. *The maneuver of avoiding any obstacle D_i is sooner or later terminated by the rule **R2**. During this maneuver, the safety requirements $d_j \geq d_{safe}\ \forall j$ are satisfied.*

Proof. Due to Assumption 2.4, the examined avoidance maneuver is initiated by the rule **R1** at some time t_{av} when the robot is at the distance C from D_i. Therefore, invoking Assumptions 2.1 and 2.5 assures that Assumptions 5.3.1 and 5.3.4 (in the simplified form (5.3.4)) of Theorem 5.4.1 are fulfilled for the convex domain $D := D_i$ and the maneuver at hand, whereas (5.4.1) is stipulated in the body of Theorem 2.1. Therefore, Theorem 5.4.1 implies that during the maneuver, the safety requirement $d_i \geq d_{\text{safe}}$ is met, whereas $d_j \geq d_{\text{safe}}$ $\forall j \neq i$ by Assumption 2.2 (see also the discussion prefacing this assumption) since $C > d_{\text{safe}}$.

Now suppose that the first claim of the lemma fails to be true. Then $d_i(t) \to d_0$ as $t \to \infty$ by Theorem 5.4.1. The computations from (5.4.8), along with the constraint on u from (1.1), show that the second derivative of d_i is bounded: $|\ddot{d}_i| \leq M < \infty$. This and the existence of $\lim_{t\to\infty} d_i(t)$ imply that $\dot{d}_i(t) \to 0$ as $t \to \infty$. Then by the first equation from (5.4.2), $\alpha(t) \to 2\pi m \pm \pi/2$ as $t \to \infty$, where m is an integer. Therefore, (5.4.3) yields $\left|\pm\dot{s} - \frac{v}{1+d_0\varkappa}\right| \to 0$ as $t \to \infty$, where s is the curvilinear abscissa of the boundary point $\boldsymbol{r}_* \in \partial D_i$ nearest to the robot's position $\boldsymbol{r}$ and $\varkappa$ is the curvature of ∂D_i at $\boldsymbol{r}_*$. Due to Assumption 2.1, $0 \leq \varkappa \leq \varkappa_+ < \infty$, where $\varkappa_+$ does not depend on $\boldsymbol{r}_*$. Hence,

$$\underline{\lim}_{t\to\infty} \pm\dot{s} \geq \frac{v}{1+d_0\varkappa_+} > 0 \Rightarrow \pm s(t) \to \infty \quad \text{as} \quad t \to \infty,$$

i.e., $\boldsymbol{r}_*$ monotonically and repeatedly circulates around the entire boundary ∂D_i since some time instant. Whence the tangential vector $\pm\mathbf{T}$ from Fig. 5.5.3(b) monotonically rotates about the origin by repetition of full rotations. Since $\alpha(t) \to 2\pi m \pm \pi/2$, the angular deviation between $\pm\mathbf{T}$ and the robot's velocity $\mathbf{v}$ annihilates as time progresses; see Fig. 5.5.3(b). Hence, $\theta(t) \to \pm\infty$ as $t \to \infty$, where $\theta(t)$ is the continuous orientation angle of the velocity (i.e., the robot itself).

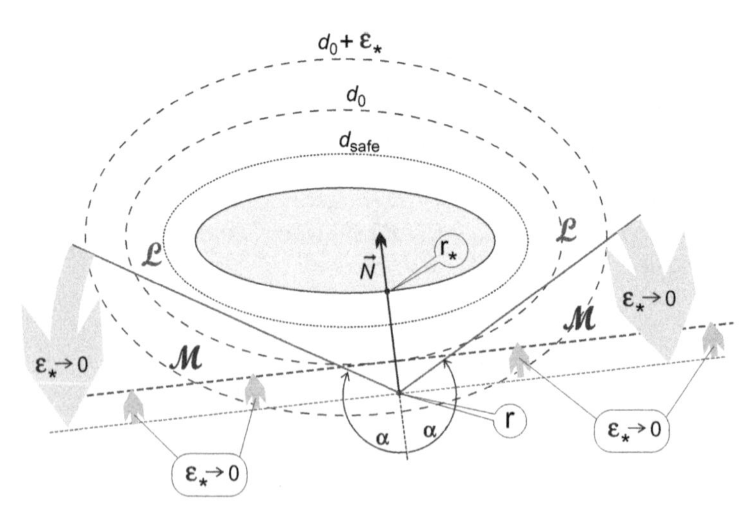

Figure 6.3 Limit as $\epsilon_* \to 0$.

On the other hand,

$$d_i(t) \xrightarrow{t\to\infty} d_0 \Rightarrow d_i \le d_0 + \epsilon/2$$

since some time $t_* \ge t_{av}$, whereas the target $\mathcal{T}$ lies outside the $[d_0+\epsilon]$-neighborhood of D_i by Assumption 2.4. Hence, the continuous angular direction $\eta(t)$ from the robot to the target evolves within some interval whose length $< \pi$. Due to the continuity of both $\theta(t)$ and $\eta(t)$, there necessarily exists a time $\tau \ge t_*$ when the robot is oriented toward the target: $\theta(\tau) = \eta(\tau) + 2\pi l$, where l is an integer. Then the avoidance maneuver is terminated by the rule **R2** at $t = \tau$, in violation of the initial hypothesis. The contradiction obtained proves the lemma. □

Lemma 2.2. *There exists $\epsilon_* > 0$ such that whenever the robot undergoes sliding motion within the maneuver of avoidance an obstacle D_i, is at the distance $d_{safe} \le d_i \le d_0 + \epsilon_*$ from D_i, and is directed toward the target, the safety requirement $d_i \ge d_{safe}$ associated with this obstacle is fulfilled for all points from the straight line segment $\mathcal{L} = [\boldsymbol{r}; \mathcal{T}]$ connecting the robot position $\boldsymbol{r}$ with the target $\mathcal{T}$.*

Proof. We examine separately two cases.

(1) $\boldsymbol{d_i \le d_0}$, where $d_i = \mathbf{dist}[\boldsymbol{r}; D_i]$. From the sliding motion equation $\dot{d}_i = -\chi[d_i - d_0]$ and (2.4), we see that

$$\dot{d}_i \ge 0 \overset{(4.2)}{\Longrightarrow} \cos\alpha \ge 0.$$

Thus, $\mathcal{L}$ forms an obtuse angle with the inner normal N to ∂D_i at the point $\boldsymbol{r}_* \in \partial D_i$ closest to $\boldsymbol{r}$; see Fig. 5.5.3(b). Since $-N = \nabla_{\boldsymbol{r}}\mathbf{dist}[\boldsymbol{r}; D_i]$ [386], we have $\langle \nabla_{\boldsymbol{r}}\mathbf{dist}[\boldsymbol{r}; D_i]; \mathcal{T} - \boldsymbol{r}\rangle \ge 0$, where the distance function is convex since so is the body D_i itself [94]. For such functions, the obtained inequality implies that the function non-strictly increases as the argument goes along the straight line from $\boldsymbol{r}$ to $\mathcal{T}$ [94]. Hence,

$$\boldsymbol{r}_\star \in \mathcal{L} \Rightarrow \mathbf{dist}[\boldsymbol{r}_\star; D_i] \ge \mathbf{dist}[\boldsymbol{r}; D_i] \ge d_{\text{safe}},$$

where the last inequality is assumed in the lemma.

(2) $\boldsymbol{d_0 \le d_i \le d_0 + \epsilon_*}$. Let $\epsilon_* < \delta$, where δ is the controller parameter from Fig. 5.5.2. The sliding motion equation and (5.2.4) yield that

$$-\gamma\epsilon_* \le \dot{d}_i \le 0 \overset{(5.4.2)}{\Longrightarrow} -\gamma\epsilon_* v^{-1} \le \cos\alpha \le 0 \Rightarrow \frac{\pi}{2} \le |\alpha| \le \frac{\pi}{2} + \arcsin\frac{\gamma\epsilon_*}{v}. \qquad (2.5)$$

Now we introduce the perpendicular $\mathcal{M}$ to the line segment $S = [\boldsymbol{r}, \boldsymbol{r}_*]$ that intersects S at the distance d_0 from the obstacle; see Fig. 6.3. Let $\epsilon_* \to 0$. Then $|\alpha| \to \pi/2, d_i \to d_0$, and $\mathcal{L}$ is asymptotically laid on $\mathcal{M}$. Due to Assumption 2.1, these convergences are uniform over all $\boldsymbol{r}$ from the stripe $d_0 \le \mathbf{dist}[\boldsymbol{r}; D_i] \le d_0 + \epsilon_*$ and α from the range given after the last $\Rightarrow$ in (2.5), as well as over all obstacles. Since $\mathcal{M}$ is separated from the forbidden area $D_{i,\text{forb}} := \{\boldsymbol{r}' : \mathbf{dist}[\boldsymbol{r}'; D_i] \le d_{\text{safe}}\}$ by positive both spatial and angular distances independent of $\boldsymbol{r}, \alpha$, and i, the segment $\mathcal{L}$ does not intersect $D_{i,\text{forb}}$ if ϵ_* is small enough.

□

Now we take $\epsilon_* > 0$ from Lemma 2.2, suppose that $\epsilon \in (0, \epsilon_*)$, and put $g := C + 2R - (d_0 + \epsilon) > 0$, where the inequality holds due to (1.5).

Corollary 2.1. *During the straight move to the target that follows the maneuver of avoidance of an obstacle D_i, the robot covers the distance $\geq g$, moves within the safe area $\{\boldsymbol{r} : d_j \geq d_{safe}\ \forall j\}$, and intersects the $[d_0+\epsilon]$-face $\mathcal{F}[d_0+\epsilon, D_i]$ of D_i.*

Proof. By Remark 5.4.2, the maneuver starts with the circular motion along an initial (for this maneuver) circle and proceeds with the sliding motion. During any motion along a circle without speed reversion, the velocity vector can be directed toward the target no more than once per full rotation. For the initial circle, this holds at the very start of the maneuver due to **R2** and Assumption 2.4. Hence, the rule **R2** cannot be activated during the circular motion and so is activated during the sliding motion. In the subsequent straight move to the target, the safety margin of D_i is never trespassed by Lemma 2.2. Since $d_i \leq d_0 + \epsilon$ at the start of this move by **R2** and the distance from D_i to any other obstacle D_j is no less than $2(C + R_{\min})$ due to (2.1), the robot covers the distance $\geq g$ without activation of the rule **R1** for any obstacle $D_j \neq D_i$: $d_j \geq C\ \forall j \neq i$. Since $C > d_{\text{safe}}$, the safety requirements for these obstacles are met during the entire straight move at hand.

Let $\mathcal{P}$ denote the path of the robot during the straight move at hand. At the starting point $\boldsymbol{r}_{\text{s}}$ of $\mathcal{P}$, $d_i \leq d_0 + \epsilon \overset{(1.5)}{<} C$ by **R2**. If this move ends at the target, $\textbf{dist}\,[\boldsymbol{r}_{\text{t}}; D_i] > d_0 + \epsilon$ at the terminal point $\boldsymbol{r}_{\text{t}}$ of $\mathcal{P}$ by the last claim from Assumption 2.4. Otherwise, this move is terminated by **R1** and so there exists j such that $\textbf{dist}\left[\boldsymbol{r}_{\text{t}}; D_j\right] \leq C$ and $d_j > C$ somewhere on $\mathcal{P}$. If $j = i$, this would mean the maximum $> C$ of the convex function $\textbf{dist}\,[\cdot; D_i]$ over the straight line segment $\mathcal{P}$ is not attained at an end point of $\mathcal{P}$, which is impossible [94]. Thus, $j \neq i$ and so

$$\begin{aligned}\textbf{dist}\,[\boldsymbol{r}_{\text{t}}; D_i] &\geq \textbf{dist}[D_i, D_j] - \textbf{dist}\left[\boldsymbol{r}_{\text{t}}; D_j\right] \\ &\overset{(2.1)}{>} 2(C + R_{\min}) - \textbf{dist}\left[\boldsymbol{r}_{\text{t}}; D_j\right] \geq C + 2R_{\min} \overset{(1.5)}{>} d_0 + \epsilon.\end{aligned}$$

Since $\textbf{dist}\,[\boldsymbol{r}_{\text{s}}; D_i] \leq d_0 + \epsilon$, we conclude that during the motion over $\mathcal{P}$, the robot passes a point $\boldsymbol{r}_\star$ such that $\textbf{dist}\,[\boldsymbol{r}_\star; D_i] = d_0 + \epsilon$ and $d_i > d_0 + \epsilon$ after passing $r_\star$. Due to Definition 2.1, $\boldsymbol{r}_\star \in \mathcal{F}[d_0 + \epsilon, D_i]$, which completes the proof.

□

Lemma 2.3. *The robot performs finitely many avoidance maneuvers.*

Proof. Suppose to the contrary that during the entire motion $\boldsymbol{r} = \boldsymbol{r}(t), t \geq 0$, the robot skirts obstacles infinitely many times: first $D_{i(1)}$, second $D_{i(2)}$, then $D_{i(3)}$, and so on. Avoidance of two successive obstacles $D_{i(m)}$ and $D_{i(m+1)}$ is separated by a straight move toward the target. By Corollary 2.1, the robot intersects the $[d_0 + \epsilon]$-face of $D_{i(m)}$ at some time τ_m during this move. Due to Definition 2.2 and the circumstances under which the rule **R1** terminates the straight move and commences avoidance of the obstacle $D_{i(m+1)}$, the obstacle $D_{i(m+1)}$ is blocking for $D_{i(m)}$. Therefore, by Assumption 2.6, $\|\boldsymbol{r}(\tau_m) - \mathcal{T}\| > \|\boldsymbol{r}(\tau_{m+1}) - \mathcal{T}\|$. Moreover, since the $[d_0+\epsilon]$-faces of compact convex sets D_i are compact [94], the discrepancy between the left- and right-hand sides of (2.3) can be lower estimated by a constant $\omega > 0$ that is independent of $\boldsymbol{r}_i, \boldsymbol{r}_j, i$, and j. Hence,

$$\|\boldsymbol{r}(\tau_m) - \mathcal{T}\| - \omega \geq \|\boldsymbol{r}(\tau_{m+1}) - \mathcal{T}\| \ \forall m = 1, 2, \ldots \Rightarrow \|\boldsymbol{r}(\tau_m) - \mathcal{T}\| \to -\infty \quad \text{as} \quad m \to \infty,$$

whereas $\|\boldsymbol{r}(\tau_m) - \mathcal{T}\| \geq 0$. The contradiction obtained proves the lemma. □

Proof of Theorem 2.1. By Lemmas 2.1 and 2.3, the concluding phase of the robot motion is necessarily a straight move to the target. Therefore, the target is reached without fail. The safety requirements are satisfied due to Lemma 2.1 and Corollary 2.1. Definition 2.3 completes the proof. □

6.3 Computer simulations of navigation with obstacle avoidance

To validate the performance of the navigation strategy **R1, R2**, we simulated a wheeled robot moving in cluttered environments with constant linear velocity $v = 0.5$ m/s. The robot's angular velocity is limited by $\overline{u} = 1$ rad/s and the desired margin is $d_0 = 2$ m. The controller parameters are as follows: $\lambda = 0.5$, $\varepsilon = 0.1$ m, $\delta = 0.1$ m, and $\gamma = 1\ \text{s}^{-1}$. The robot initially moves toward the target: $\beta(0) = 0$.

If the initial direction of the robot is not toward the target, a heading regulation phase should precede application of the guidance law **R1, R2**. In our simulations, we employed a simple regulation law $u(t) = -\overline{u} \cdot \mathbf{sgn}\,\beta$. It is clear that after some time, this law ensures that $\beta = 0$, thus making it possible to proceed with the basic algorithm **R1, R2**.

In the first simulation shown in Fig. 6.4, the robot first undergoes heading regulation and then moves straight toward the target. Switching to the obstacle avoidance mode happens when condition **R1** is met and as a result, the robot bypasses

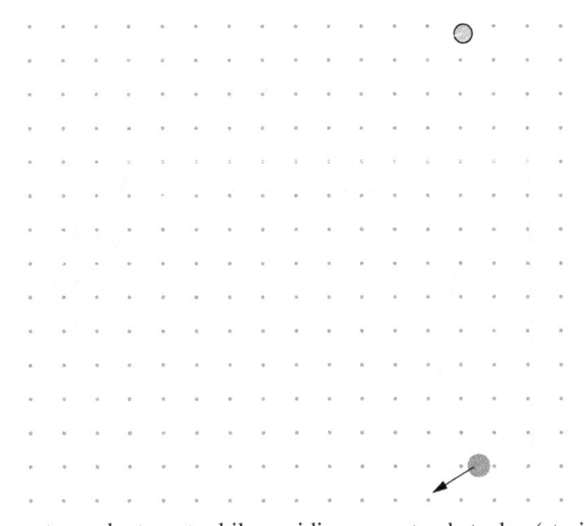

Figure 6.4 Guidance toward a target while avoiding enroute obstacles (straight lines).

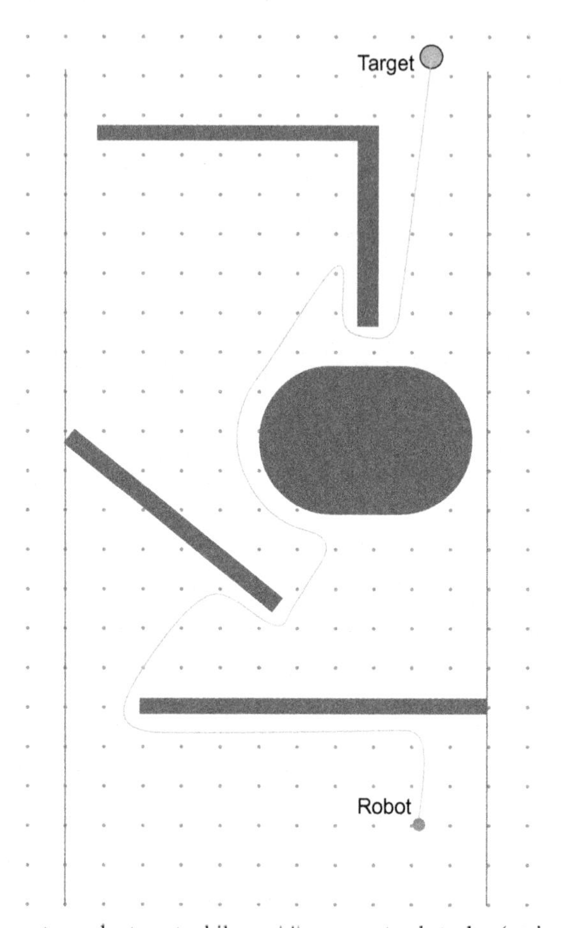

Figure 6.5 Guidance toward a target while avoiding enroute obstacles (various shapes).

the enroute obstacle by following its boundary in close range. The motion toward the target is resumed once the leaving conditions in **R2** hold. More challenging examples are illustrated in Figs. 6.5 and 6.6.

While Theorem 2.1 assumes that all obstacles D_i are convex, the presented simulations show that in practice, the proposed guidance algorithm often performs well even without the convexity assumption.

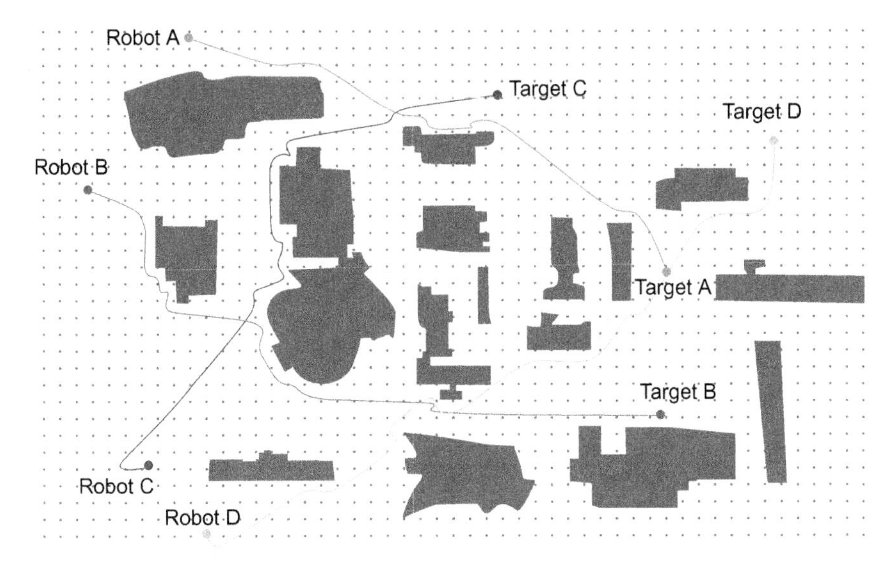

Figure 6.6 Guidance toward specific target locations in the University of New South Wales Campus.

Algorithm for reactive navigation of nonholonomic robots in maze-like environments

7

7.1 Introduction

Though autonomous navigation, guidance, and control of unmanned robotic vehicles have been topics of long-term extensive study, they still present important challenges for robotics research. Many of them emerge from uncertainty about the environment, limited perception capability and computational power of the robot, the need for quick decisions to ensure speedy motion, and restrictions on mobility of the robot due to nonholonomic constraints, bounded control range, and underactuation. This combination is well exemplified by, e.g., micro and miniature aerial robotic vehicles. They have gained a great deal of interest as a technological development aimed at achieving better surveillance in homeland security and disaster mitigation missions, which may occur in maze-like tunnels, forests, and urban structures.

Whenever shortage of computational resources is a real concern, reactive controllers, for which actuation is a simple reflex action to observation, are of particular interest. Are such controllers capable, by their own rights, of achieving a global navigation task, e.g., target reaching in a maze-like environment, under the above combination of troublesome circumstances?

As was highlighted in Chapter 3, until now a typical role of reactive controllers in motion planning algorithms was to track the curve generated by a higher-level global planner rather than to be completely responsible for the task achievement. Few counterexamples concerning nonholonomic robots include artificial potential approach [53] combined with sliding mode control for gradient climbing [50] and kinematic control based on Lyapunov-like analysis [51]. However, like any local search method, these approaches are prone to deadlocks due to, e.g., local minima, whose resolution imposes more computational burden and/or requires extended knowledge of the scene.

Another class of local path planners that are similar in spirit to that from this chapter is the Bug family algorithms. They typically drive the robot directly toward the target when possible; otherwise, the obstacle boundary is followed in close range. Motion toward the target is resumed as soon as a leaving condition is satisfied, which is designed so that the global convergence holds; see, e.g., [240, 392] and literature therein. A common problem with these strategies is that they typically take for granted the capability of carrying out maneuvers whose implementation may either

Safe Robot Navigation Among Moving and Steady Obstacles. http://dx.doi.org/10.1016/B978-0-12-803730-0.00007-X

constitute a separate engineering problem (like wall following) or be impossible (like instantaneous turns) due to kinematic, dynamic, or sensing constraints that cannot be ignored in practical settings. Implications of the impossibility to perfectly perform these maneuvers basically lie in uncharted territory.

Inspired by behaviors of animals, which are believed to use simple, local motion control rules that result in remarkable and complex intelligent behaviors [154, 393, 394], this chapter introduces and examines a navigation strategy that is aimed at reaching a steady target in a steady arbitrarily shaped maze-like environment and is composed of the following reflex-like rules:

(s.1) At considerable distances from the obstacle,
- **(a)** turn toward the target as fast as possible;
- **(b)** move toward the target when headed for it;

(s.2) In close proximity of the obstacle,
- **(c)** Follow (a,b) when moving away from the obstacle;
- **(d)** Otherwise, quickly avert the collision threat by making a sharp turn.

Studies of target pursuit in animals, ranging from dragonflies to humans, have suggested that they often use pure pursuit method (s.1) to catch both steady and moving targets. The obstacle avoidance rule (s.2) is also inspired by biological examples such as the near-wall behavior of a cockroach [154].

To address the issues of nonholonomic constraints, control saturation, and underactuation, we consider an autonomous vehicle or mobile robot of the Dubins car type, as in Chapters 4–6. It travels forward with a constant speed along planar paths of bounded curvatures and is controlled by the upper limited angular velocity. To implement (s.1), (s.2), only minor perceptual capabilities are needed, which are enough to judge whether the distance to the obstacle is small or not, to estimate the sign of its time derivative, and to determine the polar angle of the target line-of-sight in the robot's reference frame.

We show that the rules (s.1), (s.2) constitute a basically effective strategy: they do bring the robot through a simply connected arbitrarily shaped maze to the target. This holds with the fixed turn direction in (d) if the robot's initial location and the target are not deep inside the maze. Otherwise, the claim is true if this direction is sometimes randomly updated. All these are true if the robot is maneuverable enough to cope with the narrows and contortions of the maze, with constructive conditions for this being provided.

Though the rules (s.1), (s.2) do not explicitly recommend to follow walls, we show that wall following results from the interplay between (c) and (d) in the sliding mode fashion in certain circumstances. The problems of curve following are well understood for fully actuated systems, but the case of underactuated vehicles is still the area of active research; see, e.g., [206] and references therein. The proposed solutions typically ignore the control saturation and offer sophisticated nonlinear controllers based on access to extended curve parameters like the tangent or curvature. In [206], separate navigation controllers were designed for convexities and concavities of the curve. The bio-inspired simple and homogeneous rules (c,d) ensure curve tracking by an underactuated robot with saturated control without access to extended curve parameters.

By reliance on bearing-only data about the target, the proposed approach is similar to the Pledge [395] and Angulus [240] algorithms. Unlike ours, both assume access to the absolute direction, and the latter employs not one but two angles. The major distinction is that they assume the robot is able to trace paths of unlimited curvatures, e.g., broken curves, and to move exactly over and along the obstacle boundary. These assumptions are violated in the context of this chapter, which entails deficiency in the available proofs of the convergence of these algorithms.

The main results of the chapter were originally published in [203].

The body of the chapter is organized as follows. Section 7.2 describes the problem setup and the proposed navigation law. Assumptions are discussed in Section 7.3. Section 7.4 presents the main results. Section 7.5 is devoted to simulations and experiments with real robots. This chapter has an appendix, which contains the most technically demanding proofs.

7.2 Problem setup and navigation strategy

We consider a planar mobile robot that travels with a constant speed v and is controlled by the angular velocity u limited by a given constant $\overline{u}$. There also is a steady point target $\mathscr{T}$ and a single steady obstacle $D \not\ni \mathscr{T}$, which is an arbitrarily shaped compact domain. Its boundary ∂D is a Jordan piecewise analytical curve without inner corners; see Fig. 7.1(a). Modulo smoothened approximation of such corners, this requirement is typically met by all obstacles considered in robotics. The objective is to drive the robot to the target while respecting a given safety margin:

$$d(t) \geq d_{\text{safe}} > 0 \quad \forall t.$$

Here $d(t) := \mathbf{dist}\,[\boldsymbol{r}(t); D]$ is the distance $\mathbf{dist}\,[\boldsymbol{r}; D] := \min_{\boldsymbol{r}_* \in D} \|\boldsymbol{r}_* - \boldsymbol{r}\|$ to the obstacle, $\|\cdot\|$ is the Euclidian norm, $\boldsymbol{r}(t)$ is the robot location given by its abscissa $x(t)$ and ordinate $y(t)$ in the world frame. The orientation of the robot is described by the angle θ introduced in Fig. 7.1(b).

As in Chapters 4–6, the kinematics of the robot are described by the equations:

$$\begin{array}{l} \dot{x} = v\cos\theta, \\ \dot{y} = v\sin\theta, \end{array}, \quad \dot{\theta} = u \in [-\overline{u}, \overline{u}], \quad \begin{array}{l} \boldsymbol{r}(0) = \boldsymbol{r}_0 \notin D \\ \theta(0) = \theta_0 \end{array}. \tag{2.1}$$

Thus, the minimal turning radius of the robot is equal to

$$R_{\min} = \frac{v}{\overline{u}}. \tag{2.2}$$

Whenever $d(t) \leq d_{\text{range}}$, the robot has access to $d(t)$ and the signum $\mathbf{sgn}\, \dot{d}(t)$. Here $d_{\text{range}} > d_{\text{safe}}$ is a given sensor range. The robot can also measure the relative target bearing β introduced in Fig. 7.1(b).

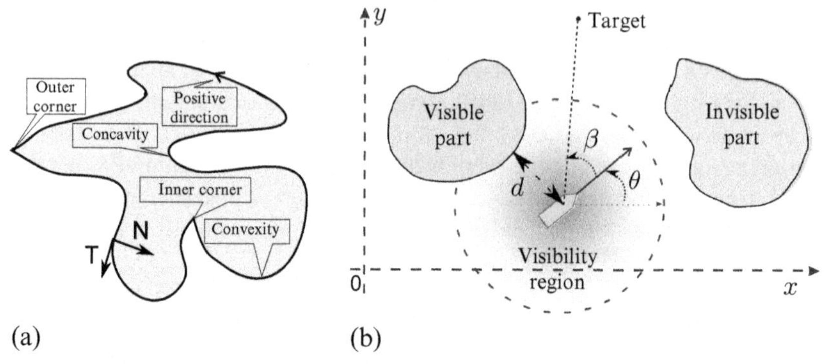

Figure 7.1 (a) Obstacle; (b) Planar mobile robot.

Mathematically, the examined navigation strategy is described by

$$u = \overline{u} \times \begin{cases} \mathbf{sgn}\,\beta & \text{if } d > d_{\updownarrow} \text{ (mode } \mathfrak{A}) \\ \begin{cases} \mathbf{sgn}\,\beta & \text{if } \dot{d} > 0 \\ -\sigma & \text{if } \dot{d} \le 0 \end{cases} & \text{if } d \le d_{\updownarrow} \text{ (mode } \mathfrak{B}) \end{cases}. \tag{2.3}$$

Here σ and $d_{\updownarrow} \in (0, d_{\text{range}})$ are controller parameters, σ can assume the values "+1" or "−1" and gives the turn direction in (d), $d_{\updownarrow}$ regulates mode switching: $\mathfrak{A} \mapsto \mathfrak{B}$ when d reduces to $d_{\updownarrow}$; $\mathfrak{B} \mapsto \mathfrak{A}$ when d increases to $d_{\updownarrow}$. When mode $\mathfrak{B}$ is switched on, $\dot{d} \le 0$; if $\dot{d} = 0$, the "turn" submode $u := -\sigma\overline{u}$ is set up. Because of discontinuities in (2.3), the solution of the closed-loop system is meant in the Filippov's sense [70].

Remark 2.1. In (2.3), β is assumed to account for not only the heading but also the sum of full turns performed by the target bearing.

Since the navigation controller (2.3) is fed by **sgn** β, but not the entire β, it suffices to know if the target is to the left of the robot and how many full turns are reckoned in β. Therefore, transitions between the left and the right side views that occur in the front-to-target and rear-to-target orientations should be distinguishable. Overall, it suffices to have access to only the quadrant containing the target line-of-sight. Likewise, it suffices to know only whether d is less or more than $d_{\updownarrow}$, but not the entire d.

In the *basic* version of the algorithm, $\sigma = \pm 1$ is fixed. As will be shown, the algorithms with $\sigma = +1$ and $\sigma = -1$ have basically identical properties except for the direction of bypassing the obstacle, which is counter-clockwise and clockwise, respectively. To find a target hidden deeply inside the maze, a modified version of the algorithm can be employed: whenever $\mathfrak{A} \mapsto \mathfrak{B}$, the parameter σ is updated. The updated value is picked randomly and independently of the previous choices from $\{+1, -1\}$, with the value $+1$ being drawn with a fixed probability $p \in (0, 1)$. This version is called the *randomized* navigation law.

7.3 Assumptions of theoretical analysis and tuning the navigation law

We start by emphasizing an already imposed requirement.

Assumption 3.1. *The boundary ∂D of the obstacle D is a Jordan piece-wise analytical curve without inner corners.*

We also assume the robot is maneuverable enough to avoid trapping in narrows of the maze; see Fig. 7.2. Specifically, when following the boundary ∂D with $d(t) \equiv d_{\text{safe}}$, the full turn can always be made without violation of the safety margin. Moreover, this can be done without crossing a center of curvature of a concavity of ∂D, the normal radius of the osculating circle at a distance $\leq R_{\min}$ from this center, and a location whose distance d from D is furnished by multiple points in D. This enhancement is aimed at reducing technicalities so far as such crossing typically requires special consideration [206]. This is elucidated by the fact[1] that at the respective points, the distance d is uncontrollable, and even if $\dot{d}(0) = 0$, no control may prevent local convergence to D. For safety reasons, we also assume that d_{safe} exceeds the unavoidable forward advancement $R_{\min}$ during the sharpest turn.

To come into details, we recall that $\varkappa(\boldsymbol{r})$ and $R_{\varkappa}(\boldsymbol{r})$ stand for the signed curvature of the boundary ∂D at the point $\boldsymbol{r} \in \partial D$ and the (unsigned) curvature radius $R_{\varkappa}(\boldsymbol{r}) := |\varkappa(\boldsymbol{r})|^{-1}$. (We refer the reader to [377, Ch. 1] for the formal definition of the curvature.) Here $\varkappa(\boldsymbol{r}) > 0$ on convexities and $\varkappa(\boldsymbol{r}) < 0$ for concavities, and $0^{-1} := +\infty$. Focal points of the boundary were introduced in Definition 5.3.1; the *focal locus* is the set of all focal points along any normal lines to the boundary.

Due to the absence of inner corners, any point $\boldsymbol{r} \notin D$ at a sufficiently small distance $\mathbf{dist}[\boldsymbol{r}; D] < d_\star$ from D does not belong to the focal locus of ∂D and $\mathbf{dist}[\boldsymbol{r}; D]$ is attained at only one point [396]. The *regular margin* $d_\star(D) > 0$ of D is the supremum of such $d_\star$'s. Therefore, $d_\star(D) = \infty$ for convex domains, whereas for non-convex D,

$$d_\star(D) \leq R_D := \inf_{\boldsymbol{r} \in \partial D : \varkappa(\boldsymbol{r}) < 0} R_\kappa(\boldsymbol{r}). \tag{3.1}$$

Here inf over the empty set is set to be $+\infty$. Since the distance to D may be increased by $2R_{\min}$ during the full turn, the above assumptions can be boiled down into

$$d_\star(D) > d_{\text{safe}} + 2R_{\min}, \quad R_D > d_{\text{safe}} + 3R_{\min}, \quad d_{\text{safe}} > R_{\min}. \tag{3.2}$$

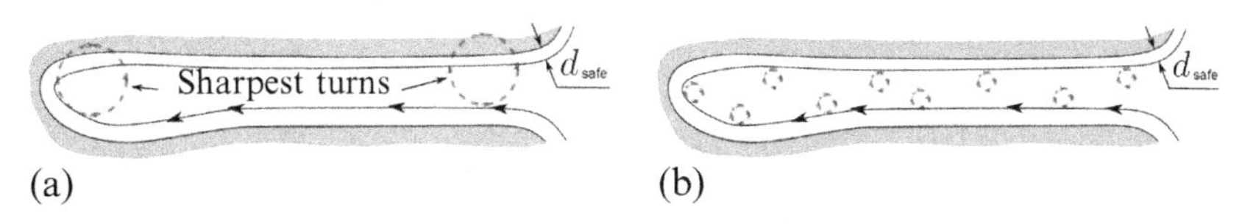

Figure 7.2 (a) Unavoidable collision; (b) Maneuverable enough mobile robot.

[1] It can be justified by retracing the proofs of Lemmas 5.3.1 and 5.3.2.

We also assume that the sensor range is large enough to avoid violation of the safety margin after detection of D:

$$d_{\text{range}} > 2R_{\min} + d_{\text{safe}}. \tag{3.3}$$

This takes into account that even the sharpest turn may decrease the distance d by $2R_{\min}$. As $d_{\text{safe}} \to R_{\min}$, the requirements (3.2) and (3.3) shape into

$$\frac{v}{\overline{u}} = R_{\min} < \min\left\{\frac{d_\star(D)}{3}; \frac{R_D}{4}; \frac{d_{\text{range}}}{3}\right\}$$

and mean that the robot speed v should not be large to cope with the maze.

The following choice of $d_\updownarrow$ is feasible thanks to (3.2) and (3.3):

$$d_{\text{safe}} + 2R_{\min} < d_\updownarrow < d_\star(D), d_{\text{range}}, R_D - R_{\min}. \tag{3.4}$$

7.4 Convergence and performance of the navigation law

To state the main theoretical results of the chapter, we need some notions. The *d-equidistant curve* $C(d)$ of D is the locus of points $\boldsymbol{r}$ at the distance $\mathbf{dist}\,[\boldsymbol{r}; D] = d$ from D; the *d-neighborhood* $\mathcal{N}(d)$ of D is the area bounded by $C(d)$:

$$C(d) := \{\boldsymbol{r} : \mathbf{dist}\,[\boldsymbol{r}; D] = d\}, \quad \mathcal{N}(d) := \{\boldsymbol{r} : \mathbf{dist}\,[\boldsymbol{r}; D] \leq d\}.$$

The symbol $[\boldsymbol{r}_1, \boldsymbol{r}_2]$ stands for the straight line segment directed from $\boldsymbol{r}_1$ to $\boldsymbol{r}_2$, whereas $(\boldsymbol{r}_1, \boldsymbol{r}_2)$ is the segment deprived of the end-points.

Let $\boldsymbol{r}_\diamondsuit, \boldsymbol{r}_* \in C(d_\updownarrow)$, and $(\boldsymbol{r}_\diamondsuit, \boldsymbol{r}_*) \cap \mathcal{N}(d_\updownarrow) = \emptyset$. The points $\boldsymbol{r}_\diamondsuit, \boldsymbol{r}_*$ divide $C(d_\updownarrow)$ into two arcs. Being concatenated with $[\boldsymbol{r}_\diamondsuit, \boldsymbol{r}_*]$, each of them gives rise to a Jordan curve encircling a bounded domain, one of which is the other united with $\mathcal{N}(d_\updownarrow)$. The smaller domain is called the *simple cave of* $\mathcal{N}(d_\updownarrow)$ *with endpoints* $\boldsymbol{r}_\diamondsuit, \boldsymbol{r}_*$. The location $\boldsymbol{r}$ is said to be *locked* if it belongs to a simple cave of $\mathcal{N}(d_\updownarrow)$ whose endpoints lie on a common ray centered at the target $\mathcal{T}$. This concept of "locking" is applicable to the target itself; see Fig. 7.3.

Remark 4.1. Let $\mathbf{co}\,D$ denote the *convex hull* of the set D, i.e., the least convex set containing D. Location $\boldsymbol{r}$ is unlocked whenever $\mathbf{dist}\,[\boldsymbol{r}; \mathbf{co}\,D] > d_\updownarrow$.

Theorem 4.1. *Suppose that* (3.2)–(3.4) *and Assumption 3.1 hold, both the robot's initial location* $\boldsymbol{r}_0$ *and the target* $\mathcal{T}$ *are unlocked in* $\mathcal{N}(d_\updownarrow)$ *and are far enough from the obstacle and each other*

$$\mathbf{dist}\,[\boldsymbol{r}_0; D] > d_\updownarrow + 2R_{\min}, \quad \|\boldsymbol{r}_0 - \mathcal{T}\| > 2R_{\min}, \quad \mathbf{dist}\,[\mathcal{T}; D] > d_\updownarrow. \tag{4.1}$$

Then the basic navigation law brings the robot to the target in finite time without violation of the safety margin.

The proof of this theorem will be given in Section 7.4.2.

The following remark shows that targets located deep inside the maze can be found by means of limited randomization.

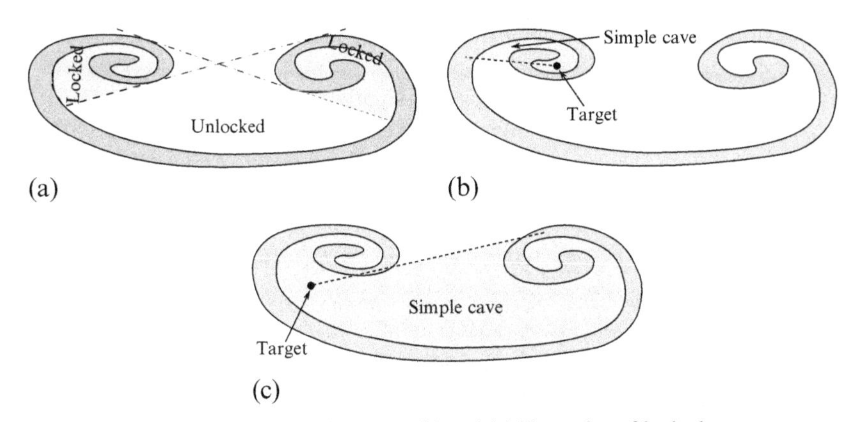

Figure 7.3 (a) Locked and unlocked targets; (b) and (c) Examples of locked targets.

Proposition 4.1.[a] *Suppose that* (3.2)–(4.1) *and Assumption 3.1 hold. With probability* 1, *the randomized navigation law drives the robot to the target* $\mathcal{T}$ *in finite time without violation of the safety margin.*

The proofs of all claims marked by [a] are given in the appendix to this chapter.

Theorem 4.1 and Proposition 4.1 remain true if in (4.1), $\mathbf{dist}\,[\boldsymbol{r}_0; D] > d_{\updownarrow} + 2R_{\min}$ is replaced by $\mathbf{dist}\,[\boldsymbol{r}_0; D] > d_{\updownarrow}$ and initially the robot is headed for the target $\beta(0) = 0$.

As $v \to 0$, (3.2)–(4.1) are reduced to $\boldsymbol{r}_0 \notin D, \boldsymbol{r}_0 \neq \mathcal{T}, \mathcal{T} \notin D$ due to (2.2) and the freedom to manipulate d_{safe} and $d_{\updownarrow}$. Therefore, these minimal assumptions imply the conclusions of Theorem 4.1 and Proposition 4.1 if the robot's speed v is small enough.

Now we disclose the tactical behavior implied by (s.1) and (s.2) from Section 7.1 and show that it includes wall following in a sliding mode. In doing so, we focus on a particular *avoidance maneuver*, i.e., the motion within uninterrupted mode $\mathfrak{B}$.

Let $\boldsymbol{\rho}(s)$ be the natural parametric representation of ∂D, where s is the curvilinear abscissa. This abscissa is cyclic: s and $s + L$ encode a common point, where L is the perimeter of ∂D. We notationally identify s and $\boldsymbol{\rho}(s)$. For any $\boldsymbol{r} \notin D$ within the regular margin $\mathbf{dist}\,[\boldsymbol{r}; D] < d_\star(D)$, the symbol $s(\boldsymbol{r})$ stands for the boundary point closest to $\boldsymbol{r}$, and $s(t) := s[\boldsymbol{r}(t)]$, where $\boldsymbol{r}(t)$ is the robot's location at time t.

To simplify matters, we first show that ∂D can be assumed C^1-smooth without any loss of generality. Indeed, if $0 < d < d_\star(D)$, the equidistant curve $C(d)$ is C^1-smooth and piece-wise C^2-smooth [396]; its parametric representation, orientation, and curvature are given by

$$s \mapsto \boldsymbol{\rho}(s) - d\mathbf{N}(s), \quad \varkappa_{C(d)}(s) = \frac{\varkappa(s)}{1 + \varkappa(s)d}. \tag{4.2}$$

The second formula holds if s is not a corner point of ∂D; such points contribute circular arcs of the radius d into $C(d)$. Therefore, by picking $\delta > 0$ small enough, expanding D to $\mathcal{N}(\delta)$, and correction $\eth := \eth - \delta$ of $\eth := d, d_{\text{safe}}, d_{\updownarrow}, d_{\text{range}}$, we keep all assumptions true and do not alter the operation of the closed-loop system. Hence, ∂D can be assumed C^1-smooth.

Writing $f(\eta_* \pm^{\approx} 0) > 0$ means there exists small enough $\Delta > 0$ such that $f(\eta) > 0$ if $0 < \pm(\eta - \eta_*) < \Delta$. The similar notations, e.g., $f(\eta_* \pm^{\approx} 0) \leq 0$, are defined likewise.

Proposition 4.2. *Suppose that* (3.2)–(3.4) *and Assumption 3.1 are valid. Let for the robot driven by the navigation law* (2.3)*, an avoidance maneuver (AM) be started with zero target bearing $\beta(t) = 0$ at $t = t_*$. Then the following claims hold:*

(i) *There exists $\tau \geq t_*$ such that the robot moves with the maximal steering angle $u \equiv -\sigma\overline{u}$ and the distance to the obstacle decreases $\dot{d} \leq 0$ until τ,*[2] *and at $t = \tau$, sliding motion along the equidistant curve $C\{\mathbf{dist}[\boldsymbol{r}(\tau); D]\}$*[3] *is started with $\sigma\dot{s} > 0$ and $\beta\dot{s} > 0$;*

(ii) *SMEC holds until β arrives at 0 at a time when $\varkappa[s(t) + \mathbf{sgn}\,\sigma^{\approx}0] > 0$, which sooner or later does hold and after which a motion to the target in a straight line*[4] *is commenced;*

(iii) *During SMT, the robot first does not approach the obstacle $\dot{d} \geq 0$ and either the triggering threshold $d_{\updownarrow}$ is ultimately trespassed and mode $\mathfrak{B}$ is switched off, or a situation is encountered where $\dot{d}(t) = 0$ and $\varkappa[s(t) + \mathbf{sgn}\,\sigma^{\approx}0] < 0$. When it is encountered, the robot starts SMEC related to the current distance;*

(iv) *There may be several transitions from SMEC to SMT and vice versa, all obeying the rules from* (4.2), (4.2)*;*

(v) *The number of transitions is finite and finally the robot does trespass the triggering threshold $d_{\updownarrow}$, thus terminating the considered avoidance maneuver;*

(vi) *Except for IT described in* (4.2)*, the robot maintains a certain direction of bypassing the obstacle: $\dot{s}$ is constantly positive if $\sigma = +1$ and negative if $\sigma = -1$.*

The proof of this proposition will be given in Section 7.4.1.

Though the curvature $\varkappa$ is concerned in (ii), (iii), the navigation controller (2.3) does not use measurement or estimation of $\varkappa$.

By (2.3), AM is commenced with $\dot{d}(t_*) \leq 0$ at $t = t_*$. The next remark shows that if $\dot{d}(t_*) = 0$, IT may have the zero duration.

Remark 4.2. If $\dot{d}(t_*) = 0$, IT has the zero duration if and only if $\sigma\dot{s}(t_*) > 0$. Then the following claims are true:

1. If $\varkappa[s(t_*) + \sigma \cdot^{\approx} 0] < 0$, SMEC is immediately started;
2. If $\varkappa[s(t_*) + \sigma \cdot^{\approx} 0] \geq 0$, the duration of SMEC is zero, and SMT continues.

The proof of Remark 4.2 will be given in Section 7.4.1.

The assumption $\beta(t_*) = 0$ of Proposition 4.2 holds for the first AM. Indeed, since initially $\mathbf{dist}[\boldsymbol{r}_0; D] > d_{\updownarrow} + 2R_{\min}$ and the robot moves over a circle of the radius $R_{\min}$, it heads for the target earlier than it arrives at the threshold $d_{\updownarrow}$ of AM. All subsequent AMs with $\beta = 0$ since any previous AM ends in course of SMT by Proposition 4.2.

[2] This part of AM is called the *initial turn* and abbreviated IT.

[3] This is abbreviated SMEC and means following the wall at the fixed distance $\mathbf{dist}[\boldsymbol{r}(\tau); D]$, which is set up at the start of SMEC.

[4] SMT= "straight motion to the target", which is sliding motion over the surface $\beta = 0$.

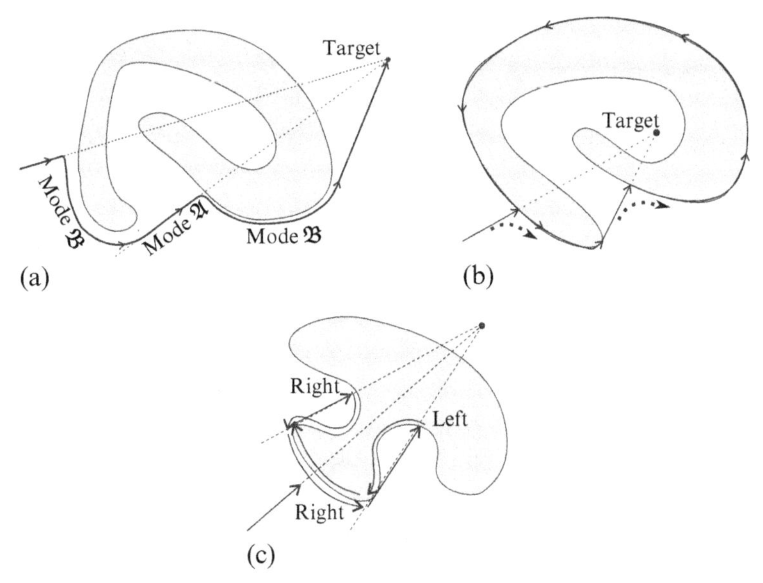

Figure 7.4 (a) Obstacle avoidance with two AM's; (b,c) Insufficiency of (b) only-right-turns and (c) cycle-left-and-right-turns options.

Dealing with a given obstacle may comprise several AM's. By (ii) and (iii), at most one both AM and SMEC is performed if D is convex.

Figure 7.4(b) shows an example where the basic navigation law fails to find a locked target. Figure 7.4(c) shows that repeatedly interchanging left and right turns is not enough either. It can be shown that periodic repetition of any finite deterministic sequence of left and right turns is not enough to find the target in an arbitrary maze. Randomization overcomes the insufficiency of deterministic algorithms and helps to cope with uncertainty about the global geometry of complex scenes.

7.4.1 Proof of Proposition 4.2 and Remark 4.2

For the sake of convenience, we put the origin of the world frame into the target $\mathcal{T}$. Let $C \not\ni \mathcal{T}$ be a regular piece-wise smooth directed curve with natural parametric representation $\varrho(s), s \in [s_-, s_+]$. The turning angle of C around a point $\boldsymbol{p} \notin C$ is denoted by $\sphericalangle_{\boldsymbol{p}} C$, and $\sphericalangle \mathbf{TANG}\,[C] := \sphericalangle_0 T_C$, where $\mathbf{T}_C(s), \mathbf{N}_C(s)$ is the Frenet frame of C at s.[1] Let $\lambda(s)$, $\zeta(s)$, and $\psi(s)$ denote the Cartesian coordinates and polar angle of $-\varrho(s)$ in this frame (see Fig. 7.5), respectively. To indicate the curve C, the symbols $\lambda, \zeta, \varkappa$, etc., may be supplied with the lower index $_C$.

[1] At the corner points, the count of $\sphericalangle_0 \mathbf{T}_C$ progresses abruptly according to the conventional rules [396].

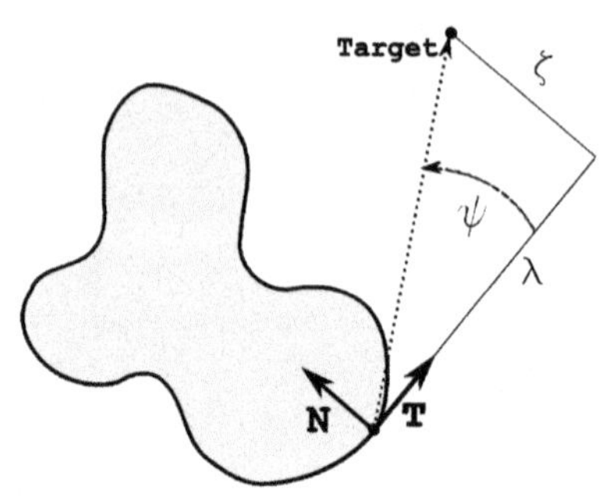

Figure 7.5 Definition of λ, ζ, and ψ.

Lemma 4.1. *The following relations hold whenever $\mathcal{T} \notin C$:*

$$\frac{d\lambda}{ds} = -1 + \varkappa\zeta, \quad \frac{d\zeta}{ds} = -\varkappa\lambda, \tag{4.3}$$

$$\frac{d\psi}{ds} = -\varkappa + \frac{\zeta}{\lambda^2+\zeta^2}, \quad \frac{d\varphi}{ds} = \frac{\zeta}{\lambda^2+\zeta^2}, \tag{4.4}$$

$$\mathfrak{r}_C := \mathbf{col}\,(\lambda, \zeta) \neq 0, \quad \sphericalangle_0 \mathfrak{r}_C = \sphericalangle_{\mathcal{T}} C - \sphericalangle \mathbf{TANG}\,[C]\,. \tag{4.5}$$

Proof. Differentiation of the equation $\mathcal{T} = \boldsymbol{\varrho} + \lambda \mathbf{T}_C + \zeta \mathbf{N}_C$ and the Frenet-Serret formulas $T_C' = \kappa \mathbf{N}_C, \mathbf{N}_C' = -\kappa \mathbf{T}_C$ [396] yield that

$$0 = \mathbf{T}_C + \lambda' \mathbf{T}_C + \lambda\varkappa \mathbf{N}_C + \zeta' \mathbf{N}_C - \zeta\varkappa \mathbf{T}_C.$$

Equating the cumulative coefficients in this linear combination of $\mathbf{T}_C$ and $\mathbf{N}_C$ to zero gives (4.3). Due to (4.3), relations (4.4) follow from the formulas [396]:

$$\psi' = \frac{\zeta'\lambda - \lambda'\zeta}{\lambda^2+\zeta^2}, \quad \varphi' = \frac{y'x - x'y}{x^2+y^2}. \tag{4.6}$$

The first relation in (4.5) holds since $\mathcal{T} \notin C$. Let $\eta(s) := \sphericalangle \mathbf{TANG}[C_{s_- \to s-0}] + \eta_0$, where η_0 is the polar angle of $\mathbf{T}_C(s_-)$. The matrix $\Phi_{\eta(s)}$ of rotation through angle $\eta(s)$ transforms the world frame into the Frenet one, and

$$\boldsymbol{\varrho}(s) = h(s)\,\mathbf{col}\,[\cos\varphi(s), \sin\varphi(s)].$$

Therefore,

$$\mathfrak{r}(s) = -\Phi_{-\eta(s)}\boldsymbol{\varrho}(s) = h(s)\,\mathbf{col}\,\{\cos[\pi + \varphi(s) - \eta(s)], \sin[\pi + \varphi(s) - \eta(s)]\}.$$

Thus, $\pi + \varphi(s) - \eta(s)$ is the piece-wise continuous polar angle of $\mathfrak{r}(s)$ that jumps according to the convention concerned by the footnote in page 133. This trivially implies (4.5). □

By (3.4) and the last inequality in (4.1), Lemma 4.1 yields that

$$\sphericalangle_0 \mathfrak{r}_{C(d_*)} = -2\pi \quad \text{for} \quad d_* \in [0, d_\updownarrow]. \tag{4.7}$$

Lemma 4.2. *The following two statements hold:*

(i) *In the domain $d \leq d_\updownarrow \wedge \dot{d} > 0 \vee d > d_\updownarrow$, the surface $\beta = 0$ is sliding, with the equivalent control $u \equiv 0$;*

(ii) *The surface $\dot{d} = 0$ is sliding in the domain*

$$d_\updownarrow - 2R_{\min} \leq d < d_\updownarrow, \quad \dot{s}\beta > 0, \quad \sigma\dot{s} > 0. \tag{4.8}$$

Proof.

(i) Let $h(t) := \|\boldsymbol{r}(t) - \mathscr{T}\|$ be the distance from the robot to the target. By (2.1) and Fig. 7.1(b),

$$\dot{h} = -v\cos\beta, \quad \dot{\beta} = h^{-1}v\sin\beta - u.$$

Therefore, $\dot{\beta} \xrightarrow{(2.3)} -\overline{u}\,\mathbf{sgn}\,\beta$ as the surface $\beta = 0$ is approached, which implies the first claim.

(ii) Let α be the polar angle of the robot velocity in the Frenet frame

$$\mathbf{T}_{\partial D}[s(t)], \mathbf{N}_{\partial D}[s(t)].$$

By (3.1), (3.4), and (4.8), $1 + \varkappa[s(t)]d(t) > 0$, and as is shown in, e.g., [397],

$$\dot{s} = \frac{v\cos\alpha}{1 + \varkappa(s)d}, \quad \dot{d} = -v\sin\alpha, \quad \dot{\alpha} = -\varkappa(s)\dot{s} + u. \tag{4.9}$$

As the state approaches a point where $\dot{d} = 0$ and (4.8) holds,

$$\sin\alpha \to 0, \quad \cos\alpha \to \mathbf{sgn}\,\dot{s}, \quad \ddot{d} \to -v^2\left[v^{-1}u\,\mathbf{sgn}\,\dot{s} - \varkappa_d\right], \tag{4.10}$$

where $\varkappa_d := \frac{\varkappa}{1+\varkappa d}$. Therefore, in view of (2.2) and (2.3), we see that

$$\begin{aligned} \ddot{d} &\xrightarrow{d>0, \dot{d}\to 0} \ddot{d}_+ := -v^2\left[R_{\min}^{-1}\mathbf{sgn}\,(\beta\dot{s}) - \varkappa_d\right] \overset{(4.8)}{=} -v^2\left[R_{\min}^{-1} - \varkappa_d\right], \\ \ddot{d} &\xrightarrow{\dot{d}<0, \dot{d}\to 0} \ddot{d}_- := v^2\left[\sigma R_{\min}^{-1}\mathbf{sgn}\,\dot{s} + \varkappa_d\right] \overset{(4.8)}{=} v^2\left[R_{\min}^{-1} + \varkappa_d\right]. \end{aligned} \tag{4.11}$$

The proof is completed by observing that by (3.4) and (4.8),

$$\begin{aligned} \ddot{d}_+ &= -v^2\frac{1 + \varkappa d - \varkappa R_{\min}}{R_{\min}(1 + \varkappa d)} < 0 \text{ since } 1 + \varkappa d > 0 \text{ and } d > d_{\text{safe}} > R_{\min}, \\ \ddot{d}_- &= v^2\frac{|\varkappa|\left[R_\varkappa + (d + R_{\min})\mathbf{sgn}\,\varkappa\right]}{R_{\min}(1 + \varkappa d)} > 0. \end{aligned} \tag{4.12}$$

□

From now on, $\sigma = +1$; $\sigma = -1$ is considered likewise.

Lemma 4.3. *If $\dot{d}(t_*) < 0$, claim (i) in Proposition 4.2 is true.*

Proof. By (2.3), initially $u \equiv -\overline{u}$. During the maximal interval $[t_*, \tau]$ on which $u \equiv -\overline{u}$ the robot moves clockwise over a circle C_{in} of the radius $R_{\min}$ and so by Remark 4.1, $\beta(t) > 0$ and

$$d(t) \geq \underbrace{\mathbf{dist}\,[\boldsymbol{r}(0); D]}_{\geq d_\ddagger} - \underbrace{\|\boldsymbol{r} - \boldsymbol{r}(0)\|}_{\leq 2R_{\min}} \overset{(3.4)}{>} R_{\min} > 0; \tag{4.13}$$

$$\dot{\alpha} \overset{(4.9)}{=} -v\left[\varkappa_d \cos\alpha + \overline{u}v^{-1}\right]$$

$$\overset{(2.2)}{\leq} -v\left[\frac{1}{R_{\min}} - \frac{|\varkappa|}{1+\varkappa d}\right] = -v\left[\frac{1}{R_{\min}} - \frac{1}{R_\varkappa + d\,\mathbf{sgn}\,\varkappa}\right].$$

While $d \leq d_\ddagger$ (in particular, while $\dot{d} \leq 0$) the expression in the last square brackets is positive. This holds by (4.13) if $\varkappa \geq 0$; otherwise, since $R_\varkappa > R_{\min} + d_\ddagger$ by (3.4). Therefore, $\dot{\alpha} \leq -\delta < 0$, i.e., the vector $\mathbf{col}\,(\cos\alpha, \sin\alpha)$ rotates clockwise. Here

$$\mathbf{sgn}\,\cos\alpha = \mathbf{sgn}\,\dot{s}, \quad \mathbf{sgn}\,\sin\alpha = -\mathbf{sgn}\,\dot{d}$$

by (4.9). Therefore, $\mathbf{col}\,(\dot{s}, \dot{d})$ evolves as is depicted in Fig. 7.6. This and the conditions (4.8) for sliding motion complete the proof. □

If some of the limits of the vector field are tangential to its discontinuity surface $\dot{d} = 0$, the arguments become more sophisticated [70]. The first lemma establishes a required technical fact, while the next lemma handles the respective particulars.

To state the first lemma, we note that whenever $d := \mathbf{dist}\,[\boldsymbol{r}; D] < R_\star(D)$, the system state (x, y, θ) is given by s, d, θ and along with $(\dot{d}, \dot{s}) \neq (0,0)$, uniquely determines $\beta \in (-\pi, \pi)$.

Lemma 4.4. *If $\lambda_{C(d_\dagger)}(s_*) \neq 0$ for $d_\dagger \in [0, d_\ddagger]$, there is $\delta > 0$ such that whenever $s_* \leq s_0 < s < s_* + \delta$ and $|d_* - d_\dagger| < \delta$, the following entailments hold with $\varsigma := \mathbf{sgn}\,\dot{s}$:*

$$\left.\begin{array}{c} \dot{s} \neq 0, \quad \dot{d} \geq 0, \quad d \geq d_*, \quad \zeta_{C(d_*)}(s_0) \geq 0 \\ \varkappa(s_* + \varsigma \overset{\approx}{} 0) < 0, \quad \dot{s}\lambda_{C(d_\dagger)}(s_0) > 0 \end{array}\right| \Rightarrow \dot{s}\beta > 0; \tag{4.14}$$

$$\left.\begin{array}{c} \dot{s} \neq 0, \quad \dot{d} \leq 0, \quad d \leq d_*, \quad \zeta_{C(d_*)}(s_0) \leq 0 \\ \varkappa(s_* + \varsigma \overset{\approx}{} 0) \geq 0, \quad \dot{s}\lambda_{C(d_\dagger)}(s_0) > 0 \end{array}\right| \Rightarrow \dot{s}\beta \leq 0. \tag{4.15}$$

In (4.15), *$\dot{s}\beta < 0$ if $\zeta_{C(d_*)}(s_0) < 0$ or $\varkappa \not\equiv 0$ on $\partial D_{s_0 \to s}$.*

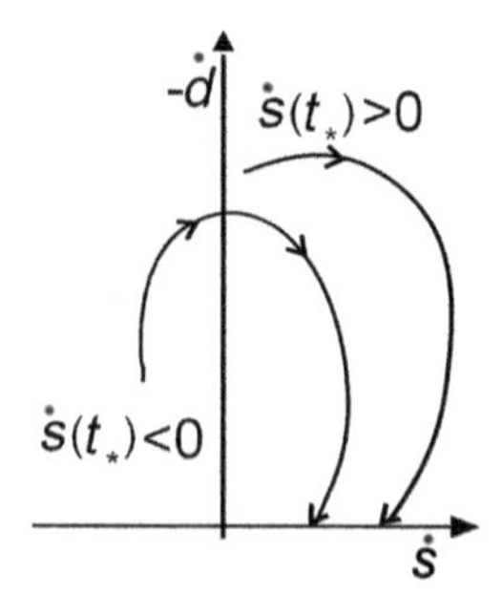

Figure 7.6 Behavior during initial turns.

Proof. We pick $\delta > 0$ so that $\lambda_{C(d_*)}(s)$ and $\varkappa(s)$ do not change the sign as s and d_* run over $(s_*, s_*+\delta)$ and $(d_\dagger - \delta, d_\dagger + \delta)$, respectively. By (4.2), the curvature $\varkappa_{C(d_*)}(s)$ does not change its sign either, which equals $\mathbf{sgn}\, \varkappa(s_* + \varsigma^{\approx}0)$.

If the conditions from (4.14) hold and $\varsigma = +$, application of the second equation from (4.3) to $C(d_*)$ yields that $\zeta_{C(d_*)}(s) > 0$. Therefore, the target polar angle in the s-related Frenet frame of $C(d_*)$ belongs to $(0, \pi/2)$. Transformation of this frame into that of the robot's path consists of a move of the origin in the negative direction along the ζ-axis (since $d \geq d_*$) and a clockwise rotation of the axes (since $\dot{d} > 0, \dot{s} > 0$). Since both operations increase the target bearing angle, $\beta > 0$. Formula (4.14) with $\varsigma = -$ and (4.15) are established likewise. □

Lemma 4.5. *Suppose that*

$$d_{safe} \leq d_* := d(t_*) \leq d_\updownarrow \quad \textit{and} \quad \dot{d}(t_*) = 0$$

at a time t_* within mode $\mathfrak{B}$. Then for $t > t_*, t \approx t_*$, the robot

(i) *performs the turn with $u \equiv -\sigma\overline{u}$ if $\sigma\dot{s}(t_*) < 0$, $d(t_*) = d_\updownarrow$, and $\beta(t_*) = 0$;*
(ii) *undergoes SMEC if $\sigma\dot{s}(t_*) > 0$ and either **(a)** $\sigma\beta(t_*) > 0$ or **(b)** $\beta(t_*) = 0$ and $\varkappa[s(t_*) + \mathbf{sgn}\, \dot{s}(t_*)^{\approx}0] < 0$;*
(iii) *moves straight to the target if $\beta(t_*) = 0, \sigma\dot{s}(t_*) > 0, \varkappa[s(t_*) + \mathbf{sgn}\, \dot{s}(t_*)^{\approx}0] \geq 0$.*

Proof. We recall that $\sigma = +$.

(i) As $t \to t_*$, (2.3) and (4.10) yield that

$$\ddot{d}|_{u=-\overline{u}} \to v^2\left[-\frac{1}{R_{\min}} + \frac{\varkappa}{1+\varkappa d_*}\right] = -\frac{1+\varkappa[d_* - R_{\min}]}{R_{\min}(1+\varkappa d_*)} < 0, \tag{4.16}$$

where $\varkappa := \varkappa[s(t_*) \pm 0]$ and the inequality holds since $d_* \geq d_{\text{safe}} > R_{\min}$ due to (3.4).

Let (i) fail to be true and $\varkappa[s(t_*) -^{\approx} 0] < 0$. If there exists an infinite sequence $\{t_i\}$ such that $t_i > t_*, d(t_i) < d_\updownarrow\ \forall i$ and $t_i \to t_*$ as $i \to \infty$, a proper decrease of every t_i yields in addition that $\dot{d}(t_i) < 0$ since $d(t_*) = d_\updownarrow$. However, then $\dot{d}(t) < 0$ for $t \geq t_i, t \approx t_*$ by (2.3), (4.16) and thus $\dot{d}(t) < 0, d(t) < d_\updownarrow$ for $t > t_*, t \approx t_*$, i.e., (i) holds in violation of the initial assumption. It follows that $d(t_* +^{\approx} 0) \geq d_\updownarrow$.

Now suppose that there is a sequence $\{t_i\}$ such that $t_i > t_*, d(t_i) = d_\updownarrow\ \forall i$, $t_i \to t_*$ as $i \to \infty$. Then $\dot{d}(t_i) = 0$ and so $\beta(t_i) < 0$ due to (4.14). By continuity, $\beta < 0$ in a vicinity of the system state at $t = t_i$. Then any option from (2.3) yields $u = -\overline{u}$ and so $u(t) \equiv -\overline{u}\ \forall t \approx t_i$ by the definition of Filippov solution. Hence,

$$d(t_i) = d_\updownarrow \wedge \dot{d}(t_i) = 0 \overset{(4.16)}{\Longrightarrow} d(t_i +^{\approx} 0) < d_\updownarrow,$$

in violation of the foregoing. Therefore, $d > d_\updownarrow$ and $u = \mathbf{sgn}\, \beta$ for $t > t_*, t \approx t_*$ by (2.3), and by Lemma 4.2, SMT is continued. Then the last relation in (4.10) (with $u := 0$) and $\varkappa[s(t_*) -^{\approx} 0] < 0$ imply the contradiction $d(t_* +^{\approx} 0) < d_\updownarrow$ to the foregoing, which proves (i).

Let $\varkappa[s(t_*) -^{\approx} 0] \geq 0$. Therefore, far as the navigation controller is first probationally set to the submode related with $\dot{d} < 0$, this submode will be maintained longer by (4.16).

(ii.a) If $d(t_*) < d_\updownarrow$, the claim is true by Lemma 4.2. Let $d(t_*) = d_\updownarrow$. If there is a sequence $\{t_i\}$ such that $t_i > t_*, d(t_i) < d_\updownarrow\ \forall i$ and $t_i \to t_*$ as $i \to \infty$; a proper decrease of every t_i yields in addition that $\dot{d}(t_i) < 0$. Let τ_i be the minimal $\tau \in [t_*, t_i]$ such that $d(t) < d_\updownarrow$ and $\dot{d}(t) < 0$ for $t \in (\tau, t_i]$. For such t, $u \equiv -\overline{u}$ by (2.3) and so $\ddot{d} > 0$ by (4.11) and

(4.12). Therefore, $\ddot{d}(\tau_i) < \dot{d}(t_i) < 0, \tau_i > t_*$, and $d(\tau_i) = d_\ddagger$, otherwise τ_i is not the minimal τ. Thus, at time τ_i, the assumptions of Lemma 4.3 hold except for $\beta(\tau_i) = 0$. In the proof of this lemma, this relation was used only to justify that $\beta > 0$, which is now true by assumption and the continuity argument. Therefore, by Lemmas 4.2 and 4.3, the sliding motion along an equidistant curve $C(d_\dagger)$ with $d_\dagger < d_\ddagger$ is commenced at time $t > \tau_i$ when $\dot{d}(t) = 0$ and is maintained while $\beta > 0$ and $\dot{s} > 0$, in violation of $d(\tau_i) = d_\ddagger \; \forall i \wedge \tau_i \xrightarrow{i\to\infty} t_*$. This contradiction proves that $d(t_* +^{\approx} 0) \geq 0$.

Now suppose there exists a sequence $\{t_i\}$ such that $t_i > t_*, d(t_i) > d_\ddagger \; \forall i$ and $t_i \to t_*$ as $i \to \infty$. Since $d(t_*) = 0$, a proper perturbation of every t_i yields in addition that $\dot{d}(t_i) > 0$. Let τ_i be the minimal $\tau \in [t_*, t_i]$ such that $d(t) > d_\ddagger$ for $t \in (\tau, t_i]$. For such t, the continuity argument gives $\beta > 0$, (2.3) yields $u \equiv \overline{u}$, and so $\ddot{d} < 0$ by (4.11) and (4.12). Hence, $\dot{d}(\tau_i) > 0, \tau_i > t_*, d(\tau_i) = d_\ddagger$ and so $d(\tau_i -^{\approx} 0) < 0$, in violation of the foregoing. This contradiction proves that $d(t_* +^{\approx} 0) \equiv 0$ indeed.

(ii.b) We first assume that $d_* < d_\ddagger$. Due to (4.11) and (4.12)

$$\ddot{d}|_{u=-\overline{u}} > 0 \quad \text{and} \quad \ddot{d}|_{u=\overline{u}} < 0 \quad \text{for} \quad t \approx t_*. \tag{4.17}$$

Therefore, it is easy to see that $\dot{d}(t_* +^{\approx} 0) \geq 0$ and $d(t_* +^{\approx} 0) \geq d_*$. Suppose that $\dot{d}(t_* +^{\approx} 0) \not\equiv 0$ and so $d(t_* +^{\approx} 0) > d_*$. For any $\delta > 0$, there is $\tau \in (t_*, t_* + \delta)$ such that $\dot{d}(\tau) > 0$. For any such τ that lies sufficiently close to t_*, (4.14) $\Rightarrow \beta(\tau) > 0$. Therefore, $u = \overline{u}$ by (2.3) and $\ddot{d}(\tau) < 0$ by (4.17). Hence, the inequality $\dot{d}(t) > 0$ is not only maintained but also enhanced as t decays from τ to t_*, in violation of the assumption $\dot{d}(t_*) = 0$ of the lemma. This contradiction shows that $\dot{d}(t_* +^{\approx} 0) \equiv 0$ and completes the proof of (ii).

It remains to consider the case where $d_* = d_\ddagger$. By the arguments from the previous paragraph, it suffices to show that $\dot{d}(t_* +^{\approx} 0) \geq 0$ and $d(t_* +^{\approx} 0) \geq d_\ddagger$. Suppose that $d(t_* +^{\approx} 0) \not\geq d_\ddagger$, i.e., there exists a sequence $\{t_i\}$ such that $t_i > t_*, d(t_i) < d_\ddagger \; \forall i$ and $t_i \to t_*$ as $i \to \infty$. Since $d(t_*) = d_\ddagger$, a proper decrease of every t_i gives $\dot{d}(t_i) < 0$ in addition. By (2.3) and (4.17), the inequality $\dot{d}(t) < 0$ is maintained and enhanced as t decreases from t_i, remaining in the domain $\{t : d(t) < d_\ddagger\}$. Since $\dot{d}(t_*) = 0$, there is $\tau_i \in (t_*, t_i)$ such that $d(\tau_i) = d_\ddagger$ and $\dot{d}(t) < 0 \; \forall t \in [\tau_i, t_i)$. Hence, $d(\tau_i -^{\approx} 0) > d_\ddagger$ and if i is large enough, there is $\theta_i > t_i$ such that $d(\theta_i) = d_\ddagger$ and $d(t) < d_\ddagger \; \forall t \in (\tau_i, \theta_i)$. Furthermore, there is $s_i \in (\tau_i, \theta_i)$ such that

$$\dot{d}(t) < 0 \; \forall t \in (\tau_i, s_i), \quad \dot{d}(s_i) = 0, \quad \dot{d}(t) \geq 0 \; \forall t \in [s_i, \theta_i].$$

Then $\beta(\theta_i) > 0$ by (4.14). We note that $\beta(t_*) = 0 \Rightarrow \zeta_{\mathscr{P}}(t_*) = 0$ for the robot's path $\mathscr{P}$ and so $\zeta_{\mathscr{P}}(t) \to 0$ as $t \to t_*$. This and (4.4) (applied to $\mathscr{P}$) imply that the sign of $\dot{\beta}$ is determined by the sign of the path curvature:

$$u = \pm\overline{u} \Rightarrow \pm\dot{\beta} < 0 \quad \forall t \approx t_*. \tag{4.18}$$

Suppose that $\exists \tau_* \in [\tau_i, s_i) : \beta(\tau_*) \geq 0$. Since $u(t) = -\overline{u} \; \forall t \in (\tau_i, s_i)$, we see that $\beta(s_i) > 0, \dot{d}(s_i) = 0, d_s := d(s_i)$. By Lemma 4.2, sliding motion along the d_s-equidistant curve is commenced at $t = s_i$ and maintained while $\beta > 0$; meanwhile, $\beta > 0$ until θ_i (if i is large enough) due to (4.14). However, this is impossible since $d_s < d_\ddagger$ and $d(\theta_i) = d_\ddagger$. This contradiction proves that $\beta(t) < 0 \; \forall t \in [\tau_i, s_i)$. The same argument and the established validity of (ii.b) for $d_* := d_s < d_\ddagger$ show that $\beta(s_i) < 0$. Since $\beta(\theta_i) > 0$, there exists $c_i \in (s_i, \theta_i)$ such that $\beta(c_i) = 0$ and $\beta(t) > 0 \; \forall t \in (c_i, \theta_i]$. If $\dot{d}(c) = 0$ for some $c \in (c_i, \theta_i)$, Lemma 4.2 assures that sliding motion along the $d(c)$-equidistant curve is started at $t = c$ and is not terminated until $t = \theta_i$, in violation of $d(\theta) = d_\ddagger$.

For any $t \in (c_i, \theta_i)$, we thus have $\dot{d}(t) > 0$. Hence, $u(t) = \overline{u}$ by (2.3), $\dot{\beta} < 0$ by (4.18), and so $\beta(c_i) = 0 \Rightarrow \beta(\theta_i) < 0$, in violation of the above inequality $\beta(\theta_i) > 0$. This contradiction proves that $d(t_* +^{\approx} 0) \geq d_{\updownarrow}$.

Now suppose that $\dot{d}(t_* +^{\approx} 0) \not\leq 0$. Then there is a sequence $\{t_i\}$ such that $t_i > t_*, \dot{d}(t_i) > 0\ \forall i$ and $t_i \to t_*$ as $i \to \infty$; a proper increase of every t_i gives $d(t_i) > d_{\updownarrow}$ in addition. By (4.14),

$$d(t) > d_{\updownarrow} \wedge \dot{d}(t) > 0 \Rightarrow \beta(t) > 0 \quad \text{for} \quad t \approx t_*$$

and so $u(t) = \overline{u}$ by (2.3) and $\ddot{d}(t) < 0$ by (4.17). Therefore, as t decreases from t_i to t_*, the derivative $\dot{d}(t) > 0$ increases while $d > d_{\updownarrow}$, in violation of the implication $d(t) = d_{\updownarrow} \Rightarrow \dot{d}(t) = 0$ for $t \in [t_*, t_i]$. This contradiction completes the proof of (ii).

(iii) Were there a sequence $\{t_i\}_{i=1}^{\infty}$ such that $\dot{d}(t_i) > 0, \beta(t_i) > 0\ \forall i$ and $t_i \to t_* + 0$ as $i \to \infty$, (2.3), (4.17), and (4.18) would imply that as t decreases from t_i to t_* for large enough i, the inequalities $\dot{d}(t) > 0, \beta(t) > 0$ would be preserved, in violation of $\dot{d}(t_*) = 0, \beta(t_*) = 0$. It follows that $\dot{d}(t) > 0 \Rightarrow \beta(t) \leq 0$ for $t \approx t_*, t > t_*$.

Now we assume the existence of a sequence such that $\dot{d}(t_i) > 0, \beta(t_i) \leq 0\ \forall i$ and $t_i \to t_* + 0$ as $i \to \infty$. For large i such that $\beta(t_i) < 0$, (2.3)$\wedge$(4.17) $\Rightarrow u(t) = -\overline{u}$, and $\dot{d}(t)$ increases and so remains positive as t grows from t_i until $\beta = 0$. By (4.18), $\overline{u}^{-1}|\beta(t_i)|$ time units later the robot becomes headed to the target, which is trivially true if $\beta(t_i) = 0$. This and (i) of Lemma 4.2 imply that then the sliding motion along the surface $\beta = 0$ is commenced. It is maintained while $\varkappa[s(t)] \geq 0$. Since $t_i \to t_*$ and $\beta(t_i) \to \beta(t_*) = 0$ as $i \to \infty$, this motion occurs for $t > t_*$, i.e., (iii) holds.

It remains to examine the case where $\dot{d}(t_* +^{\approx} 0) \leq 0$ and so $d(t_* +^{\approx} 0) \leq d_*$. Suppose first that either $\dot{d}(t_* +^{\approx} 0) \not\equiv 0$ or $\varkappa[s(t_*) +^{\approx} 0] \not\equiv 0$. Then $\beta(t_* +^{\approx} 0) < 0$ by (4.15) and $u = -\overline{u}$ at any side of the discontinuity surface $\dot{d} = 0$ by (2.3). Hence, $u(t_* +^{\approx} 0) \equiv -\overline{u}$, which yields $\ddot{d}(t_* + 0) > 0$ by (4.17), in violation of $\dot{d}(t_* + 0) = 0$. This contradiction proves that $\dot{d}(t_* +^{\approx} 0) \equiv 0$, $\varkappa[s(t_*) +^{\approx} 0] \equiv 0$. Then SMEC and SMT are initially the same, and (iii) does hold.

□

Proof of Remark 4.2. This remark is immediate from Lemma 4.5. □

Proof of Proposition 4.2.

(i) is justified by Lemmas 4.3 and 4.5.

(ii) By Lemma 4.2 and (ii), (iii) of Lemma 4.5, SMEC holds while

$$\dot{s}(t) > 0 \quad \text{and} \quad \beta(t) > 0 \vee \beta(t) = 0 \wedge \varkappa[s(t) +^{\approx} 0] < 0.$$

During SMEC $\sin\alpha \equiv 0$ and $\dot{s} \geq \frac{v}{1+\varkappa d} > 0$ by (4.9). Therefore, SMEC terminates when β arrives at zero $\beta(\theta) = 0$ and $\varkappa[s(\theta) +^{\approx} 0] > 0$. Let SMEC never terminate. By (4.7), $\mathfrak{r}_{C(d_*)}$ enters the fourth quadrant through the positive ray of the λ-axis $\exists \tau : \lambda(\tau) > 0, \zeta(\tau) = 0, \zeta(\tau +^{\approx} 0) < 0$. Then $\varkappa[s(\tau) +^{\approx} 0] > 0$ by (4.3); $\lambda(\tau) > 0 \wedge \zeta(\tau) = 0$ means the robot is headed for $\mathcal{T}$. Hence, SMEC is terminated at τ.

(vi) During SMT, $u \equiv 0$ and so (4.9) implies that

$$\begin{aligned}\dot{s} &= g(t)\cos\alpha \\ \dot{\alpha} &= f(t)\cos\alpha\end{aligned}, \quad \ddot{d} = \varkappa \frac{v^2 \cos^2\alpha}{1 + \varkappa d} \geq 0 \quad \text{if} \quad \varkappa \geq 0, \tag{4.19}$$

where $g(t) := \frac{v}{1+\varkappa(s)d} > 0, f(t) := -\varkappa g(t)$. Hence, $\cos\alpha(0) \neq 0 \Rightarrow \cos\alpha(t) \neq 0\ \forall t \geq 0$; thus **sgn** $\cos\alpha$ is constant. Therefore, $\dot{s} > 0$ during SMT if this is true initially. The same holds true for SMEC since then $\sin\alpha \equiv 0$ by (4.9) and so $\cos\alpha \equiv \pm 1$. The proof is completed by **(i)**.

(iii) By (vi) and Lemmas 4.2, 4.5, SMT continues until the threshold $d_\updownarrow$ is passed or $\dot{d}(t) = 0 \wedge \varkappa[s(t) +^{\approx} 0] < 0$, after which SMEC is started. For SMT,

$$\dot{d}(t) = 0 \wedge \varkappa[s(t) +^{\approx} 0] \geq 0 \overset{(4.19)}{\Longrightarrow} \dot{d}(t +^{\approx} 0) \geq 0$$

and so $\dot{d}(t) \geq 0$. Claim **(iv)** is evident.

(v) By (ii) and (iii), the switches SMT $\mapsto$ SMEC $\mapsto$ SMT hold when $\varkappa[s+^{\approx}0] < 0$ and $\varkappa[s+^{\approx} 0] > 0$, respectively. Therefore, between them, $s(t)$ runs the entire length of a connected component of the set $\{s : \varkappa(s) \geq 0\}$. Since there are only finitely many such components thanks to Assumption 3.1, the switches do not accumulate. Therefore, it suffices to show that AM cannot be endless.

Suppose the contrary. Then s repeatedly encircles ∂D. Consider a complete run that starts after IT at the point $s_{\max}$ furnishing $\max_{s\in\partial D} \|s - \mathscr{T}\|$. Let $\mathscr{P}$ stand for the related path of the robot, parameterized by $\boldsymbol{\rho}(s) - \mathbf{N}(s)d(s)$.

The angular discrepancy between $\mathbf{T}_{\mathscr{P}}(s)$ and $\mathbf{T}_{\partial D}(s)$ is zero during SMEC and lies in $(0, \pi/2)$ during SMT since then $\dot{d} \geq 0, \dot{s} > 0$ by (4.2), (vi). Therefore,

$$|\sphericalangle\mathbf{TANG}\,[\mathscr{P}] - \sphericalangle\mathbf{TANG}\,[\partial D]| < \pi/2,$$

where $\sphericalangle\mathbf{TANG}\,[\partial D] = 2\pi$ by Hopf's theorem [398, p. 57]. Since $\dot{s} > 0$, the point $s_{\max}$ is passed within SMEC. Hence, the vectors $\mathbf{T}_{\mathscr{P}}$ are identical at the ends of $\mathscr{P}$ and $\sphericalangle\mathbf{TANG}\,[\mathscr{P}] = 2\pi$. These ends lie on the normal to ∂D at the distance maximizer $s_{\max}$. The homotopy $\theta \in [0, 1] \mapsto \boldsymbol{\rho}(s) - \theta\mathbf{N}(s)d(s)$ transforms $\mathscr{P}$ into ∂D within the strip $\{\boldsymbol{r} \notin \operatorname{int} D : \mathbf{dist}\,[\boldsymbol{r}; D] \leq d_\updownarrow + \varepsilon\} \not\ni \mathscr{T}$, moving the end points of $\mathscr{P}$ along a ray centered at $\mathscr{T}$. It follows that $\sphericalangle_{\mathscr{T}}\mathscr{P} = \sphericalangle_{\mathscr{T}}\partial D = 0$. Therefore, by (4.5), $\sphericalangle_0\mathfrak{r}_{\mathscr{P}} = -2\pi$.

For k full runs over ∂D, we have $\sphericalangle_0\mathfrak{r}_{\mathscr{P}} = -2k\pi$. However, β is the polar angle of $\mathfrak{r}_{\mathscr{P}}$, and by (i), $\beta > 0$ during SMEC. Hence, $\mathfrak{r}_{\mathscr{P}}$ crosses the positive ray of the λ-axis in the clockwise direction with $\beta = 0$. This corresponds to a SMT, for which $\beta = \text{const}$ and after which $\beta > 0$ due to (iii) and Lemma 4.1. This contradiction completes the proof.

□

7.4.2 Proof of Theorem 4.1

We consider the basic navigation controller with right turns. Let an occurrence $\mathfrak{A}^\dagger$ of mode $\mathfrak{A}$ hold between two $\mathfrak{B}$s, start at $\boldsymbol{r}_\diamondsuit = \boldsymbol{r}(t_\diamondsuit)$, and end at $\boldsymbol{r}_* = \boldsymbol{r}(t_*)$. By (3.4),

$$\mathbf{dist}\,[\boldsymbol{r}_*; D] = \mathbf{dist}\left[\boldsymbol{r}_\diamondsuit; D\right] = d_\updownarrow$$

are attained at unique points $s_\diamondsuit, s_* \in C := \partial D$, respectively. They divide C into two arcs. Being concatenated with $[s_*, \boldsymbol{r}_*], [\boldsymbol{r}_*, \boldsymbol{r}_\diamondsuit]$, and $[\boldsymbol{r}_\diamondsuit, s_\diamondsuit]$, each of them gives rise to a Jordan curve encircling a domain, one of which is the other united with D. The smaller domain is denoted by $\mathfrak{C}_{\mathfrak{A}^\dagger}$ and the respective arc by $\gamma_{\mathfrak{A}^\dagger}$. Let $\sigma_{\mathfrak{A}^\dagger} = \pm$ be the direction (on C) of the walk from $s_\diamondsuit$ to s_* over $\gamma_{\mathfrak{A}^\dagger}$. We also recall that the proofs of the claims marked by an upper index ^{a} will be given in the appendix to this chapter.

Lemma 4.6.[a] *For any occurrence* $\mathfrak{A}^\dagger$, *we have* $\sigma_{\mathfrak{A}^\dagger} = +$.

In this sense, s goes over C in the positive direction in mode $\mathfrak{A}$ in addition to SMEC. ITs are addressed by the following.

Lemma 4.7.[a] *Let an IT start and end at* $t = t_s, t_e$*, respectively. Then*

$$\dot{s}(t_s) \geq 0 \Rightarrow \dot{s}(t) \geq 0 \quad \forall t \in [t_s, t_e];$$

otherwise, $\dot{s}$ *changes sign only once. In any case,* s *goes from* $s(t_s)$ *to* $s(t_e)$ *in the positive direction during a last phase of IT.*

Let $\angle(\mathbf{a}, \mathbf{b}) \in (-\pi, \pi]$ be the signed angle between the reference vector **a** and **b**, and $\mathscr{P}$ be the directed path of the robot. Let $\boldsymbol{r}_1, \boldsymbol{r}_2 \in \mathscr{P}$ be such that $\mathbf{dist}[\boldsymbol{r}_i; D] \leq d_\updownarrow$ and the path does not intersect itself and the normals $[\boldsymbol{r}_i, s_i]$, $s_i := s[\boldsymbol{r}_i]$ when traveling from $\boldsymbol{r}_1$ to $\boldsymbol{r}_2$. The points s_i split C into two arcs. Being concatenated with the normals and $\mathscr{P}|_{\boldsymbol{r}_1 \to \boldsymbol{r}_2}$, they give rise to Jordan loops, with one of them enveloping the other. Let **LOOP** be the inner loop, γ_{inner} be the related arc, and $\sigma = \pm$ be the direction from s_1 to s_2 along γ_{inner}.

Lemma 4.8. *If* **LOOP** *does not encircle the target,*

$$\sphericalangle_0 \mathfrak{r}_{\mathscr{P}}|_{\boldsymbol{r}_1 \to \boldsymbol{r}_2} = \sphericalangle_0 \mathfrak{r}_C|_{s_1 \xrightarrow{\sigma} s_2} + \sphericalangle_{\mathcal{T}}[\boldsymbol{r}_1, s_1] - \sphericalangle_{\mathcal{T}}[\boldsymbol{r}_2, s_2]$$
$$+ \angle[\sigma \mathbf{T}_C(s_1), \mathbf{T}_{\mathscr{P}}(\boldsymbol{r}_1)] - \angle[\sigma \mathbf{T}_C(s_2), \mathbf{T}_{\mathscr{P}}(\boldsymbol{r}_2)]. \tag{4.20}$$

Proof. Let $\sigma = +$. By Hopf's theorem [398, p. 57],

$$\sphericalangle_{\mathcal{T}}[s_1, \boldsymbol{r}_1] + \sphericalangle_{\mathcal{T}} \mathscr{P}|_{\boldsymbol{r}_1 \to \boldsymbol{r}_2} + \sphericalangle_{\mathcal{T}}[\boldsymbol{r}_2, s_2] - \sphericalangle_{\mathcal{T}} C_{s_1 \to s_2} = 0,$$
$$\sphericalangle \mathbf{TANG}\left[\mathscr{P}|_{\boldsymbol{r}_1 \to \boldsymbol{r}_2}\right] = \sphericalangle \mathbf{TANG}\left[C_{s_1 \to s_2}\right] + \sum_{i=1}^{2} (-1)^i \angle[\mathbf{T}_C(s_i), \mathbf{T}_{\mathscr{P}}(\boldsymbol{r}_i)].$$

The proof is completed by the second formula in (4.5). □

Corollary 4.1. *For* $\boldsymbol{r}_1 = \boldsymbol{r}_\Diamond, \boldsymbol{r}_2 = \boldsymbol{r}_*$, (4.20) *holds with* $\sigma = +$.

Corollary 4.2. *Let* $\boldsymbol{r}_1, \boldsymbol{r}_2$ *be successively passed within a common mode* $\mathfrak{B}$*. If* $\boldsymbol{r}_2$ *follows IT,* (4.20) *holds with* $\sigma = +$.

If (a) s does not run the entire length of C and (b) either $\boldsymbol{r}_1$ follows IT or $\mathbf{sgn}\, \dot{s} = +1$ at the start of the mode, the claim is evident. If (a) holds but (b) does not, the path may intersect $[s_1, \boldsymbol{r}_1]$. Then we apply Lemma 4.8 to $\boldsymbol{r}_1 := \boldsymbol{r}_3$, where $\boldsymbol{r}_3$ is the intersection point. It remains to note that

$$\sphericalangle_{\mathcal{T}}[\boldsymbol{r}_1, \boldsymbol{r}_3] = \sphericalangle_{\mathcal{T}} \gamma, \sphericalangle \mathbf{TANG}[\gamma] = \angle[\mathbf{T}_1, \mathbf{T}_3]$$

and so

$$\sphericalangle_0 \mathfrak{r}_{\mathscr{P}}|_{\boldsymbol{r}_1 \to \boldsymbol{r}_3} = \sphericalangle_0 \mathfrak{r}_\gamma = \sphericalangle_{\mathcal{T}}[\boldsymbol{r}_1, \boldsymbol{r}_3] - \angle[\mathbf{T}_1, \mathbf{T}_3],$$

as well as that

$$\angle[\sigma \mathbf{T}_C(s_1), \mathbf{T}_3] = \angle[\sigma \mathbf{T}_C(s_1), \mathbf{T}_1] + \angle[\mathbf{T}_1, \mathbf{T}_3].$$

If (1) is not true, the claim is proved via partition of $\mathscr{P}$ into pieces for which (1) holds.

Corollary 4.3. *Suppose the points* $\boldsymbol{r}_1$ *and* $\boldsymbol{r}_2$ *are successively passed in modes* $\mathfrak{B}$ *(maybe, different) and* $\boldsymbol{r}_2$ *is not attributed to the IT. Then* (4.20) *holds with* $\sigma = +$*, where* $\sphericalangle_0 \mathfrak{r}_{\partial D}|_{s_1 \xrightarrow{\sigma} s_2}$ *accounts for the entire motion of the projection* $s = s[\boldsymbol{r}], \boldsymbol{r} \in \mathscr{P}_{\boldsymbol{r}_1 \to \boldsymbol{r}_2}$*, including possible full runs over* ∂D.

When a SMEC ends,

$$s \in S_0 := \{s \in C : -d_{\updownarrow} \leq \zeta_{\partial D}(s) < 0, \quad \lambda_C(s) > 0\}.$$

Since C is piece-wise analytical, this set has finitely many connected components, called *exit arcs*.

Proof of Theorem 4.1. If its claim fails to be true, s repeatedly loops around C. (This includes imaginary moves of s in modes $\mathfrak{A}$.) By retracing the proof of v) in Proposition 4.2, we see that $\mathcal{P}$ can be truncated so that the first and last modes $\mathfrak{A}$ start at positions $\boldsymbol{r}_1$ and $\boldsymbol{r}_2$, respectively, from a common exit arc A, and s goes through the entire C. By definition of the exit arc, $\mathfrak{r}_C(s)$ remains in the fourth quadrant as s goes from s_1 to s_2 within the +arc and therefore the absolute value of its turning angle does not exceed $\pi/2$. This and (4.7) (where $d_* := 0$) imply that $\sphericalangle_0 \mathfrak{r}_{C|_{s_1 \to s_2}} \leq -3/2\pi$. In (4.20), $|\sphericalangle_{\mathcal{T}}[\boldsymbol{r}_i, s_i]| < \pi/2$ and $\angle[\mathbf{T}_C(s_i), \mathbf{T}_{\mathcal{P}}(\boldsymbol{r}_i)] = 0$ since $[\boldsymbol{r}_i, s_i]$ and $[\boldsymbol{r}_i, \mathcal{T}]$ are perpendicular. Overall, (4.20) implies that $\sphericalangle_0 \mathfrak{r}_{\mathcal{P}|_{\boldsymbol{r}_1 \to \boldsymbol{r}_2}} < -\frac{\pi}{2}$. The path $\mathcal{P}|_{\boldsymbol{r}_1 \to \boldsymbol{r}_2}$ starts with $\beta = 0$ and whenever $\beta = 0$, the angle β may stay constant during SMT; but after SMT, $\beta > 0$ since the robot turns right by (iii) of Proposition 4.2 and (2.3). Such behavior of β is inconsistent with the above inequality $\sphericalangle_0 \mathfrak{r}_{\mathcal{P}|_{\boldsymbol{r}_1 \to \boldsymbol{r}_2}} < -\frac{\pi}{2}$. The contradiction obtained completes the proof. □

7.5 Simulations and experiments with a real wheeled robot

In computer simulations, the control was updated every 0.02 s, $d_{\updownarrow} = 8\,\text{m}, \overline{u} = 2.5\,\text{rad/s}, v = 3\,\text{m/s}$. Figure 7.7(a) and (b) presents typical results for the randomized navigation controller, where the realizations $\sigma = +1, -1, -1, -1$ and

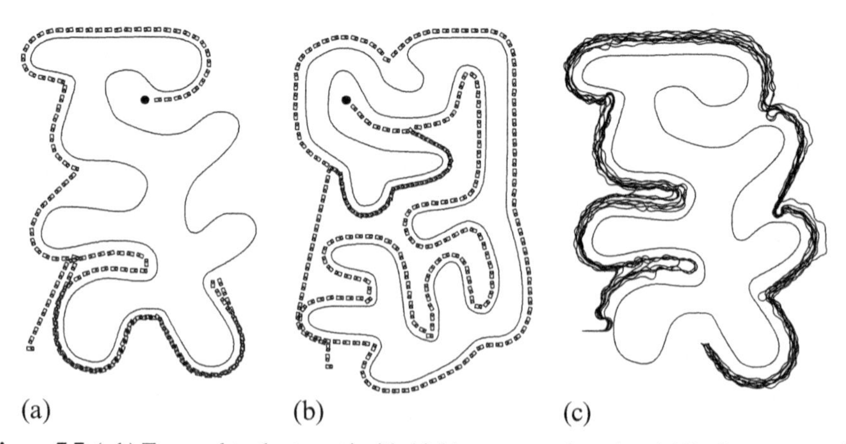

Figure 7.7 (a,b) Traversal to the target inside highly concave obstacles; (c) Performance under random noises: various paths correspond to various realizations of the noises.

$\sigma = -1, +1, +1, +1, -1$ of the random sequence of turns were observed in (a) and (b), respectively. The boundary following distance changes after bypass of any obstacle concavity.

To address performance under sensor noises, disturbances, and unmodeled dynamics, the sensor readings and system equations were corrupted by independent Gaussian white noises with the standard deviation 0.1 m for d, 0.1 rad for β, 0.5 rad/s for $\dot{\theta}$. To access $\dot{d}$, Newton's difference quotient of the noisy data d was computed for the current and previous sampling instants separated by 0.1 s. The maximum turning rate $\overline{u}$ was decreased to 0.7 rad/s, and extra dynamics were added via upper limiting the rate of change of the turning rate by 0.7 rad/s^2. The threshold where the trigger $\mathfrak{B} \mapsto \mathfrak{A}$ occurs was increased to 16 m. As can be seen from Fig. 7.7(c), the basic navigation law satisfactory guides the robot to the target for various realizations of the noise and disturbance.

The use of switching regulation often gives rise to concerns about its practical implementation and implications of the noises and unmodeled dynamics in practical setting, including possible chattering at worst. To address these issues, experiments were carried out with an Activ-Media Pioneer 3-DX mobile robot using its onboard PC and the Advanced Robot Interface for Applications (ARIA 2.7.0); see Section 1.4.1 for details. The position relative to the target was obtained through odometry, and $d_{\updownarrow}$ was taken to be 0.6 m and 0.9 m. The distance to the obstacles was accessed using both LIDAR and sonar sensors. A typical experimental result is presented in Fig. 7.8. In this experiment, as in the others, the robot reaches the target via safe navigation among the obstacles, with no visible mechanical chattering of any parts observed.

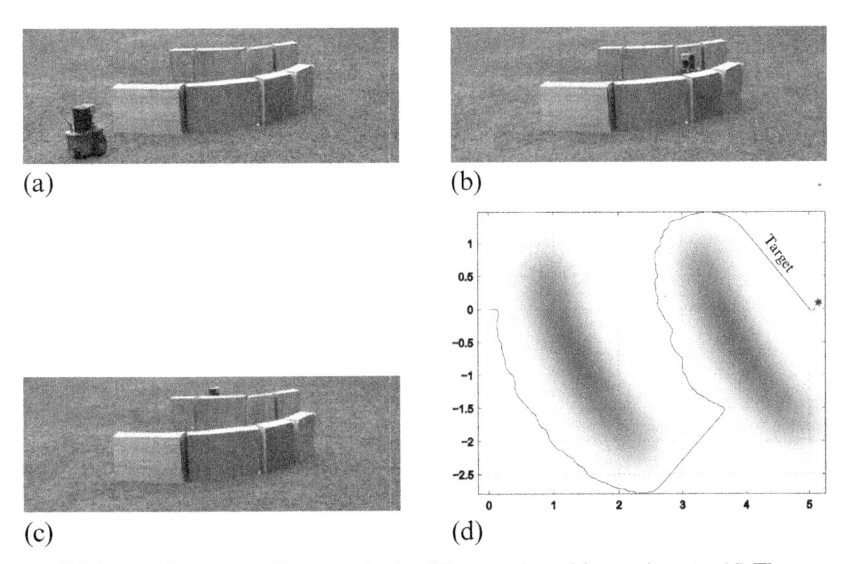

Figure 7.8 (a,b,c) Sequence of images obtained from real world experiments; (d) The trajectory of the robot, obtained through odometry.

7.A Appendix: Proofs of Proposition 4.1 and Lemmas 4.6 and 4.7

7.A.1 Technical facts and the proof of Lemma 4.7

We recall that the world frame is centered at the target $\mathscr{T}$, for a regular piece-wise smooth directed curve $C \not\ni \mathscr{T}$ with the natural parametric representation $\boldsymbol{\varrho}(s), s \in [s_-, s_+]$, the turning angle of C around a point $\boldsymbol{p} \notin C$ is denoted by $\sphericalangle_{\boldsymbol{p}} C$, and $\sphericalangle \mathbf{TANG}[C] := \sphericalangle_0 \mathbf{T}$, where $\mathbf{T}(s), \mathbf{N}(s)$ is the Frenet frame of C at s. The symbols $\lambda(s), \zeta(s)$ and $\psi(s)$ stand for the Cartesian coordinates and polar angle of $-\boldsymbol{\varrho}(s)$ in this frame; see Fig. 7.5. The directed curve traced as s runs from s_1 to s_2 is denoted by $C_{s_1 \xrightarrow{\pm} s_2}$, where the specifier $\pm$ is used for closed curves.

We start with two corollaries of Lemma 4.1, with the first of them being evident.

Corollary A.1. *Let $\zeta(s_*) = 0$ and $\varsigma = \pm$. Then*

$$\begin{aligned} &\varsigma\zeta[s_* + \varsigma^{\approx}0]\mathbf{sgn}\,\lambda[s_*] < 0 \quad \textit{if} \quad \varkappa[s_* + \varsigma^{\approx}0] > 0, \\ &\varsigma\zeta[s_* + \varsigma^{\approx}0]\mathbf{sgn}\,\lambda[s_*] > 0 \quad \textit{if} \quad \varkappa[s_* + \varsigma^{\approx}0] < 0. \end{aligned} \tag{A.1}$$

Corollary A.2. *There exist F and $d_\# > d_\updownarrow$ such that whenever $|d| \leq d_\#$, the set $S(d) := \{s \in \partial D : \zeta_{\partial D}(s) = d\}$ has no more than F connected components.*

Proof. By the last inequality in (4.1),

$$\exists d_\# : d_\updownarrow < d_\# < \mathbf{dist}\,[\mathscr{T}; D] \leq \sqrt{\zeta(s)^2 + \lambda(s)^2}.$$

Then

$$s \in S(d) \wedge |d| \leq d_\# \Rightarrow |\lambda(s)| \geq \delta := \sqrt{\mathbf{dist}\,[\mathscr{T}; D]^2 - d_\#^2} > 0.$$

Since the domain D is compact, $|\lambda'(s)| \leq M < \infty\ \forall s$. Therefore, whenever $s \in S(d)$ and $|d| \leq d_\#$, the function $\lambda(\cdot)$ does not change its sign in the δM^{-1}-neighborhood $V(s)$ of s. Since ∂D is piece-wise analytical, each set $\{s : \pm\varkappa(s) > 0\}$ and $\{s : \varkappa(s) = 0\}$ has finitely many connected components $\partial_i^{\pm}$ and ∂_ν^0, respectively. By the foregoing and (4.3), any intersection $V(s) \cap \overline{\partial_i^{\pm}}, s \in S(d), |d| \leq d_\#$ contains only one point s. Hence, the entire arc $\partial_i^{\pm}$ of the length $\left|\partial_i^{\pm}\right|$ contains no more than $\delta^{-1} M \left|\partial_i^{\pm}\right| + 1$ such points. It remains to note that $S(d)$ covers any ∂_ν^0 such that $\partial_\nu^0 \cap S(d) \neq \emptyset$. □

Observation A.1. *SMEC with $\sigma = \pm$ ends when*

$$s \in S_0 := \{s \in \partial D : -d_\# < \zeta_{\partial D}(s) < 0, \quad \pm\lambda_{\partial D}(s) > 0\}. \tag{A.2}$$

This set has no more than F connected components.

The second claim holds since $\lambda' < 0$ on S_0 due to (3.4) and (4.3).

Definition A.1. The connected components of the set (A.2) are called $\pm$*arcs*.

Lemma A.1. *Let s_* and s_b be the values of the continuously evolving s at the start and end of IT, respectively. During IT, $\sigma\dot{s} \geq 0$ if $\sigma\dot{s}(t_*) \geq 0$, and $\dot{s}$ ones changes the sign otherwise. In any case, s runs from s_* to s_b in the direction σ during a last phase of IT.*

Proof. Let $\sigma = +$. The map $\boldsymbol{r} \mapsto (s, d)$ is the orientation-reversing immersion on the disc D_{in} encircled by C_{in}. Therefore, it transforms any negatively oriented circle $C \subset D_{\text{in}}$ concentric with C_{in} into a curve ξ with $\sphericalangle\mathbf{TANG}[\xi] = 2\pi$. Then the argument from the concluding part of the proof of Lemma 4.3 shows that as the robot once runs over C_{in} in the negative direction, the vector $\mathbf{col}(\dot{s}, \dot{d})$ intersects the half-axes of the frame in the order associated with counter clockwise rotation, each only once. This implies the claim given by the first sentence in the conclusion of the lemma.

If $\dot{s}(t_*) \geq 0$, this claim yields that $s_b - s_* \geq 0$. Let $\dot{s}(t_*) < 0$. Since the robot once runs over C_{in} in the negative direction, $\dot{s} > 0$ and $\dot{d} \leq 0$ when it passes the point B from Fig. A.1, which corresponds to the second passage of $s = s_*$. Due to the order in which $\mathbf{col}(\dot{s}, \dot{d})$ intersects the half-axes, this combination of signs is possible only before $\dot{d}$ vanishes for the first time, i.e., within IT. Thus, the second occurrence of $s = s_*$ holds within IT. The proof is completed by noting that $\dot{s} > 0$ after this by the first claim of the lemma. □

Proof of Lemma 4.7. We recall that this lemma deals with the basic navigation law with only right turns. Therefore, Lemma 4.7 is immediate from Lemma A.1. □

Remark A.1. The times of switches between the modes of the discontinuous navigation law (2.3) do not accumulate.

To prove this, we first note that the projection of any robot's position $\boldsymbol{r}$ within mode $\mathfrak{B}$ onto ∂D is well defined due to (4.3). Let s_i^- and s_i^+ be its values at the start and end of the ith occurrence of the mode, respectively. By Lemma 4.7 and (vi) of Proposition 4.2, s monotonically sweeps an arc γ_i of ∂D with the ends s_i^-, s_i^+ during the concluding part of $\mathfrak{B}$.

Definition A.2. The robot's path or its part is said to be *single* if the interiors of the involved arcs γ_i are pairwise disjoint and in the case of only one arc, do not cover ∂D.

Let P and Q be the numbers of the connected components of $S_\varkappa := \{s : \varkappa(s) < 0\}$ and $S_\zeta := \{s : \zeta_{\partial D}(s) = 0\}$, respectively. They are finite due to Corollary A.2.

Lemma A.2. *Any single path accommodates no more than $(P+1)(Q+2)$ SMTs.*

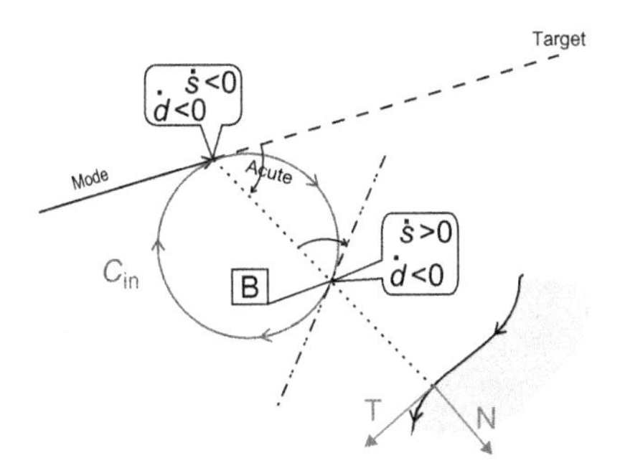

Figure A.1 Behavior during IT.

Proof. As was shown in the proof of (v) in Proposition 4.2, the number of SMTs within a common mode $\mathfrak{B}$ does not exceed $P+1$. SMT between the ith and $(i+1)$th occurrences of $\mathfrak{B}$ starts at a position $s_\dagger \in \gamma_i = [s_i^-, s_i^+]$, where $\zeta_{\partial D}(s_\dagger) = -d < 0$, and ends at the position s_{i+1}^- where $\zeta_{\partial D}(s_{i+1}^-) \geq 0$. Hence, any arc γ_i, except for the first and last ones, intersects adjacent connected components $\mathrm{Cc}_i^=$ and $\mathrm{Cc}_i^<$ of S_ζ and $\{s : \zeta_{\partial D}(s) < 0\}$, respectively, such that the left end-point of $\mathrm{Cc}_i^=$ is the right end-point of $\mathrm{Cc}_i^<$. Hence, $\mathrm{Cc}_i^= \neq \mathrm{Cc}_{i'}^= \; \forall i \neq i'$, and so the total number of the arcs γ_i does not exceed $Q+2$, which competes the proof. □

Proof of Remark A.1. Suppose to the contrary that the times t_i of switches between the modes accumulate, i.e., $t_i < t_{i+1} \to t_* < \infty$ as $i \to \infty$. At $t = t_i$, a SMT is terminated, and so $d(t_i) = d_\updownarrow, \dot{d}(t_i) \leq 0$, and $\beta(t_i) = 0$. During the subsequent AM, $d \leq d_\updownarrow$. At such distances, (4.9) implies that $|\ddot{d}| \leq M_d$ and $|\ddot{s}| \leq M_s$, where $M_d, M_s > 0$ do not depend on the system state. Since IT ends with $\dot{d} = 0$, this AM lasts no less than $M_d^{-1}|\dot{d}(t_i)|$ time units. Hence, $\dot{d}(t_i) \to 0$ as $i \to \infty$. This and (4.9) imply that $\dot{s}(t_i) - v\,\mathbf{sgn}\,\dot{s}(t_i) \to 0$ as $i \to \infty$. So far as IT lasts no less than $M_s^{-1}|\dot{s}(t_i)|$ time units if $\dot{s}$ is reversed during IT, the sign of $\dot{s}(t)$ is the same for $t_i < t < t_*$ and large enough i. Therefore, the related part of the path is single. By Lemma A.2, this part can accommodate only a finite number of SMTs, in violation of the initial hypothesis. This contradiction completes the proof. □

7.A.2 Auxiliary deterministic algorithm

In order to prove Proposition 4.1, it is convenient to start with a study of an auxiliary navigation strategy. This strategy results from the randomized navigation law by replacement of its random machinery of choosing the turn direction σ at switches $\mathfrak{A} \mapsto \mathfrak{B}$ by a deterministic rule. The first step to the proof of Proposition 4.1 will be to show that the so altered strategy achieves the navigation objective by making no more than N switches, where N does not depend on the initial state of the robot. However, the altered strategy in fact cannot be implemented since it uses unavailable data. The proof will be completed by showing that with probability 1, the initial randomized navigation law sooner or later gives rise to N successive switches identical to those generated by the altered strategy.

We recall that the symbol $[\boldsymbol{r}_1, \boldsymbol{r}_2]$ stands for the straight line segment directed from $\boldsymbol{r}_1$ to $\boldsymbol{r}_2$. In what follows, $\gamma_1 \star \gamma_2$ denotes the concatenation of directed curves γ_1, γ_2 such that γ_1 ends at the origin of γ_2.

Let an occurrence $\mathfrak{A}^\dagger$ of mode $\mathfrak{A}$ hold between two modes $\mathfrak{B}$ and let it start at $\boldsymbol{r}_\diamondsuit = \boldsymbol{r}(t_\diamondsuit)$ and end at $\boldsymbol{r}_* = \boldsymbol{r}(t_*)$. Due to (3.4),

$$\mathbf{dist}\,[\boldsymbol{r}_*; D] = \mathbf{dist}\left[\boldsymbol{r}_\diamondsuit; D\right] = d_\updownarrow$$

are attained at unique boundary points $s_\diamondsuit$ and s_*, respectively. They divide C into two arcs. Being concatenated with $\eta := [s_*, \boldsymbol{r}_*] \star [\boldsymbol{r}_*, \boldsymbol{r}_\diamondsuit] \star [\boldsymbol{r}_\diamondsuit, s_\diamondsuit]$, each of them gives rise to a Jordan curve encircling a bounded domain, one of which is the other united with D. The smaller domain is denoted $\mathfrak{C}_{\mathfrak{A}^\dagger}$; it is bounded by η and one of the above arcs $\gamma_{\mathfrak{A}^\dagger}$. Let $\sigma_{\mathfrak{A}^\dagger} = \pm$ be the direction (on ∂D) of the walk from $s_\diamondsuit$ to s_* along $\gamma_{\mathfrak{A}^\dagger}$.

We introduce the navigation law $\mathscr{A}$ that is the replica of (2.3) except for the rule to update σ when $\mathfrak{A} \mapsto \mathfrak{B}$. Now for the first such switch, σ is set to an arbitrarily pre-specified value. After any subsequent occurrence $\mathfrak{A}^{\dagger}$ of this mode,

$$\sigma := \begin{cases} \sigma_{\mathfrak{A}^{\dagger}} & \text{if } \mathfrak{C}_{\mathfrak{A}^{\dagger}} \text{ does not contain the target,} \\ -\sigma_{\mathfrak{A}^{\dagger}} & \text{if } \mathfrak{C}_{\mathfrak{A}^{\dagger}} \text{ contains the target.} \end{cases} \tag{A.3}$$

We underscore once more that this rule cannot be implemented in practice since the robot typically has no idea about whether the cave contains the target or not. The purpose of this rule is to serve subsequent theoretical analysis of the really implementable strategy: the randomized navigation law.

The main property of the auxiliary algorithm is given by the following.

Proposition A.1. *Under the law $\mathscr{A}$, the target is reached for a finite time, with making no more than N switches $\mathfrak{A} \mapsto \mathfrak{B}$, with N independent of the robot's initial state.*

The next two subsections are devoted to the proof of Proposition A.1. In doing so, the idea to retrace the arguments justifying global convergence of the algorithms like the Pledge one [395] that deal with unconstrained motion of an abstract point is troubled by two problems. First, this idea assumes that analysis can be boiled down to the study of a point moving according to self-contained rules coherent in nature with the above algorithms, i.e., those like "move along the boundary," "when hitting the boundary, turn left," etc. However, this is hardly possible, at least in full, since the robot behavior essentially depends on its distance from the boundary. For example, depending on this distance at the end of mode $\mathfrak{B}$, the robot afterward may or may not collide with a forward-horizon cusp of the obstacle. Secondly, the Pledge algorithm and the likes are maze-escaping strategies; they do not find the target inside a labyrinth when started outside of it. Novel arguments and techniques are required to justify the success of the proposed algorithm in this situation.

In what follows, we only partly reduce analysis of the robot motion to that of a kinematically controlled abstract point. This reduction concerns only special parts of the robot path and is not extended to the entire trajectory. The obstacle to be avoided by the point is introduced a posteriori with regard to the distance of the real path from the real obstacle. To justify the convergence of the abstract point to the target, we develop a novel technique based on an induction argument.

We start with study of a kinematically controlled point.

7.A.3 Symbolic path and its properties

In this subsection, "ray" means "ray emitted from the target," and we consider a domain $\mathscr{D}$ satisfying the following.

Assumption A.1. *The boundary $C := \partial \mathscr{D}$ consists of finitely many (maybe, zero) straight line segments and the remainder on which the curvature vanishes no more than finitely many times. The domain $\mathscr{D}$ does not contain the target.*

We also consider a point $\boldsymbol{r}$ moving in the plane according to the following rules:

(r.1) The point moves outside the interior of $\mathscr{D}$;
(r.2) Whenever $\boldsymbol{r} \notin \mathscr{D}$, it moves to $\mathscr{T}$ in a straight line;

(r.3) Whenever $\boldsymbol{r}$ hits $\partial \mathfrak{D}$, it proceeds with monotonic motion along the boundary, counting the angle β;
(r.4) This motion lasts until $\beta = 0$ and new SMT is possible, then SMT is indeed commenced;
(r.5) The point halts as soon as it arrives at the target.

The possibility from (r.4) means that $\mathfrak{D}$ does not obstruct the initial part of SMT. When passing the corner points of $\partial \mathfrak{D}$, the count of β obeys (4.5) and the conventional rules adopted for turning angles of tangential vector fields [396], and is assumed to instantaneously, continuously, and monotonically run between the one-sided limit values. The possibility from (r.4) may appear within this interval.

To specify the turn direction in (r.3), we need some constructions. Let the points $s_\pm \in C$ lie on a common ray and $(s_-, s_+) \cap C = \emptyset$. One of them, say s_-, is closer to the target than the other. They divide C into two arcs. Being concatenated with (s_-, s_+), each arc gives rise to a Jordan curve encircling a bounded domain. One of these domains is the other united with D. The smaller domain $\mathfrak{C}(s_-, s_+)$ is called the *cave* with the *corners* s_-, s_+. It is bounded by (s_-, s_+) and one of the above arcs $\gamma_{\mathfrak{C}}$.

To complete the rule (r.3), we note that any SMT except for the first one starts and ends at some points $s_\diamondsuit, s_* \in C$, which cut out a cave $\mathfrak{C}[s_\diamondsuit, s_*]$.

(r.3a) After the first SMT, the turn is in an arbitrarily pre-specified direction;
(r.3b) After SMT that is not the first, the point turns:

- outside $\mathfrak{C}[s_\diamondsuit, s_*]$ if the cave does not contain the target;
- inside the cave $\mathfrak{C}[s_\diamondsuit, s_*]$ if the cave contains the target.

Definition A.3. The path traced by the point obeying the rules (r.1)–(r.5), (r.3a), (r.3b) is called the *symbolic path (SP)*.

Proposition A.2. *SP arrives at $\mathcal{T}$ from any initial position. The number of performed SMTs is upper limited by a constant N independent of the initial position.*

The remainder of the subsection is devoted to the proof of this proposition. The notations $s, \mathbf{T}, \mathbf{N}, \mathfrak{r}, \lambda, \zeta, \varkappa, \psi, \varphi$ are attributed to $C = \partial \mathfrak{D}$. At the corner points of C, these variables except for s have one-sided limits and are assumed to instantaneously, continuously, and monotonically run between the one-sided limit values. An arc of C is said to be *regular* if ζ (non-strictly) does not change its sign on this arc, depending on which arc is said to be *positive/negative* (or $\pm$arc). The regular arc is *maximal* if it cannot be extended without violation of the regularity. A connected part of C and its points are said to be *singular* if ζ strictly changes the sign when passing it and, if this part contains more than one point, is identically zero on it; see Fig. A.2. The singular arc is a segment of a straight line since $\varkappa \equiv 0$ on it due to (4.3). The ends of any maximal regular arc are singular. Due to Assumption A.1 and (4.3), the boundary C has only finitely many singular parts. A boundary point $s \in C$ is said to *lie above $\mathfrak{D}$* if there exists $\delta > 0$ such that $((1-\delta)s, s) \subset \mathfrak{D}$ and $(s, (1+\delta)s) \cap \mathfrak{D} = \emptyset$. If conversely $((1-\delta)s, s) \cap \mathfrak{D} = \emptyset$ and $(s, (1+\delta)s) \subset \mathfrak{D}$, the point is said to *lie below $\mathfrak{D}$*.

Formulas (4.3) and (4.6) imply the following.

Observation A.2. *As s moves in direction $\sigma = \pm$ over a η-arc ($\eta = \pm$) of C, we have $\sigma\eta\dot{\varphi} \geq 0$. Any point of $\pm$arc that is not singular lies above/below $\mathfrak{D}$.*

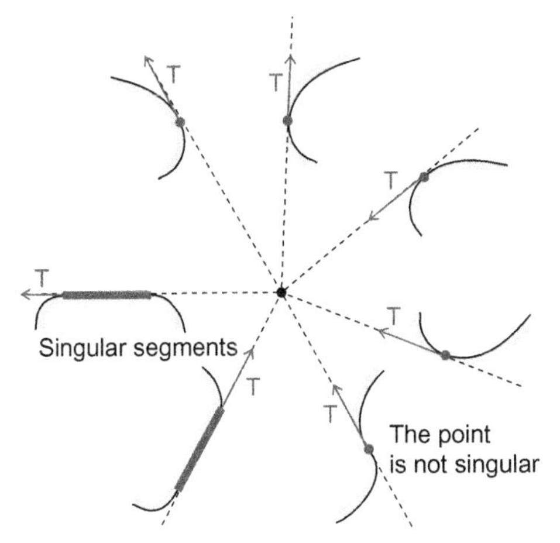

Figure A.2 Singular points.

Lemma A.3. *As s continuously moves along a regular arc, β evolves within an interval of the form $\Delta := [\pi k, \pi(k+1)]$, where k is an integer. When s reaches a singular point, β arrives at the end of Δ associated with the even or odd integer, depending on whether s moves toward or outwards the target at this moment, respectively.*

Proof. Since ζ does not change its sign, the vector $\mathfrak{r}$ does not trespass the λ-axis, whereas β is the polar angle of this vector. This gives rise to the first claim of the lemma. The second one is immediate from the first claim. □

Lemma A.4. *Whenever SP progresses along C in direction $\sigma = \pm$, we have $\sigma\beta \geq 0$.*

Proof. This is clearly true just after any SMT. During the subsequent motion along C, the inequality can be violated only at a position s where $\beta = 0$ and either s is a corner singular point or $\varkappa(s+\sigma^{\approx}0) > 0$ since $\varkappa(s+\sigma^{\approx}0) \leq 0 \Rightarrow \sigma\beta(s+\sigma^{\approx}0) \geq 0$ by the first relation from (4.4). However, at such position, motion along C is ended. □

The cave $\mathfrak{C}(s_-, s_+)$ is said to be *positive/negative* (or $\pm$cave) if the trip from s_- to s_+ over $\gamma_{\mathfrak{C}}$ is in the respective direction of C. By Observation A.2, s moves from a +arc to a −arc in this trip and so passes a singular part of C. The total number of such parts inside $\gamma_{\mathfrak{C}}$ is called the *degree* of the cave.[2]

Lemma A.5. *For any cave of degree $M = 1$, the arc $\gamma := \gamma_{\mathfrak{C}}$ consists of the positive $\gamma|_{s_- \to s_-^*}$ and negative $\gamma|_{s_+^* \to s_+}$ sub-arcs and a singular part $[s_-^*, s_+^*]$. For $s \in [s_-^*, s_+^*]$, the tangential vector $\mathbf{T}(s)$ (that is co-linear with $[\mathcal{T}, s]$ if s is the corner point) is directed outwards $\mathcal{T}$ if the cave is positive and does not contain $\mathcal{T}$ or negative and contains $\mathcal{T}$. Otherwise, this vector is directed toward $\mathcal{T}$.*

Proof. The first claim is evident. Let the cave be positive and $\mathcal{T} \notin \mathfrak{C}(s_-, s_+)$. Suppose that $\mathbf{T}(s)$ is directed toward $\mathcal{T}$. Then the same is true for $s := s_+^*$. Hence, $\zeta(s_+^* + 0) \leq 0$ and

[2] Possible singular parts at the ends of $\gamma_{\mathfrak{C}}$ are not counted.

$$\zeta(s_+^* + 0) = 0 \Rightarrow \lambda(s_+^* + 0) > 0 \Rightarrow \varkappa(s_+^* +^{\approx} 0) > 0$$

since otherwise, $\zeta(s_+^* +^{\approx} 0) \geq 0$ by (4.3), in violation of the definition of the singular part. In any case, $((1-\delta)s_+^*, s_+^*) \cap \mathcal{D} = \emptyset$ for some $\delta > 0$. Since $\mathcal{T} \notin \mathfrak{C}(s_-, s_+)$, the segment $[0, s_+^*)$ intersects $\gamma_{\mathfrak{C}}$, cutting out a smaller cave $\mathfrak{C}_{\text{sm}}$ inside $\mathfrak{C}(s_-, s_+)$. The singular part inside $\mathfrak{C}_{\text{sm}}$ is the second such part in the original cave, in violation of $M = 1$. This contradiction shows that $\mathbf{T}(s)$ is directed outwards $\mathcal{T}$.

Now suppose that $\mathcal{T} \in \mathfrak{C}(s_-, s_+)$ and $\mathbf{T}(s)$ is directed outwards $\mathcal{T}$. Let a point s_* move in the positive direction along $\gamma|_{s_+^* \to s_+}$. The ray containing s_* monotonically rotates by Observation A.2 and contains a continuously moving point $s_-^{\text{mov}} \in \gamma|_{s_-^* \to s_-}$. As s_* runs from s_+^* to s_+, the segment (s_-^{mov}, s_*) sweeps the entire cave $\mathfrak{C}[s_-, s_+]$, and so this cave does not contain $\mathcal{T}$, in violation of the assumption. This contradiction proves that $\mathbf{T}(s)$ is directed toward $\mathcal{T}$.

The second claim for negative caves and the third claim are proved likewise. □

Lemma A.6. *If SP enters a cave without the target, it leaves the cave through the other corner with $\beta \neq 0$. In this maneuver, the direction of motion along C is not changed, no point of C is passed twice, and the number of SMTs does not exceed the cave degree.*

Proof. Let SP enter the cave in the positive direction; the case of the negative direction is considered likewise. The proof will be by induction on the cave degree M.

Let $M = 1$.

- **(i)** Suppose first that the cave is positive and so s enters it through s_- moving over a +arc. By Lemma A.5, the point s moves outwards the target whenever $s \in [s_-^*, s_+^*]$, and so $\beta \geq \pi$ by Lemmas A.3 and A.4. As s moves over the subsequent −arc, ζ becomes negative and so the inequality is kept true by Lemma A.3. Thus, s leaves the cave through s_+ with $\beta \geq \pi > 0$, having made no SMT.
- **(ii)** Let the cave be negative. Then s enters it through s_+ moving over the negative arc. By Lemma A.5, the point s moves toward the target whenever $s \in [s_-^*, s_+^*]$. Since $\zeta(s_+ + 0) \leq 0$, Lemma A.4 yields that $\beta(s_+ + 0) \geq \pi$. By Lemma A.3, $\beta \geq \pi$ until s_+^* and so $\beta \geq 2\pi$ at $s \in [s_-^*, s_+^*]$ by Lemma A.5. When s passes the entire $[s_-^*, s_+^*]$, the sign of ζ reverses from − to + and so $\beta > 2\pi$ just after the passage of s_-^*. It remains to note that $\beta \geq 2\pi > 0$ while s moves over the +arc from s_-^* to s_- by Lemma A.3.

Now we suppose that the claim of the lemma is true for any cave with degree $\leq M$, and consider a cave of degree $M+1$. Let this cave be positive. Then s enters it through the lower corner s_- along a positive arc. We also consider the accompanying motion of the ray containing s. This ray contains a continuously moving point $s_+^{\circledast} \in C$ that starts at s_+. This motion is considered until a singular part of C appears on the ray segment $[s, s_+^{\circledast}]$ for the first time. Three cases are possible at this position.

- **(a)** The singular part $[s_-^*, s_+^*] \subset (s, s_+^{\circledast})$; see Fig. A.3(a), where $s_-^* = s_+^* =: s_*$. By successively applying the induction hypothesis to $\mathfrak{C}(s, s_+^*)$ and $\mathfrak{C}(s_+^*, s_+^{\circledast})$, we see that SP arrives at $s_+^{\circledast}$ in the positive direction and with $\beta > 0$. While s moves from s_+^* to s_+ over the −arc, the vector $\mathfrak{r}(s)$ is below the λ-axis and so $\beta \geq \pi > 0$ by Lemma A.3.
- **(b)** The singular point $s_+^{\circledast} = s_-^*$; see Fig. A.3(b), where $s_-^* = s_+^* =: s_*$. By successively applying the induction hypothesis to $\mathfrak{C}(s, s_\#)$ and $\mathfrak{C}(s_\#, s_+^{\circledast})$, we see that SP arrives at $s_+^{\circledast}$ in

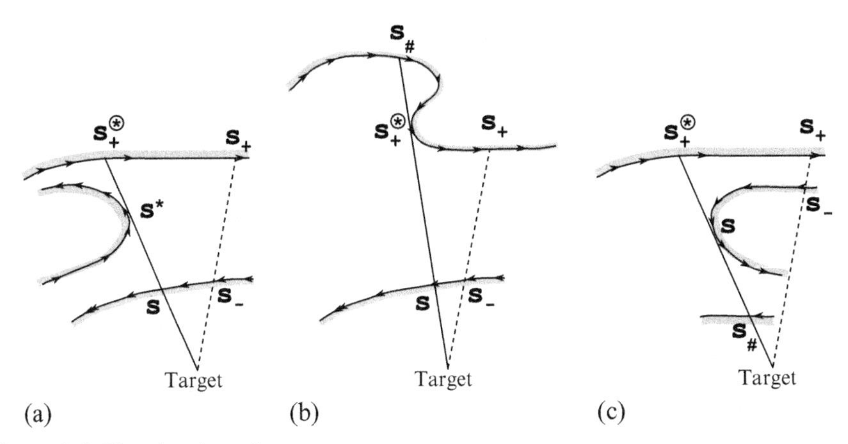

Figure A.3 First singular point.

the positive direction and with $\beta > 0$. So $\beta(s_+^{\circledast}) \geq 2\pi$ and SP proceeds along the $-$arc to s_+ with $\beta \geq \pi > 0$ by Lemma A.3, which completes the proof.

(c) The singular point s; see Fig. A.3(c). If $\beta > 0$ at this point, SP enters the cave $\mathfrak{C}[s, s_+^{\circledast}]$ of degree $\leq M$ and by the induction hypothesis, arrives at $s_+^{\circledast}$ moving in the positive direction and with $\beta > 0$. If conversely $\beta = 0$, SP undergoes SMT, which cannot be terminated at the target since it does not belong to the cave at hand. Therefore, it is terminated at some point $s_\# \in \gamma_{\mathfrak{C}}$. Since $\mathcal{T}$ does not lie in the sub-cave $\mathfrak{C}(s, s_\#)$ of the original cave, the robot turns right at $s_\#$ and thus proceeds along C in the positive direction. By applying the induction hypothesis to $\mathfrak{C}(s_\#, s_+^{\circledast})$, we see that SP arrives at $s_+^{\circledast}$ moving in the positive direction and with $\beta > 0$ in any case. The proof is completed as in the cases (a) and (b).

The case where the cave is negative is considered likewise. □

Lemma A.7. *Suppose that after SMT starting and ending at the points $s_\diamondsuit$ and s_*, respectively, the direction of the motion along C is reversed. Then the cave $\mathfrak{C}[s_\diamondsuit, s_*]$ does not contain $\mathcal{T}$ but contains the entire path traced before SMT at hand.*

Proof. Let the motion direction at $s = s_\diamondsuit$ be $+$; the case of $-$ is considered likewise. Since on arrival at s_*, the left turn is made, $\mathfrak{C}[s_\diamondsuit, s_*]$ does not contain $\mathcal{T}$ by (r.3b). Suppose that the path traced before SMT at hand is not contained by this cave, i.e., the point entered this cave before. Since this cannot be done during another SMT, the point enters the cave through either $s_\diamondsuit$ or s_*. In the first case, $s_\diamondsuit$ is passed twice in the opposite directions, in violation of Lemma A.6. In the second case, $s_\diamondsuit$ is passed with $\beta > 0$ by the same lemma and so SMT cannot be commenced. The contradiction obtained proves the initial part of SP is inside the cave. □

Lemma A.8. *If SP progresses along C in a cave not containing the target, it leaves this cave through one of its corners. During this maneuver, SP passes no point of C twice and makes no more SMTs than the degree of the cave.*

Proof. For the definiteness, let the cave be positive; the case of a negative cave is considered likewise. The proof will be by induction on the degree M of the cave.

Let $M = 1$. We employ the notations from Lemma A.5.

(α) The motion is started on $\gamma|_{s_+^* \to s_-}$ in direction $-$. The claim is evident.

(β) The motion is started on $\gamma|_{s_+ \to s_+^*}$ in direction $-$. Then the point clearly arrives at s_+^*, moving in the negative direction. Thus, the situation is reduced to (α).

(γ) The motion is started on $\gamma|_{s_-^* \to s_+}$ in the positive direction. The claim of the lemma is justified by the concluding arguments from (i) in the proof of Lemma A.6.

(δ) The motion is started on $\gamma|_{s_- \to s_-^*}$ in direction $+$. Then the point clearly arrives at s_-^*, moving in the positive direction. Thus, the situation is reduced to (γ).

Now we suppose the claim of the lemma is true for any cave with degree $\leq M$, and consider a cave of degree $M+1$. Let this cave be positive for the definiteness; the case of the negative cave is considered likewise. We also consider an auxiliary motion of the point over C from s_- into the cave and the accompanying motion of the ray containing s until one of the situations from Fig. A.3 occurs.

Case (a) from Fig. A.3.

(a.1) If the motion is started on $\gamma|_{s_+^{\circledast} \to s_+}$ in direction $+$ or on $\gamma|_{s \to s_-}$ in direction $-$, the claim of the lemma is justified by the concluding arguments from (i) in the proof of Lemma A.6.

(a.2) If the motion is started on $\gamma|_{s_-^* \to s_+^{\circledast}}$, the induction hypothesis applied to the cave $\mathfrak{C}[s_-^*, s_+^{\circledast}]$ of degree $\leq M$ ensures the point arrives at either $s_+^{\circledast}$ or s_-^*. In the first case, it arrives in direction $+$, thus reducing the situation to (a.1). In the second case, it arrives in direction $-$. If $\beta \neq 0$ at this position, the point enters the cave $\mathfrak{C}[s_-^*, s]$ in direction $-$ and afterward leaves it through s in the same direction by Lemma A.6. If $\beta = 0$, SMT is commenced, which ends at the position s with the left turn since $\mathfrak{C}[s_-^*, s]$ does not contain $\mathcal{T}$. Hence, in any case, the motion proceeds in direction $-$ from the position s, which reduces the situation to (a.1).

(a.3) The case where the motion is started on $\gamma|_{s \to s_-^*}$, is considered likewise.

(a.4) The cases where the motion starts on $\gamma|_{s_+^{\circledast} \to s_+}$ in direction $-$ or on $\gamma|_{s \to s_-}$ in direction $+$ are trivially reduced to (a.2) and (a.3), respectively.

Case (b) from Fig. A.3.

(b.1) The cases where the motion starts on $\gamma|_{s_+^{\circledast} \to s_+}$ in direction $+$ or on $\gamma|_{s \to s_-}$ in direction $-$ are considered like (a.1).

(b.2) If the start is on $\gamma|_{s \to s_\#}$, the induction hypothesis applied to $\mathfrak{C}[s, s_\#]$ ensures the point arrives at either s or $s_\#$. In the first case, it arrives in direction $-$, thus reducing the situation to (b.1). In the second case, it arrives in direction $+$ and then enters the cave $\mathfrak{C}[s_\#, s_+^{\circledast}]$. By Lemma A.6, the point leaves this cave through $s_+^{\circledast}$ in direction $+$ and with $\beta > 0$, thus reducing the situation to (b.1).

(b.3) If the motion commences on $\gamma|_{s_\# \to s_+^{\circledast}}$, the induction hypothesis applied to the cave $\mathfrak{C}[s_\#, s_+^{\circledast}]$ of degree $\leq M$ ensures the point arrives at either $s_\#$ or $s_+^{\circledast}$. In the first case, the arrival is in direction $-$, after which the situation is reduced to (b.2). In the second case, the arrival is in direction $+$. If $\beta \neq 0$ at this moment, the motion proceeds along $\gamma|_{s_+^{\circledast} \to s_+}$ in direction $+$, and the situation is reduced to (b.1). If $\beta = 0$, SMT is commenced, which ends at the position s with the left turn since the cave $\mathfrak{C}[s_+^{\circledast}, s]$ does not contain the target. Hence, the motion proceeds along $\gamma|_{s \to s_-}$ in direction $-$, and the situation is still reduced to (b.1).

(b.4) The cases where the motion starts on $\gamma|_{s_+^{\circledast}\to s_+}$ in direction $-$ or on $\gamma|_{s\to s_-}$ in direction $+$ are trivially reduced to (b.3) and (b.2), respectively.

Case (c) from Fig. A.3.

(c.1) The cases where the motion starts on $\gamma|_{s_+^{\circledast}\to s_+}$ in direction $+$ or on $\gamma|_{s\to s_-}$ in direction $-$ are considered like (a.1).

(c.2) If the start is on $\gamma|_{s_\#\to s_+^{\circledast}}$, the induction hypothesis applied to $\mathfrak{C}[s_\#, s_+^{\circledast}]$ yields that the point arrives at either $s_+^{\circledast}$ or $s_\#$. In the first case, the arrival direction is $+$ and the situation is reduced to (b.1). In the second case, the point arrives in direction $-$ and then enters $\mathfrak{C}[s_\#, s]$. By Lemma A.6, the point leaves this cave through s in direction $-$ and with $\beta > 0$. Thus, we arrive at (b.1) once more.

(c.3) If the motion commences on $\gamma|_{s_\#\to s}$, the induction hypothesis applied to the cave $\mathfrak{C}[s_\#, s]$ of degree $\leq M$ ensures the point arrives at either $s_\#$ or s. In the first case, the arrival is in direction $+$, after which the situation is reduced to (b.2). In the second case, the arrival is in direction $-$, after which the situation reduces to (b.1).

(c.4) The cases where the motion starts on $\gamma|_{s_+^{\circledast}\to s_+}$ in direction $-$ or on $\gamma|_{s\to s_-}$ in direction $+$ are trivially reduced to (c.2) and (c.3), respectively.

□

Lemma A.9. *Any part of SP where it progresses over the boundary ∂D ends with SMT.*

Proof. The proof is by retracing the proof of (v) in Proposition 4.2. □

Let K be the number of the singular parts of the boundary $\partial\mathfrak{D}$.

Lemma A.10. *If every cave examined in (r.3b) does not contain the target, SP consists of the initial $\mathcal{P}^-$ and terminal $\mathcal{P}^+$ sub-paths (some of which may contain only one point) such that each accommodates no more than K SMTs, no point of C is passed twice within $\mathcal{P}^-$, and the direction of motion along C is not altered within $\mathcal{P}^+$.*

Proof. Suppose first that the initial position lies in some cave. Among such caves, there is one enveloping the others. By Lemma A.8, SP leaves this cave and the related sub-path satisfies the properties stated in Lemma A.10. If the initial position lies outside any cave, this sub-path is taken to consist of only this position. By Lemma A.7, the direction of the motion along C is not changed on the remaining sub-path $\mathcal{P}_+$ and $\mathcal{P}_+$ does not go inside the above maximal cave.

Suppose that within $\mathcal{P}_+$, SP accommodates more than K SMTs. Any of them starts at some singular part with $\beta = 0$. Hence, SP passes some singular point with $\beta = 0$ at least twice and thus becomes cyclic. Now we consider the related minimal cyclic part CP of SP that starts and ends with commencing a SMT at a common point. Due to the constant direction, the closed curve CP is simple. It follows that $\sphericalangle$**TANG** [CP] $= \pm 2\pi$, whereas $\sphericalangle_{\mathcal{T}}\mathrm{CP} = 0$ since $W = 0$ for all bypassed caves and $\mathcal{T} \notin D$. Hence, $\sphericalangle_0 \mathfrak{r} = \mp 2\pi$ by (4.5), whereas CP starts and ends with $\beta = 0$ and so $\sphericalangle_0 \mathfrak{r} = 0$. This contradiction completes the proof. □

Lemmas A.9 and A.10 give rise to the following.

Corollary A.3. *If every cave examined in (r.3b) does not contain $\mathcal{T}$, SP arrives at $\mathcal{T}$ by making no more than 2K SMTs.*

Lemma A.11. *If SP enters a cave containing $\mathcal{T}$ over a positive arc with $|\beta| \le \pi$, it arrives at $\mathcal{T}$ not leaving the cave. During this maneuver, no point of C is passed twice and the number of SMTs does not exceed the degree of the cave.*

Proof. Let the cave be entered in direction $+$; the case of $-$ is considered likewise. The proof will be by induction on the degree M of the cave $\mathfrak{C}[s_-, s_+]$. Since s enters the cave over a positive arc, the entrance is through s_-.

Let $M = 1$. By Lemma A.5, s moves toward $\mathcal{T}$ when reaching the singular part of the cave $[s_-^*, s_+^*]$. At this position, $\beta = 0$ by Lemma A.3 and $\mathcal{D}$ does not obstruct the initial part of SMT, as was shown in the proof of Lemma A.5. Therefore, SMT is commenced. If it is not terminated at $\mathcal{T}$, the segment $[0, s_-^*)$ intersects $\gamma_{\mathfrak{C}}$, cutting out a smaller cave within the original one. The singular part inside this new cave is the second such part within the original cave, in violation of $M = 1$. Hence, $\mathcal{T}$ is reached and only one switch $\mathfrak{B} \mapsto \mathfrak{A}$ is made.

Now let the conclusion of the lemma be true for any cave with degree $\le M$, and let us inspect a cave of degree $M + 1$. As in the proof of Lemma A.6, we look at the motion of the ray containing s until a singular point appears on the segment $[s, s_+^*]$ for the first time, and handle the three possible cases from Fig. A.3 separately.

(a) The singular point $s_* \in (s, s_+^*)$; see Fig. A.3(a). The target is contained by the cave $\mathfrak{C}[s, s_*]$ of degree $\le M$, which is entered in the positive direction and by Lemma A.3, with $0 \le \beta \le \pi$. The induction hypothesis competes the proof.

(b) The singular point $s_* = s_+^*$; see Fig. A.3(b). The target is evidently contained by the cave $\mathfrak{C}[s, s_\#]$ of degree $\le M$. The proof is completed as in the previous case.

(c) The singular point $s_* = s$; see Fig. A.3(c). If at s_*, the point moves outwards $\mathcal{T}$, the arguments from the second paragraph in the proof of Lemma A.5 show that the cave does not contain $\mathcal{T}$, in violation of the assumption of the lemma. Hence, at s_*, the point moves toward $\mathcal{T}$ and so $\beta = 0$ by Lemma A.3 and $\mathcal{D}$ does not obstruct the initial part of SMT, as was shown in the proof of Lemma A.5. Thus, SMT is commenced at s_*. If it is terminated at $\mathcal{T}$, the proof is completed. Otherwise, it arrives at $s_\# \in \gamma_{\mathfrak{C}}$, as is shown in Fig. A.3(c). Evidently, the cave $\mathfrak{C}[s_\#, s]$ does not contain the target. Therefore, on reaching $s_\#$, the point turns right and continues moving in direction $+$ over a new positive arc and with $\beta \in [0, \pi]$. Therefore, the proof is completed by applying the induction hypothesis to the cave $\mathfrak{C}[s_\#, s_+^*]$ of degree $\le M$.

□

Proof of Proposition A.2. is straightforward from Corollary A.3 and Lemma A.11. □

7.A.4 Proof of Proposition A.1

Let $\mathcal{P}$ stand for the directed path traced by the robot under the navigation law $\mathcal{A}$ from Section 7.A.2. We first show that after a slight modification, this path can be viewed as SP for some domain $\mathcal{D}$ provided that $\mathcal{P}$ is single; see Definition A.2. This permits us to employ the results of Section 7.A.3.

We use the notations s_i^-, s_i^+, γ_i introduced before Definition A.2, note that for $s \in \gamma_i$, the distance d from the robot to the set D is a function $d = d_i(s)$ of s, and put:

$$\mathcal{D} := \{\boldsymbol{r} : d := \mathbf{dist}\,[\boldsymbol{r}; D] < d_\star(D) \quad \text{and} \\ \text{either} \quad s := s(\boldsymbol{r}) \in \gamma_i \wedge d \leq d_i(s) \quad \text{or} \quad s \notin \cup_i \gamma_i \wedge d \leq d_\updownarrow \}. \quad \text{(A.4)}$$

If $\sigma \dot{s} < 0$ at the start of the ith mode $\mathfrak{B}$, the abscissa s_i^- is passed twice during IT by Lemma 4.7. For every such i, the real path between these two passages is replaced by the motion along the straight line segment, which gives rise to the modified path $\mathcal{P}_*$.

Observation A.3. *Let the original path be single. Then the modified path $\mathcal{P}_*$ is SP for $\mathcal{D}_*$.*

Indeed, this path can be viewed as a trace of a point obeying the rules (r.1)–(r.5). To ensure (r.3a), the direction should be pre-specified to match that of $\mathcal{P}_*$. The property (r3.b) is satisfied due to (A.3) and the second inequality from (4.1).

Lemma A.12. *For a single path, the set* (A.4) *satisfies Assumption A.1 and its boundary has no more than N_s singular parts, where N_s is completely determined by D and $\mathcal{T}$.*

Proof. The last claim in Assumption A.1 holds by (4.1) and (A.4). The boundary $\partial\mathcal{D}$ consists of parts traced during (a) SMTs, (b) SMECs, (c) arcs of circles traced during IT's, and (d) segments of normals to ∂D resulting from path modification.

Any part (a) clearly satisfies Assumption A.1 and is either singular or does not contain singular points; their number does not exceed $(P+1)(Q+1)$ by Lemma A.2.

Since parts (b) are separated by SMTs, their number does not exceed $(P+1)(Q+1)+1$. Any part (b) lies on a d-equidistant curve $C(d)$ with $d \leq d_\updownarrow$. Due to (4.2), $\zeta_{C(d)}(s) = \zeta_{\partial D}(s) + d$, Assumption A.1 holds since the boundary ∂D is piece-wise analytical, and the singular parts of $C(d)$ are the connected components of the set from Corollary A.2. Therefore, type (b) arcs of C accommodate no more than $F[(P+1)(Q+1)+1]$ singular parts.

It remains to note that parts (c) and (d) do not contain singular points since β monotonically evolves from 0 during ITs. □

Lemma A.13. *If the robot finds the target in $\mathfrak{C}_{\mathfrak{A}^\dagger}$ after some occurrence $\mathfrak{A}^\dagger$ of mode $\mathfrak{A}$, it arrives at the target by making after this no more than N_s switches $\mathfrak{A} \mapsto \mathfrak{B}$.*

Proof. Let us consider a part $\mathcal{P}$ of the path that starts in mode $\mathfrak{B}$ preceding $\mathfrak{A}^\dagger$. Suppose first that this part is not single and truncate it from the right, leaving its maximal single sub-part $\mathcal{P}^\dagger$. The terminal position of $\mathcal{P}^\dagger$ lies on a previously passed piece of $\mathcal{P}^\dagger$. Let $\mathcal{D}^\dagger$ and $\mathcal{P}_*^\dagger$ be the related domain (A.4) and modified path. Associated with $\mathfrak{C}_{\mathfrak{A}^\dagger}$ is a cave of $\mathcal{D}^\dagger$ into which $\mathcal{P}_*^\dagger$ turns with $|\beta| \leq \pi$. By Lemma A.11, $\mathcal{P}_*^\dagger$ cannot arrive at a previously passed point, in violation of the above property. This contradiction proves that the entire path $\mathcal{P}$ is single. Then Lemmas A.11 and A.12 guarantee that $\mathcal{P}_*$ arrives at $\mathcal{T}$ by making no more than N_s SMTs. It remains to note that $\mathcal{P}$ and $\mathcal{P}_*$ arrive at $\mathcal{T}$ only simultaneously, and each occurrence of $\mathfrak{A}$ gives rise to a SMT in $\mathcal{P}_*$. □

Lemma A.14. *After no more than N_s+1 switches $\mathfrak{A} \mapsto \mathfrak{B}$, the direction in which s moves along ∂D within modes $\mathfrak{B}$ is not altered.*

Proof. Consider an occurrence $\mathfrak{A}^\dagger$ of mode $\mathfrak{A}$ after which the direction is altered and the path $\mathcal{P}$ from the start of the entire motion until the end of $\mathfrak{A}^\dagger$. Suppose that $\mathcal{P}$ is not single and truncate it from the left, leaving the maximal single part $\mathcal{P}^\dagger$. The starting point of $\mathscr{P}^\dagger$ is passed once more within $\mathscr{P}^\dagger$, both times in mode $\mathfrak{B}$. Therefore, this double point is inherited by $\mathscr{P}_*^\dagger$, where $\mathscr{D}^\dagger$ and $\mathscr{P}_*^\dagger$ are the related domain (A.4) and modified path, respectively. Associated with $\mathfrak{C}_{\mathfrak{A}^\dagger}$ is a cave $\mathfrak{C}_{\mathscr{D}^\dagger}$ of $\mathscr{D}^\dagger$; these two sets contain the target only simultaneously due to (4.1). Hence, $\mathscr{P}$ and $\mathscr{P}_*^\dagger$ acquire a common turn direction at their ends. Therefore, SP $\mathscr{P}_*^\dagger$ has converse directions of motion along the boundary at the start and end of the last involved SMT, and by Lemmas A.7 and A.8, has no double points. This contradiction proves that the entire $\mathscr{P}^\dagger$ is single. Due to Lemma A.7, the modified path $\mathscr{P}_*^\dagger$ lies in $\mathfrak{C}_{\mathscr{D}^\dagger}$ and so involves no more than N_s SMTs thanks to Lemmas A.8 and A.12. It remains to note that each occurrence of $\mathfrak{A}$ gives rise to a SMT in $\mathscr{P}_*$. □

To prove Proposition A.1, it in fact remains to show that the robot cannot pass more than N_s modes $\mathfrak{A}$ in a row, constantly not finding the target in $\mathfrak{C}_\mathfrak{A}$ and not changing the direction of the motion along ∂D. The next lemma with corollaries serves this proof. We recall that the symbol $\angle(\mathbf{a}, \mathbf{b}) \in (-\pi, \pi]$ stands for the angle from the vector $\mathbf{a}$ to $\mathbf{b}$. Let the points $\boldsymbol{r}_i, i = 1, 2$ on $\mathscr{P}$ be at the distance $\mathbf{dist}\,[\boldsymbol{r}_i; D] \leq d_\updownarrow$ and such that when traveling between them, the path does not intersect itself and except for $\boldsymbol{r}_i$, has no points in common with the normals $[\boldsymbol{r}_i, s_i]$, where $s_i := s[\boldsymbol{r}_i]$. The points s_i split ∂D into two curves. Being concatenated with the above normals and $\mathscr{P}|_{\boldsymbol{r}_1 \to \boldsymbol{r}_2}$, they give rise to Jordan loops, with one of them enveloping the other. Let γ_{inner} be the curve giving rise to the inner loop **LOOP**, and $\sigma = \pm$ be the direction from s_1 to s_2 along γ_{inner}.

Lemma A.15. *If* **LOOP** *does not encircle the target, the following relation holds:*

$$\sphericalangle_0 \mathfrak{r}_{\mathscr{P}}|_{\boldsymbol{r}_1 \to \boldsymbol{r}_2} = \sphericalangle_0 \mathfrak{r}_{\partial D}|_{s_1 \xrightarrow{\sigma} s_2} + \sphericalangle_{\mathscr{T}}[\boldsymbol{r}_1, s_1] - \sphericalangle_{\mathscr{T}}[\boldsymbol{r}_2, s_2] + \angle\left[\sigma \mathbf{T}_{\partial D}(s_1), \mathbf{T}_{\mathscr{P}}(\boldsymbol{r}_1)\right] - \angle\left[\sigma \mathbf{T}_{\partial D}(s_2), \mathbf{T}_{\mathscr{P}}(\boldsymbol{r}_2)\right]. \quad \text{(A.5)}$$

Proof. Let $\sigma = +$; $\sigma = -$ is considered likewise. By applying Hopf's theorem [398, p. 57] to **LOOP**, we see that

$$\sphericalangle_{\mathscr{T}}[s_1, \boldsymbol{r}_1] + \sphericalangle_{\mathscr{T}} \mathscr{P}|_{\boldsymbol{r}_1 \to \boldsymbol{r}_2} + \sphericalangle_{\mathscr{T}}[\boldsymbol{r}_2, s_2] - \sphericalangle_{\mathscr{T}} \partial D_{s_1 \to s_2} = 0,$$

$$\sphericalangle\mathbf{TANG}\left[\mathscr{P}|_{\boldsymbol{r}_1 \to \boldsymbol{r}_2}\right] = \sphericalangle\mathbf{TANG}\left[\partial D_{s_1 \to s_2}\right] - \angle\left[\mathbf{T}_{\partial D}(s_1), \mathbf{T}_{\mathscr{P}}(\boldsymbol{r}_1)\right] + \angle\left[\mathbf{T}_{\partial D}(s_2), \mathbf{T}_{\mathscr{P}}(\boldsymbol{r}_2)\right].$$

The proof is completed by the second formula in (4.5). □

The next claim employs the notations introduced at the beginning of Section 7.A.2.

Corollary A.4. *Suppose that* $\mathscr{T} \notin \mathfrak{C}_{\mathfrak{A}^\dagger}$ *and the value of* σ *maintained during* $\mathfrak{A}^\dagger$ *is not altered when* $\mathfrak{A}^\dagger \mapsto \mathfrak{B}$. *Then* (A.5) *holds with* $\boldsymbol{r}_1 := \boldsymbol{r}_\diamond, \boldsymbol{r}_2 := \boldsymbol{r}_*$.

This is true since in this claim and Lemma A.15, σ is the same.

Corollary A.5. *Let* $\boldsymbol{r}_1$ *and* $\boldsymbol{r}_2$ *be successively passed within a common mode* $\mathfrak{B}$, *where* $\sigma(t) \equiv \sigma = \pm$. *If* $\boldsymbol{r}_2$ *is passed after IT,* (A.5) *holds, where* $\sphericalangle_0 \mathfrak{r}_{\partial D}|_{s_1 \xrightarrow{\sigma} s_2}$ *accounts for the entire motion of the projection* $s = s[\boldsymbol{r}], \boldsymbol{r} \in \mathscr{P}_{\boldsymbol{r}_1 \to \boldsymbol{r}_2}$, *including possible full runs over* ∂D.

If (a) s does not run the entire length of ∂D and (b) either $\boldsymbol{r}_1$ is passed after IT or $\mathbf{sgn}\, \dot{s} = \sigma$ at the start of the mode, the claim is evident. If (a) holds but (b) does not,

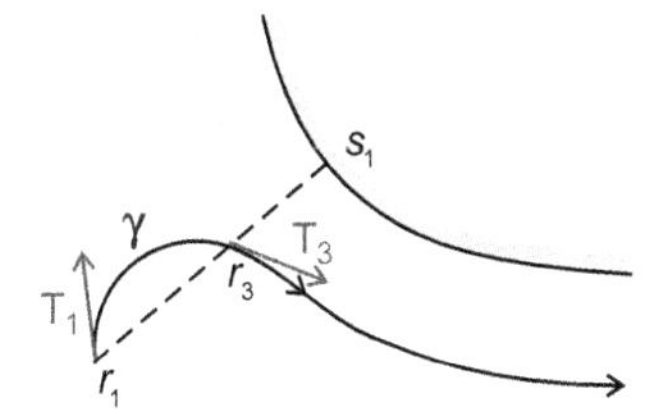

Figure A.4 Auxiliary loop.

the path may intersect $[s_1, \boldsymbol{r}_1]$ and so direct application of Lemma A.15 is impossible. Then we apply this lemma to $\boldsymbol{r}_1 := \boldsymbol{r}_3$, where $\boldsymbol{r}_3$ is the point where the robot intersects the normal for the second time during IT; see Fig. A.4. The proof is completed by noting that

$$\sphericalangle_{\mathcal{T}}[\boldsymbol{r}_1, \boldsymbol{r}_3] = \sphericalangle_{\mathcal{T}}\gamma, \quad \sphericalangle\mathbf{TANG}[\gamma] = \angle[\mathbf{T}_1, \mathbf{T}_3]$$

and so

$$\sphericalangle_0\mathfrak{r}_{\mathcal{P}}|_{\boldsymbol{r}_1\to\boldsymbol{r}_3} = \sphericalangle_0\mathfrak{r}_\gamma = \sphericalangle_{\mathcal{T}}[\boldsymbol{r}_1, \boldsymbol{r}_3] - \angle[\mathbf{T}_1, \mathbf{T}_3],$$

as well as that

$$\angle[\sigma\mathbf{T}_{\partial D}(s_1), \mathbf{T}_3] = \angle[\sigma\mathbf{T}_{\partial D}(s_1), \mathbf{T}_1] + \angle[\mathbf{T}_1, \mathbf{T}_3].$$

The claim is generalized on the case where (a) is not true by proper partition of the path, followed by summation of the formulas related to the resultant pieces.

Corollary A.6. *Let points $\boldsymbol{r}_1$ and $\boldsymbol{r}_2$ be successively passed in modes $\mathfrak{B}$ (maybe, different). Suppose that $\boldsymbol{r}_2$ is not attributed to IT and when traveling from $\boldsymbol{r}_1$ to $\boldsymbol{r}_2$, the robot constantly does not find the target in $\mathfrak{C}_{\mathfrak{A}}$ and does not change σ. Then* (A.5) *holds, where $\sphericalangle_0\mathfrak{r}_{\partial D}|_{s_1 \xrightarrow{\sigma} s_2}$ accounts for the entire motion of the projection $s = s[\boldsymbol{r}], \boldsymbol{r} \in \mathcal{P}_{\boldsymbol{r}_1\to\boldsymbol{r}_2}$, including possible full runs over ∂D.*

It is assumed that as the robot moves in mode $\mathfrak{A}$, the projection s continuously and monotonically goes over ∂D from $s_\diamondsuit$ to s_* in the direction σ.

Lemma A.16. *The robot cannot pass more than N_s modes $\mathfrak{A}$ in a row, constantly not finding the target in $\mathfrak{C}_{\mathfrak{A}}$ and not changing the direction of the motion along ∂D.*

Proof. Suppose the contrary and that $\sigma = +$; the case $\sigma = -$ is considered likewise. By Observation A.1, the ith mode $\mathfrak{A}^i$ in the row starts when s lies in an +exit arc A^i, whereas $\zeta \geq 0$ when it ends. Hence, $A^1, A^2, \ldots$ cannot repeat until s completes the full run over ∂D. However, they do repeat since the number of +arcs does not exceed F by Observation A.1, and $F \leq N_s$ by construction from the proof of Lemma A.12. Hence, the path $\mathcal{P}$ can be truncated so the first and last modes $\mathfrak{A}$ start at positions $\boldsymbol{r}_1$ and $\boldsymbol{r}_2$, respectively, lying on a common +exit arc A, whereas s encircles the entire boundary ∂D during the move over the truncated $\mathcal{P}$. By the definition of the +arc, $\mathfrak{r}_{\partial D}(s)$ evolves within the fourth quadrant as s runs from s_1 to s_2 within the +arc and therefore the absolute value of its turning angle does not exceed $\pi/2$. This and

(4.7) (where $d_* := 0$) imply that $\sphericalangle_0 \mathfrak{r}_{\partial D}|_{s_1 \to s_2} \leq -3/2\pi$. In (A.5), $|\sphericalangle_{\mathcal{T}}[\boldsymbol{r}_i, s_i]| < \pi/2$ and $\angle[\mathbf{T}_{\partial D}(s_i), \mathbf{T}_{\mathcal{P}}(\boldsymbol{r}_i)] = 0$ since the segments $[\boldsymbol{r}_i, s_i]$ and $[\boldsymbol{r}_i, \mathcal{T}]$ are perpendicular. Overall, (A.5) implies that

$$\sphericalangle_0 \mathfrak{r}_{\mathcal{P}}|_{r_1 \to r_2} < -\frac{\pi}{2}. \tag{A.6}$$

The path $\mathcal{P}|_{r_1 \to r_2}$ starts with $\beta = 0$ and whenever $\beta = 0$ is encountered, the angle β may stay constant during SMT but after this SMT β becomes positive by (A.1) (see Fig. A.2) since the robot turns right. The last claim holds thanks to (iii) of Proposition 4.2 if $\mathfrak{B}$ is not terminated during this SMT and (A.3) otherwise. Such behavior of β is inconsistent with (A.6). The contradiction obtained completes the proof. □

Proof of Proposition A.1. The proof is straightforward from (v) of Proposition 4.2 and Lemmas A.13, A.14, and A.16. □

7.A.5 Proof of Proposition 4.1

Let P_k be the probability that the robot does not arrive at $\mathcal{T}$ after making kN switches $\mathfrak{A} \to \mathfrak{B}$, where N is taken from Proposition A.1. Given a realization of σ's for the first kN switches, the probability of the $(k+1)$th event does not exceed the probability P_* that the next N realizations are not identical to those generated by the algorithm $\mathcal{A}$ for the related initial state. Here $P_* \leq \rho$, where $\rho := 1 - \min\{p, 1-p\}^N$ and p is the probability of picking $+1$ in (2.3). Therefore, the law of total probability yields that $P_{k+1} \leq \rho P_k \Rightarrow P_k \leq \rho^{k-1} P_1 \to 0$ as $k \to \infty$. It remains to note that the probability of not achieving $\mathcal{T}$ does not exceed P_k for any k.

7.A.6 Proof of Lemma 4.6

We recall that this lemma addresses the robot driven by the basic algorithm with the right turns. Therefore, in any SMEC the robot has the obstacle to the left. The proof basically follows that from the previous section and employs many facts established there. The difference is that now we do not need to introduce an auxiliary deterministic algorithm since the examined one is deterministic itself.

As before, we first consider another obstacle $\mathcal{D} \not\ni \mathcal{T}$ satisfying Assumption A.1. Let a point $\boldsymbol{r}$ move in the plane according to the following rules:

(r.1) If $\boldsymbol{r} \notin \mathcal{D}$, $\boldsymbol{r}$ moves to $\mathcal{T}$ in a straight line; $\boldsymbol{r}(0) \notin \mathcal{D}$;
(r.2) If $\boldsymbol{r}$ hits $C := \partial \mathcal{D}$, it turns right and then moves in the positive direction along the boundary, counting the angle β;
(r.3) This motion lasts until $\beta = 0$ and a new SMT is possible;
(r.4) The point halts as soon as it arrives at the target.

In this subsection, the path traced by $\boldsymbol{r}$ is called the *symbolic path (SP)*. Any SMT according to **(r.1)** except for the first one starts and ends at some points $s_\diamondsuit, s_* \in C$, which cut out a cave $\mathfrak{C}[s_\diamondsuit, s_*]$.

We start with the following specification of Observation A.2.

Observation A.4. *As* $\boldsymbol{r}$ *moves over a* $\pm$*-arc of* C*, we have* $\pm\dot{\varphi} \geq 0$*. Non-singular points of* $\pm$*-arc lie above/below* $\mathfrak{D}$*.*

Lemma A.3 evidently remains valid, whereas Lemma A.4 holds in the following specified form.

Lemma A.17. *Whenever SP lies on* C*, we have* $\beta \geq 0$*.*

It is easy to see by inspection that Lemma A.6 remains true as well, where in the case from Fig. A.3(c) the right turn at the point $s_{\#}$ is justified by not the absence of the target in the cave but the very algorithm statement. The following claim is an analog of Lemma A.7.

Lemma A.18. *Suppose that after SMT starting and ending at the points* $s_{\diamondsuit}$ *and* s_{*}*, respectively, SP enters the cave* $\mathfrak{C}[s_{\diamondsuit}, s_{*}]$*. Then this cave contains the entire path traced before SMT at hand.*

Proof. The proof is by retracing the arguments from the proof of Lemma A.7 with the only alteration: the point cannot enter the cave through $s_{\diamondsuit}$ since this violates the always positive direction of motion along the boundary. □

Now we revert to the robot at hand. The following proof employs the notations $\mathfrak{A}^{\dagger}$ and $\sigma_{\mathfrak{A}^{\dagger}}$ introduced at the beginning of Section 7.A.2.

Proof of Lemma 4.6. Suppose to the contrary that $\sigma_{\mathfrak{A}^{\dagger}} = -$. Then according to the "only-right-turns" option of the algorithm, the robot enters the cave $\mathfrak{C}_{\mathfrak{A}^{\dagger}}$ after termination of $\mathfrak{A}^{\dagger}$. We are going to show that then similar to Lemma A.7, this cave contains the entire path passed by the robot until this moment and therefore its initial location. Due to the first relation from (4.1), the last claim implies that the initial location $\boldsymbol{r}_0$ is also contained by a cave of $\mathcal{N}(d_{\updownarrow})$, in violation of the assumptions of Theorem 4.1. This contradiction will complete the proof.

Thus, it remains to show that $\mathfrak{C}_{\mathfrak{A}^{\dagger}}$ does contain the path traced so far. Suppose the contrary. Since in the mode $\mathfrak{B}$ preceding $\mathfrak{A}^{\dagger}$, the robot has the obstacle to the left, it passes to $\mathfrak{A}^{\dagger}$ from inside the cave. It follows that the moment after $\mathfrak{A}^{\dagger}$ is not the first time the robot enters the cave. Let us consider the last of these "preceding" enters and the path $\mathscr{P}$ traced by the robot since this moment until the commencement of $\mathfrak{A}^{\dagger}$. By combining Lemma A.6 with the arguments from the proof of Lemma A.13, we conclude that this path is single and $\beta > 0$ at its end, which makes mode $\mathfrak{A}^{\dagger}$ impossible. The contradiction obtained completes the proof. □

Biologically-inspired algorithm for safe navigation of a wheeled robot among moving obstacles

8

8.1 Introduction

This chapter introduces and discusses a biologically-inspired reactive algorithm for navigation toward a target while avoiding collisions with moving obstacles. We assume the mobile robot is a unicycle-like vehicle described by the standard nonholonomic model with a hard constraint on its angular velocity. As mentioned in Section 3.2, motions of many wheeled robots, missiles, and unmanned aerial vehicles can be described by this model. Unlike many works in the area of robotics, which present heuristic navigation strategies, this chapter gives a mathematically rigorous analysis of the examined navigation algorithm with a complete proof of the stated theorem. This algorithm is simple, computationally efficient, and is based on switching between moving to the target along straight lines, when possible, and a sliding mode based obstacle avoidance. The performance of the examined navigation strategy is confirmed via extensive computer simulations and experiments with real wheeled robots, including robotic wheelchairs and mobile hospital beds for head injury patients. In particular, the performance of the proposed method is compared with that of the popular velocity obstacle approach [17, 19] and found to be better under certain circumstances. In the particular case of stationary obstacles, the strategy studied in this chapter is reduced to the algorithm proposed and studied in [32].

Researchers in the area of robot navigation are finding much inspiration from biology, where the problem of controlled animal motion is a central one. Animals, such as insects, birds, or mammals, are believed to use simple, local motion control rules that result in remarkable and complex intelligent behaviors. The navigation strategy considered in this chapter also belongs to the class of biologically inspired or biomimetic navigation algorithms. In biology, a similar obstacle avoidance strategy is called "negotiating obstacles with constant curvatures"; see, e.g., [153]. An example of such a movement is a squirrel running around a tree.

The main results of the chapter were originally published in [52].[1] Preliminary versions of some of the chapter's results were also reported in [161, 399, 400].

The reminder of the chapter is organized as follows. Section 8.2 presents the system description and assumptions adopted for theoretical analysis. In Section 8.3, the navigation algorithm is introduced, while its mathematical analysis is given in

[1] Savkin and Wang [52]. Copyright ©2013 Cambridge University Press. Reprinted with permission.

Safe Robot Navigation Among Moving and Steady Obstacles. http://dx.doi.org/10.1016/B978-0-12-803730-0.00008-1

Section 8.4. Computer simulation tests are discussed in Section 8.5, whereas Section 8.6 presents experiments with a Pioneer P3-DX wheeled robot. The implementation of the proposed navigation algorithm on a real robotic wheelchair and hospital bed are covered in Sections 8.7 and 8.8, respectively.

8.2 Problem description

We consider a planar vehicle or wheeled mobile robot modeled as a unicycle. The control inputs are the robot's angular velocity and speed, both limited by given constants. The mathematical model of the robot is as follows:

$$\begin{aligned} \dot{x} &= v(t)\cos\theta, & x(0) &= x_0, \\ \dot{y} &= v(t)\sin\theta, & y(0) &= y_0, \\ \dot{\theta} &= u, & \theta(0) &= \theta_0, \end{aligned} \tag{2.1}$$

where

$$u \in [-\overline{u}, \overline{u}], \quad v(t) \in [0, \overline{v}]. \tag{2.2}$$

Here $x(t)$ and $y(t)$ are the robot's Cartesian coordinates in a world frame, $\theta(t)$ gives its orientation, and $v(t)$ and $u(t)$ are the speed and angular velocity, respectively. The angles, including $\theta(t)$, are measured in the counter-clockwise direction. The maximum angular velocity $\overline{u}$ and the maximum speed $\overline{v}$ are given.

For the sake of convenience, we introduce the following notations

$$\boldsymbol{r}(t) := \mathbf{col}\,[x(t), y(t)], \quad \mathbf{v}(t) = \dot{\boldsymbol{r}}(t) \tag{2.3}$$

for the vectors of the robot's coordinates and velocity, respectively.

We consider a quite general problem of robot navigation with collision avoidance. There is a stationary point-wise target $\mathcal{T}$ and several disjoint moving obstacles $D_1(t), \dots, D_k(t)$ in the plane. The objective is to drive the robot to the target while avoiding collisions with these obstacles. We assume that any obstacle i is a moving closed bounded planar set. The robot has access to the current distance $d_i(t)$ to obstacle i defined as

$$d_i(t) := \mathbf{dist}\,[\boldsymbol{r}(t); D_i(t)] = \min_{\boldsymbol{r}' \in D_i(t)} \|\boldsymbol{r}' - \boldsymbol{r}(t)\|.$$

Here $\|\cdot\|$ denotes the standard Euclidean vector norm, and min is achieved since $D_i(t)$ is closed. Furthermore, the robot has access to $\dot{d}_i(t)$, e.g., via numerical differentiation, and to the velocity $\mathbf{v}_i(t)$ of the mass center of obstacle i. Also, the robot measures the angles $\alpha_i^{(1)}(t)$ and $\alpha_i^{(2)}(t)$ associated with the boundary rays of the robot's vision cone of obstacle i; see Fig. 8.1(a). Finally, the robot knows the heading $H(t)$ to the target $\mathcal{T}$; see Fig. 8.1(b).

Definition 2.1. Let $d_{\text{safe}} > 0$ be a given constant. A robot navigation strategy is said to be *target reaching with collision avoidance* if there exists a time $t_f > 0$ such that $\boldsymbol{r}(t_f) = \mathcal{T}$, and $d_i(t) \geq d_{\text{safe}}$ for all $i = 1, \dots, k$ and $t \in [0, t_f]$.

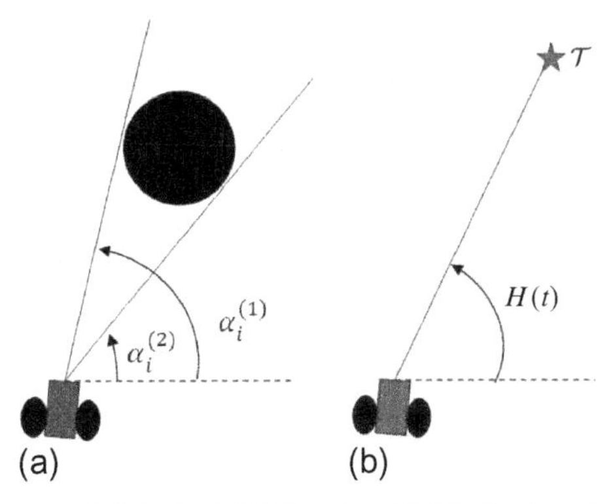

Figure 8.1 (a) Vision cone of obstacle i; (b) Heading $H(t)$ to the target.

In what follows, we assume the robot is faster than the obstacles: there exists a constant $\overline{v}_{\text{obs}}$ such that the velocity of the mass center of any obstacle always obeys the bound:

$$\|\mathbf{v}_i(t)\| \leq \overline{v}_{\text{obs}} < \overline{v}. \tag{2.4}$$

It is clear that if (2.4) does not hold, collision avoidance may be impossible.

8.3 Navigation algorithm

To describe the employed obstacle avoidance maneuver, we need some notations. Let an angle $\alpha_0 \in (0, \pi/2)$ be given. We first expand the vision cone of any obstacle i via rotation of its boundary rays through an angle of α_0; see Fig. 8.2. The expanded cone is bounded by two rays outgoing from the robot in the following respective directions:

$$\beta_i^{(1)}(t) := \alpha_i^{(1)}(t) + \alpha_0, \quad \beta_i^{(2)}(t) := \alpha_i^{(2)}(t) - \alpha_0. \tag{3.1}$$

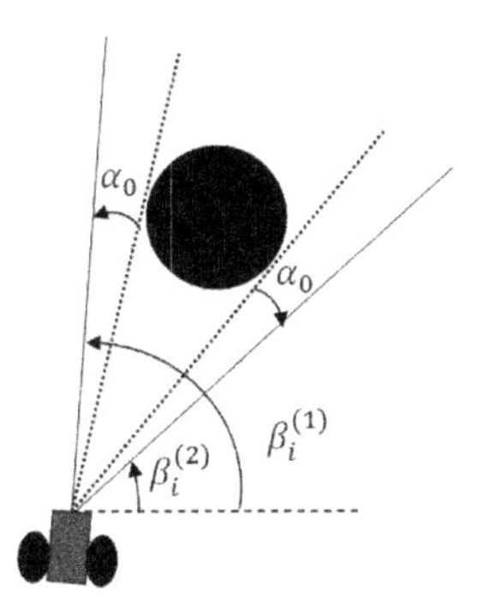

Figure 8.2 Enlarged vision cone.

These rays are spanned by the following vectors

$$\mathbf{l}_i^{(j)}(t) := (\overline{v} - \overline{v}_{\text{obs}})\,\mathbf{col}\left[\cos(\beta_i^{(j)}t), \sin(\beta_i^{(j)}(t))\right] \quad j = 1, 2, \tag{3.2}$$

where the multiplier $\overline{v} - \overline{v}_{\text{obs}}$ with the constants $\overline{v}$ and $\overline{v}_{\text{obs}}$ from (2.2) and (2.4), respectively, is introduced for the sake of convenience. These rays are also known under the name of *occlusion rays*.

Furthermore, for any two nonzero vectors $\mathbf{a}, \mathbf{b} \in \mathbb{R}^2$, the symbol $\angle(\mathbf{a}, \mathbf{b})$ stands for the angle subtended by these vectors and measured from $\mathbf{a}$ in the counterclockwise direction; $\angle(\mathbf{a}, \mathbf{b}) \in (-\pi, \pi]$. Finally, we introduce the following function:

$$f(\mathbf{a}, \mathbf{b}) := \begin{cases} 0 & \text{if} \quad \angle(\mathbf{a}, \mathbf{b}) = 0, \\ 1 & \text{if} \quad 0 < \angle(\mathbf{a}, \mathbf{b}) \le \pi, \\ -1 & \text{if} \quad -\pi < \angle(\mathbf{a}, \mathbf{b}) < 0. \end{cases} \tag{3.3}$$

Let the maneuver to avoid the moving obstacle i be started at time t_0. We first determine an index $h \in \{1, 2\}$ at which the following minimum is attained:

$$\min_{j=1,2} \left| \angle \left[\mathbf{v}_i(t_0) + \mathbf{l}_i^{(j)}(t_0); \mathbf{v}(t_0) \right] \right|. \tag{3.4}$$

By doing so, we in fact make a selection among two vectors $\mathbf{v}_i(t_0) + \mathbf{l}_i^{(j)}(t_0), j = 1, 2$ and draw the vector with lesser angular discrepancy with respect to the robot's velocity $\mathbf{v}(t_0)$. To avoid collision with the obstacle, the following navigation rule is used:

$$\begin{aligned} u(t) &= -\overline{u} f \left[\mathbf{v}_i(t) + \mathbf{l}_i^{(h)}(t), \mathbf{v}(t) \right]; \\ v(t) &= \|\mathbf{v}_i(t) + \mathbf{l}_i^{(h)}(t)\|. \end{aligned} \tag{3.5}$$

It is easy to see that these control inputs are feasible, i.e., satisfy the constraints from (2.2). Indeed, (3.3) trivially implies that $|u(t)| \le \overline{u}$. Furthermore, (3.5), (2.4), and (3.2) yield that

$$v(t) = \|\mathbf{v}_i(t) + \mathbf{l}_i^{(h)}(t)\| \le \|\mathbf{v}_i(t)\| + \|\mathbf{l}_i^{(h)}(t)\| \le \overline{v}_{\text{obs}} + (\overline{v} - \overline{v}_{\text{obs}}) = \overline{v}.$$

Therefore, the second constraint from (2.2) holds as well.

Remark 3.1. The intuition behind the obstacle avoidance rule (3.5) can be explained as follows. The navigation law (3.5) is discontinuous and, specifically, is a sliding mode law; see, e.g., [70]. Correspondingly, the closed-loop system (2.1), (3.5) is described by ordinary differential equations with discontinuous right-hand sides. The solution of such equations goes first to and then exactly over the discontinuity surface in the so-called sliding mode under certain circumstances discussed in Section 1.3; and they do occur for the navigation law at hand. In other words, it drives the robot over the switching surface of the navigation law (3.5) after an initial transient.

Meanwhile being on the discontinuity surface means that the vector $\mathbf{v}(t)$ of the robot's absolute velocity is equal to $\mathbf{v}_i(t) + \mathbf{l}_i^{(h)}(t)$. In the relative coordinate frame

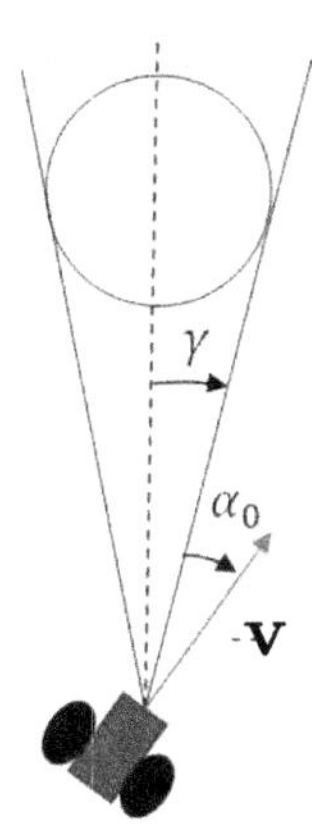

Figure 8.3 Constant avoiding angle α_0.

attached to the obstacle, the obstacle stays steady, while the robot's velocity is $[\mathbf{v}_i(t) + \mathbf{l}_i^{(h)}(t)] - \mathbf{v}_i(t) = \mathbf{l}_i^{(h)}(t)$ and therefore the robot moves in the direction of an occlusion ray. This motion is clearly an obstacle avoidance procedure, which keeps a constant avoiding angle α_0 between the instantaneous moving direction of the robot and one of the two boundary rays of the vision cone of the obstacle; see Fig. 8.3. In other words, the navigation law consists in steering toward the nearest edge of the expanded obstacle's vision cone. The idea of this obstacle avoidance scheme originates from biology where it is known under the title of "negotiating obstacles with constant curvatures"; see, e.g., [153]. An example of its application is given by a squirrel running around a tree. A similar strategy for avoiding steady obstacles was proposed and studied in [32].

The above rule takes measures against collisions with obstacles but does not cover the entire duration of the experiment, which may include periods with no collision threat. Then the robot is driven at the maximal speed along a straight line:

$$u(t) = 0, \quad v(t) = \overline{v}. \tag{3.6}$$

The algorithm of choosing this line and switching between the obstacle avoidance (3.5) and the straight motion (3.6) laws uses free constant $C > 0$ and $a_1, \ldots, a_k$, where $a_i > 0$ is associated with the moving obstacle D_i. Specifically, the switching rule is as follows:

R1: Switching from (3.6) to (3.5) occurs at any time t_0 when the distance from the robot to obstacle i reduces to the value C, i.e., $d_i(t_0) = C$ and $\dot{d}_i(t_0) < 0$;

R2: Switching from (3.5) to (3.6) occurs at any time t_* when $d_i(t_*) \leq 1.1a_i$ and the robot is oriented toward the target, i.e., $\theta(t_*) = H(t_*)$.

To exclude uncertainty in choosing between (3.5) and (3.6), C should exceed $1.1a_i$ for all i.

8.4 Mathematical analysis of the navigation strategy

In this section, we show that the proposed navigation strategy is target reaching with collision avoidance, in the sense of Definition 2.1. This is proved under a number of assumptions, and we start with their exposition.

Assumption 4.1. *Every obstacle i is a disk of radius $R_i > 0$ moving with the constant velocity $\mathbf{v}_i \neq 0$. Furthermore, $v_i := \|\mathbf{v}_i\| < \overline{v}$ for all $i = 1, \ldots, k$, and the vectors $\mathbf{v}_i$ and $\mathbf{v}_j$ are not collinear whenever $i \neq j$.*

The next assumption employs the following constants

$$F_i := \frac{(v_i + \overline{v})R_i}{(R_i + d_{\text{safe}})^2\sqrt{1 - \frac{R_i^2}{(R_i + d_{\text{safe}})^2}}} \qquad i = 1, 2, \ldots, k. \tag{4.1}$$

Assumption 4.2. *The inequality $F_i < \overline{u}$ holds for all $i = 1, \ldots, k$.*

Assumption 4.3. *The parameters of the switching rule are such that for any i,*

$$a_i = \frac{R_i}{\cos\alpha_0} - R_i, \tag{4.2}$$

$$C \geq \frac{\pi\overline{v}}{\overline{u} - F_i} + 1.1a_i.$$

Let $S_{ij}(t)$ denote the distance between obstacles i and j at time t, and let $S_i(t)$ be the distance between obstacle i and the target $\mathscr{T}$ at time t:

$$S_{ij}(t) := \min_{\boldsymbol{q}_i \in D_i(t), \boldsymbol{q}_j \in D_j(t)} \|\boldsymbol{q}_i - \boldsymbol{q}_j\|, \quad S_i(t) := \min_{\boldsymbol{q} \in D_i(t)} \|\boldsymbol{q} - \mathscr{T}\|.$$

Assumption 4.4. *The following inequalities hold for all $i \neq j$ and t:*

$$S_{ij}(t) \geq 2C + \frac{\pi\overline{v}}{\overline{u} - F_i}, \quad S_i(t) > 1.1a_i.$$

Assumption 4.5. *The parameter $\alpha_0 < \pi/2$ in (3.1) and the safety margin $d_{safe} > 0$ are chosen so that*

$$\alpha_0 \geq \arccos\left(\frac{R_i}{R_i + d_{safe}}\right) \quad \forall i = 1, 2, \ldots, k.$$

Remark 4.1. It obviously follows from Assumption 4.5 and (4.2) that $a_i \geq d_{\text{safe}}$.

Assumption 4.6. *At the initial time $t = 0$ the robot is moving toward the target in accordance with (3.6), and $d_i(0) > C$ for all i.*

Theorem 4.1. *Suppose that Assumptions 4.1–4.6 hold. Then the navigation strategy **R1, R2** is target reaching strategy with collision avoidance.*

8.4.1 Proof of Theorem 4.1

We start with a lemma, which in fact plays a decisive role in the proof of the theorem. To formulate it, we note that the statements of the switching rules **R1** and

R2 include descriptions of triggering events; every of these events will be denoted by the respective notation **R1** or **R2**.

Lemma 4.1. *If the event **R1** occurs at some time t_0, then the event **R2** necessarily occurs at a later time $t_* > t_0$.*

Proof. Suppose the contrary. The navigation law (3.5) steers the vector $\mathbf{v}_i + \mathbf{l}_i^{(h)}(t)$ toward the robot velocity vector $\mathbf{v}(t)$. An elementary geometric argument assures that whenever $d_i(t) \geq d_{\text{safe}}$, the angular velocity of the vector $\mathbf{v}_i + \mathbf{l}_i^{(h)}(t)$ is less than F_i, where F_i is defined by (4.1). In view of Assumption 4.2, the rule (3.5) guarantees that $\mathbf{v}_i + \mathbf{l}_i^{(h)}(t) = \mathbf{v}(t)$ for $t > t_0 + \frac{\pi}{\bar{u} - F_i}$. Meanwhile, Assumption 4.4 ensures that during the time interval when the robot is engaged in bypassing obstacle i and is governed by the navigation law (3.5), $d_j(t) > C$ for all other obstacles $j \neq i$. So switching to bypass of any other obstacle cannot occur.

Now we focus on a time interval where $\mathbf{v}_i + \mathbf{l}_i^{(h)}(t) \equiv \mathbf{v}(t)$, and consider the process in the relative coordinate frame attached to obstacle i. In this frame, obstacle i stays still, whereas the robot moves with the velocity $\mathbf{l}_i^{(h)}(t)$.

It follows that

$$\dot{d}_i(t) = -(\overline{v} - \overline{v}_{\text{obs}}) \cos[\gamma(t) + \alpha_0], \tag{4.3}$$

where $\gamma(t) + \alpha_0$ is the angle between the robot's velocity and the line connecting the robot with the center of obstacle i; see Fig. 8.3. By (4.3), the distance $d_i(t)$ decays if and only if $\gamma(t) + \alpha_0 < \frac{\pi}{2}$. This and elementary analysis yield that $d_i(t)$ decays when the robot is outside the circle of radius $\frac{R_i}{\cos \alpha_0}$ concentric with the obstacle, and $\dot{d}_i(t) = 0$ when the robot is exactly on this circle. Hence the robot's trajectory converges to this circle from outside; see Fig. 8.4. So Assumption 4.5 ensures that the safety constraint $d_i(t) \geq d_{\text{safe}}$ is satisfied. Meanwhile the circle's radius $\frac{R_i}{\cos \alpha_0} = R_i + a_i$ thanks to (4.2) and so $d_i(t) \leq 1.1a_i$ for large enough t. Convergence to the circle, which does not enclose the target by Assumption 4.4, obviously implies existence of a time $t_* \geq t_0$ such that $d_i(t) \leq 1.1a_i$ and the robot is oriented toward the target. But this means that the event **R2** occurs at $t = t_*$, in violation of the starting hypothesis of the proof. This contradiction completes the proof. □

Proof of Theorem 4.1. Let $I_i(t)$ be an interval of a straight line that goes from the target to some point on the circle of radius $R_i + 1.1a_i$ concentric with obstacle i. By Assumption 4.1, the obstacles move in different directions. So there exists a time

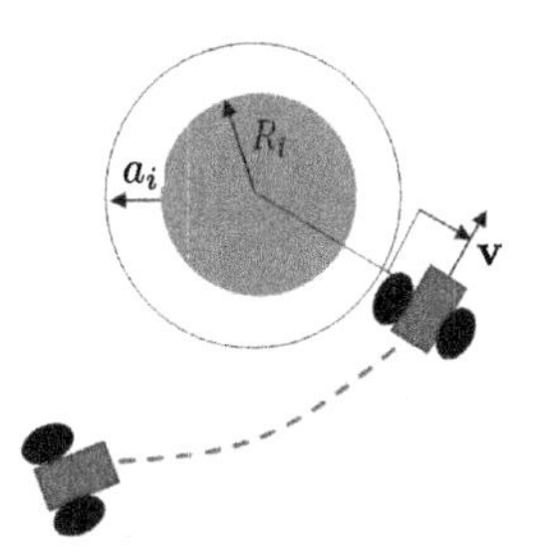

Figure 8.4 Asymptotical behavior of the robot.

$t_{**} > 0$ since which $t \geq t_{**}$ the distance between any interval $I_i(t)$ and any obstacle $j \neq i$ is greater than C. It follows that after the time t_{**}, motion to the target in a straight line cannot be switched off by virtue of **R1**. This and Lemma 4.1 imply that the target is inevitably reached. □

8.5 Computer simulations

In this section, we present computer simulation results for a wheeled mobile robot navigating in a dynamic environment with moving obstacles. The robot is governed by the biomimetic algorithm **R1, R2** described in Section 8.3. The simulations are performed in Matlab. The parameters used for simulations are given in Table 8.1.

First, we examine the simplest case: the environment with only one moving obstacle, as is shown in Fig. 8.5. Figure 8.5(a) depicts the initial position of the robot, obstacle, and target, as well as the moving direction of the obstacle. The paths of the robot and obstacle are shown as circled lines. The time profiles of the speed and angular velocity of the robot are shown in Fig. 8.5(c) and (d).

Figure 8.6 addresses a more challenging scenario where the robot is in a dynamic environment with six moving obstacles. Figures 8.6(a), (c), and (e) display the crucial moments of the robot's maneuvers to avoid each of the obstacles. Figures 8.6(b), (d), and (f) show the evolution of distance between the robot and the closest obstacle. The red (dark gray in print versions) ovals in Fig. 8.6(a), (c), and (e) mark the positions where the distances between the robot and obstacles are minimal; the corresponding minimal distance values are similarly marked in Fig. 8.6(b), (d), and (f). In this experiment, the robot successfully reaches the target and avoids every moving obstacle encountered along its path.

The performance of the biologically-inspired algorithm from Section 8.3 was compared with that of the popular Velocity Obstacle Approach (VOA) [17, 19] in the scenario illustrated in Fig. 8.7: a long obstacle is moving perpendicularly to the line connecting the robot and target. Figures 8.7(c) and (d) address the moment when the robot is just about to bypass the lower end of the obstacle. As it can be observed, the biomimetic algorithm avoids the obstacle by keeping a constant avoiding angle between the robot's moving direction and the lower end of the obstacle. Meanwhile

Table 8.1 **Parameters used for simulations**

Parameter	Value	Comments
Δt	0.1 s	Sampling intervals
$\overline{v}$	0.7 m/s	Maximum linear lelocity
$\overline{u}$	$\frac{\pi}{3}$ rad/s	Maximum angular velocity
α_0	$\frac{\pi}{10}$ rad	Avoidance angle
C	1.5 m	Switching distance

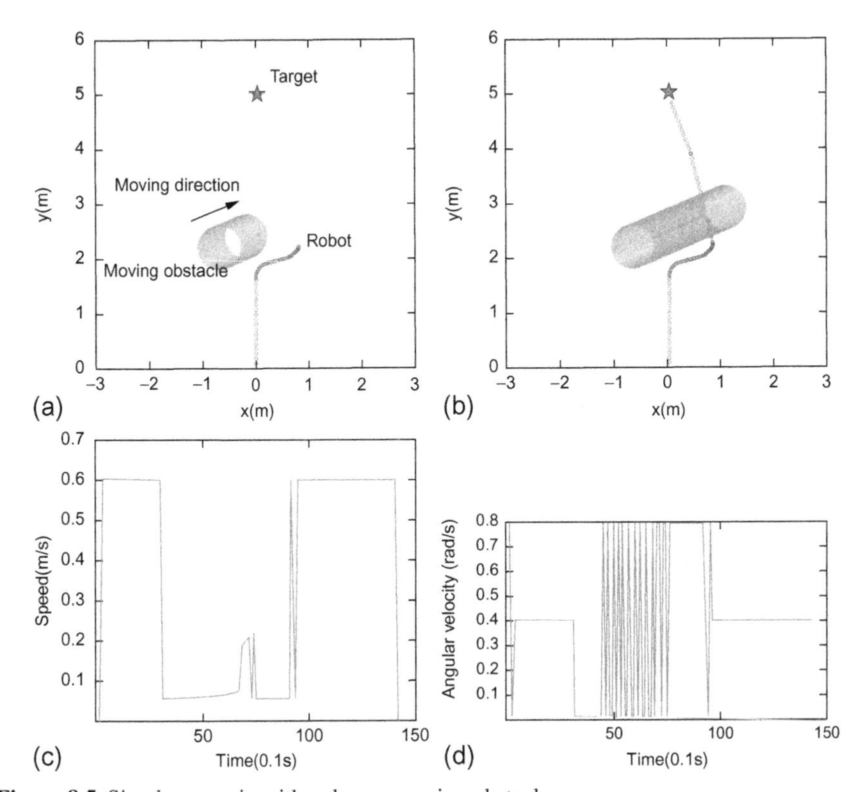

Figure 8.5 Simple scenario with only one moving obstacle.

VOA drives the robot toward the lower end of the obstacle in a straight line and then steers it toward the target position. The complete paths of the robot are shown in Fig. 8.7(e) and (f). The overall maneuver time for the biomimetic algorithm is 17.3 s and for VOA is 18.7 s. Hence, the biomimetic algorithm outperforms VOA in this scenario in terms of the maneuver time.

8.6 Experiments with a laboratorial wheeled robot

The experiments were conducted with an ActivMedia Pioneer 3-DX wheeled robot described in Section 1.4.1. To implement the biomimetic algorithm on this robot, we used the C++ programming language and Active Media Robotics Application, ARIA, which is an object oriented C++ library for controlling ActivMedia mobile robots, running in the Linux operating system.

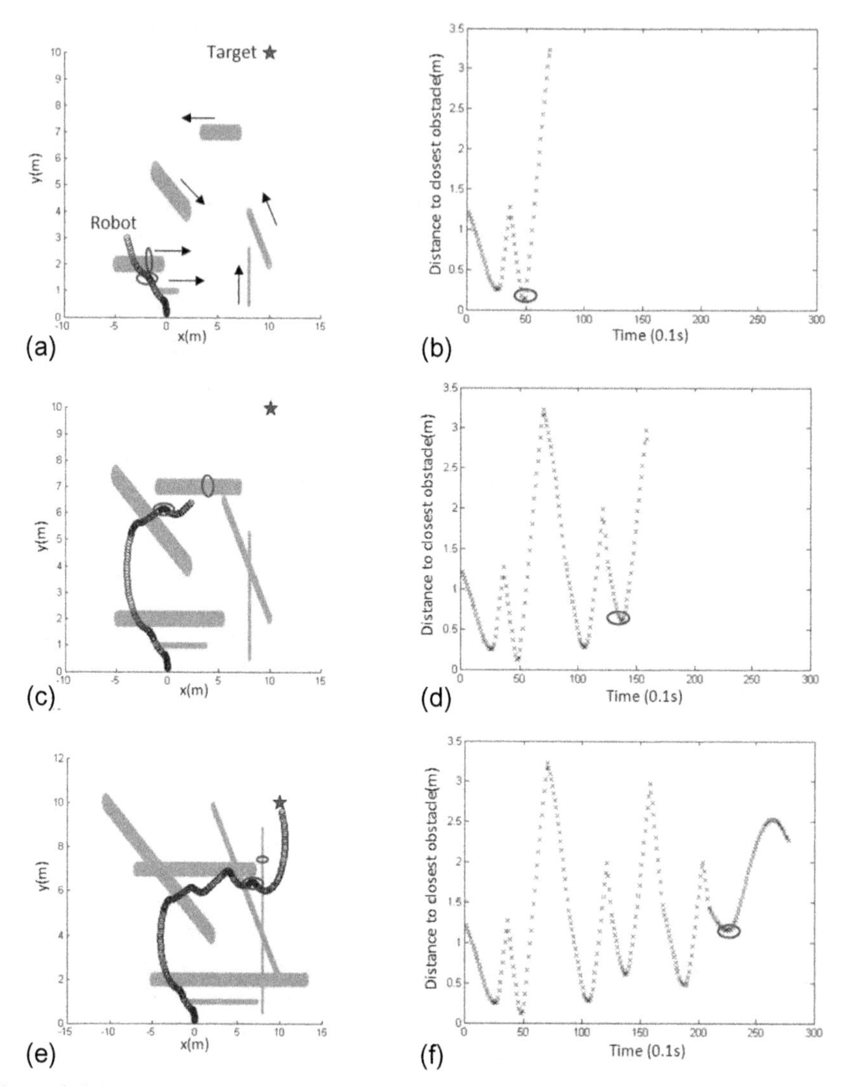

Figure 8.6 Robot navigation in a cluttered dynamic environment.

An odometry system was used to evaluate the position of the robot, obstacle, and target. Based on these data, the robot's heading $H(t)$ to the target was computed. The relative distance $d(t)$ to the obstacle and the angle $\theta(t)$ were determined based on the readings of laser and sonar range finders mounted to the robot. Therefore, the coordinates of the obstacle's center can be computed by

$$x_{ob} = x_R + d\cos(\theta), \quad y_{ob} = y_R + d\sin(\theta)$$

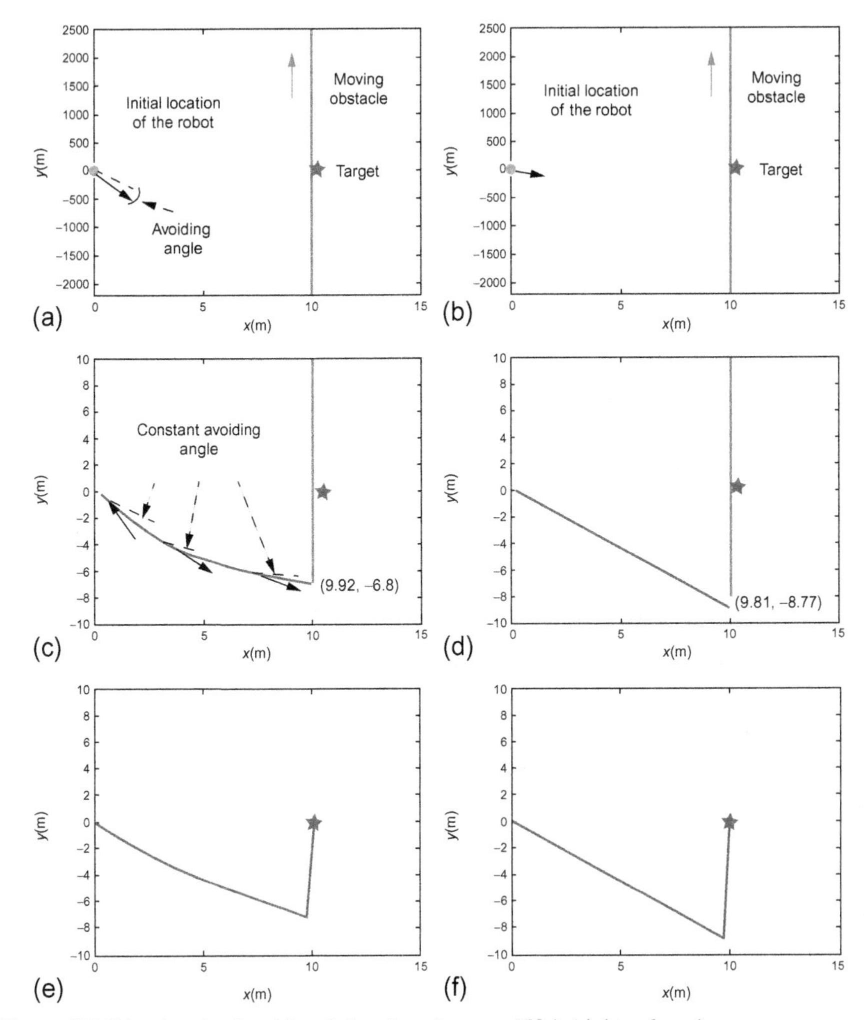

Figure 8.7 Biomimetic algorithm (left column) versus VOA (right column).

and thus the velocity of the obstacle is estimated by

$$v_{\rm ob} = \frac{\sqrt{[y_{\rm ob}(t) - y_{\rm ob}(t - \Delta t)]^2 + [x_{\rm ob}(t) - x_{\rm ob}(t - \Delta t)]^2}}{\Delta t},$$

$$\theta_{ob} = \arctan\left(\frac{y_{ob}(t) - y_{ob}(t - \Delta t)}{x_{ob}(t) - x_{ob}(t - \Delta t)}\right).$$

The first experiment is in an environment with three moving obstacles, as is shown in Fig. 8.8. Figures 8.8(a)–(d) display characteristic snapshots of the experiment. The moving directions of persons (which played the role of moving obstacles) are

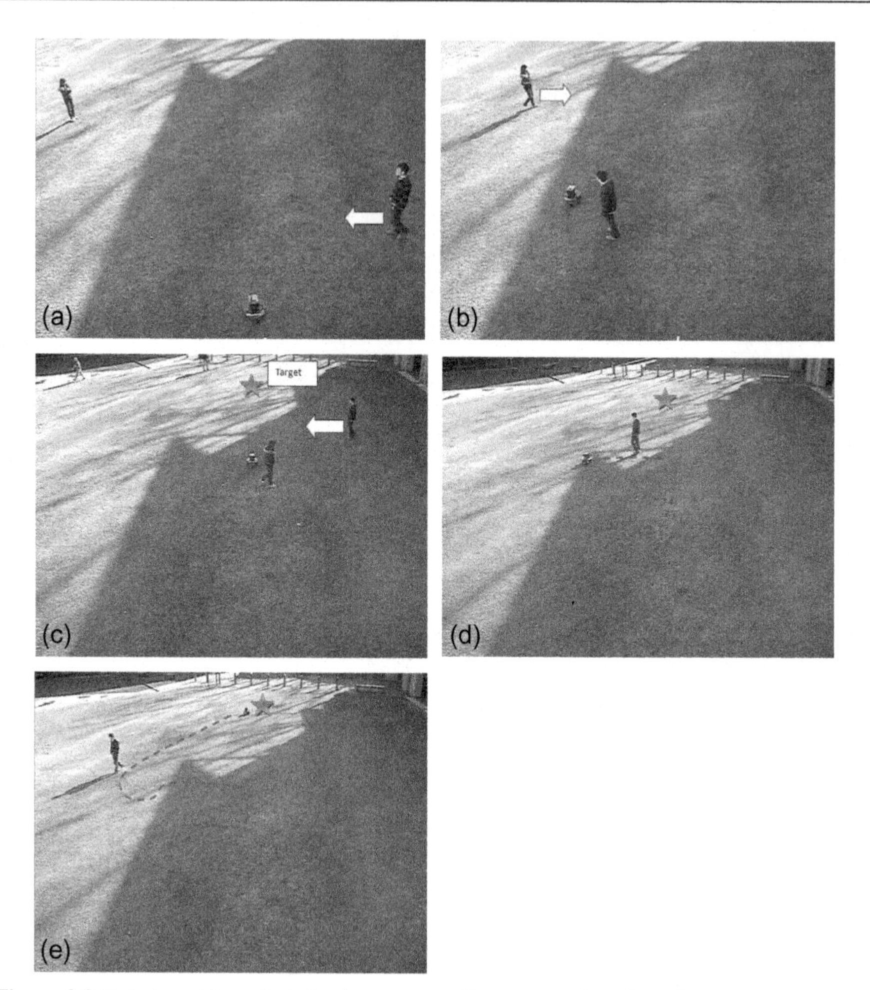

Figure 8.8 Robot avoiding obstacles in a scene with three moving obstacles.

indicated by white arrows, and the trajectory of the robot is depicted by a dashed line. The overall trajectory is displayed in Fig. 8.8(e).

The second experiment deals with a more challenging scenario: the robot should pass through a narrow corridor cluttered with both stationary and moving obstacles, as is illustrated in Fig. 8.9. In doing so, the robot should both avoid the obstacles and not to collide with the walls. Figures 8.9(a)–(c) show the moments when the robot avoids obstacles, while keeping a certain distance to the walls. The complete path taken by the robot is drawn in Fig. 8.9(d).

Figure 8.10 shows a situation where the robot is in a corridor with two moving obstacles, one of which is a girl and the other is another Pioneer 3-DX wheeled

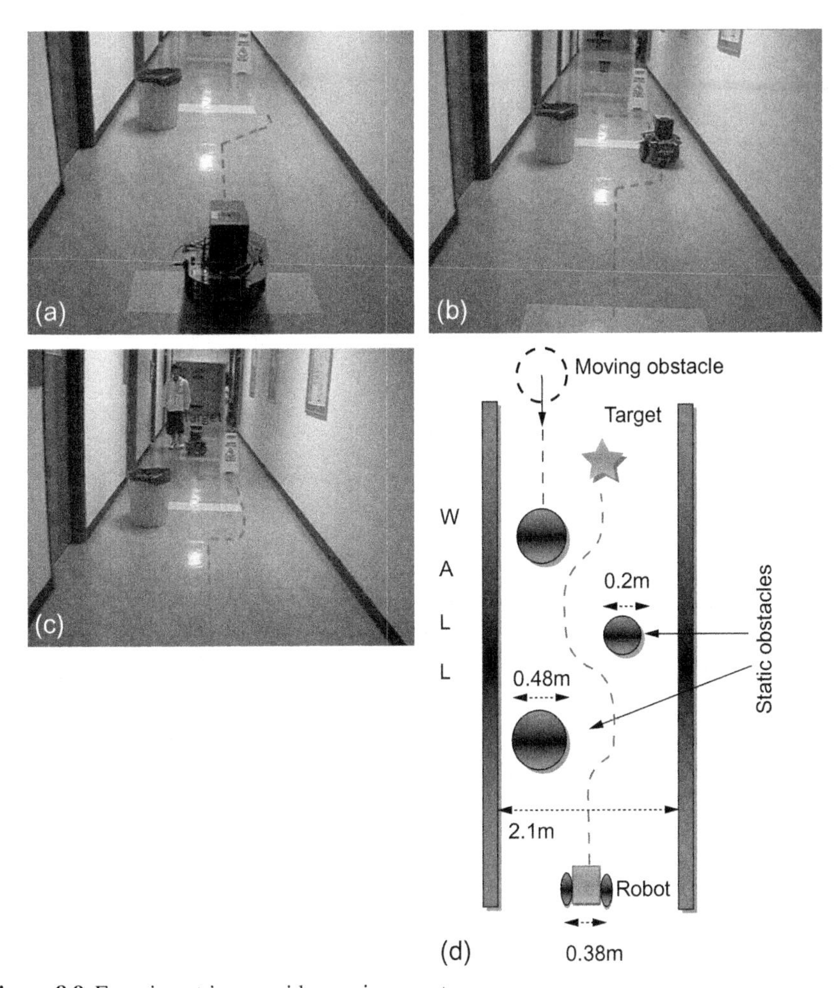

Figure 8.9 Experiment in a corridor environment.

robot. Figure 8.10(a) gives a snapshot at the moment when the robot bypasses the first moving obstacle, which comes toward the robot along a straight line. Figures 8.10(b) and (c) correspond to the moments when the robot avoids the stationary obstacle and after this, immediately faces and successfully avoids the second moving obstacle. The overall path is shown in Fig. 8.10(d).

Though the theoretical analysis of Section 8.4 was based on the assumption that vision cones of the obstacles are infinite, the real life experiments deal with finite vision cones. In particular, the vision cone of a wall is in fact composed by the rays emitted from the robot to points on the wall at a distance less than some value D.

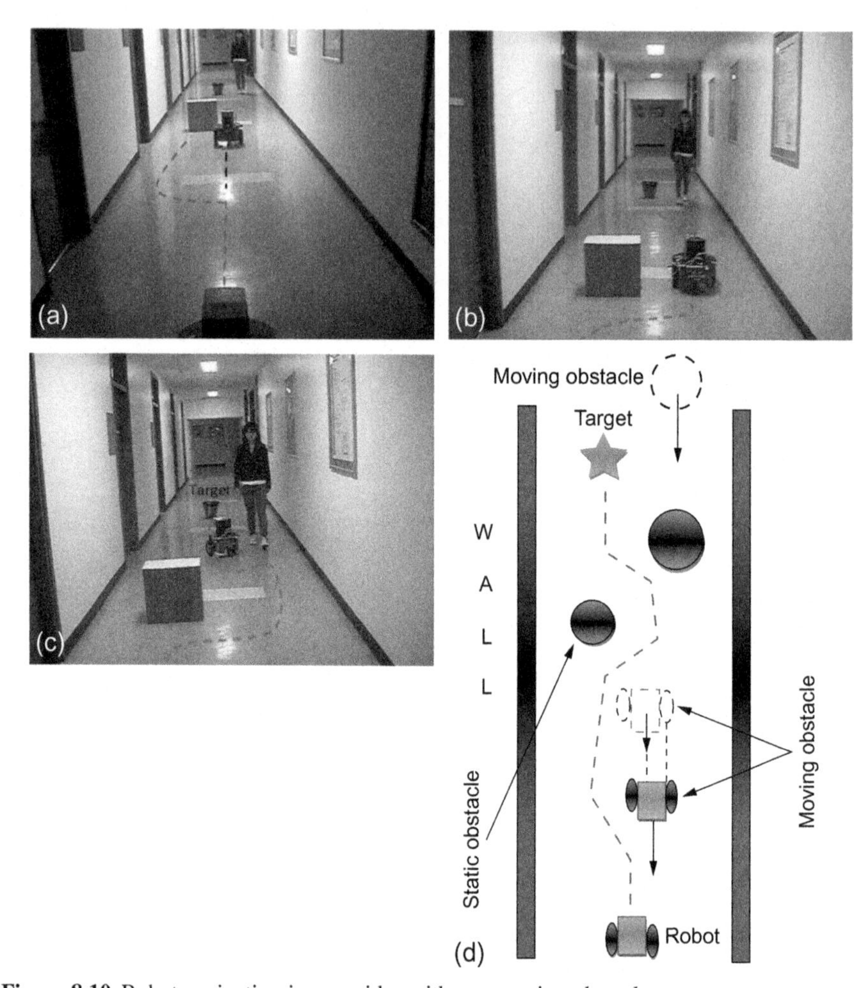

Figure 8.10 Robot navigation in a corridor with two moving obstacles.

8.7 Algorithm implementation with a robotic wheelchair

Numerous people with mobility impairments often suffer from real life problems such as losing social connections (which leads to social isolation, depression, and anxiety) or emotional damage (which leads to fear and loss of self-esteem) [401]. The statistics have shown that the population of the impaired people was growing over the past decades [402]. Clinical studies provide a convincing evidence that utilization of independent mobility aids helps the users to ease the problem and significantly improve his/her lifestyle [403]. Wheelchairs are considered as one of the most commonly used mobility aids to assist the movements of the users [401], and a large number of patients benefit from it [404].

The primary objective for operating a wheelchair is to safely deliver the user to the target location by detecting and avoiding the en-route obstacles. The methods for controlling a wheelchair have evolved from purely manual control through user-wheelchair interaction (semi-autonomous navigation) to autonomous control (autonomous navigation).

Manual control of the wheelchair is achieved by pushing from the back of the wheelchair by an assistant or the manual steering of the wheels by the user. It is the earliest and simplest way to control the movements of the wheelchairs. The problem with it is that proper steering of the wheelchairs requires a great amount of forces, especially in emergency cases where sharp turning is required. Furthermore, it is exhausting to manually drive the wheelchairs for a long period of time. So despite its simplicity, manual control approach is most inefficient.

One way to solve the problem is to motorize a wheelchair and endow it with a certain level of autonomy in motion control, which makes controlling wheelchairs more efficient and comfortable for the users. Various control strategies have been proposed for wheelchairs to perform different "local" tasks such as: tracking control [405], door passing [406],wall following [407], and motion control under uncertainties and external disturbances [408]. However at the "strategic" level, the motorized semi-autonomous wheelchairs need "signaling" commands from the user, which can be delivered by various means (the user-wheelchair interface), e.g., hand gesture [409], head position and movements [410], oral motion [411], etc. Impaired people with different disabilities can choose the most suitable method. Unfortunately, the common problem with this approach is that the overall performance may suffer much from inevitable errors and imperfections in signaling, to say nothing of external noises. Furthermore, the safety of the users is heavily dependent on their own reaction, which necessitates them to stay vigilant and watch out for potential dangers all the time. Finally, this approach requires a fair amount of practices and training for the users to operate the wheelchairs properly and safely. A clinical survey shows that forty percent of the users found it difficult, up to complete inability, to properly perform steering tasks using semi-autonomous wheelchairs [412].

Intelligent autonomous wheelchairs with navigation abilities have several advantages over the conventional wheelchairs and the semi-autonomous wheelchairs. First of all, the navigation process does not require any effort from a human labor (assistant) or the user. Instead, a navigation algorithm autonomously guides the wheelchair to the target location. Second, properly designed navigation algorithms are capable to easily adapt to a variety of environments and make proper decisions to ensure the user's safety, especially in complex cluttered dynamic scenes. Finally, the users do not need to be trained in order to control the wheelchair appropriately.

8.7.1 Experimental results with an intelligent autonomous wheelchair

For experiments, we used the intelligent wheelchair system described in Section 1.4.2. To endow it with the capacity of autonomy, the navigation law described in Section 8.3

was implemented on the wheelchair. All experiments were concerned with realistic scenarios that are commonly faced by wheelchair users.

The first experiment demonstrates the wheelchair avoiding two static (the chairs) and a moving (a human) obstacles. The initial setup of the experiment is depicted in Fig. 8.11(a). The wheelchair bypasses the static obstacles with a reasonable margin, as can be seen in Fig. 8.11(b) and (c). Figure 8.11(d) corresponds to the moment when the wheelchair avoids the moving obstacle, whereas it successfully reaches its destination at Fig. 8.11(e). Figure 8.11(f) displays the complete path for the wheelchair in this experiment.

In the scenario from Fig. 8.12, the wheelchair moves through a corridor and encounters two pedestrians along its path to the target position. The snapshots in Fig. 8.12(a) and (b) show that the wheelchair avoids the both pedestrians with a satisfactory space. Figure 8.12(c) depicts the complete path for the experiment.

A more complicated scenario is addressed in Fig. 8.13, where the wheelchair has to avoid a variety of obstacles in a cornered corridor. In this scenario, the target position is not sensible from the starting point of the wheelchair; so an intermediate sub-goal was assigned, as is depicted in Fig. 8.13(e). As can be seen, the wheelchair successfully avoids a number of enroute pedestrians and other obstacles (a wall, trash bin, couches). Figures 8.13(a)–(d) show the progression of the wheelchair over its path to the sub-goal and then to the target; the overall path is depicted in Fig. 8.13(e).

8.8 Algorithm implementation with a robotic motorized hospital bed

The patients with diagnosis of head injuries need to be transferred from an ambulance car to the neurosurgical unit for definitive medical treatments immediately. The whole transfer process is time critical, every minute in the process is directly related to the patient's life. At the same time, studies have shown that the transfer time for transportation of these patients inside of a hospital is often below the acceptable level; see, e.g., [413–415].

The patients cannot be transferred from one location to another directly using conventional hospital beds. They need to be switched to different beds with wheels and pushed to their destinations by hospital employees. The whole transportation process suffers from low efficiency and potential safety risks as hospital employees need to push the bed through a number of corridors and transfer the patient between an ambulance vehicle, the Intensive Care Unit and an operation theatre. It is very difficult for hospital employees to control the movements of the hospital beds and simultaneously take care of the safety of the patients during the entire process. Reports and studies have shown that a human-based mistake is a major contributing factor for the potential and unexpected risks of hospital beds transportation [414, 416]. Furthermore, moving the patients from one bed to another put them into potential danger of the secondary injury, which is another concern of the current transportation process.

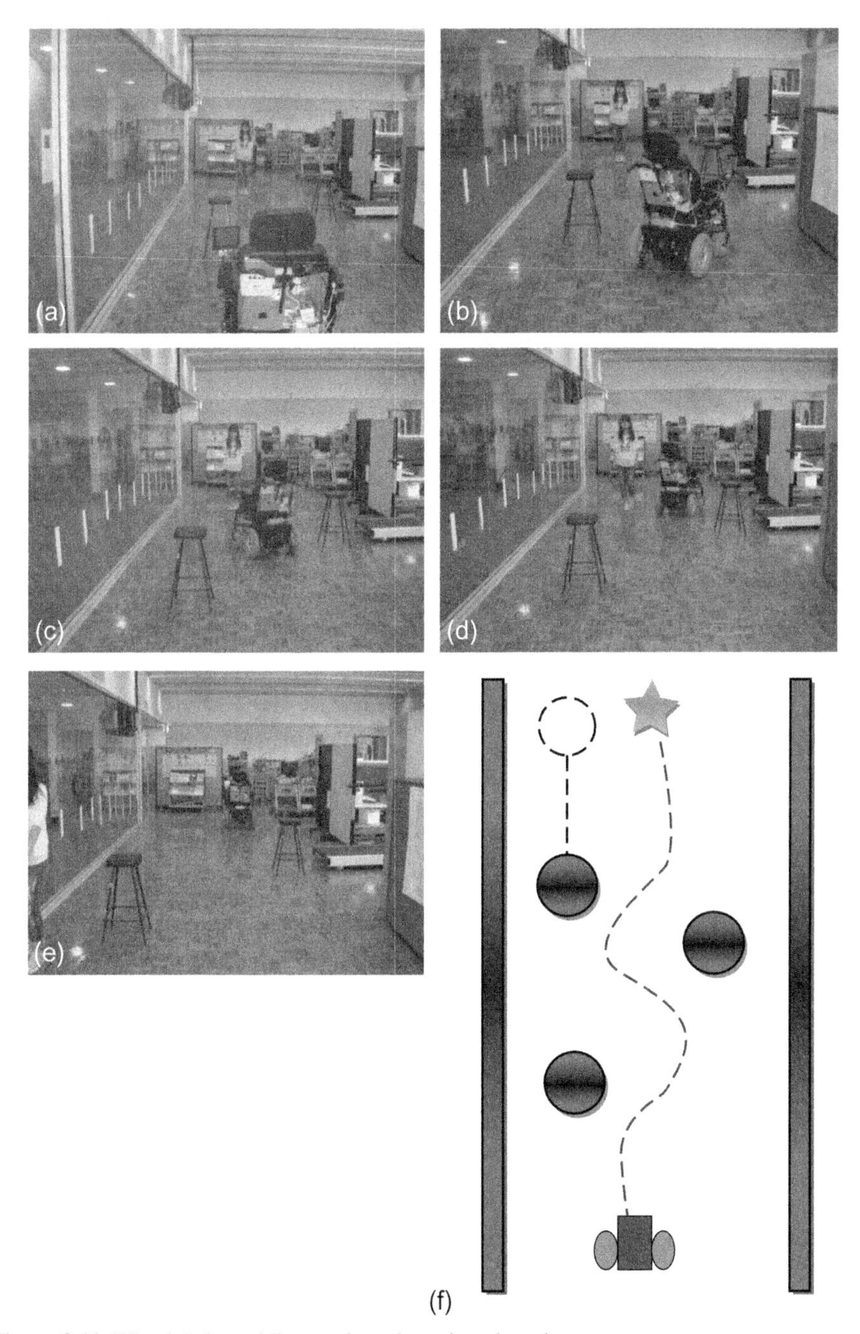

Figure 8.11 Wheelchair avoiding static and moving obstacles.

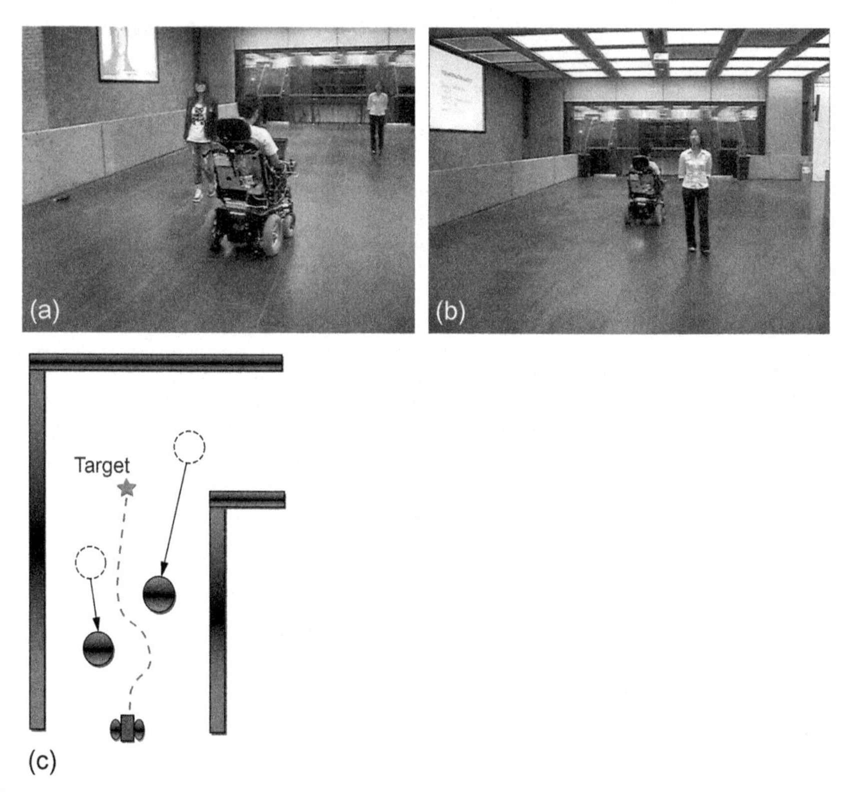

Figure 8.12 Wheelchair avoiding pedestrians in a corridor.

Utilization of intelligent hospital beds with autonomous navigation ability is able to increase the efficiency and safety level of the current transportation techniques. A properly designed navigation algorithm is able to guide the hospital bed from its current location to the destination location while avoiding en-route obstacles autonomously. Numerous navigation strategies had been proposed over the past decades, and many of them had been implemented on various robotics, such as electric-powered wheelchairs, to achieve autonomous navigation tasks. The results have shown that these navigation algorithms can dramatically increase the safety levels of robotic systems and users in the real world environments. Therefore, there is no doubt that the navigation algorithms are able to help the intelligent hospital beds to increase their safety level and efficiency during transportation. The overall transportation process is simplified by utilizing intelligent robotic hospital beds, and patients no longer need to be switched between different beds for transportation. This reduces the overall transportation time and potentially increases the safety level of the transportation process.

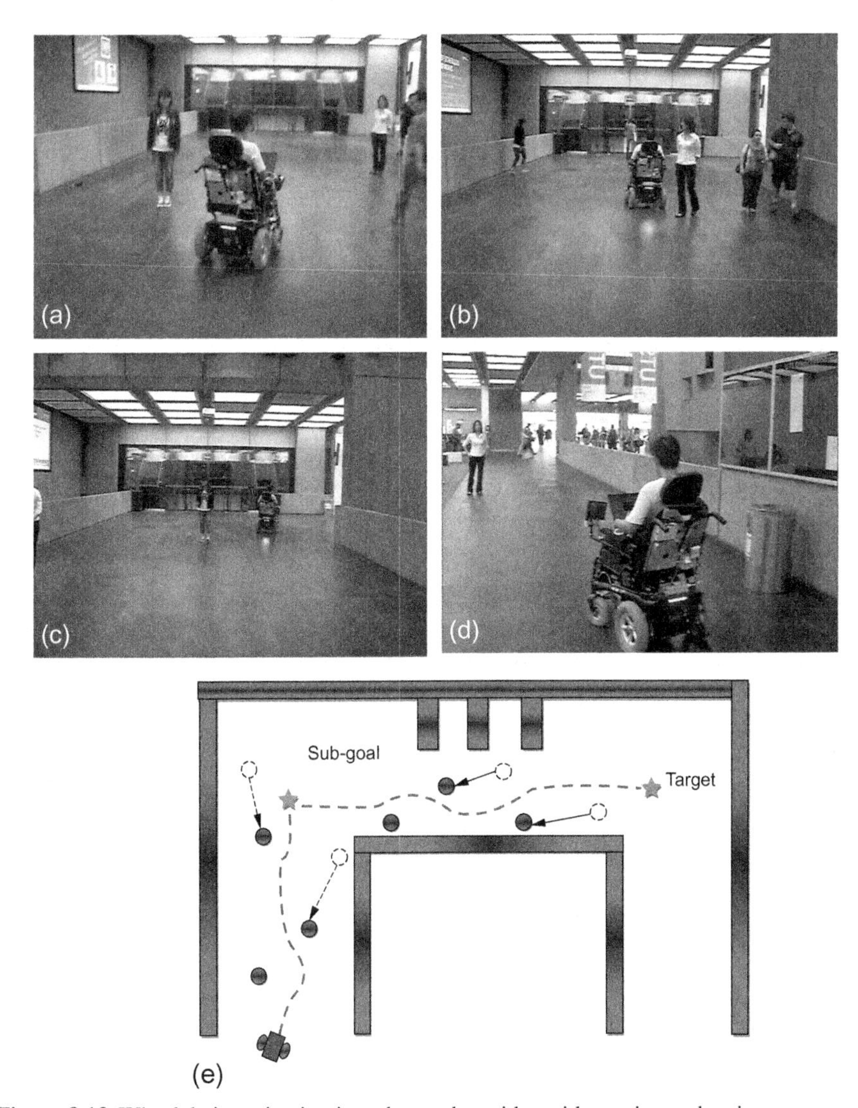

Figure 8.13 Wheelchair navigating in a cluttered corridor with moving pedestrians.

8.8.1 Experimental results with Flexbed

The reported experiments are carried out with the intelligent robotic hospital bed Flexbed described in Section 1.4.3. The objective is to demonstrate the capability of the proposed law to successfully navigate Flexbed in real world dynamic environments.

In the first experiment, we examine two key metrics that evaluate safety during obstacle avoidance, which are as follows:

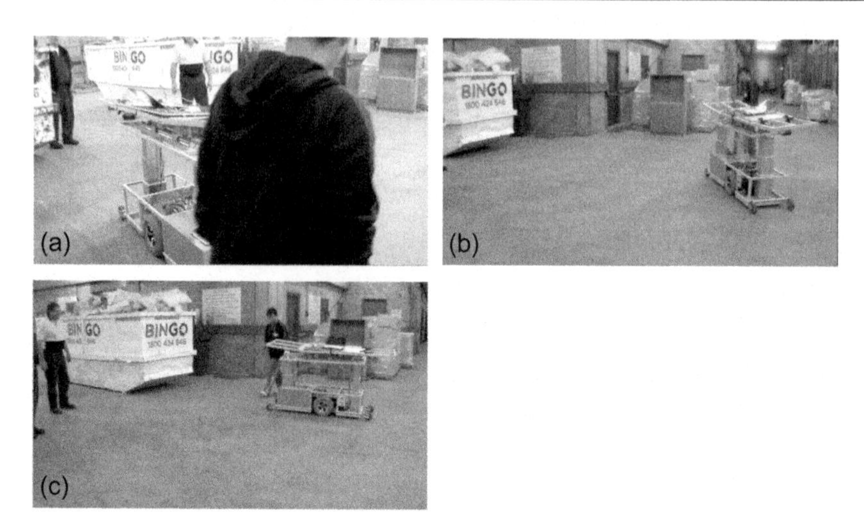

Figure 8.14 Characteristic moments of an experiment with Flexbed.

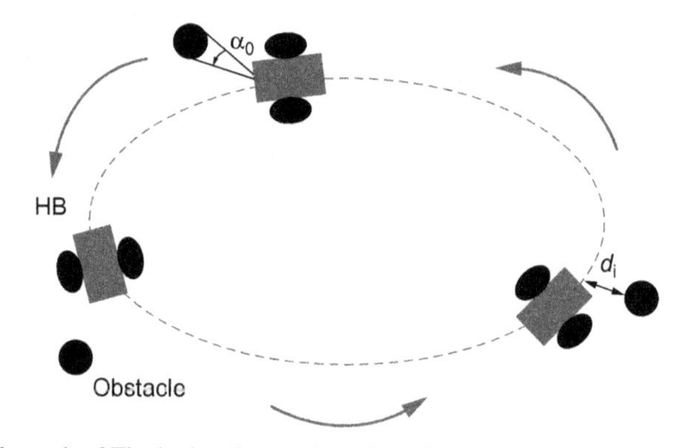

Figure 8.15 The path of Flexbed and a moving obstacle.

- The angular difference between the instantaneous moving direction of Flexbed and the closet boundary of the enlarged vision cone;
- The minimum distance between Flexbed and the obstacle.

The experimenter intentionally moves so that it is always within the sensing range of Flexbed and is thus treated as an obstacle during the entire duration of the experiment. The characteristic snapshots are shown in Fig. 8.14; whereas the paths of both Flexbed and obstacle are given in Fig. 8.15. The angular difference between the instantaneous moving direction of Flexbed and the closest vision cone boundary is shown in Fig. 8.16(a); it can be observed that there is a constant avoiding angle between them. (The mean value of the avoiding angle equals -0.3287 rad, with a standard deviation of 0.0551.) The safety of Flexbed is further confirmed by the minimum distance to the

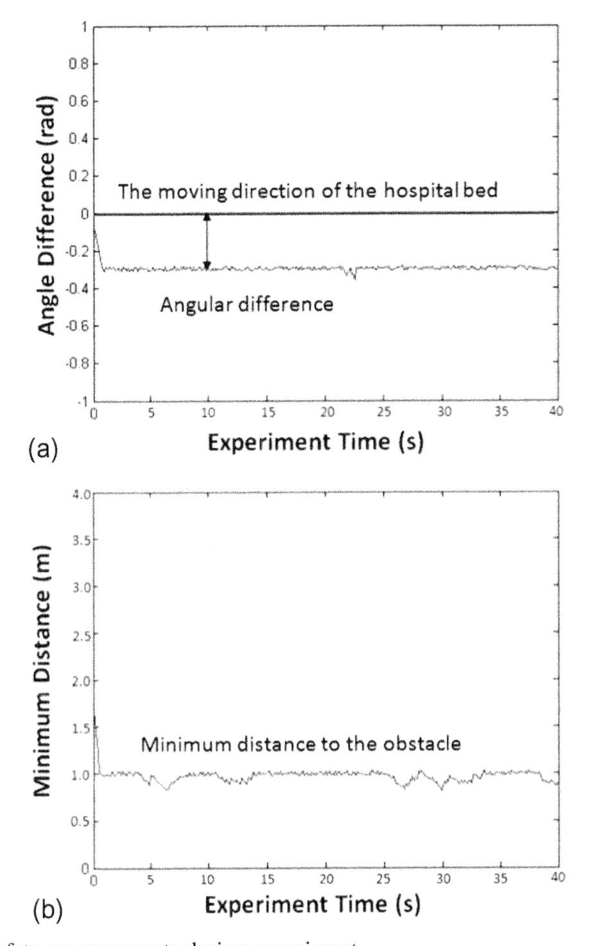

Figure 8.16 Safety measurements during experiment.

obstacle d_i, which is always greater than a safety distance of 0.816 m, as is shown in Fig. 8.16(b).

In the following experiments, we examine the navigation ability of Flexbed and start with a relatively simple scenario. Specifically, Flexbed encounters two moving obstacles in a narrow environment. Figures 8.17(b) and (c) correspond to the moments where Flexbed avoids these two obstacles; it eventually arrives at the destination location, as can be seen in Fig. 8.17(d).

We extend this experiment to a more complicated scenario where a number of moving obstacles are present in the same environment, and Flexbed has to safely arrive at the destination location while avoiding collisions with them. The snapshots of this experiment are shown in Fig. 8.18.

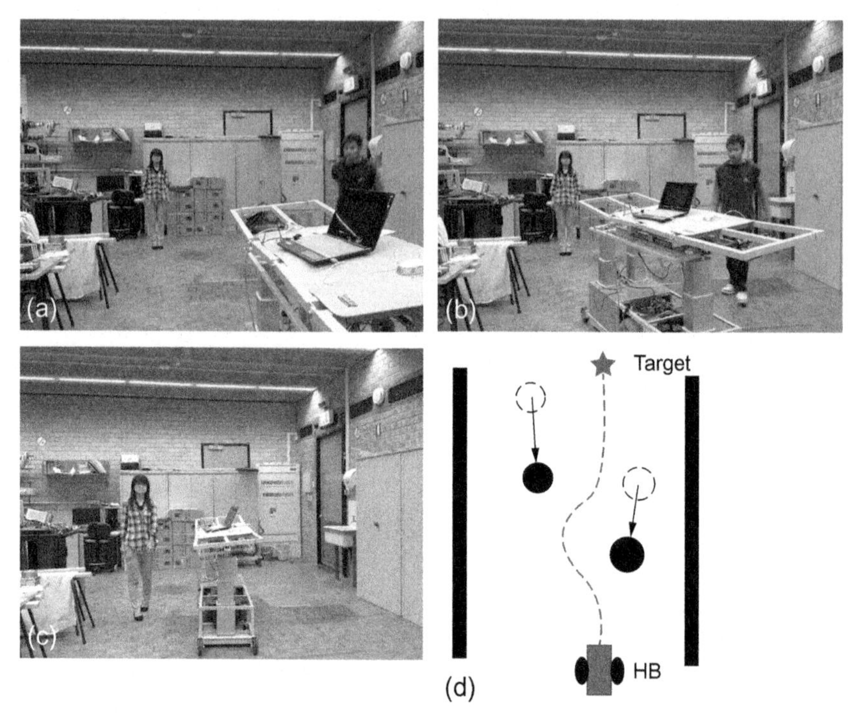

Figure 8.17 Flexbed avoids two moving obstacles.

In the last experiment, we examine the performance of Flexbed in a scenario with a "hidden" obstacle, i.e., an obstacle that appear suddenly and may be considered as a serious potential problem to the safety of Flexbed. This scenario is concerned in Fig. 8.19. After Flexbed senses the "hidden" obstacle within the group of two obstacles, it turns more to the left in order to avoid the entire group with a larger radius. This provides an evidence that Flexbed is able to adapt to sudden changes in the environment of operation and to prevent collisions with "hidden" obstacles.

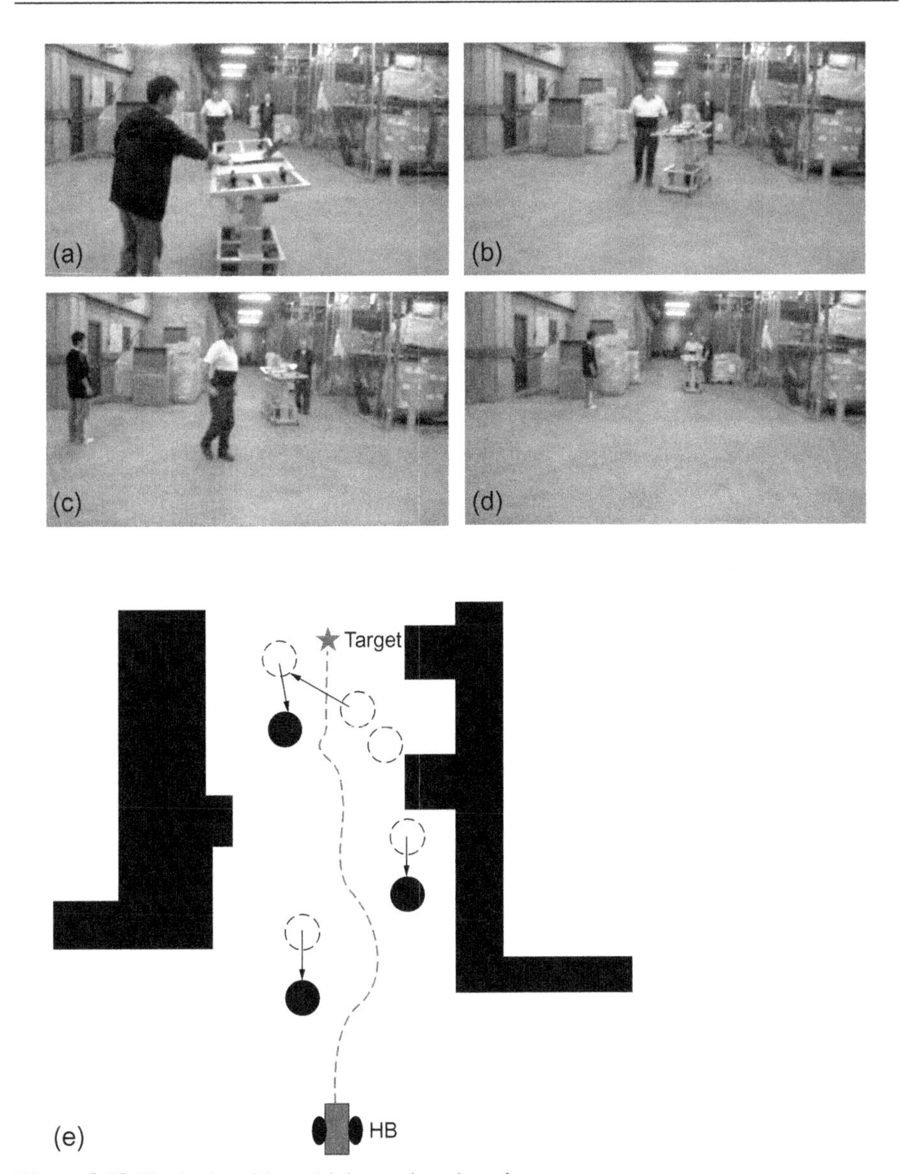

Figure 8.18 Flexbed avoids multiple moving obstacles.

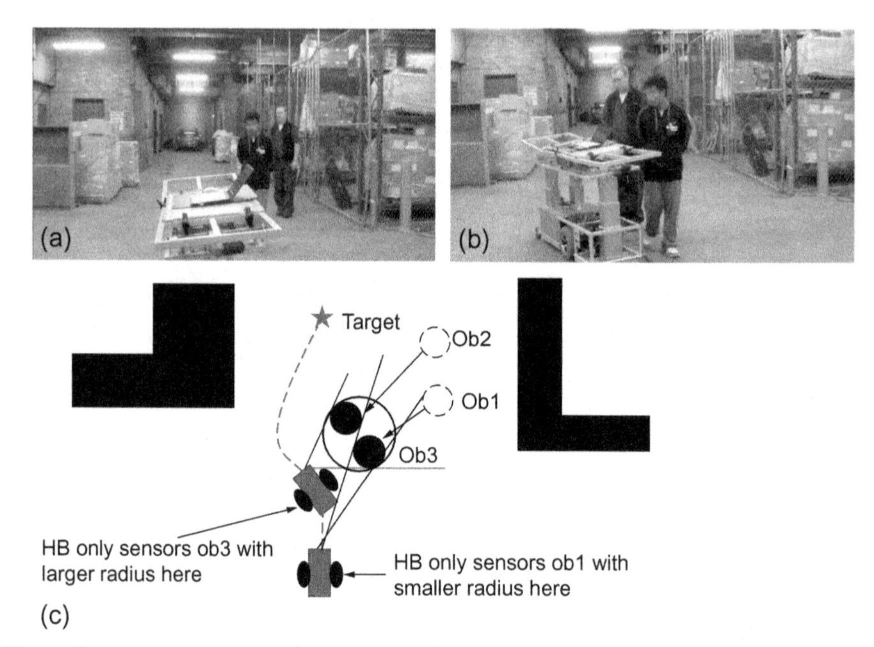

Figure 8.19 Flexbed avoids a “hidden” obstacle.

Reactive navigation among moving and deforming obstacles: Problems of border patrolling and avoiding collisions

9.1 Introduction

Avoidance of collisions with obstacles is a key component of safe navigation. A typical objective is to reach a target through the obstacle-free part of the environment. This may involve bypassing an obstacle, especially a long one, in close range with a safety margin, as in Chapters 4–6. This maneuver is similar to border patrolling, which mission is of self-interest for many applications, as was detailed in Chapter 5.

Unlike the previous chapters and an overwhelming majority of available works in the area of mobile robotics, the obstacles are not assumed to be rigid or even solid in this chapter. They are continuums of arbitrary and time-varying shapes, which may rotate, twist, wring, skew, wriggle, or be deformed in any other way. This, for example, covers scenarios with reconfigurable rigid obstacles, rigid obstacles with large moving parts, forbidden zones between moving obstacles, like inter-vehicle those in a dense platoon, flexible underwater obstacles, like fishing nets, schools of big fish, bunches of cables, etc., virtual obstacles, like areas corrupted with hazardous chemicals, vapor, radiation, high turbulence, etc., or on-line estimated areas of operation of a hostile agent.

In this troublesome scenario, this chapter addresses the problem of reactive navigation of a mobile robot described by the standard model of the Dubins car type. In other words, we consider a nonholonomic under-actuated robotic vehicle moving with a constant speed along planar paths of bounded curvature without reversing direction. As was discussed in Section 3.2, this model describes many mechanical systems such as wheeled robots, UAVs, unmanned underwater vehicles, missiles, etc. The robot is strongly perceptually deficient, may observe only a small part of the obstacle, may not distinguish between its points, and so may be unable to estimate many of its parameters, like size, center, edge, full velocity, etc. It has access only to the distance to the nearest obstacle within the sensor range, the time derivative of this measurement, and the angle-of-sight at the target.

As in Chapters 6 and 7, the examined navigation strategy consists of properly switching between moves to the target in straight lines, when possible, and sliding mode based bypassing enroute obstacles at a pre-specified distance. When used in

Safe Robot Navigation Among Moving and Steady Obstacles. http://dx.doi.org/10.1016/B978-0-12-803730-0.00009-3

isolation, the controller governing the second of these maneuvers solves the problem of patrolling the border of a moving domain with a given safety margin based on only the distance to the border. Therefore, it can be used for pure border or perimeter surveillance, which is of independent self-interest. Unlike the prior research in the area, where only static domains were examined and a certain focus was on a multiple vehicle scenario (see Section 3.6 for details), this chapter deals with the problem of patrolling a moving and deforming border with a single robot scenario.

When applied to border patrolling problems, the examined controller needs sensors with only a bounded range exceeding the safety margin. If the sensor range is large enough, it is rigorously shown that the basin of convergence of the closed-loop system is not local and so this controller is suitable for non-local navigation of the robot at the required distance to the moving domain based on range-only measurements. In some cases (e.g., for several scenarios with convex rigid moving domains), the convergence holds from any remote initial location of the robot.

As was discussed in Section 3.4.1, reliance on range-only measurements is of self-interest since the extensive literature on navigation and guidance mostly assumes the robot has access to both the line-of-sight angle (bearing) and the distance between the robot and the target (range). There is also a relatively large body of research that deals with bearing-only measurements; see, e.g., [417, 418] and literature therein. At the same time, not many publications address guidance toward an unpredictable target on the basis of range-only measurements. However, this is of interest in many areas, e.g., wireless networks, unmanned vehicles, and surveillance services, since many sensors typical for these areas, like sonar or range-only radars, provide only the relative distance between the pursuer and the target via, e.g., measurement of the time-of-flight of an acoustics pulse [419] or the strength of the signal radiated by the target [420]. In fact, this chapter offers an algorithm for approaching and then following a not necessarily point-wise target with a predefined margin based on range-only measurements.

The navigation strategy considered in this chapter develops some ideas set forth in Chapter 6. However, in Chapter 6, only the case of static rigid domains and relatively sparse scenes with convex obstacles was examined. In this chapter, we show that those ideas remain viable for much more general scenarios with moving and deforming domains. We first offer a mathematically rigorous justification of the proposed approach for the problem of patrolling a border of such a domain. This approach originates from the equiangular navigation and guidance law [32, 126] and employs sliding mode regulation; see Section 1.3. The latter gives rise to concerns about the implications of noise and un-modeled dynamics in a practical setting. They are addressed via experimental studies and extensive computer simulations in the second part of the chapter, where border patrolling is used as a part of an integrated navigation strategy to safely guide the robot to the target through an environment cluttered with moving non-convex obstacles. The applicability of the proposed strategy is demonstrated via experiments with real robots and extensive computer simulations. In doing so, its performance has been compared with that of the popular velocity obstacle approach [17, 19] and found to be better under certain circumstances.

The main results of the chapter were originally published in [60]. A preliminary version of some of these results was also reported in the conference paper [421].

The remainder of the chapter is organized as follows. Sections 9.2 and 9.4 describe the problem of border patrolling and the main assumptions, respectively. Section 9.3 introduces the navigation law for border patrolling and offers its informal discussion, while Section 9.5 presents the main theoretical results of the chapter, which are concerned with convergence of this law. Section 9.6 illustrates these results, along with the assumptions and controller parameters requirements, via their specification in five particular scenarios. The first three of them deal with patrolling the boundary of an arbitrarily shaped rigid body that is steady, or moves with an unknown constant velocity, or undergoes a translational motion with unknown and time-varying velocity and acceleration, respectively. The fourth scenario is about escorting a convoy of several vehicles by an under-actuated robot, while the fifth one deals with escort of a single bulky cigar-shaped vehicle. Section 9.7 introduces an algorithm for guidance toward a target in dynamic environments. The related computer simulations and experimental studies are discussed in Sections 9.8 and 9.9, respectively.

9.2 System description and border patrolling problem

As in Chapters 4–8, we consider a planar under-actuated nonholonomic mobile robot of the Dubins car type [146]. It travels in a given plane with a constant speed v and is controlled by the angular velocity u limited by a given constant $\overline{u}$. In the two-dimensional space, the position of the robot is given by the abscissa x and ordinate y of its characteristic point[1] in the world frame, while its orientation is described by the angle θ from the abscissa axis to the robot centerline; see Fig. 9.1. The kinematics of such robots are classically described by the following equations:

$$\begin{array}{l}\dot{x} = v\cos\theta \\ \dot{y} = v\sin\theta\end{array}, \quad \dot{\theta} = u \in [-\overline{u}, \overline{u}], \quad \begin{array}{l}x(0) = x_0 \\ y(0) = y_0\end{array}, \quad \theta(0) = \theta_0. \tag{2.1}$$

These equations capture the capacity of the robot to travel with the given speed in the forward direction along planar curves whose curvature radius exceeds a given threshold. In the case at hand, this minimal turning radius equals

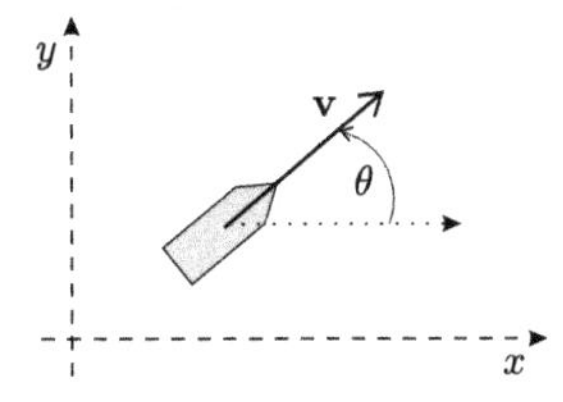

Figure 9.1 Planar robotic vehicle.

[1] For example, the middle of the rear axle for a four wheeled robot.

$$R_{\min} = \frac{v}{\overline{u}}. \tag{2.2}$$

Apart from the robot, there is an unknown moving domain $D = D(t)$ in the plane. This domain is not necessarily a rigid body: it may perpetually change the shape and undergo any displacements and deformations, including stretches, bends, twists, relative motions of one parts with respect to others, etc. We do not limit ourselves to a particular type of domain kinematics and do not specify whether the body D is a rigid, elastic, or plastic solid or fluid, etc. Moreover, the domain D may represent a virtual structure that obeys no laws governing physical continuums.[2]

To cover a larger range of scenarios, we consider a perceptually deficient robot: it has access only to the current distance $d(t)$ to the current border $\partial D(t)$ and the rate $\dot{d}(t)$ at which this measurement evolves over time. The distance to the moving and deforming domain $D(t)$ is given by $d(t) := \mathbf{dist}\,[\boldsymbol{r}(t); D(t)]$, where $\boldsymbol{r}(t) := \mathbf{col}\,[x(t), y(t)]$ is the vector of the vehicle's Cartesian coordinates,

$$\mathbf{dist}\,[\boldsymbol{r}; D] := \min_{\boldsymbol{r}' \in D} \|\boldsymbol{r} - \boldsymbol{r}'\|,$$

and $\|\cdot\|$ denotes the standard Euclidean norm. This min is achieved if D is closed, which is assumed for $D := D(t)$.

The objective is to design a navigation and guidance law that allows the robot to advance on the (a priori) unknown moving domain $D(t)$ and then to patrol its border $\partial D(t)$ at the pre-specified distance d_0 from it and at the given speed v. In doing so, a prescribed safety margin $d_{\text{safe}} \in (0, d_0)$ should always be respected: $d(t) \geq d_{\text{safe}}\ \forall t$. In some applications, the robot should also constantly overtake the nearest boundary point so the full scan of the border is unavoidably completed in due time. This will be referred to as the *full scan* border patrolling.

We first derive a solution for the border patrolling problem. Then we use this solution as a part of a navigation strategy that drives the robot to a given target while avoiding en-route moving obstacles with a pre-specified safety margin.

The border patrolling problem, as is stated above, in fact consists of two parts: advancement to the body and patrolling itself. The first part is basically the problem of target reaching. While most works in the area assume access to the angle-of-sight at the target for solution of this problem, we ensure advancement to the target at the pre-specified distance based on range-only measurements.

9.3 Navigation for border patrolling

We examine the navigation law (5.2.3), which was applied in Chapter 5 to the problem of patrolling the border of a steady rigid domain. We recall that this law is described by the equation

[2] For example, $D(t)$ may be the current estimate of the zone where a hostile agent is operating.

$$u(t) = \overline{u} \cdot \mathbf{sgn} \left\{ \dot{d}(t) + \chi[d(t) - d_0] \right\}, \tag{3.1}$$

where $\chi(\cdot)$ is a linear function with saturation (see Fig. 9.2(a))

$$\chi(z) := \begin{cases} \gamma z & \text{if } |z| \leq \delta, \\ \mu \mathbf{sgn}(z) & \text{if } |z| > \delta, \end{cases} \qquad \mu := \gamma\delta \tag{3.2}$$

and $\gamma, \delta > 0$ are the controller parameters. We stress that this navigation controller generates only feasible controls $u \in [-\overline{u}, \overline{u}]$.

The idea behind the switching law (3.1) is based on the expectation, confirmed by rigorous analysis in Section 9.5, that the major part of the robot's trajectory corresponds to sliding motion along the switching surface $\dot{d} + \chi(d - d_0) = 0$. To illustrate the geometrical meaning of this equation, we assume the body D is rigid for simplicity. Then $|\dot{d}| \leq v_r$, where v_r is the speed of the robot in the coordinate frame attached to the body. Therefore, for the sliding mode to be realistic, the values of $\chi(\cdot)$ should not exceed v_r, i.e., $\mu \leq v_r$ in (3.2). This is ensured by proper tuning of the controller parameters based on analysis of v_r; see Sections 9.4 and 9.5 for details. Under the above condition, the angle α subtended by the normal to the body boundary and the vector of the relative velocity $\mathbf{v}_r$ of the robot (see Fig. 9.2(b)) obeys the equation $\alpha = \arccos\left[-v_r^{-1}\chi(d - d_0)\right]$ in sliding mode. According to this equation, this angle is obtuse for $d > d_0$ and acute for $d < d_0$, and thus the robot is driven to the required distance d_0 in any case.

In the instance of steady body, the angle α is maintained constant $\alpha = \arccos\left[\pm\frac{\mu}{v}\right]$ in the saturation mode, i.e., when the robot is far enough from the body $|d - d_0| > \delta$. The geometry of motion with a constant angle between the heading and the direction to the steady target (with negligible size) is described by the so-called equiangular spiral; see, e.g., [422]. This pattern of target reaching was discovered in, e.g., insects flying toward a candle and peregrine falcons plummeting toward their prey [152, 393]. Inspired by this animal behavior, the equiangular navigation and guidance law $\alpha =$ const for reaching a steady point-wise target was proposed and studied in [32, 126]. In this chapter, the applicability of this law is justified for the general case of moving and deforming domains of substantial sizes.

As the robot approaches the desired distance $|d - d_0| \leq \delta$, the linear mode of the navigation controller (3.1), (3.2) is activated, where $\alpha = \arccos[-\gamma(d - d_0)/v_r]$.

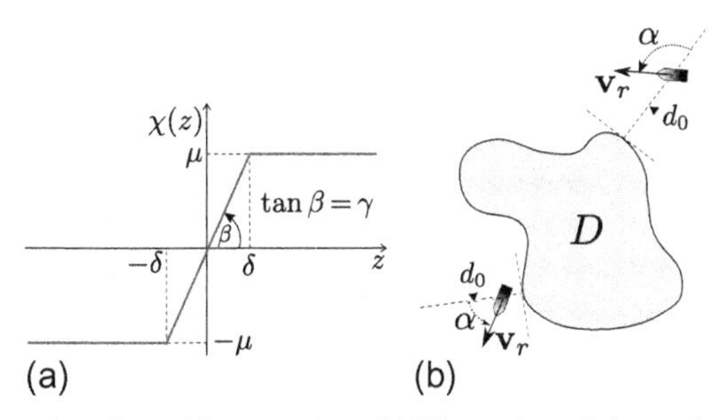

Figure 9.2 (a) Linear function with saturation; (b) Illustration of the navigation law.

Hence, for $d = d_0$, the heading is directed parallel to the body boundary $\alpha = \pi/2$, thus driving the robot along the d_0-equidistant curve. The speed of "descent" to this curve $\dot{d}$ is proportional $\dot{d} = -\gamma(d - d_0)$ to the distance $d - d_0$ to it, which is similar to some biologically inspired landing strategies of fixed-wing aircrafts [423, Section X.C].

Practical implementation of the proposed navigation law requires only range measurements, including access to the derivative $\dot{d}$. If it is not directly measured, e.g., by the use of the Doppler effect, numerical differentiation of d may be employed. Estimation of derivatives from noise-corrupted data is a well-established discipline that offers a variety of approaches. Apart from classic numerical differentiation, they include approximation of the transfer function of the ideal differentiator, optimal differentiation based on stochastic models, observers with sliding modes and large gains, difference methods, approximation and Tikhonov regularization, etc. (see, e.g., [388, 389] for recent surveys). The implications of the numerical differentiation errors will be addressed in Sections 9.8 and 9.9 by means of computer simulations and experimental studies. The problem of chattering, which often accompanies the use of switching controllers like (3.1), will be addressed in Section 9.9 via experimental studies as well.

For the navigation objective to be achievable, the robot should be capable of tracking the d_0-equidistant curve of the body boundary. However, this is impossible irrespective of the robot's capabilities if this curve contains cusp singularities, so far as any path of the unicycle (2.1) is everywhere smooth. Such singularities typically occur whenever the boundary contains concavities and the required distance d_0 exceeds the critical value, which is equal to the minimal curvature radius of the concavity parts of the boundary [390]; see Fig. 9.3. In this case, the objective of maintaining the distance $d = d_0$ cannot be achieved by any means and should be replaced by a realistic one, e.g., to follow a non-equidistant path that smoothly connects the regular parts of the equidistant curve. In this chapter, we do not consider the related developments, including path planning, which may be viewed as concerned with a separate problem, and assume that if the body boundary contains concavities, the required distance d_0 is so small that the related equidistant curve is smooth. This also ensures uniqueness of the minimum-distance point [422].

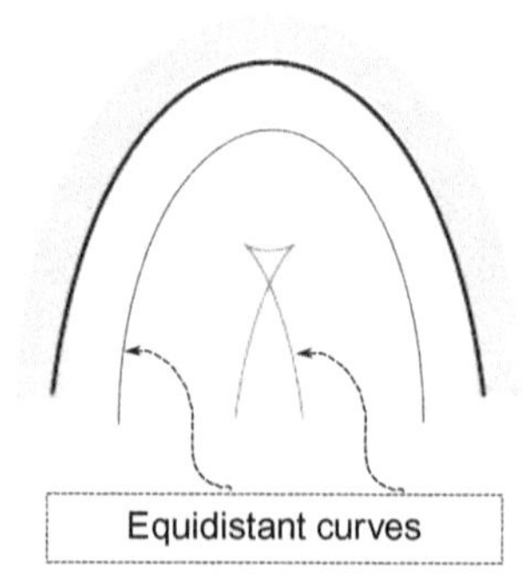

Figure 9.3 Singularities of equidistant curves.

The preliminary analysis in the previous paragraph tacitly assumed that the body is rigid and steady. In this case, the non-local convergence of the navigation controller (3.1), (3.2) was theoretically justified in Chapter 5. In Section 9.5, we extend these results to a much wider class of moving and deforming domains. To this end, we first establish conditions necessary for the navigation objective to be achievable in this general case (Lemma 4.1). Assuming these unavoidable conditions do hold in a slightly enhanced form (Assumption 4.5), we show that this is enough for the navigation law (3.1), (3.2) to ensure the local stability of the output d near d_0 (Lemma 5.1). Non-local convergence is established in Theorem 5.1 at the expense of extending the above conditions from the vicinity of the equidistant curve to the estimated area of the transient localization.

9.4 Main assumptions

To qualitatively describe the properties of the moving continuum $D(t)$, we use the Lagrangian formalism [424]: we introduce a *reference configuration* $D_* \subset \mathbb{R}^2$ of the body and the *configuration map* $\Phi(\cdot, t) : \mathbb{R}^2 \to \mathbb{R}^2$ that transforms D_* into the current configuration $D(t) = \Phi[D_*, t]$. We limit ourselves by only a few conventions typical for the continuum mechanics and listed in the following.

Assumption 4.1. *The reference domain D_* is compact and has a smooth boundary ∂D_*. The configuration map $\Phi(\cdot, t)$ is defined on an open neighborhood of D_* and is smooth, the determinant of its Jacobian matrix is everywhere positive* $\det \Phi_{\boldsymbol{r}}' > 0$, *and* $\Phi(\boldsymbol{r}_1, t) \neq \Phi(\boldsymbol{r}_2, t)$ *whenever* $\boldsymbol{r}_1 \neq \boldsymbol{r}_2$.

These assumptions imply that topological properties of $D(t)$ do not change as time progresses (the body $D(t)$ remains homeomorphic to itself), e.g., the body does not split into separate parts or intersect with itself, etc.[3] Furthermore, the boundary $\partial D(t)$ is smooth at any time.

To proceed, we need some notations, which partly refer to the Eulerian formalism [424] in the description of the body motion:

- $\Phi^{-1}(\cdot, t)$—the inverse map;
- $\mathbf{V}(\boldsymbol{r}, t) := \Phi_t'[\Phi^{-1}(\boldsymbol{r}, t)]$—the velocity vector field;
- $\mathbf{A}(\boldsymbol{r}, t) := \Phi_{tt}''[\Phi^{-1}(\boldsymbol{r}, t)]$—the acceleration vector field;
- $\mathbf{V}_r'[\boldsymbol{r}, t]$—the spatial velocity gradient tensor;
- $\mathcal{E} := \frac{1}{2}\left[\mathbf{V}_r'[\boldsymbol{r}, t] + \mathbf{V}_r'[\boldsymbol{r}, t]^{\mathrm{T}}\right]$—the strain rate tensor;
- $\omega = \omega(\boldsymbol{r}, t)$—the vorticity, or angular velocity of the rigid-body-rotation, i.e.,

$$\begin{pmatrix} 0 & \omega \\ -\omega & 0 \end{pmatrix} = \frac{1}{2}\left[\mathbf{V}_r' - \mathbf{V}_r'^{\mathrm{T}}\right];$$

- $\langle \cdot ; \cdot \rangle$—the standard inner product;
- $\mathbf{T}(\boldsymbol{r}, t), \mathbf{N}(\boldsymbol{r}, t)$—the Frenet frame of $\partial D(t)$ at $\boldsymbol{r} \in \partial D(t)$; the domain $D(t)$ is to the left to the tangent vector $\mathbf{T}$, the normal vector $\mathbf{N}$ is directed inward $D(t)$;
- $W_{T,t}(\boldsymbol{r}, t) := \langle \mathbf{W}(\boldsymbol{r}, t); \mathbf{T}(\boldsymbol{r}, t) \rangle$—the tangential component of the vector field $\mathbf{W}$ at the point $\boldsymbol{r} \in \partial D(t)$;

[3] This does hold for many solids but may be a restriction for, e.g., fluids or virtual bodies.

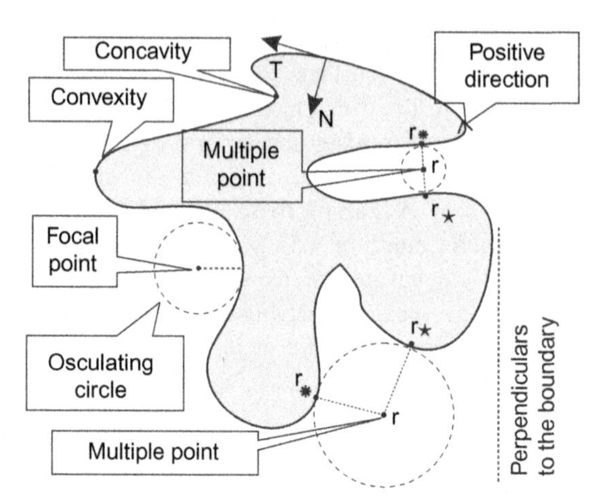

Figure 9.4 Various types of boundary points.

- $W_{N,t}(\boldsymbol{r}, t)$—the normal component at the point $\boldsymbol{r} \in \partial D(t)$;
- $\kappa(\boldsymbol{r}, t)$—the signed curvature of $\partial D(t)$ at the point $\boldsymbol{r}$;
- $\sigma[\boldsymbol{r}, t] := \langle \mathcal{E}\mathbf{T}; \mathbf{N}\rangle + \omega = \langle \mathbf{V}'_{\boldsymbol{r}}\mathbf{T}; \mathbf{N}\rangle$—the normal component of the velocity rate-of-change under an infinitesimally small shift along the boundary $\partial D(t)$;
- $\boldsymbol{r}(t)$—the current position of the robot;
- $\boldsymbol{r}_*(t)$—the point of $\partial D(t)$ nearest to $\boldsymbol{r}(t)$;
- $\mathcal{R}_\theta$—the matrix of counter-clockwise rotation through angle θ.

The signed curvature κ is positive and negative on the convex and concave parts of $\partial D(t)$, respectively, so that the Frenet-Serrat equations are satisfied, i.e.,

$$\frac{\mathrm{d}\mathbf{T}}{\mathrm{d}s} = \kappa \mathbf{N}, \quad \frac{\mathrm{d}\mathbf{N}}{\mathrm{d}s} = -\kappa \mathbf{T}, \tag{4.1}$$

where s is the natural parameter on $\partial D(t)$; see Fig. 9.4.

We also recall the following useful relations [424, Ch.4]:

$$\mathbf{A} = \mathbf{V}'_{\boldsymbol{r}}\mathbf{V} + \mathbf{V}'_t, \quad \Phi''_{\boldsymbol{r}t} = \mathbf{V}'_{\boldsymbol{r}}[\Phi, t]\Phi'_{\boldsymbol{r}}. \tag{4.2}$$

The second technical assumption serves asymptotic (as $t \to \infty$) analysis of the closed-loop system.

Assumption 4.2. *The scalars $V_N[\boldsymbol{r}, t]$, $V_T[\boldsymbol{r}, t]$, $A_N[\boldsymbol{r}, t]$, $\sigma[\boldsymbol{r}, t]$, $\kappa[\boldsymbol{r}, t]$ remain bounded as $\boldsymbol{r} \in \partial D(t)$ and $t \to \infty$.*

It is worth noting that these quantities are bounded over $\boldsymbol{r} \in \partial D(t)$ and t within any finite time horizon since all of them, along with $\Phi(\boldsymbol{r}, t)$, continuously depend on $\boldsymbol{r}$ and t, and the reference domain D_* is compact by Assumption 4.1.

Due to restricted maneuverability, the robot is not always capable of border patrolling. Briefly, the d_0-equidistant curve to be traced should not move too fast and be sharply contorted due to the limited turning radius (2.2). The following discussion details these requirements and provides evidence that the majority of them are nearly unavoidable.

Definition 4.1. A point $\boldsymbol{r} \notin D$ is said to be *multiple* (for D) if there exist two different *minimum-distance* points $\boldsymbol{r}_*, \boldsymbol{r}_\star \in D$, i.e., such that $\mathbf{dist}\,[\boldsymbol{r}; D] = \|\boldsymbol{r} - \boldsymbol{r}_*\| = \|\boldsymbol{r} - \boldsymbol{r}_\star\|$, and *focal* if there exists a minimum-distance point $\boldsymbol{r}_* \in D$ such that $1 + \mathbf{dist}\,[\boldsymbol{r}; D]\,\kappa = 0$, where κ is the curvature of ∂D at $\boldsymbol{r}_*$.

These notions are illustrated in Fig. 9.4.

As is shown by Lemma 5.3.1, the following assumption is nearly unavoidable in the case of the static $D(t) \equiv D$ domain.

Assumption 4.3. *There are neither multiple nor focal points at the d_0-equidistant curve of the domain $D(t)$.*

For any point $\boldsymbol{r}$ on the equidistant curve $\mathscr{E}\mathscr{C}$, there are no focal points between $\boldsymbol{r}$ and the nearest point $\boldsymbol{r}_* \in D(t)$ [425, Ch.16]. Assumption 4.3 excludes the marginal case where the end $\boldsymbol{r}$ of the segment $[\boldsymbol{r}, \boldsymbol{r}_*]$ is focal. Furthermore, this assumption implies that $1 + \varkappa d_0 > 0$ on $\mathscr{E}\mathscr{C}$. Now we require that this remains true as $t \to \infty$.

Assumption 4.4. *The following relation holds:*

$$\lim_{t \to \infty} \inf_{\boldsymbol{r} \in \partial D(t) : \kappa[\boldsymbol{r}, t] < 0} (1 + d_0 \kappa[\boldsymbol{r}, t]) > 0. \tag{4.3}$$

This necessarily holds if either $D(t)$ is convex or d_0 is small enough [425, Ch.16].

The following proposition discloses conditions necessary for the navigation objective to be realistic.

Proposition 4.1. *Let the robot travel so that $d(t) \equiv d_0$. For any time t, the parameters of the domain motion at the point $[\boldsymbol{r}_*(t), t]$ satisfy the following relations:*

$$|V_N| \le v, \quad \overline{u} \ge \frac{|\mathcal{A}|}{\sqrt{v^2 - V_N^2}}, \quad \textit{where}$$

$$\mathcal{A} = \mathcal{A}(\boldsymbol{r}_*, t, d_0) := A_N + \frac{2\sigma\xi + \kappa\xi^2 - d_0\sigma^2}{1 + \kappa d_0}$$

$$\textit{and} \quad \xi := -V_T \pm \sqrt{v^2 - V_N^2}. \tag{4.4}$$

If the robot overtakes the nearest boundary point, then additionally

$$\sqrt{v^2 - V_N^2} \ge \pm (V_T + d_0\sigma). \tag{4.5}$$

In (4.4) *and* (4.5), *the sign* $+$ *is taken if the robot moves so that the domain is to the left, and* $-$ *is taken otherwise.*

The proof of this proposition is given in Section 9.4.1.

Thus, conditions (4.4) must be satisfied, otherwise the navigation objective can be achieved by no means. Since the nearest boundary point $\boldsymbol{r}_*(t)$ is not known in advance, it is reasonable to require that (4.4) holds for all boundary points at any time. The next assumption enhances this a bit by substituting the uniformly strict inequality in place of the non-strict one.

Assumption 4.5. *There exist $\lambda_v, \lambda_a \in (0, 1)$ such that*

$$|V_N| \le \lambda_v v, \quad \frac{|\mathcal{A}|}{\sqrt{v^2 - V_N^2}} \le \lambda_a \overline{u} \tag{4.6}$$

at any time t, point $\boldsymbol{r}_* \in \partial D(t)$, and with both signs $\pm$.

For any finite time horizon, such λ_v, λ_a exist if and only if the left-hand sides of (4.6) are strictly less than v and $\overline{u}$, respectively, for all concerned $\boldsymbol{r}_*$ and t. By introducing λ_v, λ_a, we ensure the discrepancies between the right- and left-hand sides do not vanish as $t \to \infty$.

The strict inequalities contained in (4.6) allow not only to maintain the distance to the domain $\dot{d} = 0$ but also to make it both descending $\dot{d} < 0$ and ascending $\dot{d} > 0$. In other words, they in fact guarantee that the output d is locally controllable. As will be shown, the above assumptions are enough for local stability of the closed-loop system in a vicinity of the desired equidistant trajectory. To establish non-local convergence to this trajectory, we need to ensure controllability during the transient. To this end, we introduce two definitions.

Definition 4.2. An interval $[d_-, d_+]$ is said to be *regular* if there exist $\lambda_v, \lambda_a \in (0, 1)$ such that Assumptions 4.3–4.5 hold with d_0 replaced $d_0 := d$ by any d from this interval.

Since in (4.4) $\mathcal{A}$ is the (first degree)/(first degree) rational function in d_0, simple calculus shows that it suffices to check Assumption 4.5 only for $d = d_-, d_+$. Similarly, Assumption 4.4 holds for all $d \in [d_-, d_+]$ if and only if

$$\lim_{t \to \infty} \inf_{\boldsymbol{r} \in \partial D(t): \kappa[\boldsymbol{r}, t] < 0} (1 + d_+ \kappa[\boldsymbol{r}, t]) > 0.$$

Remark 4.1. Any interval $[d_-, d_+]$, where d_-, d_+ are close enough to d_0, is regular due to uniform continuity of the concerned expressions from (4.3) and (4.6) in d_0.

However, we need more: this interval can be expanded to include all distances $d(t)$ that may be encountered during the transient. To flesh out this, we introduce the following.

Definition 4.3. The $\pm$-*initial circle* is the path of the robot driven by the maximal actuation $u \equiv \pm\overline{u}$ from its initial position; the corresponding motion of the robot during the first one and a half full turns over the initial circle is said to be $\pm$-*initial* (see Fig. 9.5). The distance from the robot to $D(t)$ observed within any initial motion is said to be *launching*.

Now we summarize the above discussion by imposing the cumulative assumption, which covers Assumptions 4.3–4.5.

Assumption 4.6. *Let $\mathfrak{D}$ be the set of all launching distances d, along with $d := d_0$, and*

$$d_- := \min_{d \in \mathfrak{D}} d, \quad d_+ := \max_{d \in \mathfrak{D}} d.$$

The interval $[d_-, d_+]$ is regular and $d_- > d_{safe}$.

The last requirement ensures that any initial motion (which lasts $\frac{3\pi}{\overline{u}}$ time units) does not involve collisions with $D(t)$. The last assumption enhances this.

Assumption 4.7. *During the first $\frac{3\pi}{\overline{u}}$ time units, the body $D(t)$ and the union of the two initial circles remain in disjoint steady half-planes; see Fig. 9.5.*

In other words, the robot starts far enough from the domain.

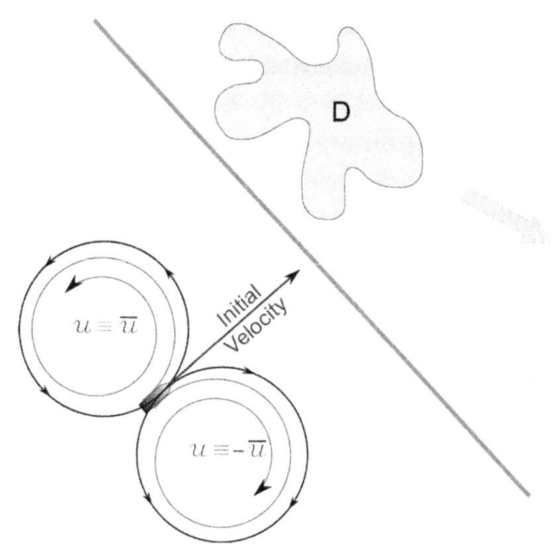

Figure 9.5 Initial circles.

9.4.1 Proof of Proposition 4.1

In this subsection, the Frenet frame $[\mathbf{T}(\boldsymbol{r},t), \mathbf{N}(\boldsymbol{r},t)]$ and the variables attributed to the motion of the domain, such as $\mathbf{V}(\boldsymbol{r},t), \mathbf{A}(\boldsymbol{r},t)$, etc., are considered only for $\boldsymbol{r} := \boldsymbol{r}_*(t)$, where $\boldsymbol{r}_*(t)$ is the point of $D(t)$ that is closest to $\boldsymbol{r}(t)$. With a slight abuse of notations, the resultant argument $[\boldsymbol{r}_*(t), t]$ is replaced by t and dropped if t is clear from the context. We denote by $\mathbf{Nr}[\mathbf{W}] := \mathbf{W}/\|\mathbf{W}\|, \mathbf{Pr}_{\mathbf{W}}$, and $\mathbf{W}^{\perp}$ the normalization of the vector $\mathbf{W}$, the orthogonal projection on the line spanned by $\mathbf{W}$, and the line perpendicular to $\mathbf{W}$, respectively.

Lemma 4.1. *The velocity* $\mathbf{v}$ *of the robot has the form*

$$\mathbf{v} = \underbrace{\left[\overset{\bullet}{s} + d\mu \right]}_{\xi} \mathbf{T} + \mathbf{V} - \dot{d}\,\mathbf{N}, \tag{4.7}$$

where $\overset{\bullet}{s}$ *is the speed of the relative motion of* $\boldsymbol{r}_*$ *along the boundary* ∂D, *i.e.,*

$$\dot{\boldsymbol{r}}_* - \mathbf{V} = \overset{\bullet}{s}\mathbf{T},$$

and

$$\varsigma := \left\langle \mathbf{V}_r'\mathbf{T}, \mathbf{N} \right\rangle + \kappa \overset{\bullet}{s} = \langle \mathcal{E}\mathbf{T}, \mathbf{N} \rangle + \omega + \kappa \overset{\bullet}{s}. \tag{4.8}$$

The Frenet frame evolves so that

$$\frac{\mathrm{d}\mathbf{T}}{\mathrm{d}t} = \varsigma \mathbf{N}, \quad \frac{\mathrm{d}\mathbf{N}}{\mathrm{d}t} = -\varsigma \mathbf{T}. \tag{4.9}$$

Proof. To prove (4.7) and (4.9) at a particular time instant τ, we reset the reference configuration to be $D(\tau)$. This does not alter the velocity and acceleration fields and their descendants but converts $\Phi(\boldsymbol{r},t)$ into $\Phi(\boldsymbol{r},t|\tau) := \Phi\left[\Phi^{-1}(\boldsymbol{r},\tau),t\right]$. Let $\boldsymbol{\rho}(s)$ stand for a regular parametric representation of $\partial D(\tau)$, where s is the arc length, or natural parameter (which ascends so that $D(\tau)$ is to the left). Then

$$\boldsymbol{r}(t) = \Phi\left\{\boldsymbol{\rho}[s(t)],t\middle|\tau\right\} - d\mathbf{N} \tag{4.10}$$

and $\boldsymbol{r}_*(t) = \Phi\left\{\boldsymbol{\rho}[s(t)],t\middle|\tau\right\}$, whence $\dot{\boldsymbol{r}}_*(\tau) = \dot{s}(\tau)\mathbf{T}(\tau) + \mathbf{V}(\tau) \Rightarrow \overset{\bullet}{s} = \dot{s}(\tau)$. Since the map $s \mapsto \Phi\left\{\boldsymbol{\rho}[s],t\middle|\tau\right\}$ provides parametric representation of $\partial D(t)$, we have

$$\mathbf{T}\{\Phi[\boldsymbol{\rho}(s),t|\tau]\} = \mathbf{Nr}\left\{\Phi_{\boldsymbol{r}}'\left\{\boldsymbol{\rho}[s],t\middle|\tau\right\}\mathbf{T}[\boldsymbol{\rho}(s),\tau]\right\},$$
$$\mathbf{T} = \mathbf{Nr}\left\{\Phi_{\boldsymbol{r}}'\left\{\boldsymbol{\rho}[s(t)],t\middle|\tau\right\}\mathbf{T}[\boldsymbol{\rho}(s(t)),\tau]\right\}.$$

In what follows, $\mathbf{d}$ stands for the differential. Since $\Phi(r,\tau|\tau) \equiv r$ and for infinitesimally small increments,

$$\mathbf{d}\,\mathbf{Nr}[\mathbf{W}] = \frac{1}{\|\mathbf{W}\|}\mathbf{Pr}_{\mathbf{W}^{\perp}}[\mathbf{dW}],$$

we see that for an infinitesimal time increment $t = \tau + \mathbf{d}t$,

$$\begin{aligned}
\mathbf{dT} &= \mathbf{Pr}_{\mathbf{N}(\tau)}\left[\mathbf{d}\Phi_{\boldsymbol{r}}'\left\{\boldsymbol{\rho}[s(t)],t\middle|\tau\right\}\mathbf{T}\left\{\boldsymbol{\rho}[s(t)],\tau\right\}\right] \\
&= \mathbf{Pr}_{\mathbf{N}(\tau)}\Bigg[\Phi_{\boldsymbol{r}}'\left\{\boldsymbol{\rho}[s(\tau+\mathbf{d}t)],\tau+\mathbf{d}t\middle|\tau\right\}\mathbf{T}\left\{\boldsymbol{\rho}[s(\tau+\mathbf{d}t),\tau]\right\}. \\
&\quad - \Phi_{\boldsymbol{r}}'\left\{\boldsymbol{\rho}[s(\tau+\mathbf{d}t)],\tau\middle|\tau\right\}\mathbf{T}\left\{\boldsymbol{\rho}[s(\tau+\mathbf{d}t),\tau]\right\} + \mathbf{d}\,\underbrace{\Phi_{\boldsymbol{r}}'\left\{\boldsymbol{\rho}[s(t)],\tau\middle|\tau\right\}}_{=I}\mathbf{T}\left\{\boldsymbol{\rho}[s(t),\tau]\right\}\Bigg] \\
&\overset{(4.1)}{=\!=} \mathbf{Pr}_{\mathbf{N}(\tau)}\left[\Phi_{rt}''\left\{\boldsymbol{\rho}[s(\tau)],\tau\middle|\tau\right\}\mathbf{T}(\tau) + \kappa(\tau)\dot{s}(\tau)\mathbf{N}[\tau]\right]\,\mathbf{d}t \\
&\overset{(4.2)}{=\!=} \mathbf{Pr}_{\mathbf{N}(\tau)}\left[\mathbf{V}_{\boldsymbol{r}}'(\tau)\mathbf{T}(\tau) + \kappa(\tau)\dot{s}(\tau)\mathbf{N}(\tau)\right]\,\mathbf{d}t \\
&= \underbrace{\left[\left\langle\mathbf{V}_{\boldsymbol{r}}'(\tau)\mathbf{T}(\tau),\mathbf{N}(\tau)\right\rangle + \kappa(\tau)\dot{s}(\tau)\right]}_{\varsigma}\mathbf{N}(\tau)\,\mathbf{d}t,
\end{aligned}$$

which implies the first equation from (4.9). The second one is implied by the relations $\mathbf{N} = \mathscr{R}_{\frac{\pi}{2}}\mathbf{T}, -\mathbf{T} = \mathscr{R}_{\frac{\pi}{2}}\mathbf{N}$, where $\mathscr{R}_{\frac{\pi}{2}}$ is the matrix of counter-clockwise rotation through angle $\pi/2$. Equation (4.7) is immediate from (4.9) and (4.10). □

By invoking that $\|\mathbf{v}\| = v$ for the unicycle, we arrive at

$$(\xi + V_T)^2 + (V_N - \dot{d})^2 = v^2. \tag{4.11}$$

Lemma 4.2. *The following relation holds*

$$\ddot{d} = \xi\varsigma - u\left[\xi + V_T\right] + A_N + \overset{\bullet}{s}\left\langle\mathbf{V}_{\boldsymbol{r}}'\mathbf{T};\mathbf{N}\right\rangle \tag{4.12}$$

$$= -u\left[\xi + V_T\right] + A_N + \frac{2\sigma\xi + \kappa\xi^2 - d\sigma^2}{1+\kappa d}; \tag{4.13}$$

$$\overset{\bullet}{s} = \frac{\xi - d\sigma}{1+\kappa d}. \tag{4.14}$$

Proof. To prove the required relations at a particular time instant τ, we employ the notations introduced in the proof of the previous lemma and consider an infinitesimal time increment $t = \tau + \mathbf{d}t$. To avoid hash in notations of the differential $\mathbf{d}$ and distance d to the domain, we denote the latter by η in this proof. Due to (4.7), (4.9),

$$\begin{aligned}\mathbf{dv} &= (\mathbf{d}\xi)\mathbf{T} + \xi\varsigma\mathbf{N}\mathbf{d}t + \mathbf{dV} - \ddot{\eta}\mathbf{N}\mathbf{d}t + \dot{\eta}\varsigma\mathbf{T}\mathbf{d}t \\ &= [(\mathbf{d}\xi) + \dot{\eta}\varsigma\mathbf{d}t]\,\mathbf{T} + [\xi\varsigma - \ddot{\eta}]\,\mathbf{N}\mathbf{d}t + \mathbf{dV}.\end{aligned}$$

Here

$$\begin{aligned}\mathbf{dV} &= \mathbf{V}[\boldsymbol{r}_*(\tau + \mathbf{d}t), \tau + \mathbf{d}t] - \mathbf{V}[\boldsymbol{r}_*(\tau), \tau] \\ &= \left\{\mathbf{V}_r'[\boldsymbol{r}_*(\tau), \tau]\dot{\boldsymbol{r}}_*(\tau) + \mathbf{V}_t'[\boldsymbol{r}_*(\tau), \tau]\right\}\mathbf{d}t \\ &\overset{(4.2)}{=\!=} \left\{\mathbf{V}_r'\left[\mathbf{V} + \overset{\bullet}{s}\mathbf{T}\right] + \mathbf{A} - \mathbf{V}_r'\mathbf{V}\right\}\mathbf{d}t = \left\{\mathbf{A} + \overset{\bullet}{s}\mathbf{V}_r'\mathbf{T}\right\}\mathbf{d}t.\end{aligned}$$

Summarizing, we have

$$\mathbf{dv} = [(\mathbf{d}\xi) + \dot{\eta}\varsigma\mathbf{d}t]\,\mathbf{T} + [\xi\varsigma - \ddot{\eta}]\,\mathbf{N}\mathbf{d}t + \left\{\mathbf{A} + \overset{\bullet}{s}\mathbf{V}_r'\mathbf{T}\right\}\mathbf{d}t.$$

On the other hand, (2.1) implies that

$$\mathbf{dv} = u\mathcal{R}_{\frac{\pi}{2}}\mathbf{v}\mathbf{d}t \overset{(4.7)}{=\!=} u\mathcal{R}_{\frac{\pi}{2}}\,[\xi\mathbf{T} + \mathbf{V} - \dot{\eta}\mathbf{N}]\,\mathbf{d}t = u\,[\xi\mathbf{N} + V_T\mathbf{N} - V_N\mathbf{T} + \dot{\eta}\mathbf{T}]\,\mathbf{d}t.$$

Equating these two expressions for $d\mathbf{v}$ yields that

$$\begin{aligned}u\,[\dot{\eta} - V_N] &= \frac{\mathrm{d}\xi}{\mathrm{d}t} + \dot{\eta}\varsigma + A_T + \overset{\bullet}{s}\left\langle\mathbf{V}_r'\mathbf{T};\mathbf{T}\right\rangle, \\ u\,[\xi + V_T] &= \xi\varsigma - \ddot{\eta} + A_N + \overset{\bullet}{s}\left\langle\mathbf{V}_r'\mathbf{T};\mathbf{N}\right\rangle \Rightarrow (4.12).\end{aligned}$$

By (4.7), (4.8),

$$\xi = \overset{\bullet}{s} + \eta\varsigma, \quad \varsigma = \sigma + \kappa\overset{\bullet}{s} \Rightarrow \overset{\bullet}{s} = \frac{\xi - \eta\sigma}{1 + \kappa\eta}, \quad \varsigma = \frac{\sigma + \kappa\xi}{1 + \kappa\eta}.$$

Therefore, (4.14) does hold and (4.13) is immediate from (4.12). □

Corollary 4.1. *Whenever the robot travels with $d \equiv d_0$, the parameter ξ is determined by the formula from* (4.4) *and*

$$\overline{u}^2 v^2 \geq V_N^2\overline{u}^2 + \left[A_N + \frac{2\sigma\xi + \kappa\xi^2 - d_0\sigma^2}{1 + \kappa d_0}\right]^2; \tag{4.15}$$

$$\overset{\bullet}{s} = \frac{-V_T \pm \sqrt{v^2 - V_N^2} - d_0\sigma}{1 + \kappa d_0}, \tag{4.16}$$

where + is taken if the robot moves so that the domain is to the left, and − otherwise.

Proof. By (4.11), $(\xi + V_T)^2 + V_N^2 = v^2$, which implies the first claim of the lemma. This claim and (4.14) justify (4.16). To prove (4.15), we note that (4.13) shapes into

$$u\,[\xi + V_T] = A_N + \frac{2\sigma\xi + \kappa\xi^2 - d_0\sigma^2}{1 + \kappa d_0},$$

and take into account that $|u| \leq \overline{u}$. □

Proof of Proposition 4.1. The proof is straightforward from Corollary 4.1 and (4.14) since $1 + d_0\sigma > 0$ in (4.14) and overtaking means that $\overset{\bullet}{s} > 0$ if the robot has the domain to the left, and that $\overset{\bullet}{s} < 0$ otherwise. □

9.5 Main results concerning border patrolling problem

To properly tune the parameters of the controller (3.1), (3.2), two real numbers $\eta_v > 0, \eta_a > 0$ should first be picked such that

$$\lambda_v + \eta_v < 1, \quad \lambda_a + \eta_a < 1, \tag{5.1}$$

where $\lambda_v, \lambda_a < 1$ are the parameters associated with the regular interval at hand by Definition 4.2. We also note that for any choice of the sign in $\pm$, the function

$$\Omega(\boldsymbol{r}_*, t, d, z) := \frac{A_N + \frac{2\sigma\overline{\xi} + \kappa\overline{\xi}^2 - d\sigma^2}{1+\kappa d}}{\sqrt{v^2 - (V_N + z)^2}},$$
$$\text{where} \quad \overline{\xi} := -V_T \pm \sqrt{v^2 - (V_N + z)^2}, \tag{5.2}$$

is continuous in $z \approx 0$ uniformly over $\boldsymbol{r}_* \in \partial D(t), t \geq 0, d \in [d_-, d_+]$. For $z := 0$, its absolute value equals the left-hand side of the second inequality from (4.6). Hence, there exists $z_* > 0$ such that for the above $\boldsymbol{r}_*, t, d$,

$$|\Omega(\boldsymbol{r}_*, t, d, z)| < (\lambda_a + \eta_a)\overline{u} \quad \forall z \in [-z_*, z_*]. \tag{5.3}$$

Now we are in a position to state the first result concerning the border patrolling problem.

Theorem 5.1. *Let Assumptions 4.1, 4.2, 4.6, and 4.7 be satisfied and the robot be driven by the navigation rule* (3.1), (3.2), *where the parameters $\gamma > 0$ and $\delta > 0$ are chosen so that*

$$\mu = \gamma\delta \leq \min\{\eta_v v, z_*\}, \quad \gamma\mu < v\overline{u}[1 - (\lambda_a + \eta_a)]\sqrt{1 - (\lambda_v + \eta_v)^2}. \tag{5.4}$$

Then the following statements hold:

- **i.** *The navigation law* (3.1) *asymptotically drives the robot at the desired distance to $D(t)$, i.e., $d(t) \to d_0$ and $\dot{d}(t) \to 0$ as $t \to \infty$;*
- **ii.** *The safety margin is always respected: $d(t) \geq d_{safe}$ $\forall t$;*
- **iii.** *The motion splits into the* transient *and sliding motion as follows:*
  - **a.** *The transient holds while $S := \dot{d} + \chi(d - d_0) \neq 0$, necessarily terminates and lasts no longer than $\frac{3\pi}{\overline{u}}$ time units; during this phase, the robot is driven by a constant control $u \equiv \pm\overline{u}$ along an initial circle;*
  - **b.** *After the transient, a sliding motion over the surface $S = 0$ is commenced and afterward maintained, which results in monotonic and exponentially fast convergence of the robot to the desired distance d_0 to the domain;*
  - **c.** *During the sliding motion, the angle $\alpha(t)$ between the robot's velocity $\mathbf{v}(t)$ and the tangent vector $\mathbf{T}[\boldsymbol{r}_*(t), t]$ is constantly acute, i.e., the robot encircles the domain $D(t)$ so that it is to the left.*

The proof of this theorem will be given in Section 9.5.1.

The choice of the controller parameters δ, γ from (3.2) recommended by this theorem is always possible under the assumptions made. Indeed, since $\lambda_v, \lambda_a < 1$ by Assumption 4.5, inequalities (5.1) can be satisfied by picking $\eta_v \in (0, 1 - \lambda_v), \eta_a \in (0, 1 - \lambda_a)$. After this, (5.3) can be ensured by picking small enough z_*, as has been shown. It remains to choose μ and γ small enough to ensure (5.4) and put $\delta := \gamma^{-1}\mu$.

Remark 5.1. It follows from **iii** that during the entire maneuver, the distance from the robot to the moving body remains in the interval $[d_-, d_+]$ from Assumption 4.6. Therefore, the sensor range may be upper limited by d_+: the navigation law (3.1) can be implemented and achieves the navigation objective even if the sensor provides the distance d only when $d \leq d_+$.

Remark 5.2. Theorem 5.1 is true for the navigation law (3.1) with the reversed sign

$$u(t) = -\overline{u} \cdot \mathbf{sgn} \left\{\dot{d}(t) + \chi[d(t) - d_0]\right\} \tag{5.5}$$

provided that in (iii.c), "acute" and "to the left" are replaced by "obtuse" and "to the right," respectively.

The proof of this remark will be given in Section 9.5.1.

The following lemma shows that Assumptions 4.6 and 4.7 are responsible only for transition of the robot from a remote location to motion along the required equidistant curve based on range-only measurements. If the robot is initially close to this curve and nearly tracks it, this navigation law ensures stable maintenance of the required distance under much milder and partly unavoidable Assumptions 4.1–4.5.

Lemma 5.1. *Suppose that only Assumptions 4.1–4.5 are true and the controller parameters are chosen so that* (5.4) *is satisfied with* λ_v, λ_a *and* z_* *taken from* (4.6) *and* (5.3), *where only* $d = d_0$ *is considered. Then the following claims hold:*

i. *For proper initial states, the navigation law* (3.1) *does generate the desired motion: whenever at some time instant* $d = d_0, \dot{d} = 0$ *and the angle* α *from (iii.c) is acute, these features remain true afterward;*

ii. *This motion is locally stable: for any* $\varepsilon_0 > 0$ *there exists* $c = c(\varepsilon_0) > 0$ *and* $\tau = \tau(\varepsilon_0) > 0$ *such that*

$$\boxed{|d(t_0) - d_0| < c \; |\dot{d}(t_0)| < c, \; |\alpha(t_0)| < \frac{\pi}{2}}$$
$$\Downarrow \qquad\qquad \Downarrow$$
$$\overbrace{\begin{array}{c} |d(t) - d_0| < \varepsilon_0 \\ |\dot{d}(t)| < \varepsilon_0 \\ t \geq t_0 \end{array}} \qquad \overbrace{\begin{array}{c} d(t) = d_0 \\ \dot{d} = 0 \\ \forall t \geq t_0 + \tau \end{array}} \tag{5.6}$$

where $\tau(\varepsilon_0) \to 0$ *as* $\varepsilon_0 \to 0$.

The proof of this lemma will be given in Section 9.5.1.

By the first entailment in (5.6), local stability requires sensors that give access to d only at the distances $d \leq d_0 + \varepsilon_0$, where $\varepsilon_0 > 0$ may be arbitrarily small.

When dealing with non-local convergence, requirements to the robot's initial state and the desired distance (like Assumptions 4.6 and 4.7) are unavoidable since the navigation law (3.1) is unable to ensure global convergence even in the simplest case of a point-wise steady domain [33].

Lemma 5.1 remains true for the navigation law with the reversed sign discussed in Remark 5.2 provided that in **i** of the lemma, "acute" and replaced by "obtuse" and in (5.6), $|\alpha| < \frac{\pi}{2}$ is reversed into $|\alpha| > \frac{\pi}{2}$.

The next remark addresses the full scan patrolling. In the necessary condition (4.5) for such a mission to be realistic, the boundary point and sign are not known in advance. Therefore, it is natural to require that this condition be satisfied for all boundary points and with any sign:

$$\sqrt{v^2 - V_N^2} \geq |V_T + d_0\sigma| \quad \forall \boldsymbol{r}_* \in \partial D(t), t. \tag{5.7}$$

Since $V_N^2 + V_T^2 = \|\mathbf{V}\|^2$, this can be shaped into

$$v^2 \geq \|\mathbf{V}\|^2 + 2V_T d_0 \sigma + d_0^2 \sigma^2 \quad \forall \boldsymbol{r}_* \in \partial D(t), t.$$

Remark 5.3. Suppose the assumptions of Theorem 5.1 are true and (5.7) holds uniformly over the infinite time horizon and with the strict inequality sign:

$$\inf_{\boldsymbol{r}_* \in \partial D(t), t \geq 0} \left[\sqrt{v^2 - V_N^2} - |V_T + d_0\sigma| \right] > 0.$$

Then since some time instant, the robot constantly overtakes the nearest boundary point. Moreover, its speed exceeds that of this point by a positive time-invariant constant.

The proof of this remark will be given in Section 9.5.1.

Our last remark concerns the choice of the parameter $z_* > 0$, which requires analyzing (5.3). Since this analysis is usually performed after that of "almost necessary" conditions (4.6), it is worth of noting that (5.3) is identical to the second condition in (4.6) written for $\lambda_a := \lambda_a + \eta_a$ and $v := \sqrt{v^2 - 2V_N z - z^2}$. Therefore, it suffices to pick z_* so that the latter condition is satisfied with any $z \in [-z_*, z_*]$ and feasible V_N. Moreover, since $|V_N| \leq v$ by (4.6), it suffices to ensure that this condition holds for all speed values from $\left[\sqrt{v^2 - 2vz_* - z_*^2}, \sqrt{v^2 + 2vz_* - z_*^2}\right]$ or even from the larger and simpler interval $\left[v - \frac{z_*}{\sqrt{2}-1}, v + z_*\right]$.[4] This may be easier than a direct analysis of (5.3), although it provides less conservative conditions.

9.5.1 Proofs of Theorem 5.1, Lemma 5.1, and Remarks 5.2 and 5.3

The first lemma reveals conditions under which the discontinuous navigation law (3.1) exhibits sliding motion.

[4] The square roots are well-defined if $z_* \in (0, [\sqrt{2} - 1]v)$.

Lemma 5.2. *Within the domain $d \in [d_-, d_+]$, the surface $S := \dot{d} + \chi(d - d_0) = 0$ is sliding in the sub-domain*

$$\xi + V_T > 0 \tag{5.8}$$

and two side repelling in the sub-domain

$$\xi + V_T < 0 \tag{5.9}$$

if (5.4) *holds. On this surface within the above domain,*

$$|\xi + V_T| \geq v\sqrt{1 - (\lambda_v + \eta_v)^2} > 0. \tag{5.10}$$

Proof. Whenever $S := \dot{d} - \nu = 0$, where $\nu := -\chi(d - d_0)$, we have by (3.2),

$$|\dot{d}| = |\nu| \leq \mu; \quad |\dot{\nu}| \leq \gamma|\dot{d}| = \gamma|\chi(d - d_0)| \leq \gamma\mu. \tag{5.11}$$

Due to (4.11),

$$|\xi + V_T| = \sqrt{v^2 - (V_N - \dot{d})^2} \geq \sqrt{v^2 - (|V_N| + \mu)^2} \\ \overset{(a)}{\geq} v\sqrt{1 - (\lambda_v + \eta_v)^2} \Rightarrow (5.10), \tag{5.12}$$

where (a) holds due to (4.6) and (5.4). Furthermore,

$$\dot{S} = \frac{\mathrm{d}}{\mathrm{d}t}\left[\dot{d} + \chi(d - d_0)\right] = \ddot{d} - \dot{\nu} \\ \overset{(4.13)}{=\!=} -u[\xi + V_T] + A_N + \frac{2\sigma\xi + \kappa\xi^2 - d\sigma^2}{1 + \kappa d} - \dot{\nu},$$

where $\xi = -V_T \pm \sqrt{v^2 - (V_N - \dot{d})^2}$ by (4.11). Therefore, by invoking (5.2), we see that

$$\dot{S} = [\xi + V_T]\left[-u + \Omega(\boldsymbol{r}_*, t, d, \dot{d}) - \frac{\dot{\nu}}{\xi + V_T}\right]. \tag{5.13}$$

Here due to (5.3),

$$\left|\Omega(\boldsymbol{r}_*, t, d, \dot{d}) - \frac{\dot{\nu}}{\xi + V_T}\right| \leq (\lambda_a + \eta_a)\overline{u} + \frac{\gamma\mu}{|\xi + V_T|} \\ \overset{(5.10)}{\leq} (\lambda_a + \eta_a)\overline{u} + \frac{\gamma\mu}{v\sqrt{1 - (\lambda_v + \eta_v)^2}} \overset{(5.4)}{<} \overline{u}. \tag{5.14}$$

Therefore, the signs taken by $\dot{S}$ for $u = \overline{u}$ and $u = -\overline{u}$, respectively, are opposite. In the case (5.8), the sign is opposite to $\mathbf{sgn}\, u \overset{(3.1)}{=} \mathbf{sgn}\, S$; in the case (5.9), the signs are equal. This implies the conclusion of the lemma. □

Lemma 5.3. *If the equation $\dot{d} + \chi(d - d_0) = 0$ becomes true at some time t_0 when $d \in [d_-, d_+]$, then monotonically and exponentially fast $d \to d_0, \dot{d} \to 0$ as $t \to \infty$. Furthermore, $d(t) \in [d_-, d_+]$ and* (5.8) *holds for all $t \geq t_0$.*

Proof. Lemma 5.2 guarantees that first, $\xi + V_T > 0$ at $t = t_0$ and second, this inequality is still valid and sliding motion occurs while $d \in [d_-, d_+]$. During this motion, $\dot{y} = -\chi(y)$ for $y := d - d_0$, where $\chi(y) \cdot y > 0 \ \forall y \neq 0$ and $\chi(0) = 0$. It follows that any solution d of the sliding mode differential equation monotonically converges to d_0. At the same time, $d_0 \in [d_-, d_+]$. Hence, d will never leave the interval $[d_-, d_+]$, the sliding mode will never be terminated, the inequality $\xi + V_T > 0$ will never be violated, and $d \to d_0, \dot{d} \to 0$ as $t \to \infty$. Application of Lyapunov's first method [426, Ch. 6] to the equation $\dot{y} = -\chi(y)$ at the equilibrium point $y = 0$ shows that the convergence is exponentially fast. □

Remark 5.4. Lemmas 5.2 and 5.3 evidently remain true for the navigation law (5.5) provided that formulas (5.8) and (5.9) are interchanged in their formulations.

Lemma 5.4. *Under the assumptions of Theorem 5.1, both relations* $\dot{d} + \chi(d - d_0) = 0$ *and* $\xi + V_T > 0$ *become true at some time* $t_0 \in \left[0, \frac{3\pi}{\overline{u}}\right)$, *and* $d \in [d_-, d_+]$ *at the first such time.*

Proof. If these relations are true initially, the claim is true as well. Otherwise, $\dot{d} + \chi(d - d_0) \neq 0$ for $t > 0, t \approx 0$, and until the first time t_0 when the equation becomes true, the robot moves with the constant control $u \equiv \pm\overline{u}$.

Let $u \equiv \overline{u}$ for the definiteness. Now we analyze the motion of the robot driven by the constant control $u \equiv \overline{u}$ during the time interval $\left[0, \frac{3\pi}{\overline{u}}\right]$. This motion is along the +initial circle by Definition 4.3. By Assumption 4.7, this circle and the domain $D(t)$ are constantly separated by a steady straight line for $t \in \left[0, \frac{3\pi}{\overline{u}}\right]$. Therefore, the polar angle of the vector $\mathbf{h}(t)$ directed from $\boldsymbol{r}_*(t)$ to $\boldsymbol{r}(t)$ continuously evolves over an interval whose length does not exceed π; see Fig. 9.6. Meanwhile, the polar angle of $\mathbf{v}(t)$ continuously runs over an interval with length 3π. Hence, there unavoidably exist two time instants $t_i \in \left[0, \frac{3\pi}{\overline{u}}\right], i = 1, 2$ such that $\mathbf{v}(t_i)$ and $(-1)^i\mathbf{h}(t_i)$ are co-linear and identically directed for $i = 1, 2$. Thanks to (4.7),

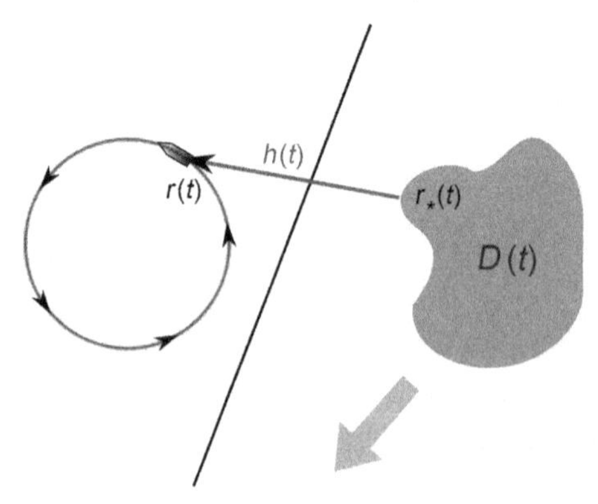

Figure 9.6 Vector $\mathbf{h}(t)$.

$$(-1)^i\dot{d}(t_i) = v + (-1)^i V_N(t_i) \geq v - |V_N| \overset{(4.6)}{\geq} (1-\lambda_v)v$$
$$\overset{(5.1)}{>} \eta_v v \overset{(5.4)}{\geq} \mu \overset{\text{Fig. 9.2(a)}}{\geq} |\chi[d(t_i) - d_0]|.$$

Thus, the continuous function of time S assumes values of opposite signs at $t = t_1, t_2$. It follows that S inevitably arrives at zero within $\left[0, \frac{3\pi}{\overline{u}}\right]$. Lemma 5.2 implies that (5.9) cannot be true at this moment. Therefore, (5.8) is true due to (5.10), which completes the proof. □

Proof of Theorem 5.1. This theorem is immediate from Lemmas 5.3–5.4 since (5.8) implies that the angle between **v** and **T** is acute due to (4.7). □

Proof of Remark 5.2 is based on Remark 5.4 and consists of merely retracing the above arguments. □

Proof of Remark 5.3. This remark is straightforward from Theorem 5.1 and (4.11), (4.14). □

To prove Lemma 5.1, we suppose that only Assumptions 4.1–4.5 are true.

Lemma 5.5. *Let* $\mathfrak{D} = [d_-, d_+]$ *be a regular interval (not necessarily that from Assumption 4.6) and let the controller parameters be chosen so that* (5.4) *is satisfied with* λ_v, λ_a *and* z_* *taken from* (4.6) *and* (5.3)*, where* $d \in \mathfrak{D}$ *are considered. Then there exist* $\varepsilon > 0, \zeta_v > 0$ *and* $\zeta > 0$ *such that*

$$d \in [d_-, d_+] \wedge |S| < \varepsilon \Rightarrow |\xi + V_T| \geq \zeta_v, \tag{5.15}$$
$$d \in [d_-, d_+] \wedge |S| < \varepsilon \wedge \xi + V_T > 0 \Rightarrow \dot{S}S \leq -\zeta|S|. \tag{5.16}$$

Proof. The proof is basically by retracing the arguments from the proof of Lemma 5.2. Whenever the premises in (5.15) are true, (5.11) is altered into

$$|\dot{d}| \leq |\nu| + \varepsilon \leq \mu + \varepsilon; \quad |\dot{\nu}| \leq \gamma|\dot{d}| \leq \gamma(|\nu| + \varepsilon) \leq \gamma(\mu + \varepsilon).$$

Hence, by taking $\varepsilon > 0$ small enough, we transform (5.12) into

$$|\xi + V_T| \geq v\sqrt{1 - \left(\lambda_v + \eta_v + \frac{\varepsilon}{v}\right)^2} \overset{\varepsilon\approx 0}{\geq} \zeta_v := v\frac{1}{2}\sqrt{1 - (\lambda_v + \eta_v)^2},$$

i.e., (5.15) does hold. Similarly in (5.14), the estimated quantity does not exceed

$$(\lambda_a + \eta_a)\overline{u} + \frac{\gamma(\mu + \varepsilon)}{v\sqrt{1 - (\lambda_v + \eta_v)^2}} \overset{\varepsilon\approx 0}{<} \overline{u}.$$

By bringing the pieces together, we arrive at (5.16). □

Note that (5.15) and (5.16) remain true as ε decreases. Therefore, without any loss of generality, we may assume that $\varepsilon < \mu$.

Corollary 5.1. *Let*

$$[d_-, d_+] \supset \mathfrak{D}_\varepsilon := \left[d_0 - \frac{\varepsilon}{\gamma}, d_0 + \frac{\varepsilon}{\gamma}\right].$$

If $|S| < \varepsilon, d \in \mathfrak{D}_\varepsilon$, *and* $\xi + V_T > 0$ *at some time instant, the system reaches the surface* $S = 0$ *for no more than* $\frac{4S^2(0)}{\gamma}$ *time units, not violating the inclusion* $d \in \mathfrak{D}_\varepsilon$, *and then undergoes sliding motion over this surface.*

Proof. Thanks to Lemma 5.5, S^2 decreases while the system remains in the domain $d \in [d_-, d_+] \wedge |S| < \varepsilon$. In this domain,

$$-\varepsilon - \chi(d - d_0) < \dot{d} < -\chi(d - d_0) + \varepsilon,$$

where

$$-\varepsilon - \chi(d - d_0) > 0 \quad \forall d < d_0 - \frac{\varepsilon}{\gamma}$$

and

$$-\chi(d - d_0) + \varepsilon < 0 \quad \forall d > d_0 + \frac{\varepsilon}{\gamma}.$$

Hence, d cannot leave the interval $\left[d_0 - \frac{\varepsilon}{\gamma}, d_0 + \frac{\varepsilon}{\gamma}\right] \subset [d_-, d_+]$. Therefore, the system remains in that domain and $\dot{S}S \le -\zeta|S|\ \forall t \ge t_0$. It follows that $S^2(t) - S^2(t_0) \le -4\zeta(t - t_0)$ while the sliding surface $S = 0$ is not reached, which implies the claim of the lemma. □

Proof of Lemma 5.1. By Remark 4.1, there is a regular interval of the form $[d_-, d_+] = [d_0 - \varepsilon_*, d_0 + \varepsilon_*]$, where $\varepsilon_* > 0$. Lemma 5.5 assigns some $\varepsilon > 0$ to this interval. By decreasing $\varepsilon > 0$ if necessary, we can ensure the assumption of Corollary 5.1 holds and $\varepsilon < \varepsilon_0(\gamma + 2)$. The constant c in (5.6) is picked so that $c < \varepsilon/(\gamma + 1)$. Whenever the premises from (5.6) hold,

$$|S| \le |\dot{d}| + |\chi(d - d_0)| \le c + \gamma|d - d_0| \le (1 + \gamma)c < \varepsilon$$

and $d \in \left[d_0 - \frac{\varepsilon}{\gamma}, d_0 + \frac{\varepsilon}{\gamma}\right]$. Then for $t \ge t_0$, Corollary 5.1 guarantees that $|d(t) - d_0| \le \varepsilon/\gamma < \varepsilon_0$ and

$$|\dot{d}| = |S - \chi(d - d_0)| \le |S| + |\chi(d - d_0)| \le \varepsilon + \gamma|d - d_0| \le 2\varepsilon < \varepsilon_0,$$

i.e., the first conclusion from (5.6) does hold. The second one is immediate from Corollary 5.1. □

9.6 Illustrative examples of border patrolling

The foregoing assumptions and parameter requirements employ some knowledge of the body $D(t)$ that is rarely available in the requested details. Typically, only some qualitative features and estimates are known a priori, which determine a whole class of body motions. For practical design, this means the need to extend the assumptions and requirements on any motion from this class. Now we offer illustrative examples of such extensions. For rigid bodies, D (without (t)) will stand for the snapshot of the body at an arbitrary time instant; the boundary ∂D is assumed to be smooth in the examples 9.6.1–9.6.3.

9.6.1 Steady rigid body

The velocity **V** and acceleration **A** vector fields are identically zero. As can be easily seen by inspection, Assumptions 4.1–4.5 mean the following properties hold:

(p) there are no multiple points at the distance d_0 from D and

$$R_{\varkappa}(\boldsymbol{r}_*, d_0) := \frac{1}{|\kappa(\boldsymbol{r}_*)|} + d_0 \mathbf{sgn}\, \kappa(\boldsymbol{r}_*) > R_{\min} \quad \forall \boldsymbol{r}_* \in \partial D. \tag{6.1}$$

Here $R_{\min}$ is the minimal turning radius (2.2) of the robot and $R_{\varkappa}(\boldsymbol{r}, d_0)$ is the curvature radius of the d_0-equidistant curve to be tracked. Inequality (6.1) means the robot is capable of tracking this curve, which is an unavoidable requirement. By Lemma 5.1, this is enough for the navigation law (3.1) to ensure local stability if in (3.1), μ, γ are chosen so that

$$\mu < v, \qquad \max_{\boldsymbol{r}_* \in \partial D} \frac{R_{\min}}{R_{\varkappa}(\boldsymbol{r}_*, d_0)} + \frac{\gamma \mu}{\overline{u}\sqrt{v^2 - \mu^2}} < 1, \tag{6.2}$$

which is always possible under the condition (6.1). For non-local convergence, Assumptions 4.6 and 4.7 are needed. They and (5.4) mean that the union of the initial circles can be separated from D by a line, **(p)** holds for $d_0 := d$ and all $d \in [d_-, d_+]$, and (6.1), (6.2) are true with $d_0 := d_-, d_+$, where d_- and d_+ are the minimal and maximal distance, respectively, from the points of this union to D.

If $\kappa(\boldsymbol{r}_*) \geq 0 \ \forall \boldsymbol{r}_* \in \partial D$ (i.e., if D is convex), (6.1) and (6.2) are true for all large enough d_0 (provided that small enough μ is taken). This implies that the controller at hand navigates the robot to the domain from any remote enough initial location based on range-only measurements.

Until now, we tacitly assumed that D is known. Now let only the lower estimates of the boundary curvature radius $R_\kappa(\boldsymbol{r}_*) = |\kappa(\boldsymbol{r})|^{-1}$ be known, which are separate for convex $R_\kappa(\boldsymbol{r}_*) \geq R_+ \ \forall \boldsymbol{r}_* \in \partial D : \kappa(\boldsymbol{r}_*) \geq 0$ and concave $R_\kappa(\boldsymbol{r}_*) \geq R_- \ \forall \boldsymbol{r}_* \in \partial D : \kappa(\boldsymbol{r}_*) \leq 0$ parts of the boundary. Then the focus on the worst-case scenario within these bounds yields that (6.1) should be replaced by $R_\pm > R_{\min} \mp d_0$ and (6.2) by

$$\mu < v, \qquad \frac{R_{\min}}{\min\{R_+ + d_0, R_- - d_0\}} + \frac{\gamma \mu}{\overline{u}\sqrt{v^2 - \mu^2}} < 1.$$

Then the foregoing remains true.

9.6.2 Rigid body moving with a constant speed V in a priori unknown and fixed direction

Let $\alpha(\boldsymbol{r}_*)$ be the polar angle of the domain's velocity **V** in the Frenet frame $\mathbf{T}(\boldsymbol{r}_*), \mathbf{N}(\boldsymbol{r}_*)$. Since $\mathbf{A} \equiv 0, \sigma \equiv 0$, trivial manipulations reduce (4.6) to the requirement that for all $\boldsymbol{r}_* \in \partial D$,

$$\mu := \frac{v}{V} \geq \lambda_v^{-1} |\sin \alpha(\boldsymbol{r}_*)|,$$

$$\lambda_a^{-1} R_{\min} \leq R_\kappa(\boldsymbol{r}_*, d_0) \frac{\mu\sqrt{\mu^2 - \sin^2 \alpha(\boldsymbol{r}_*)}}{\left(|\cos \alpha(\boldsymbol{r}_*)| + \sqrt{\mu^2 - \sin^2 \alpha(\boldsymbol{r}_*)}\right)^2}.$$

Elementary calculus shows that the right-hand side of the last inequality decreases as $|\cos \alpha(\boldsymbol{r}_*)|$ increases. By putting here the worst velocity scenario (for which $|\cos \alpha(\boldsymbol{r}_*)| = 1$), we see that (6.1) still holds and for all $\boldsymbol{r}_* \in \partial D$,

$$V \leq \lambda_v v < v, \quad \frac{(V+v)^2}{R_\varkappa(\boldsymbol{r}_*, d_0)} \leq \lambda_a v\overline{u} < v\overline{u}. \tag{6.3}$$

Thus, Assumptions 4.1–4.5 hold for all headings of the velocity **V** if and only if **(p)** and (6.1), (6.3) are true. Lemma 4.1 and the foregoing imply that these requirements are "almost necessary" for the robot to be capable of border patrolling irrespective of the direction in which the body moves. By Lemma 5.1, they are simultaneously sufficient for the navigation law (3.1) to ensure local stability if δ, γ in Fig. 9.2(a) are chosen so that (5.4) holds, where $\eta_\nu \in (0, 1 - \lambda_\nu)$, $\nu = v, a$ and

$$z_* \leq \min_{\boldsymbol{r}_* \in \partial D} \sqrt{(\lambda_a + \eta_a) v\overline{u} R_\varkappa(\boldsymbol{r}_*, d_0)} - V - v. \tag{6.4}$$

The expression $\frac{(V+v)^2}{R_\varkappa(\boldsymbol{r}_*, d_0)}$ in (6.3) is the centrifugal acceleration of the point tracing the d_0-equidistant curve at the speed v in the case where the velocity of the nearest boundary point is tangential to the boundary and directed opposite to that of the point; see Fig. 9.7. The expression $v\overline{u}$ gives the maximal linear acceleration of the robot (2.1). Therefore, the second inequality in (6.3) means the robot is able to produce accelerations required for the curve tracking in any direction.

Now suppose that Assumption 4.7 holds. Under the circumstances, Assumption 4.6 means that **(p)** is valid for $d_0 := d$ with any $d \in [d_-, d_+]$, and (6.1), (6.3) are true with $d_0 := d_-, d_+$, where d_- and d_+ are the minimal and maximal distances, respectively,

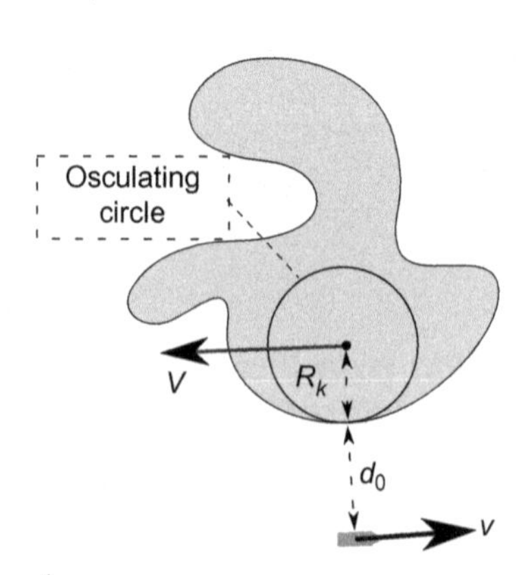

Figure 9.7 Worst-case scenario.

from the points of the union of the initial circles to the body within the first $\frac{3\pi}{\overline{u}}$ time units of its motion. The parameter z_* involved in the controller design should be taken as the minimum among the right-hand sides of (6.4) calculated for $d_0 := d_-$ and $d_0 := d_+$, respectively. Under these assumptions, non-local convergence (i.e., the conclusion of Theorem 5.1) holds.

For a convex domain D, the argument similar to that from the previous subsection shows that the convergence holds whenever initially the robot is distant enough from the domain.

Now let V be unknown but can be estimated $V \leq \overline{V}$ by a known constant. It is easy to see that the foregoing remains true if V is formally replaced by $\overline{V}$ everywhere.

9.6.3 Translational motion of a rigid body

Suppose that only upper bounds $V_+ \geq \|\mathbf{V}(t)\|, A_+ \geq \|\mathbf{A}(t)\|$ on the body velocity $\mathbf{V}(t)$ and acceleration $\mathbf{A}(t)$ are unknown. Since the motion is translational, $\mathbf{V}(\boldsymbol{r}, t) = \mathbf{V}(t)$, $\frac{\partial \mathbf{V}}{\partial \boldsymbol{r}} \equiv 0, \sigma \equiv 0$ and (4.6) takes the form

$$|V_N| \leq \lambda_v v, \qquad \frac{\left| A_N + \frac{(V_T \mp \sqrt{v^2 - V_N^2})^2}{R_\kappa(\boldsymbol{r}_*, d_0)} \mathbf{sgn} \kappa \right|}{\sqrt{v^2 - V_N^2}} \leq \lambda_a \overline{u}.$$

The first inequality holds for all feasible velocities if and only if

$$V_+ \leq \lambda_v v < v. \tag{6.5}$$

Given the value of V_N, the worst-case scenario for the second inequality is attained when $A_N = A_+ \mathbf{sgn} \kappa$ and $V_T = \mp\sqrt{V_+^2 - V_N^2}$. Then this inequality takes the form

$$\frac{A_+}{\sqrt{v^2 - V_N^2}} + \frac{\Psi(|V_N|)}{R_\kappa(\boldsymbol{r}_*, d_0)} \leq \lambda_a \overline{u}, \quad \text{where } 0 \leq |V_N| \leq V_+ \quad \text{and}$$

$$\Psi(z) := \frac{\left(\sqrt{V_+^2 - z^2} + \sqrt{v^2 - z^2} \right)^2}{\sqrt{v^2 - z^2}}. \tag{6.6}$$

For $z \in [0, V_+]$, $A := \sqrt{V_+^2 - z^2}, B := \sqrt{v^2 - z^2}$, we have

$$\begin{aligned}
(v^2 - z^2)\frac{d\Psi}{dz}(z) &= 2(A+B)\left(-\frac{z}{A} - \frac{z}{B}\right)B + (A+B)^2 \frac{z}{B} \\
&= (A+B)^2 z \frac{A - 2B}{AB} \overset{V_+ < v}{<} 0,
\end{aligned}$$

i.e., $\Psi(|V_N|)$ decreases as $|V_N|$ increases, while the first addend in the left-hand side of the inequality from (6.6) conversely increases. This complicates analytically finding the maximum of the entire left-hand side.

Anyhow, the second inequality from (4.6) holds for all feasible scenarios if and only if for all $\boldsymbol{r}_* \in \partial D$,

$$\max_{z \in [0, V_+]} \left[\frac{A_+}{\sqrt{v^2 - z^2}} + \frac{\Psi(z)}{R_\kappa(\boldsymbol{r}_*, d_0)} \right] \le \lambda_a \overline{u} < \overline{u}. \tag{6.7}$$

By separately maximizing the first and second addend, we see that (6.7) is implied by the more constructive condition

$$\frac{A_+}{\sqrt{1 - \frac{V_+^2}{v^2}}} + \frac{(V_+ + v)^2}{R_\kappa(\boldsymbol{r}_*, d_0)} \le \lambda_a \overline{u} v < \overline{u} v.$$

Note that for $A_+ := 0$, (6.5), complemented by either the last inequality or inequality (6.7), reduces to (6.3).

Overall, we see that Assumptions 4.1–4.5 hold for all feasible scenarios if and only if **(p)**, (6.1), (6.5), and (6.7) are true. Moreover, these requirements are "almost necessary" for the robot to be capable of performing border patrolling irrespective of the scenario. By Lemma 5.1, they are simultaneously sufficient for the navigation law (3.1) to ensure locally stable patrolling if δ, γ in Fig. 9.2(a) are chosen so that (5.4) holds, where $\eta_\nu \in (0, 1 - \lambda_\nu)$, $\nu = v, a$ and $z_* \in (0, V_+)$ is chosen so that (6.7) is true for $V_+ := V_+ + z_*$ and $\lambda_a := \lambda_a + \eta_a$.

Analysis similar to that from the previous subsection shows that for non-local convergence (i.e., for the conclusion of Theorem 5.1 to be valid) it suffices that Assumption 4.7 is true, **(p)** is valid for $d_0 := d$ with any $d \in [d_-, d_+]$, and (6.1), (6.7) are true with $d_0 := d_-, d_+$, where d_- and d_+ are the minimal and maximal distances, respectively, from the points of the union of the initial circles to the body within the first $\frac{3\pi}{\overline{u}}$ time units of its motion. The parameter z_* involved in the controller design should be taken as the minimum among the values calculated in accordance with the above recommendations for $d_0 := d_-$ and $d_0 := d_+$, respectively.

For a convex domain D, (6.7) remains true as d_0 grows. Therefore, the argument similar to that from the previous section shows the convergence holds whenever initially the robot is remote enough from the moving domain.

9.6.4 Escorting a convoy of unicycle-like vehicles

A team of N planar vehicles travels in a plane; the vehicles move with a common constant speed V in a convoy one after another. The first of them acts as a leader; every other vehicle traces the path of its predecessor at a constant distance[1] Δs from it. There also is a mobile escorting robot. It has to constantly encircle the entire convoy with the margin d_0, not intervening between the vehicles; see Fig. 9.8(a).

The leader motion obeys the unicycle-like equations:

$$\dot{X} = V \cos \alpha, \quad \dot{Y} = V \sin \alpha, \quad \dot{\alpha} = \omega. \tag{6.8}$$

[1] The distance is measured along the path.

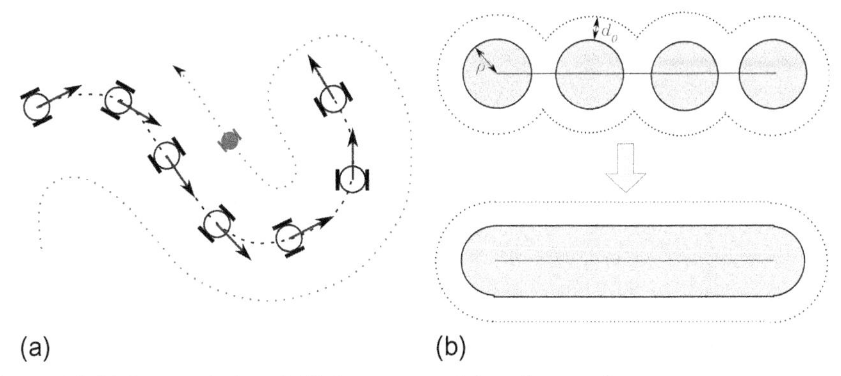

Figure 9.8 (a) Escort mission; (b) Enhancement of the safety requirement.

Here X and Y are its abscissa and ordinate, respectively, in the world frame, α gives the orientation of the linear velocity vector, and the angular velocity ω is time-varying. The escorting robot is interpreted as that discussed in Section 9.2; its governing equations are given by (2.1).

We assume that any vehicle in the convoy can be approximated by a disc of the radius ρ. Correspondingly, the escorting robot should be at the distance d_0 from the union of N discs, each of the radius ρ, whose centers are uniformly distributed over the path $\mathscr{P}|_{t-\tau}^{t}$ tracked by the leader during the last $\tau := (N-1)\Delta s/V$ time units. In this requirement, we upgrade "N discs" into "all discs"; see Fig. 9.8(b), where the dotted line is the locus of desirable positions of the escorting robot and $N = 4$. In other words, we inject more safety by requiring that the convoy is escorted at the distance d_0 from the union $D(t)$ of all ρ-discs centered at $\mathscr{P}|_{t-\tau}^{t}$. This guarantees that the robot does not intervene between the vehicles.

By doing so, the problem is put into the framework of Section 9.4. Conversely to the preceding examples, now the body $D(t)$ is not rigid and may wriggle during the motion.

The speed V and the maneuver of the leader or the navigation law driving it are not known to the escorting robot. At the same time, it knows ρ and estimates of the linear and rotational velocities of the leader:

$$0 < \underline{V} \leq V \leq \overline{V}, \quad |\omega| \leq \overline{\omega}. \tag{6.9}$$

Pursuing its own objective, the leader takes into account the escorting robot by avoiding maneuvers that make the escort impossible or entail a singular situation, like, e.g., those from Fig. 9.9(a) and (b). (In (a), the robot can neither enter nor exit the inner dotted loop while maintaining the distance d_0; in (b), the robot cannot pass the cusp P due to incapability of instantly turning.) This means [425, Ch.16] that for any path $\mathscr{P}|_{t-\tau}^{t}$ traced by the leader during τ time units, any point at the distance $\leq d_0 + \rho$ from $\mathscr{P}|_{t-\tau}^{t}$

(i) has only one minimum distance point on $\mathscr{P}|_{t-\tau}^{t}$ and
(ii) is not focal for this path.

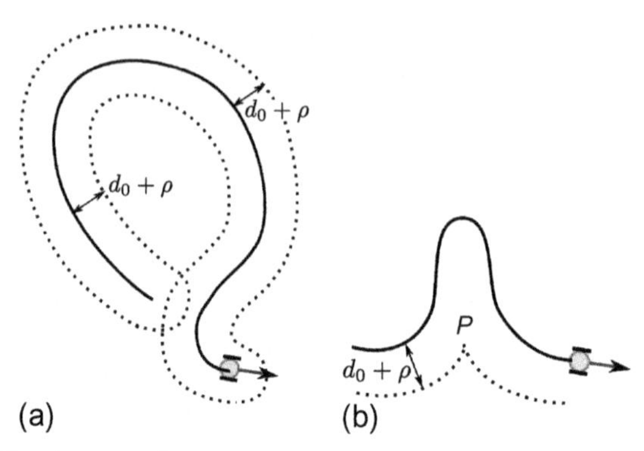

Figure 9.9 (a,b) Singular cases for escort.

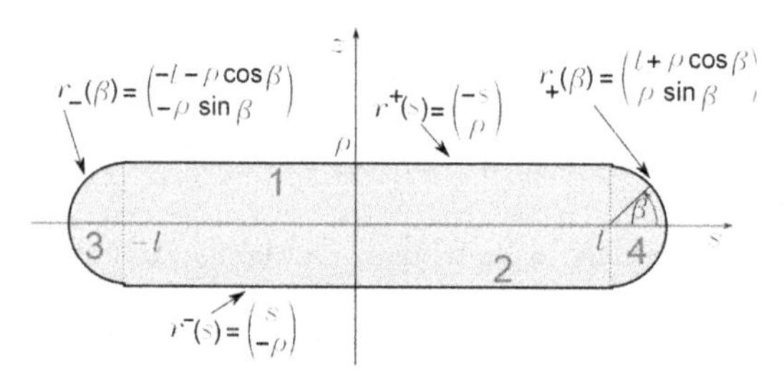

Figure 9.10 Boundary parametrization and enumeration of its parts.

(These are true if $\rho + d_0$ is small enough [425, Ch.16].) Due to (6.9), the tight lower bound on the path curvature radius is $\underline{V}/\overline{u}$. Therefore, **(ii)** brings the following assumption

$$\rho + d_0 < \frac{\underline{V}}{\overline{\omega}}.$$

These assumptions imply that the convoy does not collide with itself and that $D(t)$ has a smooth boundary [425, Ch.16].

Let the escorting robot be driven by the navigation law (3.1). (We assume that the sensor system permits the robot to access the distance $d(t)$ from it to $D(t)$.) To check the assumptions from Section 9.4, the reference configuration of $D(t)$ is taken to be that related to the straight path. We parameterize its boundary as is shown in Fig. 9.10, where $l := 1/2V\tau$. The "$\pm$-side" $S^{\pm}$ of the boundary $\partial D(t)$ (i.e., the part parameterized by $\boldsymbol{r}^{\pm}(s)$) moves via extension of the head and erasure of the tail, whereas their points remain on the ρ-equidistant curve $\mathfrak{C}^{\pm}$ of the leader's path and move with the constant speed V along this curve; see Fig. 9.11. So for $\boldsymbol{r} \in S^{\pm}$, the velocity does depend on time $\mathbf{V}(\boldsymbol{r}) = \mp V\mathbf{T}(\boldsymbol{r})$. Let $\varkappa^{\pm}(\boldsymbol{r})$ denote the curvature of $\mathfrak{C}^{\pm}$ at the point $\boldsymbol{r}$. Then for $\boldsymbol{r} \in S^{\pm}$,

$$V_T = \mp V, \quad V_N = 0, \quad \sigma = \left\langle \mathbf{V}_{\boldsymbol{r}}'\mathbf{T};\mathbf{N}\right\rangle \overset{(4.1)}{=} \mp V\varkappa^{\pm}(\boldsymbol{r}),$$
$$A_N \overset{(4.2)}{=} \left\langle \mathbf{V}_{\boldsymbol{r}}'\mathbf{V};\mathbf{N}\right\rangle = V^2\varkappa^{\pm}(\boldsymbol{r}), \quad \xi \overset{(4.4)}{=} \pm V \overset{\star}{\pm} v,$$

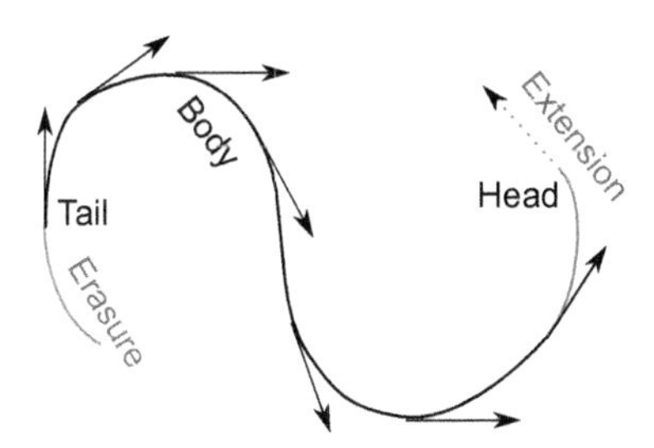

Figure 9.11 Motion of the boundary sides.

where $\star$ signals that the sign in this $\pm$ is independent of the signs in the others. So dropping $\boldsymbol{r}$ in $\varkappa^{\pm}(\boldsymbol{r})$, we have by (4.4),

$$\begin{aligned}\mathcal{A} &= V^2\varkappa^{\pm} + \frac{\mp 2V\varkappa^{\pm}\xi + \varkappa^{\pm}\xi^2 - d_0 V^2(\varkappa^{\pm})^2}{1+\varkappa^{\pm}d_0} \\ &= \varkappa^{\pm}\frac{V^2 \mp 2V\xi + \xi^2}{1+\varkappa^{\pm}d_0} = \varkappa^{\pm}\frac{(\xi \mp V)^2}{1+\varkappa^{\pm}d_0} = \frac{v^2}{(\varkappa^{\pm})^{-1}+d_0}.\end{aligned}$$

Hence the second inequality from (4.6) now takes the form

$$R_{\min} = \frac{v}{\overline{u}} \le \lambda_a \left|(\varkappa^{\pm})^{-1} + d_0\right| \tag{6.10}$$

and means that the minimal turning radius of the escorting robot (2.2) is less than the curvature radius of the d_0-equidistant curve of the side $S^{\pm}$, which is the expression in $|\dots|$. Let $\boldsymbol{r}_0(t) := \mathbf{col}\,[X(t), Y(t)]$ denote the position of the leader. By parameterizing $S^{\pm}$ as follows

$$\boldsymbol{r}(s) = \boldsymbol{r}_0(t_s^{\mp}) \pm \rho\mathbf{e}_{\perp}[\alpha(t_s^{\mp})], \quad t_s^{\mp} := t - \frac{\tau}{2} \mp \frac{s}{V}, \quad s \in [-l, l],$$

it is straightforward to compute the curvature radius $\varkappa^{\pm}[\boldsymbol{r}(s)]^{-1} = \rho \mp \frac{V}{\omega}$. This converts (6.10) into

$$\frac{v}{\overline{u}} \le \lambda_a \left| \pm\frac{V}{\omega} + \rho + d_0 \right|,$$

which holds for all maneuvers obeying (6.9) if and only if

$$\frac{v}{\overline{u}} \le \lambda_a \left[\frac{V}{\overline{\omega}} - (\rho + d_0) \right] < \frac{V}{\overline{\omega}} - (\rho + d_0). \tag{6.11}$$

Now we proceed to analysis of the second inequality from (4.6) for the front and rear half-circle parts of the body; see Fig. 9.10. Putting

$$\mathbf{e}(\alpha) := \mathbf{col}\,[\cos\alpha, \sin\alpha], \quad \mathbf{e}_{\perp}(\alpha) := \mathbf{col}\,[-\sin\alpha, \cos\alpha], \quad t_*^{+} := t, \quad t_*^{-} := t - \tau,$$

we observe that

$$\underbrace{\Phi\left[\boldsymbol{r}_{\pm}(\beta),t\right]}_{r_{\pm}(\beta,t)} = \boldsymbol{r}_0\left(t_*^{\pm}\right) \pm \rho \left\{ \begin{array}{c} \mathbf{e}\left[\alpha\left(t_*^{\pm}\right)\right]\cos\beta \\ +\mathbf{e}_{\perp}\left[\alpha\left(t_*^{\pm}\right)\right]\sin\beta \end{array} \right\};$$

$$\mathbf{T}\left[\boldsymbol{r}_{\pm}(\beta),t\right] = \mp\mathbf{e}[\alpha(t_*^{\pm})]\sin\beta \pm \mathbf{e}_{\perp}[\alpha(t_*^{\pm})]\cos\beta,$$

$$\mathbf{N}\left[\boldsymbol{r}_{\pm}(\beta),t\right] = \mp\mathbf{e}[\alpha(t_*^{\pm})]\cos\beta \mp \mathbf{e}_{\perp}[\alpha(t_*^{\pm})]\sin\beta;$$

$$\mathbf{V}\left[\boldsymbol{r}_{\pm}(\beta,t),t\right] = \left[V \mp \rho\omega(t_*^{\pm})\sin\beta\right]\mathbf{e}\left[\alpha\left(t_*^{\pm}\right)\right] \pm \rho\omega(t_*^{\pm})\cos\beta\mathbf{e}_{\perp}\left[\alpha\left(t_*^{\pm}\right)\right];$$

$$V_T\left[\boldsymbol{r}_{\pm}(\beta),t\right] = \mp V\sin\beta + \rho\omega(t_*^{\pm}),$$

$$V_N\left[\boldsymbol{r}_{\pm}(\beta),t\right] = \mp V\cos\beta, \quad \sigma\left[\boldsymbol{r}_{\pm}(\beta,t),t\right] = \omega(t_*^{\pm}),$$

$$A_N\left[\boldsymbol{r}_{\pm}(\beta,t),t\right] = \mp V\omega(t_*^{\pm})\sin\beta + \rho\omega^2(t_*^{\pm}).$$

The inequality $|V_N| \leq \lambda_v v$ from (4.6) should be true for all $\beta \in [-\pi/2, \pi/2]$. This means that the escorting robot is faster than the convoyed vehicles: $V < \lambda_v v < v$. Furthermore,

$$\xi\left[\boldsymbol{r}_{\pm}(\beta,t),t\right] = \pm V\sin\beta - \rho\omega(t_*^{\pm}) \overset{\star}{\pm} \sqrt{v^2 - V^2\cos^2\beta}, \quad \kappa\left[\boldsymbol{r}_{\pm}(\beta,t),t\right] = \rho^{-1},$$

$$\mathcal{A} = \mp V\omega(t_*^{\pm})\sin\beta + \rho\omega^2(t_*^{\pm}) + \frac{2\omega\xi + \rho^{-1}\xi^2 - d_0\omega^2}{1 + \rho^{-1}d_0}$$

$$= \mp V\omega(t_*^{\pm})\sin\beta + \frac{(\xi + \rho\omega)^2}{\rho + d_0}.$$

Hence the second condition from (4.6) takes the form

$$\frac{\left| \mp V\omega(t_*^{\pm})\sin\beta + \frac{\left(V\sin\beta \overset{\star}{\pm}\sqrt{v^2 - V^2\cos^2\beta}\right)^2}{\rho + d_0} \right|}{\sqrt{v^2 - V^2\cos^2\beta}} \leq \lambda_a \overline{u}.$$

Due to arbitrary choice of the signs in $\pm$'s and ω within the range (6.9), this is transformed into

$$\frac{V\overline{\omega}|\sin\beta| + \frac{\left(V|\sin\beta| + \sqrt{v^2 - V^2 + V^2\sin^2\beta}\right)^2}{\rho + d_0}}{\sqrt{v^2 - V^2 + V^2\sin^2\beta}} \leq \lambda_a \overline{u}.$$

Via checking the derivative sign, it is easy to see that the left-hand side is an increasing function of $V|\sin\beta|$, which ranges over $[0, \overline{V}]$ as β and V run over $[-\pi/2, \pi/2]$ and $[\underline{V}, \overline{V}]$, respectively. So the inequality holds for all these β and V if and only if it is true for $\beta := \pi/2$ and $V := \overline{V}$:

$$\overline{V}\overline{\omega} + \frac{\left(\overline{V} + v\right)^2}{\rho + d_0} \leq \lambda_a \overline{u} v < \overline{u} v. \tag{6.12}$$

Summarizing, we see that Assumptions 4.1–4.5 hold for all feasible scenarios if and only if $\overline{V} < v$ and (6.11), (6.12) are true. Moreover, these requirements are “almost necessary" for the robot to be capable of escorting the convoy irrespective of the scenario within the range (6.9). By Lemma 5.1, they are simultaneously sufficient for the navigation law (3.1) to ensure locally stable escort at the distance d_0 if δ, γ in Fig. 9.2(a) are chosen so that (5.4) holds. In (5.4), $\eta_\nu \in (0, 1 - \lambda_\nu), \nu = v, a$ and according to the paragraph preceding Section 9.5.1, $z_* > 0$ is chosen so that the following system of linear and quadratic (in z_*) inequalities is satisfied

$$z_* \leq \eta_a \overline{u} \left[\frac{\underline{V}}{\overline{\overline{\omega}}} - (\rho + d_0) \right], \quad z_* < \frac{v}{\sqrt{2} - 1},$$

$$\frac{2(\overline{V} + v) z_* + z_*^2}{\rho + d_0} + (\lambda_a + \eta_a) \frac{z_*}{\sqrt{2} - 1} < \eta_a \overline{u} v.$$

This system is consistent and is satisfied by all small enough z_*.

Analysis similar to that from the previous section shows that for non-local convergence (i.e., for the conclusion of Theorem 5.1 to be valid) it suffices that Assumption 4.7 is true and (6.12), (6.12) are valid with $d_0 := \max\{d_0, d_+\}$ and $d_0 := \min\{d_0, d_-\}$, respectively. Here d_- and d_+ are the minimal and maximal distances, respectively, from the points of the union of the initial circles to $D(t)$ within the first $\frac{3\pi}{\overline{u}}$ time units of motion. In the controller design, the parameter z_* should be taken as the minimum among the values calculated in accordance with the above recommendations for $d_0 := \max\{d_0, d_+\}$ and $d_0 := \min\{d_0, d_-\}$, respectively.

9.6.5 Escorting a bulky cigar-shaped vehicle

The body $D(t)$ is a rigid and cigar-shaped vehicle, as is shown in Fig. 9.12. It moves forward with the constant speed V in the direction of its centerline, which is time-varying. The motion is described by the unicycle-like equations (6.8), where X and Y are the abscissa and ordinate, respectively, of the vehicle's center in the world frame, and α gives the centerline orientation. The speed V and the maneuver of the body or the navigation law driving it are not known. At the same time, the shape parameters

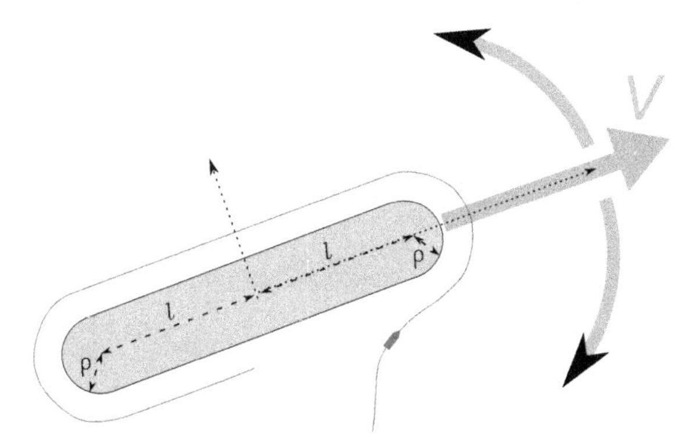

Figure 9.12 Escorting a bulky vehicle.

l and ρ from Fig. 9.7(b) are known, and the speed, as well as the rotational velocity and acceleration of D can be upper bounded by known constants:

$$0 \le V \le \overline{V}, \quad |\omega| \le \overline{\omega}, \quad |\dot{\omega}| \le \aleph. \tag{6.13}$$

In the reference frame attached to the body, the boundary ∂D can be parameterized as is shown in Fig. 9.10. In this frame,

$$\mathbf{V} = V \begin{pmatrix} 1 \\ 0 \end{pmatrix} + \omega \mathcal{R}_{\frac{\pi}{2}} \boldsymbol{r}, \quad \mathbf{A} = \omega V \begin{pmatrix} 0 \\ 1 \end{pmatrix} + \dot{\omega} \mathcal{R}_{\frac{\pi}{2}} \boldsymbol{r} - \omega^2 \boldsymbol{r}.$$

Elementary computation results in Table 9.1, where No. refers to the serial number of the boundary part introduced in Fig. 9.10, and in $\pm$ and $\mp$, the upper and lower signs correspond to the parts with odd and even numbers, respectively. By substituting data from this table into (4.6), we see that the first inequality from (4.6) takes the form

$$\max_{|s| \le l} |\omega s| \le \lambda_v v, \qquad \max_{|\beta| \le \pi/2} | \mp V \cos \beta - \omega l \sin \beta| \le \lambda_v v$$

and after calculating these maxima, shapes into $\sqrt{V^2 + \omega^2 l^2} \le \lambda_v v$. This inequality is satisfied for all scenarios within the range (6.13) if and only if

$$V_\omega := \sqrt{\overline{V}^2 + \overline{\omega}^2 l^2} \le \lambda_v v < v. \tag{6.14}$$

Similarly the second inequality from (4.6) written for the straight parts of the boundary takes the form

$$\frac{\left| \pm V\omega - \omega^2(\rho + d_0) + \dot{\omega} s \overset{\star}{\pm} 2\omega \sqrt{v^2 - \omega^2 s^2} \right|}{\sqrt{v^2 - \omega^2 s^2}} \le \lambda_a \overline{u}.$$

This inequality holds for all V and $\dot{\omega}$ within the range (6.13) if and only if

$$\frac{\overline{V}|\omega| + \omega^2(\rho + d_0) + \aleph |s|}{\sqrt{v^2 - \omega^2 s^2}} + 2|\omega| \le \lambda_a \overline{u}.$$

Here the left-hand side increases as $|\omega|$ or $|s|$ grows. So the second inequality from (4.6) holds for the straight parts of the boundary and all feasible scenarios if and only if

$$\frac{\overline{V}\overline{\omega} + \overline{\omega}^2(\rho + d_0) + \aleph l}{\sqrt{v^2 - \overline{\omega}^2 l^2}} + 2\overline{\omega} \le \lambda_a \overline{u} < \overline{u}. \tag{6.15}$$

For the circular parts, we have due to (4.4),

$$\begin{aligned} \mathcal{A} = A_N + \frac{2\omega\xi + \rho^{-1}\xi^2 - d_0\omega^2}{1 + \rho^{-1} d_0} &= \omega^2 \rho - \dot{\omega} l \sin \beta \\ + \omega \underbrace{[\mp V \sin \beta + \omega l \cos \beta]}_{=:V_\beta} &+ \frac{2\omega\rho\xi + \xi^2 - \rho d_0 \omega^2}{\rho + d_0} \\ = \omega V_\beta - \dot{\omega} l \sin \beta + \frac{(\xi + \rho\omega)^2}{\rho + d_0}. \end{aligned}$$

Table 9.1 Some geometrical, kinematic, and dynamical parameters

No.	T	N	σ	V_T
1,2	$\mp\begin{pmatrix}1\\0\end{pmatrix}$	$\mp\begin{pmatrix}0\\1\end{pmatrix}$	ω	$\mp V + \rho\omega$
3,4	$\pm\begin{pmatrix}-\sin\beta\\ \cos\beta\end{pmatrix}$	$\mp\begin{pmatrix}cos\beta\\ \sin\beta\end{pmatrix}$	ω	$\omega\rho + \omega l\cos\beta \mp V\sin\beta$

No.	V_N	A_N	κ
1,2	$\omega x'$	$\mp\omega V + \dot{\omega}x' + \omega^2\rho$	0
3,4	$\mp V\cos\beta - \omega l\sin\beta$	$\omega^2\rho - \dot{\omega}l\sin\beta$ $\mp\omega V\sin\beta + \omega^2 l\cos\beta$	$\frac{1}{\rho}$

In Table 9.1,

$$V_N = \mp V\cos\beta - \omega l\sin\beta$$

and so

$$V_N^2 + V_\beta^2 = V^2 + \omega^2 l^2, \quad \sqrt{v^2 - V_N^2} = \sqrt{\Delta^2 + V_\beta^2},$$

where $\Delta := \sqrt{v^2 - V^2 - \omega^2 l^2}$ is well defined by (6.14). Hence with regard to (4.4),

$$\xi + \omega\rho = -\omega\rho - \omega l\cos\beta \pm V\sin\beta \pm \sqrt{\Delta^2 + V_\beta^2} + \omega\rho = -V_\beta \pm \sqrt{\Delta^2 + V_\beta^2}.$$

Overall, the examined second inequality from (4.6) now looks as follows:

$$\frac{\left|\omega V_\beta - \dot{\omega}l\sin\beta + \frac{\left(V_\beta \mp \sqrt{\Delta^2 + V_\beta^2}\right)^2}{\rho + d_0}\right|}{\sqrt{\Delta^2 + V_\beta^2}} \leq \lambda_a \overline{u}.$$

This holds for all scenarios satisfying (6.13) if and only if

$$\max_{\substack{|\omega|\leq\overline{\omega}\\ 0\leq V\leq\overline{V}\\ |\beta|\leq\frac{\pi}{2}}} \left[\frac{\aleph l|\sin\beta|}{\sqrt{\Delta^2 + V_\beta^2}} + \frac{\left|\omega V_\beta + \frac{\left(V_\beta \mp \sqrt{\Delta^2 + V_\beta^2}\right)^2}{\rho + d_0}\right|}{\sqrt{\Delta^2 + V_\beta^2}}\right] \leq \lambda_a \overline{u}.$$

It can be shown that the first and second addends differently behave as β runs over its range $[-\pi/2, \pi/2]$, which troubles explicit computation of the maximum. To proceed analytically, we inject some non-conservatism by upper estimating the maximum of the sum by the sum of the maxima of the addends, thus substituting the following stronger condition in place of the above one

$$\max_{\substack{|\omega|\le\overline{\omega}\\ 0\le V\le\overline{V}\\ |\beta|\le\frac{\pi}{2}}} \underbrace{\frac{\aleph l|\sin\beta|}{\sqrt{\Delta^2+V_\beta^2}}}_{\Upsilon} + \max_{\substack{|\omega|\le\overline{\omega}\\ 0\le V\le\overline{V}\\ |\beta|\le\frac{\pi}{2}}} \underbrace{\frac{\left|\omega V_\beta + \frac{\left(V_\beta\mp\sqrt{\Delta^2+V_\beta^2}\right)^2}{\rho+d_0}\right|}{\sqrt{\Delta^2+V_\beta^2}}}_{\Xi} \le \lambda_a\overline{u}. \tag{6.16}$$

Given β, a larger value Υ is provided by the sign in $\mp$ that makes the signs of the addends in $V_\beta = \mp V\sin\beta + \omega l\cos\beta$ opposite. Hence the first maximum equals

$$\max_{\substack{|\omega|\le\overline{\omega}\\ 0\le V\le\overline{V}\\ 0\le\beta\le\frac{\pi}{2}}} \underbrace{\frac{\aleph l\sin\beta}{\sqrt{\Delta^2+(|\omega|l\cos\beta - V\sin\beta)^2}}}_{\Psi}.$$

It is easy to check that $\Psi'_\beta \ge 0\ \forall\beta\in[0,\pi/2]$. So the maximum over β is attained at $\beta=\pi/2$, and the entire maximum equals $\frac{\aleph l}{\sqrt{v^2-\overline{\omega}^2 l^2}}$.

As for Ξ, we note that the expression V_β with $+$ in $\pm$ equals $V_{-\beta}$ with $-$, whereas the range $[-\pi/2,\pi/2]$ of β is symmetric. This permits us to neglect one of the sign options by considering only $V_\beta = \omega l\cos\beta + V\sin\beta$. Then $\Xi = \Xi(V_\beta, V, \omega)$, where V_β ranges over $[-V, \sqrt{V^2+\omega^2 l^2}]$ for $\omega\ge 0$ and over $[-\sqrt{V^2+\omega^2 l^2}, V]$ for $\omega<0$. For $V_\beta\in[-V,V]$, the maximum of Ξ over all possible combinations of the signs in Ξ and $\pm\omega, \pm V_\beta$ evidently equals

$$\frac{|\omega||V_\beta| + \frac{\left(|V_\beta|+\sqrt{\Delta^2+V_\beta^2}\right)^2}{\rho+d_0}}{\sqrt{\Delta^2+V_\beta^2}}.$$

For the remaining parts ($V_\beta\in[V,\sqrt{V^2+\omega^2 l^2}]$ for $\omega\ge 0$ and $V_\beta\in[-\sqrt{V^2+\omega^2 l^2}, -V]$ for $\omega<0$) the same is true with respect to the maximum over two signs in Ξ. Thus we see that the second maximum in (6.16) is equal to

$$\max_{\substack{0\le\omega\le\overline{\omega}\\ 0\le V\le\overline{V}\\ 0\le V_\beta\le\sqrt{V^2+\omega^2 l^2}}} \frac{\omega V_\beta + \frac{\left(V_\beta+\sqrt{\Delta^2+V_\beta^2}\right)^2}{\rho+d_0}}{\sqrt{\Delta^2+V_\beta^2}}$$

$$= \max_{\substack{0\le\omega\le\overline{\omega}\\ 0\le V\le\overline{V}\\ 0\le V_\beta\le\sqrt{V^2+\omega^2 l^2}}} \left[\omega\frac{V_\beta}{\sqrt{\Delta^2+V_\beta^2}} + \frac{1}{\rho+d_0}\left(\frac{V_\beta^2}{\sqrt{\Delta^2+V_\beta^2}} + 2V_\beta + \sqrt{\Delta^2+V_\beta^2}\right)\right].$$

It is easy to check that in the last expression, all addends increase as V_β grows. Thus the maximum over V_β is attained at $V_\beta = \sqrt{V^2+\omega^2 l^2}$ and the entire maximum is equal to

$$\max_{\substack{0\le\omega\le\overline{\omega}\\ 0\le V\le\overline{V}}} \frac{\omega\sqrt{V^2+\omega^2 l^2}+\frac{\left(\sqrt{V^2+\omega^2 l^2}+v\right)^2}{\rho+d_0}}{v}.$$

Overall, the second condition from (4.6) for the circular parts is implied by the inequality (where V_ω is taken from (6.14)):

$$\frac{\aleph l}{\sqrt{v^2-\overline{\omega}^2 l^2}}+\frac{\overline{\omega}V_\omega+\frac{(V_\omega+v)^2}{\rho+d_0}}{v}\le\lambda_a\overline{u}<\overline{u}. \tag{6.17}$$

Bringing the pieces together, we see that Assumptions 4.1–4.5 hold for all feasible scenarios if (6.14), (6.15), and (6.17) are true. Moreover, these requirements are sufficient for the navigation law (3.1) to ensure locally stable patrolling if δ,γ in Fig. 9.2(a) are chosen so that (5.4) holds, where $\eta_\nu\in(0,1-\lambda_\nu), \nu=v,a$ and, according to the paragraph preceding Section 9.5.1, $z_*\in\left(0,\frac{v}{\sqrt{2}-1}\right)$ is chosen so that (6.15) and (6.17) are true with $\lambda_a:=\lambda_a+\eta_a$ and under substitution in place of v an arbitrary value from the interval $\left[v-\frac{z_*}{\sqrt{2}-1},v+z_*\right]$. This evidently holds if z_* is chosen small enough so that

$$\frac{\overline{V}\overline{\omega}+\overline{\omega}^2(\rho+d_0)+\aleph l}{\sqrt{\left(v-\frac{z_*}{\sqrt{2}-1}\right)^2-\overline{\omega}^2 l^2}}+2\overline{\omega}\le(\lambda_a+\eta_a)\overline{u},$$

$$\frac{\aleph l}{\sqrt{\left(v-\frac{z_*}{\sqrt{2}-1}\right)^2-\overline{\omega}^2 l^2}}+\frac{\overline{\omega}V_\omega+\frac{(V_\omega+v+z_*)^2}{\rho+d_0}}{v-\frac{z_*}{\sqrt{2}-1}}\le(\lambda_a+\eta_a)\overline{u}.$$

Analysis similar to that from the previous subsections also shows that for non-local convergence (i.e., for the conclusion of Theorem 5.1 to be valid) it suffices that Assumption 4.7 is true, (6.15) is true with $d_0:=\max\{d_0,d_+\}$, and (6.17) is valid with $d_0:=\min\{d_0,d_-\}$. The parameter z_* should be taken as the minimum among the values calculated in accordance with the above recommendations for $d_0:=\min\{d_-,d_0\}$ and $d_0:=\max\{d_+,d_0\}$, respectively.

9.7 Navigation in an environment cluttered with moving obstacles

In this section, we follow the lines of Chapter 6 and apply the algorithm of border patrolling to a more general problem of safe guidance of an autonomous wheeled robot toward a target in an a priori unknown cluttered environment. Like in Chapter 6, this algorithm is used to bypass the enroute obstacles with a predefined safety margin, whereas the robot is directed straight to the target when possible. The rules regulating switches between the regimes of obstacle avoidance and straight moves to the target, respectively, are basically borrowed from Chapter 6. The novelty as compared with Chapter 6 is that the obstacles are not steady but may undergo arbitrary motions, including rotations and deformations.

Specifically, we assume that apart from the robot, there is a steady point target $\mathcal{T}$ and several constantly disjoint moving obstacles $D_1(t), \dots, D_k(t)$ in the plane. The objective is to drive the robot to the target through the obstacle-free part of the plane: $\boldsymbol{r}(t) \notin \bigcup_{i=1}^k D_i(t) \quad \forall t$. Moreover, a given safety margin must be respected:

$$\mathbf{dist}\,[\boldsymbol{r}(t); D_i(t)] \geq d_{\text{safe}} \geq 0 \quad \forall t, i.$$

Our proposed navigation controller switches between the obstacle avoidance law (3.1) and straight moves to the target:

$$u(t) = 0. \tag{7.1}$$

In (3.1), $d_0 > d_{\text{safe}}$ is now a tunable parameter of the controller (the desired distance to the obstacle during bypassing it) and $d(t)$ is replaced by $d_i(t) := \mathbf{dist}\,[\boldsymbol{r}(t); D_i(t)]$ for i chosen by the controller. There are two more controller parameters:

- $\epsilon > 0$—participates to decision of obstacle avoidance termination, which is allowed only if the robot comes close enough to the obstacle: $d_i \leq d_0 + \epsilon$;
- $C > d_0 + \epsilon$—the trigger threshold: the distance to an obstacle at which the avoidance mode is switched on.

The switch (7.1) $\mapsto$ (3.1) occurs when the distance d_i to the obstacle $D_i(t)$ reduces to C, with this i being employed $d := d_i$ in (3.1). The converse (3.1) $\mapsto$ (7.1) is carried out when $d_i(t) \leq d_0 + \epsilon$ and the robot is headed for the target. For this rule to be well-defined, (3.1) must not be activated simultaneously for several obstacles. To offer constructive conditions for this to hold, we introduce the following.

Definition 7.1. The *averaged span* of obstacle i is $\mathcal{R}_i^{\text{av}} = \min \mathcal{R}_i$, where min is over all $\mathcal{R}_i$ such that during any time interval of duration $(3\pi)/\overline{u}$ the body $D_i(t)$ remains in some steady disc of the radius $\mathcal{R}_i$.

By **iii** of Theorem 5.1, since the start of the avoidance maneuver at the distance C from obstacle i and until the commencement of the sliding motion toward the d_0-equidistant curve, the robot moves inside the disc of the radius $2R_{\min}$ centered at its initial position, where $R_{\min}$ is the minimal turning radius (2.2). Meanwhile the obstacle moves within some disc of the radius $\mathcal{R}_i^{\text{av}}$. So at this phase, the distance d_i from the robot to the obstacle constantly lies in the interval

$$d_i \in \left[C - 2(R_{\min} + \mathcal{R}_i^{\text{av}}); C + 2(R_{\min} + \mathcal{R}_i^{\text{av}})\right]. \tag{7.2}$$

Afterwards d_i monotonically goes to $d_0 < C$ by Theorem 5.1. Thus while the navigation rule (3.1) is active for obstacle i, the robot is at the distance $\leq C + 2(R_{\min} + \mathcal{R}_i^{\text{av}})$ from $D_i(t)$. So to exclude simultaneous activation of this rule for two obstacles, it suffices that they are separated by a distance $> 2(C + R_{\min} + \mathcal{R}_{\max}^{\text{av}})$, where $\mathcal{R}_{\max}^{\text{av}} := \max_i \mathcal{R}_i^{\text{av}}$. With regard to the available freedom in the choices of

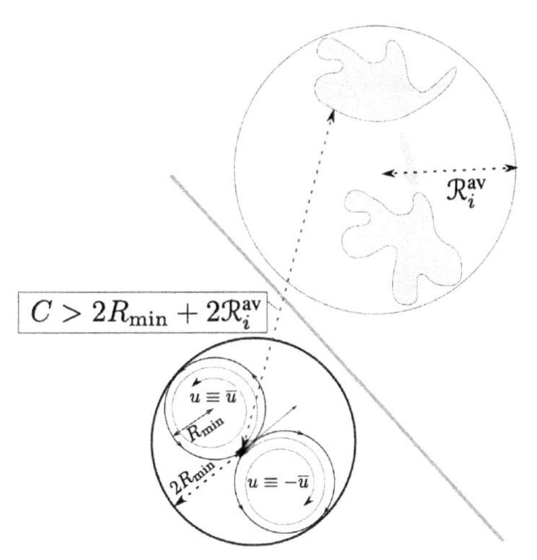

Figure 9.13 Disjoint discs.

C, d_0, ϵ $(C > d_0 + \epsilon, \epsilon > 0, d_0 > d_{safe})$, this gives rise to the following requirement to the distance $d_{i,j}(t)$ between any two obstacles $i \neq j$ at any time t:

$$d_{i,j}(t) \geq d_{obs} > 2(R_{min} + \mathscr{R}^{av}_{max} + d_{safe}). \tag{7.3}$$

Then C, d_0 and ϵ may and should be chosen so that

$$\frac{d_{obs}}{2} - (R_{min} + \mathscr{R}^{av}_{max}) > C > d_0 + \epsilon, \quad \epsilon > 0, \quad d_0 > d_{safe}.$$

Since the navigation law (3.1) is still employed, the conditions for its convergence should be satisfied. We suppose that Assumptions 4.1 and 4.2 hold for any obstacle. If $C \geq 2(R_{min} + \mathscr{R}^{av}_{max})$, the afore-mentioned two discs do not intersect (see Fig. 9.13) and so can be separated by a straight line; thus Assumption 4.7 holds. Such a choice of C is possible only if $d_{obs} > 6R_{min} + 6\mathscr{R}^{av}_{max}$ in addition to (7.3), which is assumed. Accordingly, C should be chosen so that

$$\frac{d_{obs}}{2} - (R_{min} + \mathscr{R}^{av}_{max}) > C > \max\left\{d_0 + \epsilon; 2(R_{min} + \mathscr{R}^{av}_{max})\right\}. \tag{7.4}$$

As for Assumption 4.6, we note that due to (7.2), all launching distances for any maneuver of bypassing obstacle i are contained by the interval from (7.2) and so by

$$\left[C - 2(R_{min} + \mathscr{R}^{av}_{max}); C + 2(R_{min} + \mathscr{R}^{av}_{max})\right]. \tag{7.5}$$

So Assumption 4.6 can be satisfied if within the range (7.4), there is a value of C such that $C - 2(R_{min} + \mathscr{R}^{av}_{max}) > d_{safe}$, and the interval (7.5) is regular for any obstacle, provided that d_0 is chosen within this interval. Overall, we arrive at the following recommendations on the choices of C, d_0, ϵ:

$$2(R_{\min} + \mathscr{R}^{\text{av}}_{\max}) > \epsilon > 0,$$
$$\frac{d_{\text{obs}}}{2} - (R_{\min} + \mathscr{R}^{\text{av}}_{\max}) > d_0 > d_{\text{safe}},$$
$$\min\left\{\frac{d_{\text{obs}}}{2} - (R_{\min} + \mathscr{R}^{\text{av}}_{\max}); d_0 + 2(R_{\min} + \mathscr{R}^{\text{av}}_{\max})\right\} > C$$
$$> \max\left\{d_0 + \epsilon; 2(R_{\min} + \mathscr{R}^{\text{av}}_{\max}) + d_{\text{safe}}\right\}. \quad (7.6)$$

This is possible if and only if $d_{\text{obs}} > 6R_{\min} + 6\mathscr{R}^{\text{av}}_{\max} + 2d_{\text{safe}}$.

To properly choose the controller parameters γ and δ, it suffices to pick the auxiliary parameters $z_*, \lambda_a, \lambda_v, \eta_a, \eta_v$ in (7.3) common for all obstacles. This is possible: for every obstacle, λ_a, λ_v in (4.6) are first increased (within $(0, 1)$) to a common value, then common $\eta_v \in (0, 1 - \lambda_v), \eta_a \in (0, 1 - \lambda_a)$ are picked. Finally, z_* is computed for all obstacles and the minimal result is put in (7.3).

For the proposed guidance law to be implementable, the robot should have access to the angle-of-sight $\beta(t)$ to the target $\mathscr{T}$, as well as to the distance d_i and its derivative $\dot{d}_i$ whenever $d_i \leq C + 2R_{\min} + 2\mathscr{R}^{\text{av}}_i$. We also assume that initially the robot is above the trigger threshold of all obstacles $d_i(0) > C \; \forall i$ and that they are always far enough from the target $\textbf{dist}\,[\mathscr{T}; D_i(t)] > d_0 + \epsilon \; \forall t, i$. Since the controller (3.1) is still used, conditions (4.6) and (5.4) should be also replicated.

In the remainder of this chapter, the convergence and performance of the proposed guidance law for target reaching with obstacle avoidance is demonstrated via extensive simulations and experimental studies. They are presented in the next two sections.

9.8 Simulations

Matlab and MobileSim 0.5 (a simulator for ActivMedia robots) were used to illustrate the performance of the proposed guidance law (PGL); the simulation parameters are given in Table 9.2, where T_s is the sampling period. The derivative $\dot{d}$ was approximated by Newton's difference quotient $\dot{d}(t) \approx T_s^{-1}[d(t) - d(t - T_s)]$. The robot was navigated through environment with four moving nearly rectangular shaped rigid obstacles, and has access only to the current distance to the obstacles $d(t)$. Figure 9.14(a) shows the initial states of the robot and obstacles and the position of the target, Fig. 9.14(c), (e), and (g) display the crucial moments of the maneuver when the distance from the robot to a particular obstacle achieves its minimum. The paths of the robot and obstacles are depicted as circled lines. Figures 9.14(d), (f), and (h) show the evolution of the distance from the robot to each of the obstacles. The rectangles in Fig. 9.14(c),

Table 9.2 Robot and controller parameters employed in simulations

$T_s = 0.1$ s	$\overline{u} = 0.8$ rad/s	$d_0 = 1.2$ m	$\varepsilon = 0.1$ m
$v = 1$ m/s	$\gamma = 1.5$ 1/s	$d_{\text{safe}} = 1$ m	$C = 1.5$ m

(e), and (g) illustrate the states of the robot and nearest obstacle at the above crucial moments; the minimal distance is similarly marked in Fig. 9.14(d), (f), and (h). In this experiment, the robot successfully arrives at the target with avoiding collisions with the moving obstacles.

The performance of PGL was compared with that of the popular Velocity Obstacle approach (VOA) [17, 19] via simulation in MatLab in the scenario where a long obstacle perpendicularly intervenes between the robot and the target, which is hidden just behind the obstacle; see Fig. 9.15(a) and (b). Figures 9.15(c) and (d) display the situations at the moments when the robot transverses the path of the obstacle. The entire paths of the robot during target reaching with obstacle avoidance are depicted in Figs. 9.15(d) and (f). (We considered VOA with two choices of velocity per maneuver, with the second choice being made at the favorable moment of bypassing the obstacle.) PGL first drives the robot directly to the target in a straight line until the trigger threshold is trespassed and then undertakes a relatively small de-tour by

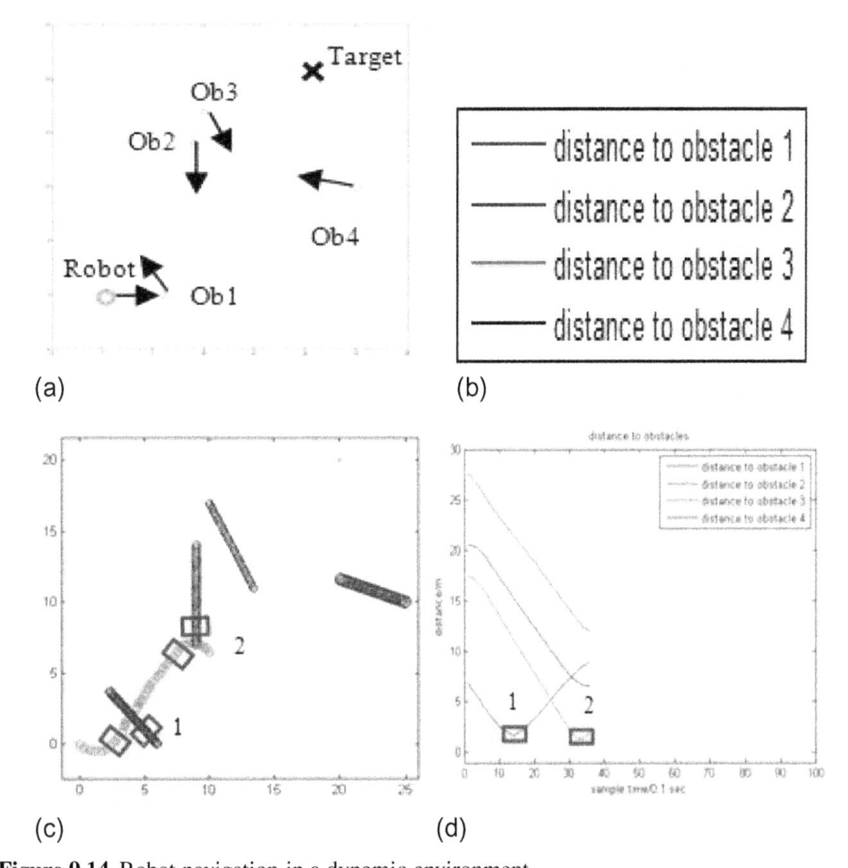

Figure 9.14 Robot navigation in a dynamic environment.

(Continued)

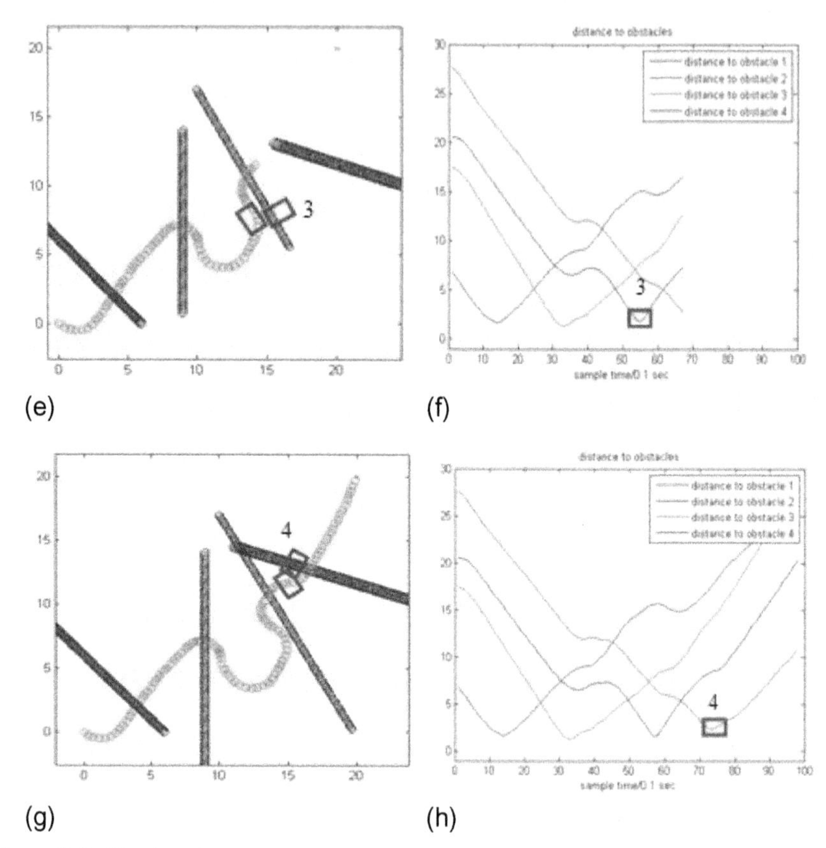

Figure 9.14 Continued

following the obstacle boundary. VOA basically drives the robot to the target along two straight lines with bypass of the obstacle in a close range at the rear. Thus the both guidance laws do drive the robot to the target. However PGL does this faster: the ratio of the maneuver times is as follows $\frac{T_{\text{PGL}}}{T_{\text{VOA}}} = \frac{14.51}{18.61} \approx 0.78$. Thus in this experiment, PGL outperforms VOA.

In the next simulation, the robot moves to the target (the red (dark gray in print versions) circle in Fig. 9.16) with bypassing a large cross rotating about a moving pivot. The translational and rotational directions of the obstacle, its initial state and that of the robot are shown in Fig. 9.16(a). The robot first moves toward the target until its distance to the obstacle reduces to the trigger threshold, as is shown in Fig. 9.16(b). Then it turns left to avoid the obstacle, as shown in Fig. 9.16(c). Finally, it bypasses the cross and arrives at the target, as is shown in Fig. 9.16(d).

Figure 9.17 displays the results of simulations concerned with the problem of escorting a convoy of unicycle-like vehicles, which was described in Section 9.6.4. In this experiment, the leader moves with the constant speed $V = 0.3$ m/s and time-varying turning rate not exceeding $\overline{\omega} = 0.55$ rad/s, the convoy is composed of four

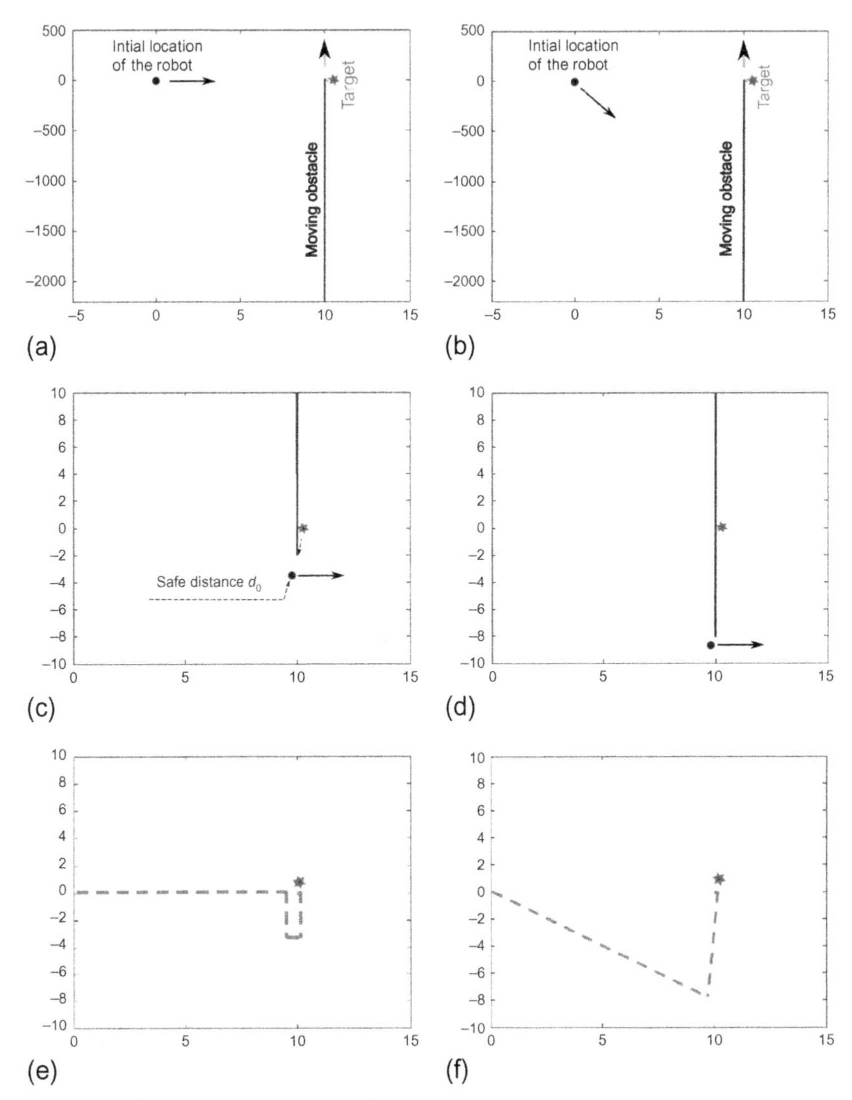

Figure 9.15 PGL (left column) versus VOA (right column).

vehicles, each approximated by a disc with the radius $\rho = 0.1$ m, moving at the distance $\Delta s = 0.133$ m one after another so that the total length of the convoy equals 1.2 m, and the desired escort distance was taken to be $d_0 = 1.5$ m. The escorting robot starts motion at a remote location. As is shown in Fig. 9.17(b), after a transient, the robot escorts the convoy with maintaining the desired distance 1.5 m from it with the error not exceeding 0.247 m.

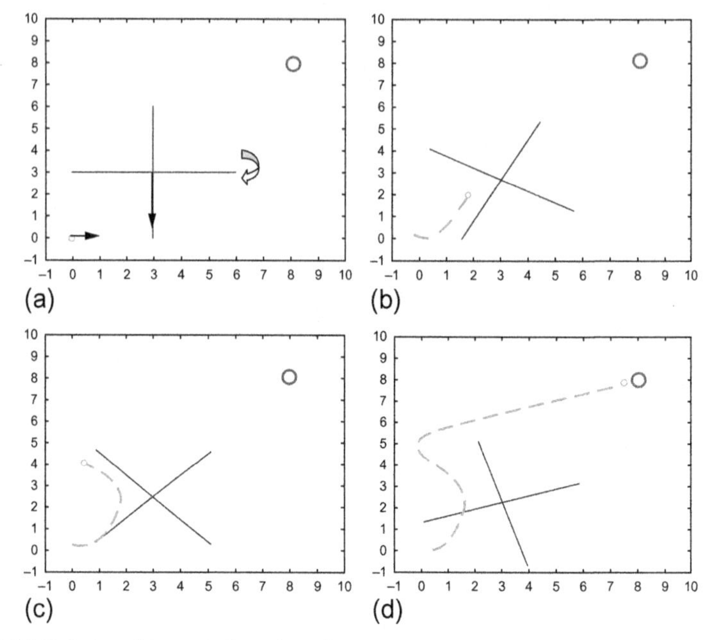

Figure 9.16 Robot avoids a rotating obstacle.

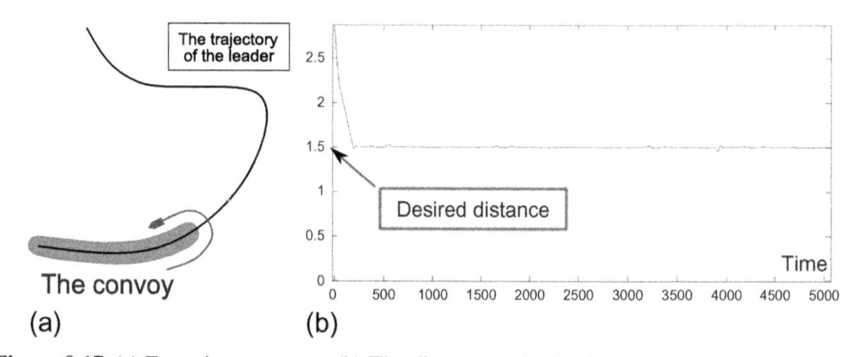

Figure 9.17 (a) Escorting a convoy; (b) The distance to the border.

9.9 Experimental results

The performance and applicability of PGL was examined on a Pioneer 3-DX robot described in Section 1.4.1.

Figure 9.18(a) shows an environment with five static obstacles (four boxes and one human) and black target. In Fig. 9.18, the trajectory of the robot at various stages of progression is depicted by the dashed lines; the dashed red (light gray in

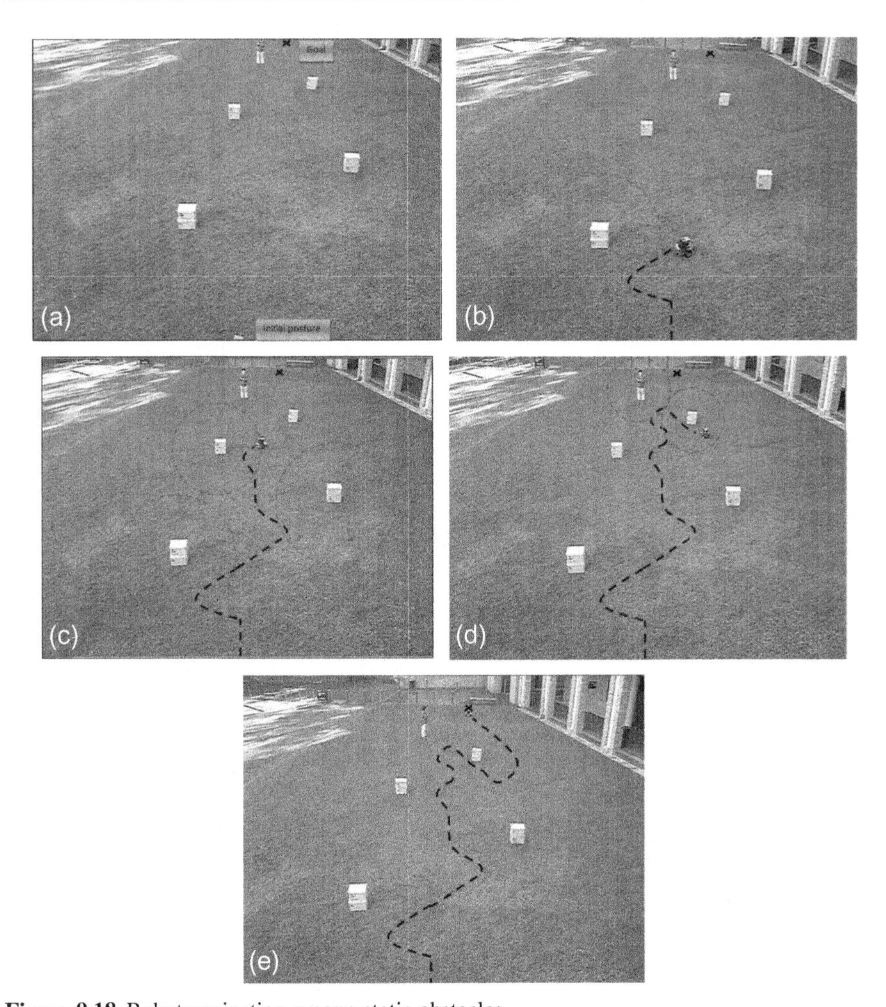

Figure 9.18 Robot navigation among static obstacles.

print versions) circles display the trigger curves where the robot is switched to the avoidance mode. In this experiment, like in the others, the robot reaches the target via safe navigation among the obstacles, with no visible chattering of the mechanical parts being observed.

Figure 9.19 concerns another experiment, where both static (the boxes) and moving (the humans) obstacles are involved. The target is positioned at the black cross. The overall path of the robot is shown in Fig. 9.19(e), the crucial moments during progression along the path are displayed in Fig. 9.19(b)–(d). The robot copes with moving obstacles as well.

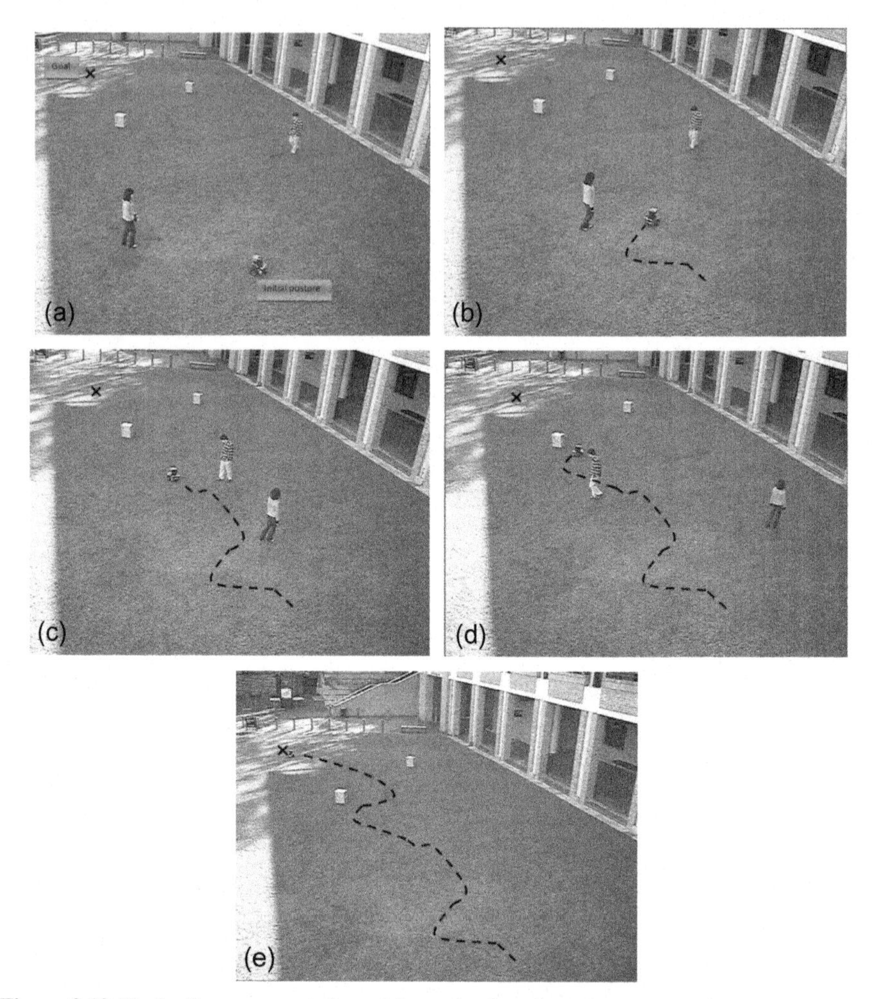

Figure 9.19 Navigation among static and dynamic obstacles.

In the last experiment, the influence of the parameters on the overall performance was studied. To this end, both the speed of the robot and the trigger threshold were decreased. As is shown in Fig. 9.20(b), this may result in bypassing an obstacle with a smaller turning maneuver, which carries a potential to decrease the deviation from the optimal motion in the straight line. However Fig. 9.20(c) displays the opposite: due to the reduced speed, the robot accompanies the moving obstacle for a longer time, which engages the robot in a long sideways maneuver. The reduced speed also caused an increase of the overall time of target reaching.

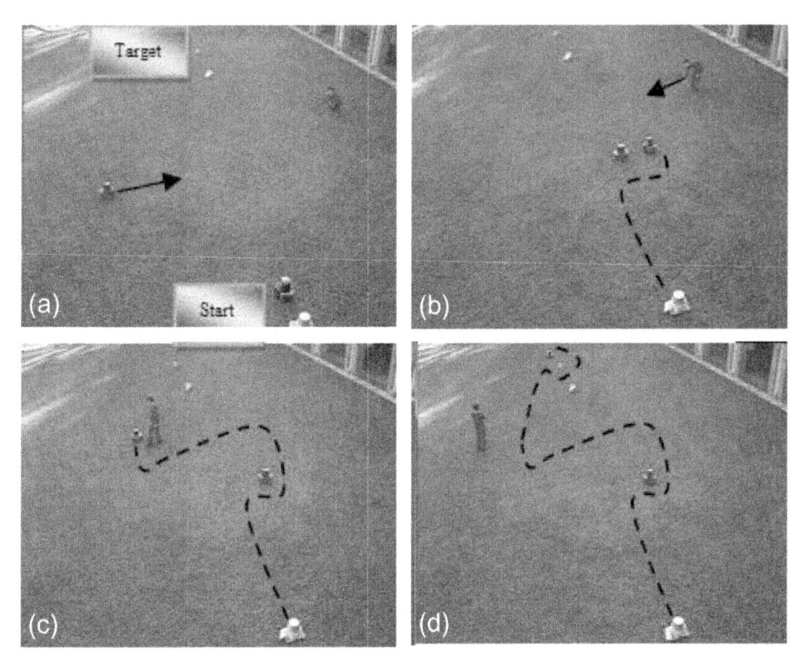

Figure 9.20 The effect of changing parameters.

Seeking a path through the crowd: Robot navigation among unknowingly moving obstacles based on an integrated representation of the environment

10.1 Introduction

Collision-free navigation of a mobile robot in cluttered environments is a fundamental problem of robotics, and this problem has been the subject of extensive research.

According to detailed discussions in Chapters 1 and 3, existing navigation approaches can be generally classified as global or local. Global path planning is based on a complete model of the environment and tries to find the best possible solution; this approach basically assumes the environment is known a priori. On the other hand, local navigation algorithms use onboard sensors to locally observe small fragments of an unknown environment at each time. However, a problem is that these strategies are often heuristic and are not based on mathematical modeling of the vehicles and their nonholonomic constraints. Many of the proposed obstacle avoidance techniques, such as the dynamic window, the curvature velocity, and the lane curvature methods, along with the boundary following method from Chapter 6 and the tangent graph based navigation algorithm from Chapter 4, consider the case of stationary obstacles. The problem of avoiding collisions with moving obstacles is much harder. For this problem, a number of approaches such as velocity obstacles, collision cones, or inevitable collision states have been proposed. However, these approaches assume deterministic knowledge of the obstacle velocity and a moderate rate of its change, and tend to be computationally expensive. Furthermore, nonholonomic constraints on a robot's motion were rarely taken into account, and fully actuated robots were mostly studied in this area until now, which is a severe limitation in practice. The obstacles were predominantly interpreted as rigid bodies, often of the simplest shapes (e.g., discs or polygons), and the sensory data were assumed to be enough to determine the location of obstacle characteristic points concerned with its global geometry (e.g., the disc center or angularly most distant polygon vertex) and to provide access to its full

Safe Robot Navigation Among Moving and Steady Obstacles. http://dx.doi.org/10.1016/B978-0-12-803730-0.00010-X

velocity. Finally, mathematically rigorous performance justification of the proposed algorithms was rarely given.

A local range-only based navigation algorithm for nonholonomic robots was presented in Chapter 9, where the obstacles were not assumed to be rigid or even solid. They may have arbitrary and time-varying shapes, and may rotate, twist, wring, skew, or be deformed in another way. However, the main results of Chapter 9 were not directly devoted to navigation among obstacles, their main focus was on a border patrolling problem. New algorithms for navigation in dynamic environments were proposed in [427, 428]. However, the algorithms of [427, 428] do not take into account nonholonomic constraints. Moreover, the papers [427, 428] do not give any mathematically rigorous analysis of the proposed navigation algorithms.

Until now, a typical obstacle-avoidance algorithm in a complex environment with many obstacles was based on selection of the "closest" or "most dangerous" obstacle in order to focus on avoidance of collision with only this obstacle in a nearest future. It was assumed that until the threat of collision with this obstacle is eliminated, no other obstacle becomes "threatful," and moreover, this "no-new-threat" situation prolongs a bit longer so that after the "dangerous" obstacle is bypassed the robot can move in a desired direction until it meets another "dangerous" obstacle. However, this approach raises many difficult issues, such as how to identify the "most dangerous" obstacle, how to separate obstacles, how to handle obstacles that are hidden or are outside the sensing field, how to deal with situations where there is more than one "dangerous" obstacle at the same time, etc.

In this chapter, we advocate another approach that is based on an integrated representation of the environment, which dismisses many of the above issues and thus overcomes the related problems. This approach is partially inspired by the paradigm shift from binary interaction models to an integrated treatment of multiple interactions, which is typical for social interactions in human crowds or animal swarms, as was suggested in [429] for analysis of pedestrian behaviors in crowds. The presented approach is based on an integrated representation of the information about the environment in which the combined effect of close multiple stationary and moving obstacles is implicitly included in the representation of the sensing field of the robot. An advantage of this approach is that it involves no need to separate obstacles or to approximate their shapes by discs or polygons. Moreover, it does not require any information on the obstacles' velocities or other derivatives of measurements. The proposed approach results in a very efficient and intelligent robot's behavior. Instead of being repelled by a crowd of obstacles, as it happens for many other navigation algorithms, the robot seeks a path through this crowd.

In this chapter, we still examine a Dubins-car-like mobile robot described by the standard nonholonomic model with a hard constraint on its angular velocity. As was discussed in Section 3.2, the motions of many wheeled robots, missiles, and unmanned aerial and underwater vehicles can be described by this model. Unlike many papers on the area of robotics, which present heuristic navigation strategies, we offer a mathematically rigorous analysis of the presented navigation algorithm with a complete proof of the stated theorem. The proposed real-time algorithm is quite simple and computationally efficient. Its performance is confirmed by extensive computer simulations and experiments with a Pioneer P3-DX mobile wheeled robot. In doing so, the performance of this algorithm is compared with that of the popular velocity obstacle approach [17, 19]. The navigation algorithm developed in this chapter was successfully applied to collision-free assisted navigation of an intelligent semi-autonomous wheelchair [430].

The main results of the chapter were originally published in [431]. A preliminary version of some of these results was also reported in the conference paper [432].

The remainder of the chapter is organized as follows. Section 10.2 presents the system description and the problem statement. Section 10.3 introduces the navigation algorithm. Computer simulations for the proposed navigation algorithm are given in Section 10.5, while Section 10.6 presents experiments with a Pioneer P3-DX mobile nonholonomic robot. Mathematical analysis of the algorithm is given in Section 10.4.

10.2 Problem description

We consider a planar vehicle or wheeled mobile robot modeled as a Dubins car. The control input is the angular velocity limited by a given constant. The mathematical model of the robot is as follows:

$$\begin{aligned} \dot{x} &= v\cos\theta, \\ \dot{y} &= v\sin\theta, \\ \dot{\theta} &= u \end{aligned} \tag{2.1}$$

where

$$u \in [-\overline{u}, \overline{u}]. \tag{2.2}$$

Here $x(t)$ and $y(t)$ are the robot's Cartesian coordinates in the world frame, $\theta(t)$ gives its heading, and $v > 0$ and $u(t)$ are the speed and angular velocity, respectively. The angle $\theta(t) \in (-\pi, \pi]$ is measured in the counter-clockwise direction from the x-axis. The maximum angular velocity $\overline{u}$ and the speed v are given, while the angular velocity $u(t)$ is updated at discrete times $0, \delta > 0, 2\delta, 3\delta, \ldots$.

We consider a general problem of robot navigation with collision avoidance. We assume the environment is dynamic and unknown to the robot. This environment consists of an arbitrary number of moving or stationary components, which can be deformable. In other words, the environment is described by a time-varying planar set $E(t) \subset \mathbb{R}^2$, possibly with many components. This set represents the part of the plane impenetrable by the robot; $E(t)$ is not known to the robot a priori.

We also consider a stationary target $\mathcal{T}$ modeled as a small circle of radius $R_{\mathcal{T}}$. The direction $\theta_0(t) \in (-\frac{\pi}{2}, \frac{\pi}{2})$ from the robot to the center of the target $\mathcal{T}$ is assumed be known to the robot. Furthermore, let $d_{\text{safe}} > 0$ be a given distance. The goal is to navigate the robot toward the target $\mathcal{T}$ while satisfying the safety constraint $d(t) \geq d_{\text{safe}}$ for all t. Here $d(t)$ is the current distance from the robot to the set $E(t)$:

$$d(t) := \mathbf{dist}\,[\boldsymbol{r}(t); E(t)] := \min_{\boldsymbol{r}' \in E(t)} \|\boldsymbol{r}' - \boldsymbol{r}(t)\|,$$

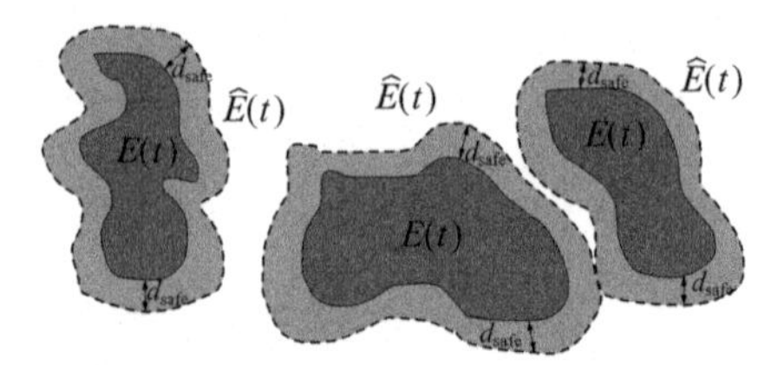

Figure 10.1 Original environment $E(t)$ (interior darker region) and the enlarged environment $\hat{E}(t)$ (the grey larger region that covers $E(t)$).

where $\|\cdot\|$ denotes the standard Euclidean vector norm and $\boldsymbol{r}(t)$ is the current location of the robot:

$$\boldsymbol{r}(t) := \mathbf{col}\,[x(t), y(t)]. \tag{2.3}$$

Definition 2.1. A robot navigation strategy is said to be *safe collision free* if $d(t) > d_{\text{safe}}$ for all t. Furthermore, a robot navigation strategy is said to be *target reaching* if $\boldsymbol{r}(t_f)$ belongs to the target circle at some finite time $t_f \geq 0$.

Consider the enlarged environment $\hat{E}(t)$ that is defined as the d_{safe}-neighborhood of $E(t)$; see Fig. 10.1. It is obvious that the robot's navigation strategy is safe collision free in the sense of Definition 2.1 if and only if the robot is always outside the enlarged environment $\hat{E}(t)$.

Assumption 2.1. *The set $\hat{E}(t)$ is closed and has a piecewise analytic boundary.*

Let $d_{\text{sen}} > 0$ be a given constant defining the sensing ability of the robot. We introduce the binary function $M(\alpha, t) \in \{0, 1\}$ defined for all $t \geq 0$ and all $\alpha \in (\theta(t) - \frac{\pi}{2}, \theta(t) + \frac{\pi}{2})$ as follows:

- $M(\alpha, t) = 1$ if the ray emitted at time t in the direction of α from the robot's position hits the enlarged environment $\hat{E}(t)$ at a point $\boldsymbol{p}$ whose distance $d_p(t)$ from the robot does not exceed

$$d_p(t) := \|\boldsymbol{r}(t) - \boldsymbol{p}\| \leq d_{\text{sen}} \cos\alpha; \tag{2.4}$$

- $M(\alpha, t) = 0$ otherwise; see Fig. 10.2.

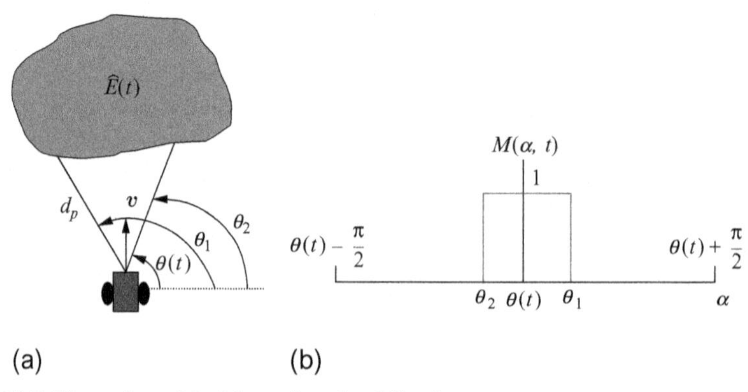

Figure 10.2 Illustration of the binary function $M(\alpha, t)$.

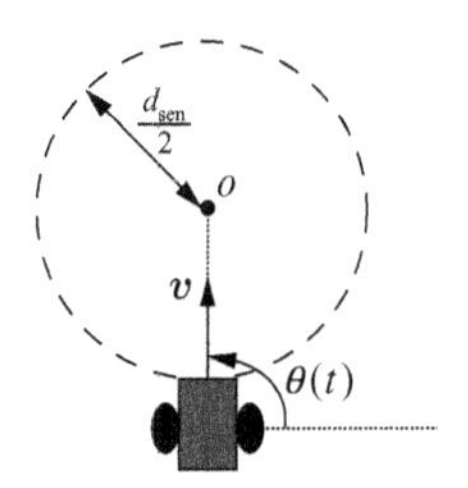

Figure 10.3 Disc $\mathcal{D}(t)$ of diameter d_{sen}.

Available measurements. The only information about the environment $E(t)$ available to the robot is the binary function $M(\alpha, t)$ at discrete sampling times $t = 0, \delta, 2\delta, 3\delta, \ldots$. Also, the robot has access to its heading $\theta(t)$ and to the direction $\theta_0(t)$ to the target's center.

It is easy to understand a geometrical sense of the condition (2.4). Indeed, it follows from elementary geometry that (2.4) defines a disc $\mathcal{D}(t)$ of diameter d_{sen} centered at the point O that is ahead of the robot's position by $\frac{d_{\text{sen}}}{2}$ in the direction of its current heading $\theta(t)$; see Fig. 10.3. Therefore, the binary function $M(\alpha, t)$ informs the robot that some points of the unknown dynamic environment have crossed the boundary of $\mathcal{D}(t)$ in the direction defined by the angle α.

Definition 2.2. The disk $\mathcal{D}(t)$ is called the *sensored disk* (at time t).

Remark 2.1. The information available to the robot is based on an integrated representation of the environment and is very easy to obtain in practice. Unlike many other approaches to collision avoidance with moving obstacles, the adopted approach does not require separating obstacles, and approximating their shapes by discs or polygons. We do not need to estimate obstacles' velocities or any other measurements' derivatives. The obstacles are not assumed to be rigid. They may rotate, twist, skew, wriggle, or be deformed in another way. This, for example, covers scenarios with reconfigurable rigid obstacles, rigid obstacles with large moving parts, flexible underwater obstacles, like fishing nets, schools of big fish, virtual obstacles, like areas corrupted with hazardous chemicals, radiation, high turbulence, etc.

Now we illustrate the rationale behind the choice of the sensored disc $\mathcal{D}(t)$ as the key component describing the sensor information available to the robot. To this end, we consider a point-wise obstacle $D(t)$ moving with a time-varying velocity $v_D(t)$ satisfying the only constraint

$$\|v_D(t)\| \le v. \tag{2.5}$$

Now we are in a position to state the following proposition.

Proposition 2.1. *Suppose that the robot moves in a straight line with the control input* $u(t) = 0$ *for* $t \in [0, t_0]$, *where* $t_0 > 0$ *is given. Any obstacle* $D(t)$ *moving for* $t \in [0, t_0]$ *from a given location* $\mathbf{p} = \mathbf{col}\,[x_D(0), y_D(0)]$ *while respecting the velocity bound* (2.5) *does not collide with the robot over the time interval* $t \in [0, t_0]$ *if and only if the initial location* $\mathbf{p}$ *does not belong to the initial sensored disc* $\mathcal{D}(0)$ *with the diameter* $d_{sen} = 2vt_0$, *which is shown in Fig. 10.3.*

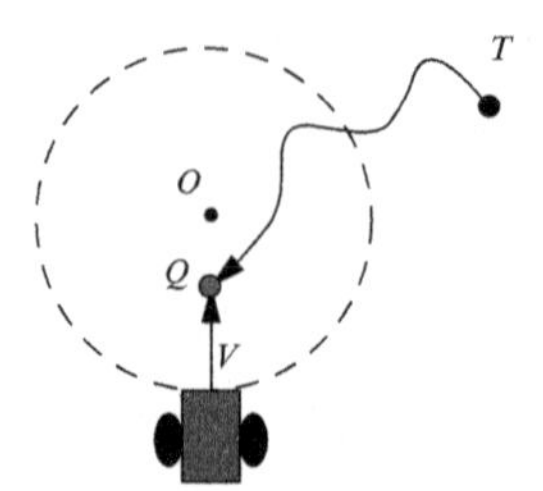

Figure 10.4 Illustration to the proof of Proposition 2.1.

Proof. First, we prove that if the obstacle's initial location $\boldsymbol{p}$ belongs to the disc $\mathcal{D}(0)$ with diameter $D_s = 2vt_0$, then the robot and the obstacle will collide for some obstacle velocity satisfying (2.5). Indeed, if $\boldsymbol{p}$ belongs to the disc $\mathcal{D}(0)$, then the distance d_p from $\boldsymbol{p}$ to the center of the disc $\mathcal{D}(0)$ satisfies $d_p \leq vt_0$. Therefore, if the obstacle $D(t)$ moves straight to the center of $\mathcal{D}(0)$ with the speed $\frac{d_p}{t_0}$, the constraint (2.5) is satisfied and $D(t)$ will be at the center of the disc at time t_0. On the other hand, the robot will also reach the center of the disc $\mathcal{D}(0)$ at time t_0. Hence, the robot and obstacle will collide at time t_0.

Now we prove that if the obstacle's initial location does not belong to the disc $\mathcal{D}(0)$ with diameter $d_{\text{sen}} = 2vt_0$, then the robot and obstacle do not collide for $t \in [0, t_0]$ whenever (2.5) holds. Indeed, suppose to the contrary that the robot and $D(t)$ collide at some time $t_1 \leq t_0$ though the obstacle velocity satisfies (2.5). The collision point Q clearly lies on the interval between the robot's initial position and the center of the disk $\mathcal{D}(0)$; see Fig. 10.4. An elementary geometrical argument assures that since the obstacle's initial position is outside of $\mathcal{D}(0)$, the distance from the obstacle's initial position to Q is greater than the distance from the robot's initial position to Q. Therefore, the obstacle's speed should be greater than the robot's speed, in violation of (2.5). This contradiction completes the proof. □

10.3 Navigation algorithm

We will need the following assumption.

Assumption 3.1. *Initially the robot is headed for the target $\theta(0) = \theta_0(0)$ and the sensored disk is free from obstacles:*

$$M(\alpha, 0) = 0 \quad \textit{for all} \quad \alpha \in \left(\theta(0) - \frac{\pi}{2}, \theta(0) + \frac{\pi}{2}\right).$$

Furthermore, the target's relative bearing is always acute:

$$\theta_0(t) \in \left(\theta(t) - \frac{\pi}{2}, \theta(t) + \frac{\pi}{2}\right) \quad \textit{for all} \quad t \geq 0.$$

To proceed, we introduce the function $m(t)$ as follows:

$$m(t) := \begin{cases} 0 & \text{if } M(\alpha, t) = 0 \ \forall \alpha \in \left(\theta(t) - \frac{\pi}{2}, \theta(t) + \frac{\pi}{2}\right), \\ 1 & \text{otherwise.} \end{cases} \tag{3.1}$$

If $m(t) = 1$ and $M(\alpha, t) = 0$ for some $\alpha \in (-\frac{\pi}{2}, \frac{\pi}{2})$, then Assumption 2.1 implies that the set

$$\left\{\alpha \in (\theta(t) - \frac{\pi}{2}, \theta(t) + \frac{\pi}{2}) : M(t, \alpha) = 0\right\}$$

consists of a finite number of non-overlapping open intervals (A_i^-, A_i^+), $A_i^- < A_i^+$. They are the intervals of directions in which the environment is far enough from the robot's current position. Let $j(t)$ be the index of the interval that is closest to the robot's current heading $\theta(t)$. In other words, if the environment is far enough in the direction of the robot's heading $M[\theta(t), t] = 0$, then $\theta(t) \in (A_i^-, A_i^+)$ for some i and $j(t) := i$. If conversely $M[\theta(t), t] = 1$, the heading $\theta(t)$ does not belong to any (A_i^-, A_i^+), and in this case,

$$j(t) := \arg\min_i \{|A_i^-|, |A_i^+|\}, \tag{3.2}$$

i.e., $j(t)$ is the index i furnishing the minimum of $\min\{|A_i^-|, |A_i^+|\}$. The middle $C(t)$ of the closest interval will be of particular interest to us:

$$C(t) := \frac{A_{j(t)}^- + A_{j(t)}^+}{2}. \tag{3.3}$$

These definitions are illustrated in Fig. 10.5. In Fig. 10.5(a), there are three "close" obstacles and four intervals (A_i^-, A_i^+) shown in Fig. 10.5(b), where $A_1^- = \theta(t) - \frac{\pi}{2}$,

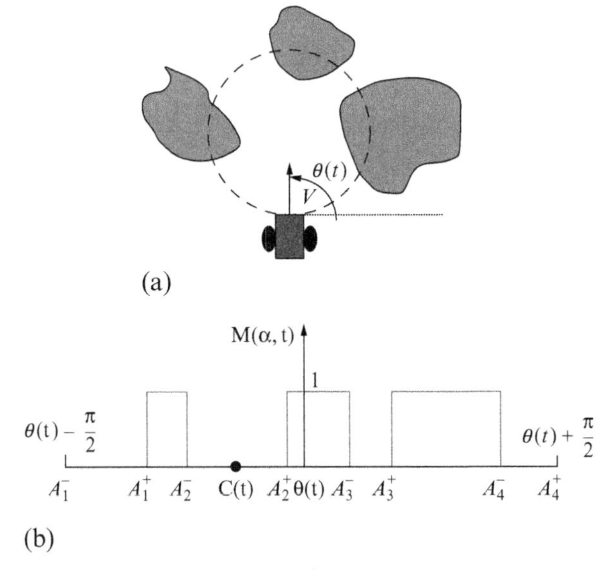

Figure 10.5 Illustration of the intervals (A_i^-, A_i^+) and the direction $C(t)$.

$A_4^+ = \theta(t) + \frac{\pi}{2}$. Furthermore, $M[\theta(t), t] = 1$, and the interval (A_2^-, A_2^+) is closest to $\theta(t)$. Therefore,

$$C(t) := \frac{A_2^- + A_2^+}{2}.$$

Our definition of $C(t)$ relies on Assumption 2.1. If contrary to Assumption 2.1 the boundary is not piecewise analytic, an infinite number of intervals (A_i^-, A_i^+) may appear, with the lengths $A_i^+ - A_i^-$ converging to 0. Of course, it is a purely mathematical technicality without any practical significance.

Now we introduce the following reactive navigation law, which is a rule for updating the robot's angular velocity $u(t)$ at sampling times $0, \delta, 2\delta, 3\delta, \ldots$:

$$\textbf{if } m(k\delta) = 0 \quad \textbf{then} \quad u(t) := \overline{u}\,\textbf{sgn}\,[\theta_0(t) - \theta(t)] \quad \forall t \in [k\delta, (k+1)\delta); \tag{3.4}$$

$$\textbf{if } m(k\delta) = 1 \quad \textbf{then} \quad u(t) := \overline{u}\,\textbf{sgn}\,[C(t) - \theta(t)] \quad \forall t \in [k\delta, (k+1)\delta). \tag{3.5}$$

Here $C(t)$ is defined by (3.3) and **sgn** $(\cdot)$ is the standard signum function.

Remark 3.1. The intuition behind the navigation law (3.4) is pretty straightforward. When the robot does not sense the environment in front of itself $m(t) = 0$, it makes its heading as close as possible to the direction to the target $\theta_0(t)$. When the robot senses the environment $m(t) = 1$, it makes its heading as close as possible to the middle of the closest interval of possible vacant headings $C(t)$.

10.4 Mathematical analysis of the navigation strategy

This section offers a theoretical analysis of the proposed algorithm under a set of simplifying assumptions.

We now suppose the environment $E(t)$ consists of several disjoint stationary or moving obstacles $D_1(t), \ldots, D_n(t)$. These can be deformable. We also consider the enlarged obstacles $\hat{D}_1(t), \ldots, \hat{D}_n(t)$ that are defined as the d_{safe}-neighborhoods of the respective obstacles $D_j(t)$.

Assumption 4.1. *The enlarged obstacles $\hat{D}_1(t), \ldots, \hat{D}_n(t)$ are convex closed sets with piecewise analytical boundaries.*

The maximum displacement of the set $\hat{D}_i(t)$ over the time interval $[t_1, t_2)$ is defined as

$$\rho_i(t_1, t_2) := \max_{\boldsymbol{r}_2 \in \hat{D}_i(t_2)} \min_{\boldsymbol{r}_1 \in \hat{D}_i(t_1)} \|\boldsymbol{r}_2 - \boldsymbol{r}_1\|.$$

We also introduce the following upper bound on the displacement speed of the environment:

$$\overline{v}_E := \max_{i=1,\ldots,n} \sup_{t_2 > t_1 \geq 0} \frac{\rho_i(t_1, t_2)}{(t_2 - t_1)}. \tag{4.1}$$

Assumption 4.2. *The following inequalities hold:*

$$\frac{v \sin(\overline{u}\delta)}{\overline{u}} > \overline{v}_E \delta, \tag{4.2}$$

$$d_{sen} > 2(v + \overline{v}_E)\delta, \tag{4.3}$$

$$(d_{sen} - 2v\delta)\sin(\overline{u}\delta) > 2\overline{v}_E \delta, \tag{4.4}$$

$$(d_{sen} - 2v\delta)(1 - \cos(\overline{u}\delta)) > 2\overline{v}_E \delta. \tag{4.5}$$

Remark 4.1. Assumption 4.2 is not very restrictive. Indeed, for small δ, $\sin(\overline{u}\delta)$ is close to $\overline{u}\delta$, and so (4.2) becomes close to the requirement that the robot is faster than the obstacles: $v > \overline{v}_E$. It is obvious that if this inequality does not hold, in general, it is impossible to obtain mathematically rigorous conditions for the robot to surely avoid collisions for any navigation strategy. However, the last illustrative example from Section 10.5 will show that the proposed navigation strategy works well even in many scenarios with obstacles moving faster than the robot. Furthermore, the requirements (4.3), (4.4), and (4.5) can be satisfied by taking a large enough sensing parameter d_{sen}.

We stress that Assumption 4.2 and requirements (4.2), (4.3), and (4.4) are needed for mathematically rigorous proofs of the theoretical results presented in this chapter. Computer simulations and experiments with a real robot show that in practice, the proposed algorithm produces good results even in situations where these assumption and requirements do not hold.

We define the distance between enlarged obstacles $\hat{D}_i(t)$ and $\hat{D}_j(t)$, $i \neq j$ as

$$d_{ij}(t) := \min_{\boldsymbol{r}_i \in \hat{D}_i(t), \boldsymbol{r}_j \in \hat{D}_j(t)} \|\boldsymbol{r}_i - \boldsymbol{r}_j\|.$$

Assumption 4.3. *For all $i \neq j, t \geq 0$ the inequality $d_{ij}(t) > d_{sen} + 2v\delta$ holds.*

Assumption 4.4. *At any time t no point of the enlarged environment $\hat{E}(t)$ occupies a position in the direction opposite to the robot's current velocity vector at a distance less than $2v\delta$ from the robot's position.*

Theorem 4.1. *Suppose that Assumptions 3.1, 4.1–4.4 hold. Then the robot navigation strategy* (3.4), (3.5) *is safe collision free and target reaching.*

10.4.1 Proof of Theorem 4.1

In this subsection, we suppose that Assumptions 3.1, 4.1–4.4 hold, and examine the robot navigated by the strategy (3.4), (3.5). Thanks to (4.3), $d_{\text{sen}} - 2v\delta > 0$. Hence, we can introduce a disc $\mathcal{D}_0(t)$ of the diameter $d_{\text{sen}} - 2v\delta$ concentric with $\mathcal{D}(t)$.

Lemma 4.1. *Suppose that at a time $t = k\delta$ the disc $\mathcal{D}_0(k\delta)$ does not contain any point of the enlarged environment $\hat{E}(k\delta)$. Then the disc $\mathcal{D}_0((k+1)\delta)$ contains no point of $\hat{E}((k+1)\delta)$ either.*

Proof. We consider separately two cases.

1. $\boldsymbol{u(k\delta) = 0}$. Due to Assumptions 4.1 and 4.3, the intersection of the sensored disc $\mathcal{D}(k\delta)$ and the enlarged environment $\hat{E}(k\delta)$ is a convex closed set. This and the navigation rule (3.4), (3.5) imply that $\mathcal{D}(k\delta)$ and $\hat{E}(k\delta)$ do not overlap. Therefore, for any $t \in [k\delta, (k+1)\delta]$, the open disc of radius $\frac{d_{\text{sen}} - \overline{v}_E(t-k\delta)}{2}$ centered at O, the center of $\mathcal{D}(k\delta)$, does not contain points of $\hat{E}(t)$. Since the robot moves straight toward the point O with the speed v and $v > \overline{v}_E$ by (4.4), we conclude that the disc $\mathcal{D}_0(t)$ does not contain any point of $\hat{E}(t)$ for any $t \in [k\delta, (k+1)\delta]$.

2. $\boldsymbol{u(k\delta) \neq 0}$. It immediately follows from (3.4), (3.5) that $u(k\delta)$ is equal to either $\overline{u}$ or $-\overline{u}$. Let $u(k\delta) = \overline{u}$; the case $u(k\delta) = -\overline{u}$ is considered likewise. As was shown in the first part of the proof, the intersection of $\mathcal{D}(k\delta)$ and $\hat{E}(k\delta)$ is a convex closed set. This, the navigation rule (3.4), and the assumption of the lemma that the disc $\mathcal{D}_0(k\delta)$ does not contain any point of $\hat{E}(k\delta)$, imply that if $u(k\delta) = \overline{u}$ then the intersection of $\mathcal{D}_0(k\delta)$ and $\hat{E}(k\delta)$ lies outside of the set consisting of the right half of the disc $\mathcal{D}_0(k\delta)$ and the half-strip of the width $d_{\text{sen}} - 2v\delta$; see Fig. 10.6. Furthermore, any point of the environment from the intersection of $\mathcal{D}_0(k\delta)$ and $\hat{E}(k\delta)$, at any time $t \in [k\delta, (k+1)\delta]$, lies outside of the set consisting of the right half of the disc of diameter

$$d_{\text{diam}}(t) := d_{\text{sen}} - 2v(t - k\delta) - 2\overline{v}_E(t - k\delta) \tag{4.6}$$

and the half-strip of the width $d_{\text{diam}}(t)$; see Fig. 10.6. Now it follows from standard high school geometry and (4.3), (4.4), (4.5), that the right half of the disc $\mathcal{D}_0((k+1)\delta)$ lies inside the set consisting of the right half of the disc of diameter $d_{\text{diam}}(t)$ and the half-strip of the width $d_{\text{diam}}(t)$; see Fig. 10.6. Assumption 4.4 and the fact that $\hat{E}(k\delta)$ intersects with the right-half plane of $\mathcal{D}(k\delta)$ imply that the left half of the disc $\mathcal{D}_0((k+1)\delta)$ does not intersect with $\hat{E}(t)$ for any $t \in [k\delta, (k+1)\delta]$. □

Proof of Theorem 3.1. First, we prove that the navigation strategy is safe collision free. By the mathematical induction method, we immediately obtain from Assumption 4.1 and Lemma 4.1 that if the robot is navigated by the strategy (3.4), the disc $\mathcal{D}_0(k\delta)$ does not contain any point of $\hat{E}(k\delta)$ at any time $k\delta$ with $k = 0, 1, 2, \ldots$. Moreover, it was shown in the proof of Lemma 4.1 that if the disc $\mathcal{D}_0(k\delta)$ does not contain any point of $\hat{E}(k\delta)$, then any point of the environment from the intersection of

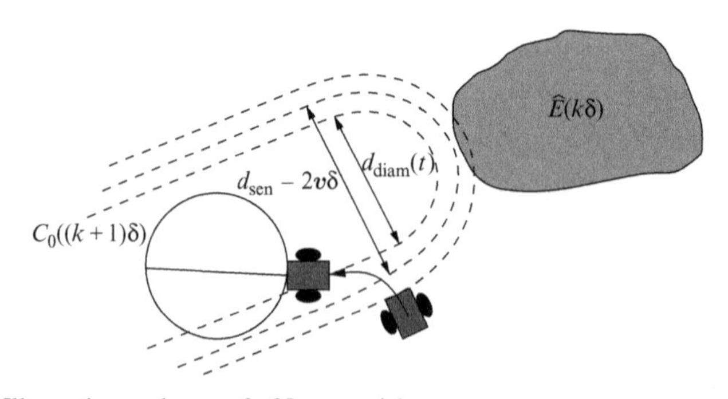

Figure 10.6 Illustration to the proof of Lemma 4.1.

$\mathcal{D}_0(k\delta)$ and $\hat{E}(k\delta)$ at any time $t \in [k\delta, (k+1)\delta]$ lies outside of the set consisting of the right half of the disc of diameter $d_{\text{diam}}(t)$ and the half-strip of the width $d_{\text{diam}}(t)$, where $d_{\text{diam}}(t)$ is defined by (4.6); see Fig. 10.6. On the order hand, it follows from (4.2) that the robot lies inside this set at any time $t \in [k\delta, (k+1)\delta]$.

Now we prove that the navigation strategy is target reaching. Indeed, let $b(t)$ be the distance between the robot and center of the target $\mathcal{T}$ at time t. Then

$$\dot{b}(t) = -v\cos(\theta_0(t) - \theta(t)); \tag{4.7}$$

see, e.g., [126]. It immediately follows from Assumption 3.1 that $-\frac{\pi}{2} < \theta_0(t) - \theta(t) < \frac{\pi}{2}$. From this and (4.7), we obtain that $\dot{b}(t) < 0$ for all t. Therefore, $b(t)$ is monotonically decreasing and converging to some $b_0 \geq 0$. If $b_0 > 0$, then the robot converges to some circle of radius b_0 concentric with the target. However, this is impossible because the navigation strategy consists of switching between two values of angular velocity. Hence, $b_0 = 0$ and the distance between the robot and the target's center converges to 0. Since the target is a circle, the robot will reach the target area in a finite time. This completes the proof of the theorem. □

10.5 Computer simulations

In this section, we present computer simulation results for a wheeled mobile robot navigating in a cluttered environment in accordance with the proposed algorithm. The simulations are performed in MobotSim and Matlab; the sampling period is $\delta = 0.1$ s.

In the first simulation, we show the basic ability of the proposed navigation algorithm to avoid static obstacles in cluttered environments. Unlike many existing navigation algorithms, which restrict the shapes of the obstacles for the sake of computational simplicity, the shapes of the obstacles are random and irregular in this simulation, as is shown in Fig. 10.7(a). It can be observed from the overall path taken by the robot and depicted in Fig. 10.7(a) that the robot is able to bypass the static obstacles through the vacancy between them. The values of the function m and the control u are displayed in Fig. 10.7(b) and (c), respectively.

In Fig. 10.8, the mobile robot faces a dynamic environment with four moving obstacles. The current heading of the robot is shown as a black arrow and the intended direction (middle of vacancy) is depicted as a red (gray in the printed version) arrow, the dashed line represents the sensing range of the robot. Once the robot senses one (Fig. 10.8(a)) or more (Fig. 10.8(b) and (c)) obstacles, it tries to steer its heading toward a safe direction to avoid them. In the "trapped" situations such as in Fig. 10.8(b) and (c), the robot decides to go through the gap between the obstacles in Fig. 10.8(b) (since the space is sufficient, i.e., greater than $2d_{\text{safe}}$) or to go around the obstacles in Fig. 10.8(c) (since the space is too small). The overall path is shown in Fig. 10.8(d).

The simulations concerned in Figs. 10.9–10.11 demonstrate the superior performance of the proposed navigation algorithm over the velocity obstacle approach

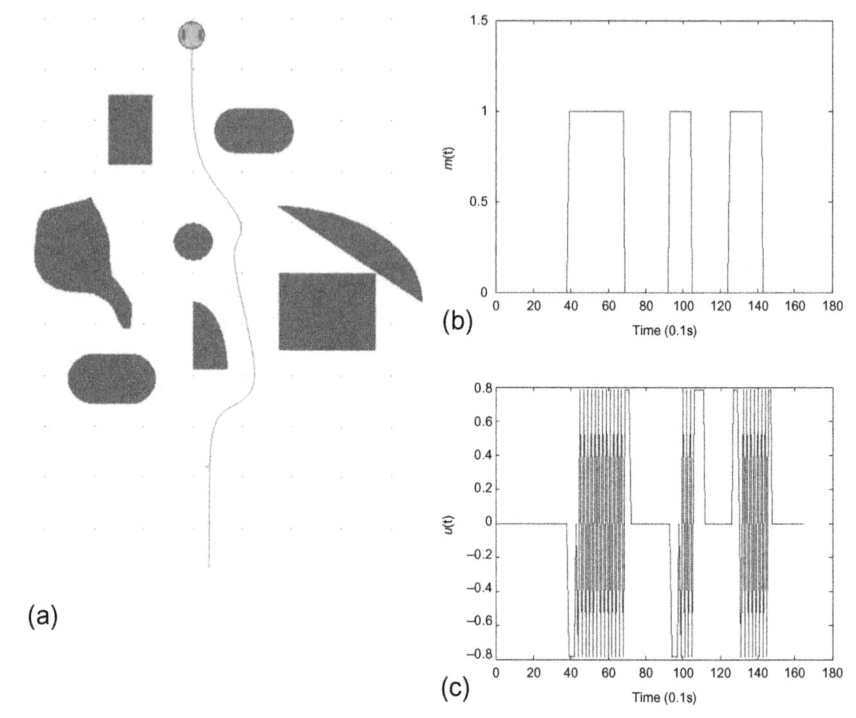

Figure 10.7 Robot navigating in a cluttered static environment.

(VOA) [19], which is one of the most well-known navigation algorithms, at least in some scenarios. In these simulations, the velocity obstacles built by the VOA are depicted as gray circles. Some potentially safe areas are forbidden by these velocity obstacles because of non-conservatism of estimation, while the proposed navigation algorithm efficiently utilizes paths through these areas and therefore bypasses the obstacles with a shorter path. In other words, with the proposed algorithm, the robot efficiently seeks a free path through a group of obstacles, whereas with the VOA, the robot avoids a crowd of obstacles, which results in a much longer path. The performance of the proposed navigation algorithm is shown in figures in left-hand side column and that of VOA is shown in right-hand side column.

Finally, the simulation in Fig. 10.12 demonstrates the ability of the proposed algorithm to avoid obstacles with various speeds, including obstacles that move faster than the robot. In Fig. 10.12(a), six obstacles (labeled from Ob1 to Ob6) with different sizes and moving speeds are depicted. The maximum speed of the obstacles for Ob1 to Ob6 are 0.75 m/s, 1.3 m/s, 0.35 m/s, 1.5 m/s, 0.6 m/s, and 0.7 m/s, respectively. The progression of the robot to avoid these obstacles is shown in Fig. 10.12(b)–(f).

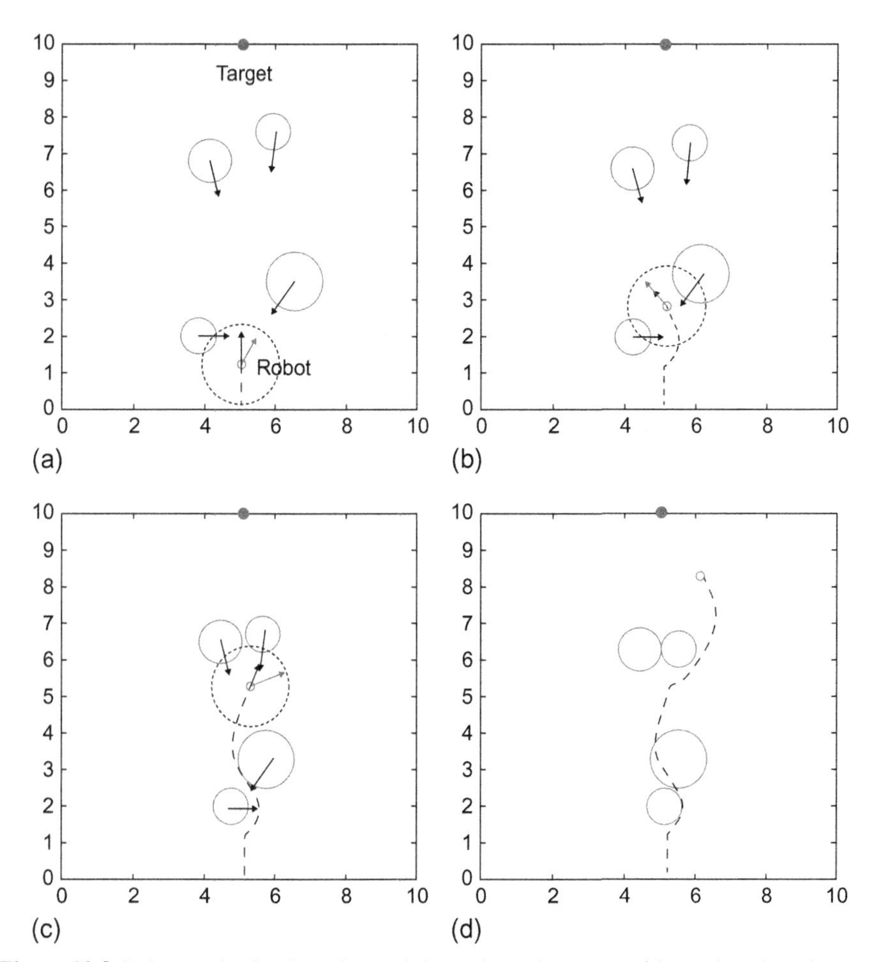

Figure 10.8 Robot navigating in a cluttered dynamic environment with moving obstacles.

10.6 Experiments with a real robot

Experiments with the proposed navigation algorithm were carried out with a Pioneer 3-Dx robot described in Section 1.4.1

Figure 10.13 shows a relatively simple scenario: the robot faces a static environment with a number of obstacles. Figures 10.13(a)–(c) correspond to moments when the robot bypasses a particular obstacles; the overall path is depicted in Fig. 10.13(d).

A "trapped" situation is shown in Fig. 10.14. Two obstacles move toward the robot so that the robot finds itself trapped between them. The robot senses that the space

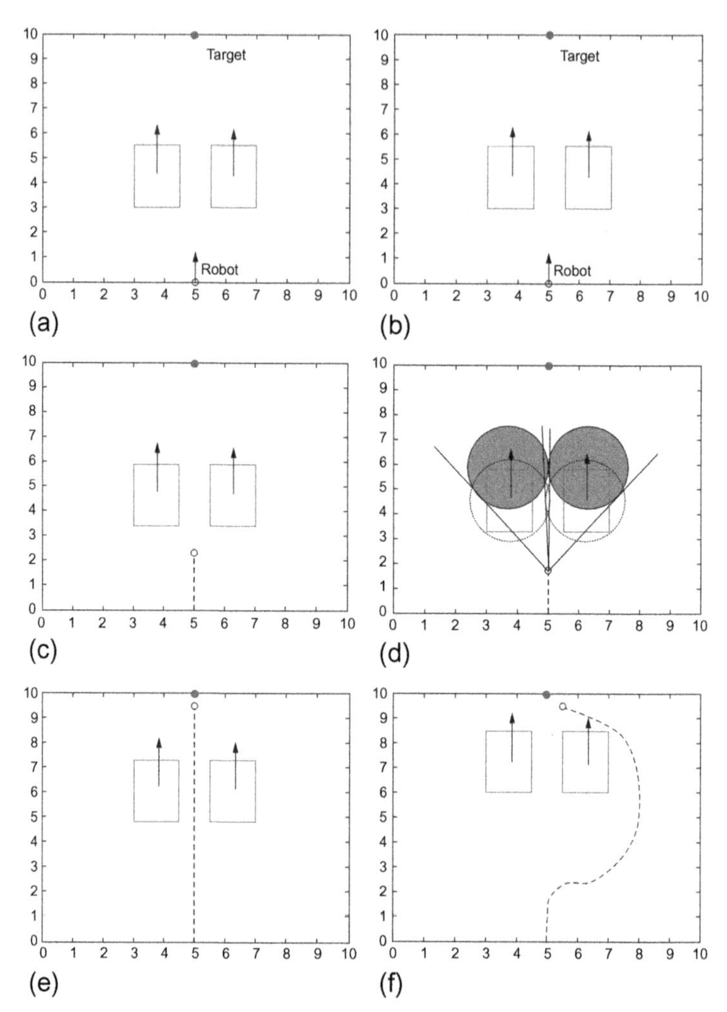

Figure 10.9 Performance comparison: two square obstacles moving side by side.

between the obstacles is large enough, and takes the respective path. The overall path for the entire duration of the experiment is shown in Fig. 10.14(d) and is quite efficient.

The performance of the proposed navigation algorithm in an environment with both static and dynamic obstacles is examined in Fig. 10.15, where the robot has to avoid two static and two moving obstacles. The robot safely completes its journey in Fig. 10.15(c), and its overall path is depicted in Fig. 10.15(d). More complicated scenarios are shown in Figs. 10.16 and 10.17.

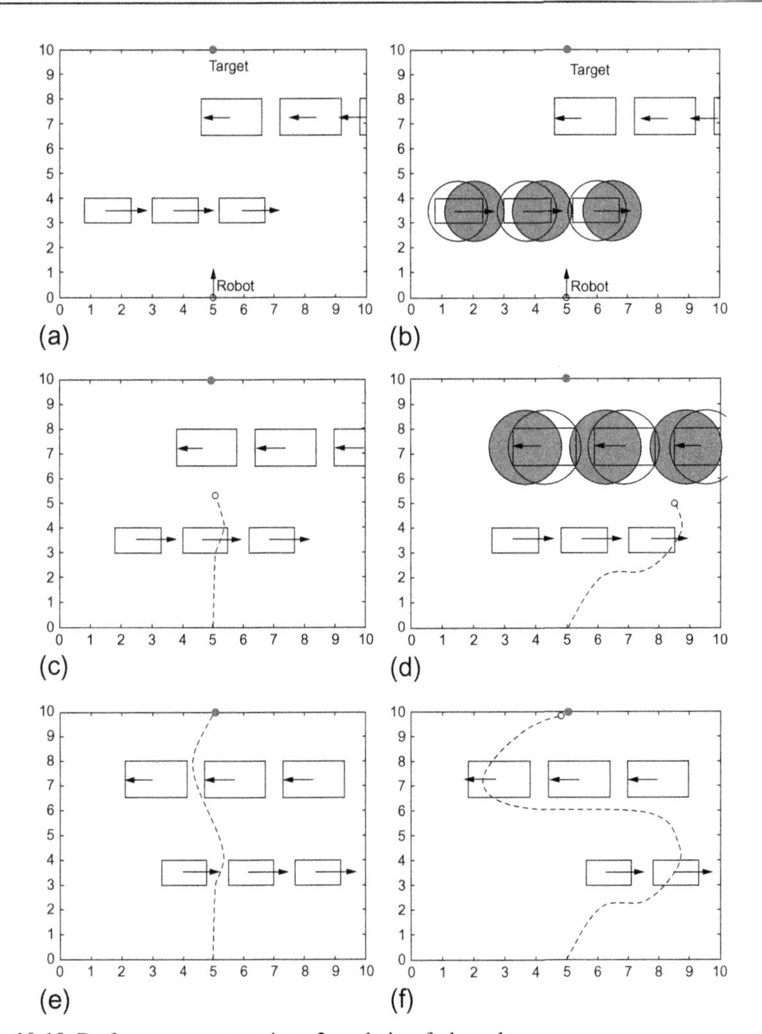

Figure 10.10 Performance comparison 2: a chain of obstacles.

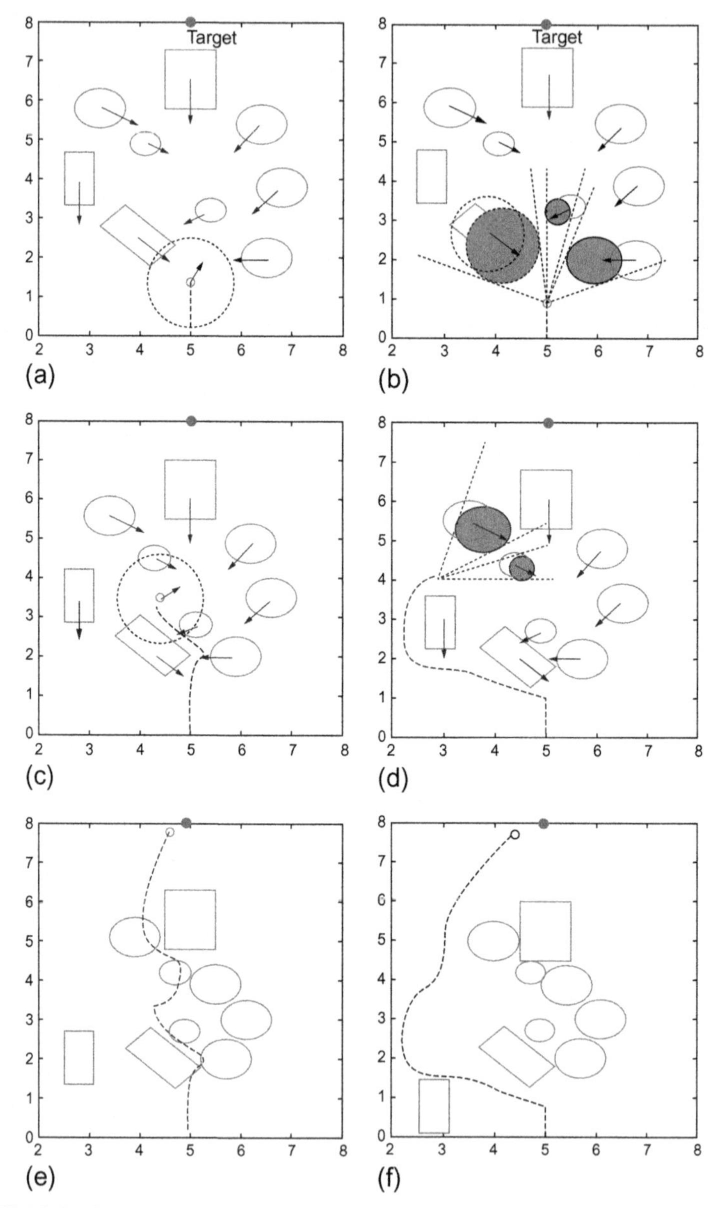

Figure 10.11 Performance comparison 3: robot navigating in a crowded small environment.

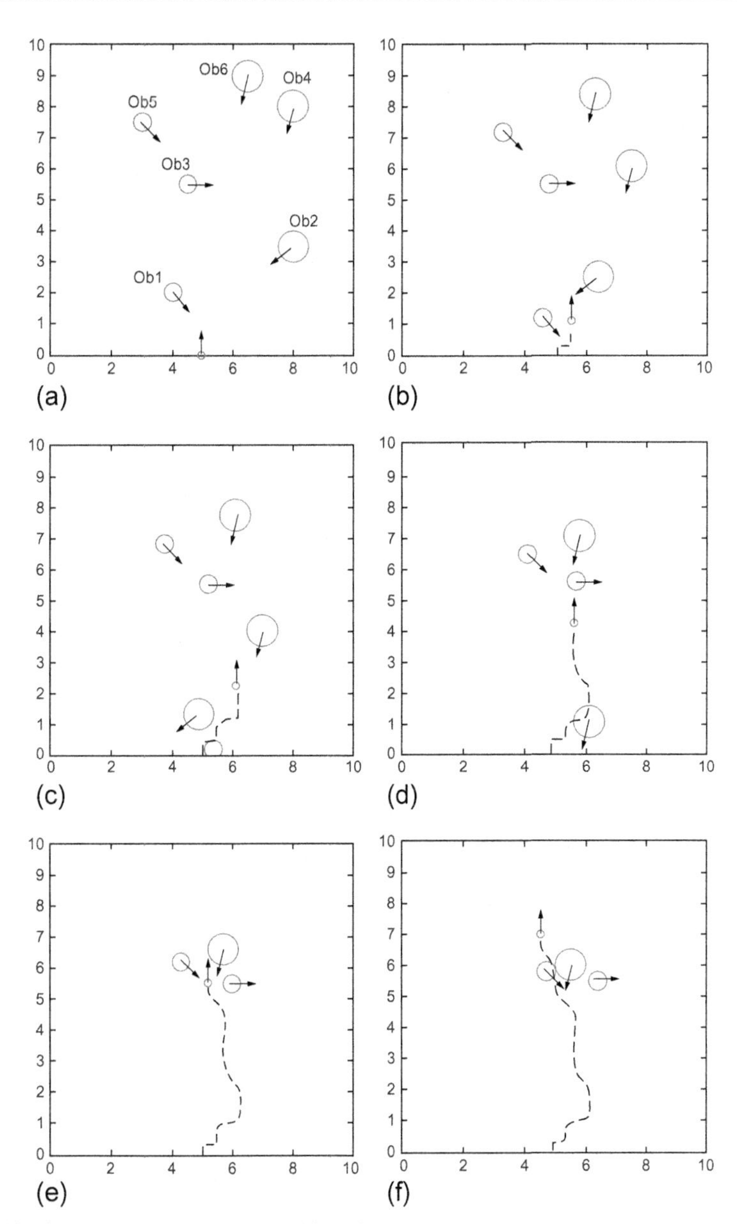

Figure 10.12 Robot avoids obstacles with various speeds.

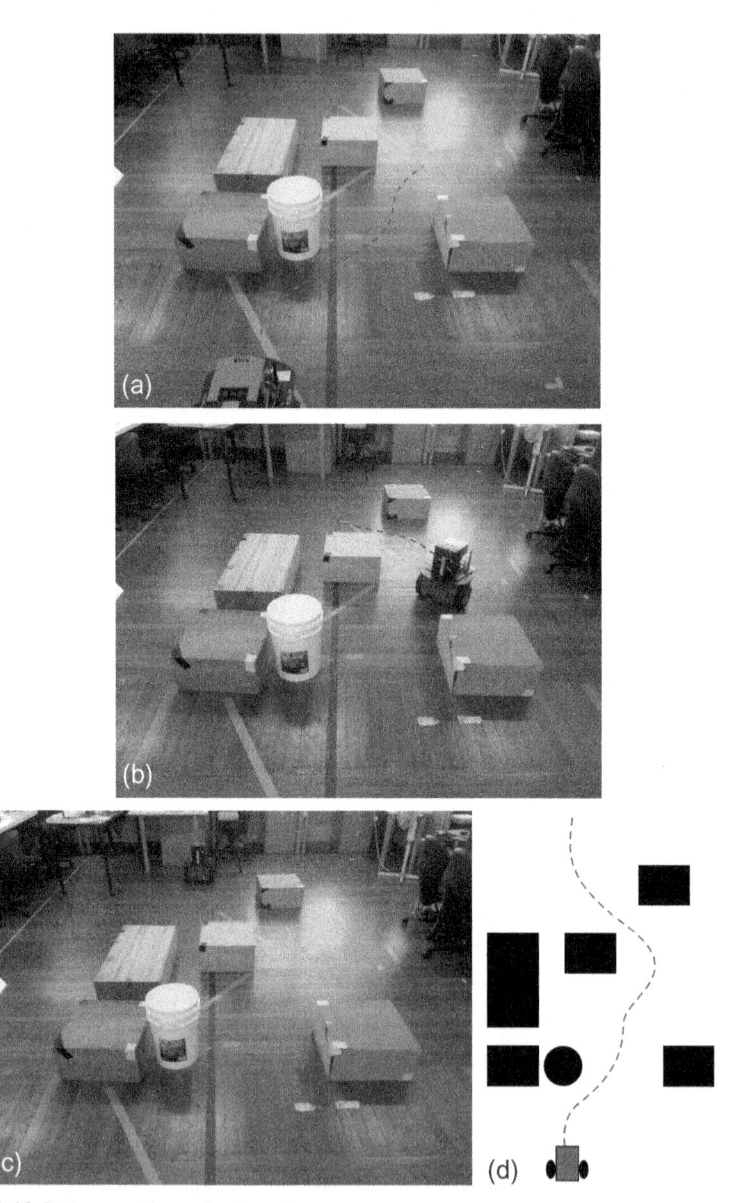

Figure 10.13 Robot avoids static obstacles.

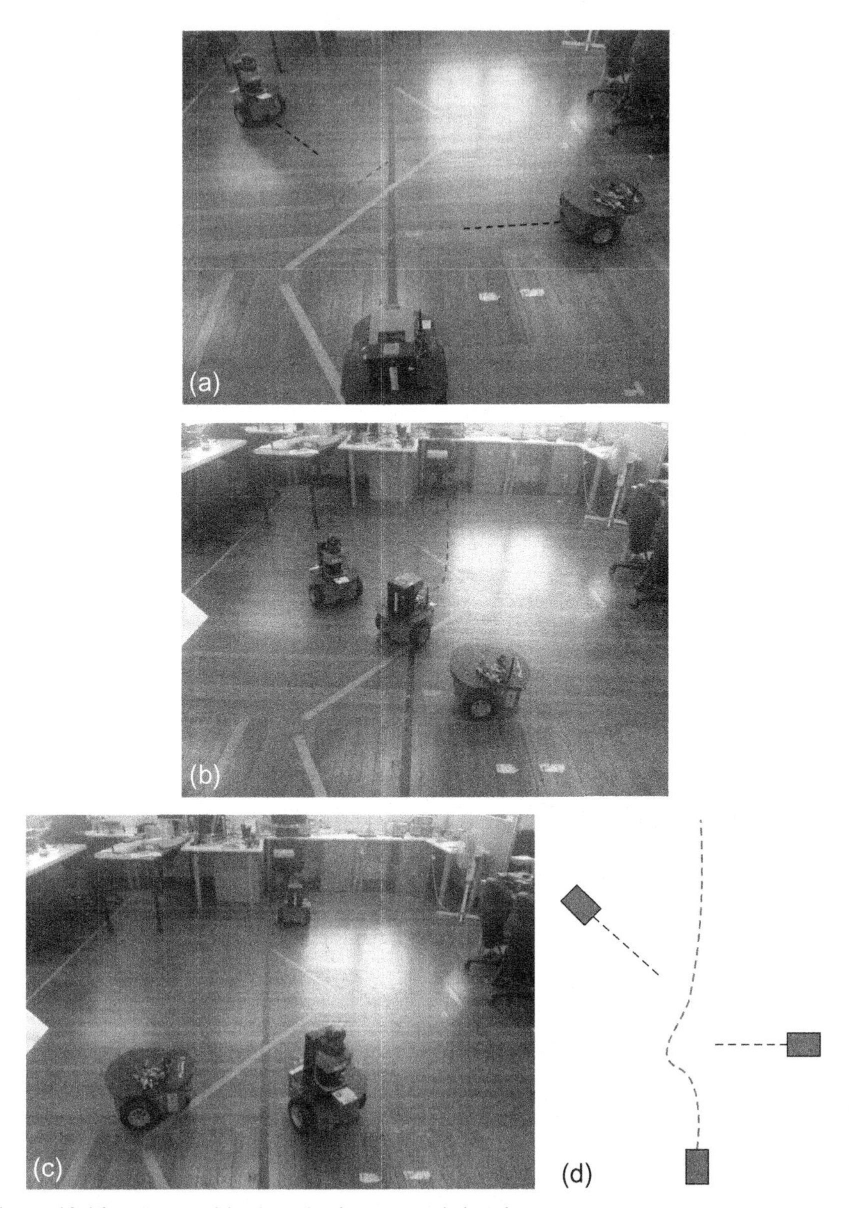

Figure 10.14 Robot avoids obstacles in a trapped situation.

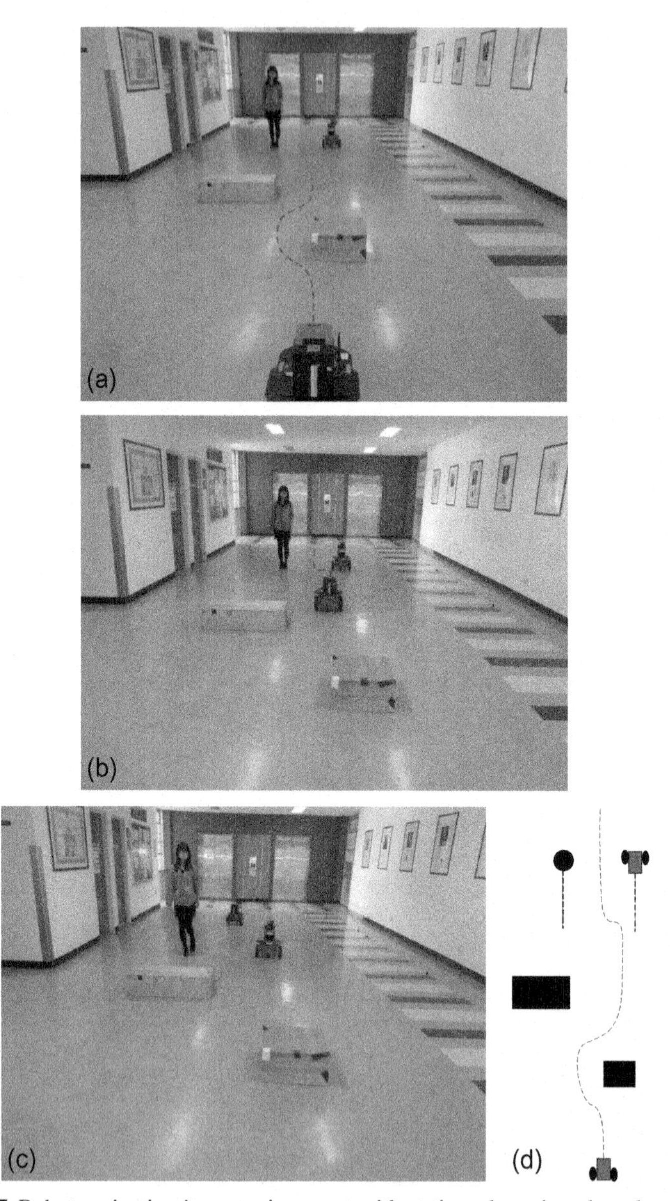

Figure 10.15 Robot navigating in an environment with static and moving obstacles.

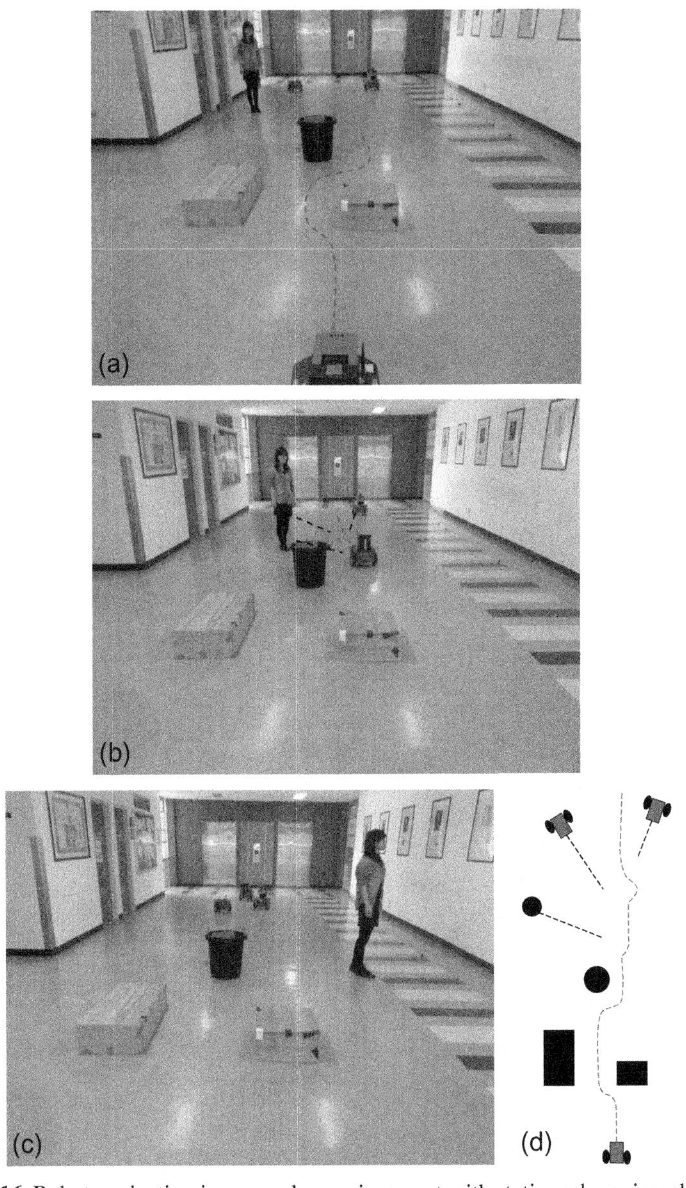

Figure 10.16 Robot navigating in a complex environment with static and moving obstacles.

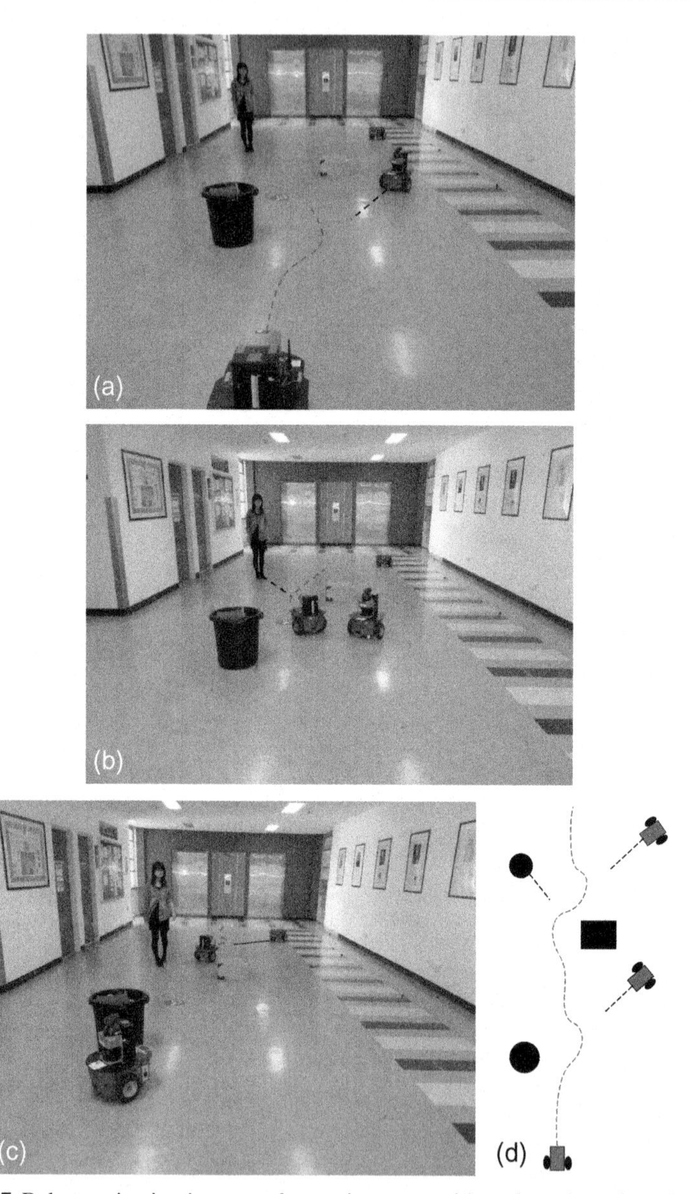

Figure 10.17 Robot navigating in a complex environment with static and moving obstacles 2.

A globally converging reactive algorithm for robot navigation in scenes densely cluttered with moving and deforming obstacles

11

11.1 Introduction

The capability to operate in dynamic and a priori unknown environments is a key requirement to mobile robots. Various scenarios gave rise to many algorithms under various sets of assumptions. However this issue still represents a real challenge in many cases, often because of uncertainties and deficiencies in available knowledge.

In this chapter, we revert to the problem of navigating a mobile robot through an environment with obstacles, which has been already treated in Chapters 4, 6–10. Unlike Chapters 4, 6–9, this chapter deals with a much more challenging scenario. Specifically, whereas all obstacles were assumed steady in Chapters 4, 6, and 7, now we consider moving obstacles. Moving obstacles were treated in Chapters 8 and 9. However, the theoretical results of Chapter 9 were confined to the border patrolling problem and were not directly extended on robot's navigation among obstacles. The last issue was theoretically treated in Chapter 8, but only for rigid, disk-shaped, slow enough, regularly moving, and relatively sparse obstacles; e.g., assumptions underlying the theoretical results of Chapter 8 imply that the velocity of any obstacle is constant and any given bounded region becomes free of obstacles sooner or later.

On the contrary, the theoretical part of this chapter admits that the obstacles are arbitrarily shaped and may undergo arbitrary motions, including transitions, rotations, and deformations, not necessarily regular. Furthermore they may constantly be a threat at the most part of the workspace so that its "threat-free" part is miserable and is not enough for even a mild advance in the desired direction. This situation is illustrated by an example, whose detailed consideration will be offered in Section 11.6.3. It deals with a scene with many straight line segments rotating with a common angular velocity about pivots evenly distributed over a planar strip, as is shown in Fig. 11.1(a). The part of the strip with no collision threat is composed of many small compact disconnected domains depicted in Fig. 11.22(a) in white. It is clear that motion through the strip cannot be confined to only "threat-less" white domains but conversely, should be mostly through the grey "threatful" areas.

The scenario considered in this chapter is similar to that from Chapter 10. Moreover, this chapter develops the approach of Chapter 10 by employing an

Safe Robot Navigation Among Moving and Steady Obstacles. http://dx.doi.org/10.1016/B978-0-12-803730-0.00011-1

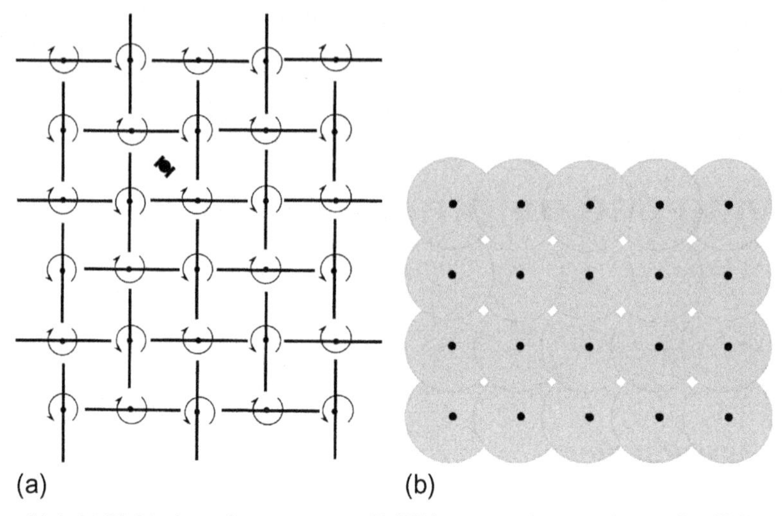

Figure 11.1 (a) Field of rotating segments; (b) White areas where no threat of collisions with the obstacles occurs.

integrated representation of the environment. However, Chapter 10 deduced achievement of the global navigation objective from some assumptions about a posteriori, resultant behavior of the robot; by doing so, the success was put in dependence on circumstances that may happen or not in the experiment. Conversely, no assumptions of such kind are imposed in this chapter, which thus provides a completed delineation of environments where success of the proposed navigation algorithm is rigorously guaranteed. Furthermore, the assumptions underlying the main theoretical result of Chapter 10 assumed convex obstacles and implied that the scene a relatively sparse: no more than one obstacle can be an "immediate" threat for the robot at any time instant. The theoretical developments of this chapter deal with non-convex obstacles and admit more densely cluttered environments, where the robot may simultaneously have many threats. Finally, this chapter offers another navigation strategy.

The navigation algorithm examined in this chapter belongs to the class of local, reactive methods; see Sections 3.1 and 3.4 for their general discussion. We briefly recall that contrary to global planners, which are computationally expensive, hardly suit real-time implementation, and are highly troubled by data incompleteness and unpredictability of the scene, up to failure in generation of a plan, local planners iteratively re-plan a short-horizon portion of the path, thus reducing the computational burden toward implementability in real time and truncating the needed data to those about a nearest, in both space and time, fraction of the scene. At the same time, this makes the ultimate result of iterations an open issue. The principal both strong and weak points of local planners attain apotheosis at reactive controllers, which directly convert the current observation into the current control. In more details, basic samples of local techniques were reviewed in Chapters 1 and 3.

Because of inevitable failure scenarios, the deficiency of the previous research on local planners is the lack of global convergence results that guarantee achieving the primary objective in dynamic environments [49]. At best, rigorous analysis examined an isolated bypass of an obstacle during which the other obstacles were neglected until the bypass end, with an idea that thereafter, the robot focuses on the main goal. However in densely cluttered dynamic scenes, bypasses may be systematically intervened by companion obstacles so that no bypass is completed, whereas the robot almost constantly performs obstacle avoidance. Ultimate goal was left, by and large, beyond the scope of theoretical analysis, especially for cluttered unpredictable environments, like a dense crowd of people. However it is in these cases that rigorous quantitative delineation between failure and success scenarios is highly important since by its own right, any experimentation is not convincing enough due to horizonless diversity of feasible scenarios. Another deficiency is that moving obstacles were viewed as rigid bodies undergoing only translational motions and often of the simplest shapes (e.g., discs [45, 51] or polygons [53, 54]). Finally, assumed awareness about the obstacles often meant access to their possibly "invisible" parts (e.g., to determine the disc center [45, 51] or angularly most distant polygon vertex [54]) or full velocity [45, 51, 54].

Conversely, this chapter offers a purely reactive navigation algorithm with the capacity of being supplied with firm, mathematically rigorous guarantees of achieving a required global objective in environments densely cluttered with unpredictably moving and deforming obstacles. This objective is perpetual drift in the desired direction in spite of possibly almost continual obstacle avoidance. The obstacles are not rigid: they have arbitrary time-varying shapes and may rotate, twist, wring, skew, wriggle, etc. As in Chapter 9, this covers scenarios with reconfigurable rigid obstacles, forbidden zones between moving obstacles, flexible obstacles, like a fluttered curtain or fishing net, virtual obstacles, like areas contaminated with hazardous chemicals or on-line estimated areas of operation of a hostile agent.

The considered navigation law uses omnidirectional vision of the scene up to the nearest reflection point and, apart from access to the desired azimuth, assumes no further sensing capacity or knowledge of the scene configuration. Like some other algorithms, it starts with finding points where the distance reading abruptly jumps as the angle of measurement continuously varies, which are interpreted as edges of visible facets of obstacles. In the vein of [32] and Chapter 8, these facets are angularly expanded. Finally the motion direction is determined via compromising between bearings at the edges of the extended facets and the desired azimuth.

Apart from Chapter 10, the navigation strategy considered in this chapter develops some ideas set forth in [32] and Chapter 8. However, in [32] and Chapter 8, only rigid and fully visible obstacles are examined, which are static in [32] and disk-shaped in Chapter 8, and the scene is assumed to be so sparse that bypasses of obstacles are sufficiently isolated from one another. Now we show that being properly developed, those ideas remain viable for much more general scenarios with dense scenes and deforming obstacles with arbitrary shapes. We first offer a mathematically rigorous justification of the proposed approach. In doing so, we start with conditions necessary for a mobile robot to be capable of obstacle avoidance in scenarios like motion within a dense crowd of people. Then we show that under a slight enhancement of these conditions, obstacle avoidance is ensured by the proposed navigation controller provided that it is properly tuned. Success in following the desired azimuth is rigorously proved for a general problem setup. By illustrating this result in several particular scenarios, it is better displayed that the algorithm does cope with densely cluttered dynamic scenes. The applicability of the proposed strategy is confirmed via extensive computer simulations. In doing so, its performance has been compared with that of the velocity obstacle approach [17, 19] and found to be better under certain circumstances.

The main results of this chapter were originally published in [433]. A preliminary version of some of these results was also reported in the conference paper [434].

The remainder of the chapter is organized as follows. Sections 11.2 and 11.3 describe the problem setup and the navigation strategy, whereas the main results concerned with collision avoidance and the main navigation objective are given in Sections 11.4 and 11.5, respectively. In Section 11.6, they are illustrated in special scenarios. Section 11.7 discusses simulations.

11.2 Problem setup

A planar point-wise robot travels in a plane. The robot is controlled by the linear velocity $\mathbf{v}(t)$ whose magnitude does not exceed a given constant $\overline{v} > 0$. The objective is to follow the azimuth given by a unit vector $\mathbf{f}$. However, pure following is impossible since the scene is cluttered with untraversable disjoint obstacles $D_1(t), \dots, D_N(t)$, and the robot should always be in the obstacle-free part of the plane

$$\boldsymbol{r}(t) := \mathbf{col}\,[x(t), y(t)] \notin \bigcup_i D_i(t) \quad \forall t,$$

where $x(t)$ and $y(t)$ are the abscissa and ordinate of the robot in the world frame.

The obstacles undergo general motions, including rotations and deformations, but do not split into parts, merge together, or collide with each other. We assume that any obstacle is bounded by a smooth Jordan curve. Thus holes in obstacles are ignored since they cannot affect the robot's navigation.

Remark 2.1. To take the robot's size into account, a standard hint is to treat the obstacle as if it is enlarged and bounded by a proper equidistant curve. This automatically smoothes the outer cusps of the obstacle if they do exist. The inner cusps can be smoothed via approximation.

We assume a reference frame attached to the robot.[1] For any polar angle α in this frame, the robot measures the distance $d(\alpha, t)$ to the nearest obstacle in the direction given by α (see Fig. 11.2(a)): the point with the relative polar coordinates $[\alpha, d]$ belongs and does not belong to an obstacle if $d = d(\alpha, t)$ and $0 \leq d < d(\alpha, t)$, respectively. If there is no obstacle in the examined direction, $d(\alpha, t) := \infty$. The desired direction of motion $\mathbf{f}$ is also accessed via its relative polar angle $\beta(t)$.

[1] Its details and evolution over time do not affect both the proposed navigation law and its analysis. So they are not discussed.

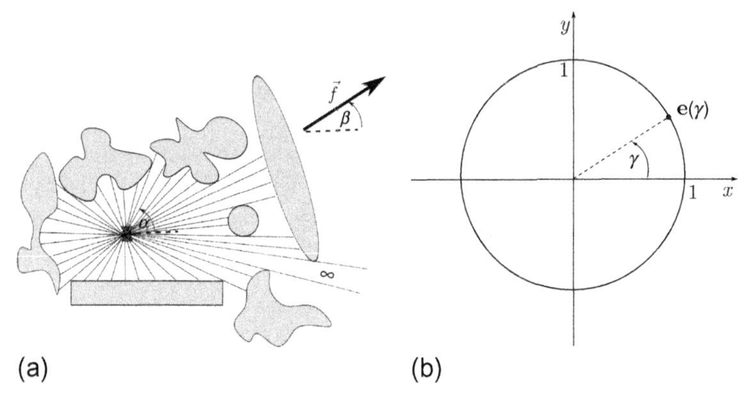

Figure 11.2 (a) Sensing capacity of the robot; (b) Vector $\mathbf{e}(\gamma)$ in the relative frame of the robot.

The robot is able to classify enroute obstacles into disjoint classes $C_1, \ldots, C_K$, e.g., into "definitely static" and "probably moving" or into "slow" and "speedy", etc. In the case of no classification capacity, $K = 1$ and the robot has to treat all obstacles in a common way. If conversely the robot recognizes the obstacles, every class contains only one element.

The robot should move along the given azimuth on the basis of these data. We do not assume any predictability in the motion of the obstacles. The only assumption is that they are slower than the robot; this assumption will be justified by Lemma 4.1.

We close the section with a list of notations that will be often used in this chapter:

$D_1(t), \ldots, D_N(t)$—moving obstacles;
$C_1, \ldots, C_K$—disjoint classes of the obstacles;
$\mathbf{v}$—the velocity of the robot;
$\overline{v}$—an upper bound on its speed;
d—the distance from the robot to the nearest obstacle;
$d(\alpha, t)$—the distance to the nearest obstacle in the direction given by the polar angle α (see Fig. 11.2(a));
$\mathbf{f}$—the unit vector in the desired direction of motion;
$\overline{v}_o^i$—an upper bound on the speeds of all obstacles from the class C_i;
$\mathbf{e}(\gamma) := \mathbf{col}\,[\cos\gamma, \sin\gamma]$—the point on the unit circle with the polar angle γ in the relative frame of the robot; see Fig. 11.2(b);
$\Delta_i(\cdot)$—a decaying function that determines the extent of expansion of the facets for obstacles from class C_i;
$\mathbf{V}_j(\boldsymbol{r}, t)$—the velocity of the point $\boldsymbol{r} \in D_j(t)$ at time t;
$[\mathbf{T}_j(\boldsymbol{r}, t), \mathbf{N}_j(\boldsymbol{r}, t)]$—the Frenet frame of the boundary $\partial D_j(t)$ at the boundary point $\boldsymbol{r}$;
$W_{T,t,j}(\boldsymbol{r}, t)$—the tangential component $W_{T,t,j}(\boldsymbol{r}, t) := \langle \mathbf{W}; \mathbf{T}_j(\boldsymbol{r}, t)\rangle$ of a vector $\mathbf{W}$ at the boundary point $\boldsymbol{r} \in \partial D_j(t)$;
$W_{N,t,j}(\boldsymbol{r}, t)$—the normal component;

In these notations, some indices may be dropped if they are clear from the context.

11.3 The navigation algorithm

Angles γ will be treated as cyclic $\gamma \pm 2\pi = \gamma$ variables associated with respective points $\mathbf{e}(\gamma)$ on the unit circle; see Fig. 11.2(b).

At time t, the sensory data are given by the function $d(\alpha) = d(\alpha, t)$ of the polar angle α and the relative bearing $\beta = \beta(t)$ of the desired direction given by the unit vector $\mathbf{f}$. For any class C_i of obstacles, a continuous decaying function of the distance $d \geq 0$ is pre-specified to be thereafter used to set the extent of facet's enlargement:

$$\Delta_i(d) \in \left[0, \frac{\pi}{2}\right), \quad \Delta_i(d_1) \geq \Delta_i(d_2) \quad \forall d_1 \leq d_2. \tag{3.1}$$

At time t, the algorithm carries out the following consecutive steps to generate the current control $\mathbf{v}(t)$:

- *Computation of facets.* Via finding the discontinuities of the distance function $d(\alpha) = d(\alpha, t)$, the maximal angular intervals $A_k = (\alpha_k^-, \alpha_k^+)$ on which $d(\cdot)$ is continuous and finite are determined.
  The *facets* are lines associated with intervals A_k, with the kth facet being given by its distance profile (equation in polar coordinates) $d = d(\alpha), \alpha \in A_k$. This is illustrated in Fig. 11.3(a) for a scene with three grey convex obstacles, where four facets are depicted as black lines, and angular intervals are extended into cones shown in green, yellow, pink, and blue (different tones of gray in print versions). As can be seen, even a convex obstacle (the large rectangle in the case from Fig. 11.3(a)) may give rise to several distinct facets.
- *Determining parameters* of every facet k. They are the distance

  $$d_k^{\min} := \inf_{\alpha \in A_k} d(\alpha)$$

  to the facet and the class $C_i, i = i(k)$ of the concerned obstacle.
- *Enlargement of the facets.* The *enlarged facet* k is given by its angular range $\hat{A}_k$ and distance profile $d_k(\cdot)$

$$\hat{A}_k := \left[\alpha_k^- - \delta_k, \alpha_k^+ + \delta_k\right], \quad \text{where} \quad \delta_k := \Delta_{i(k)}(d_k^{\min}), \tag{3.2}$$

$$d_k(\alpha) := d[p^{A_k}(\alpha)] \quad \forall \alpha \in \hat{A}_k. \tag{3.3}$$

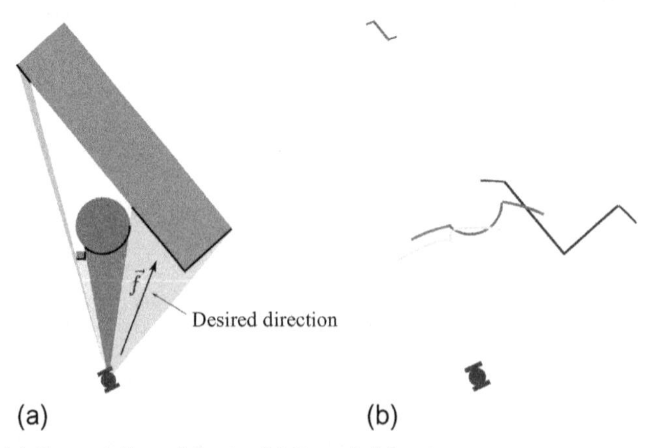

Figure 11.3 (a) Computation of facets; (b) Extended facets.

Here $p^{A_k}(\alpha)$ is the point of the angular interval A_k nearest to α. In other words,

$$p^{A_k}(\alpha) = \begin{cases} \alpha & \text{if } \alpha \in A_k, \\ \text{the nearest end-point of } A_k & \text{otherwise.} \end{cases}$$

Thanks to (3.1), the closer the facet, the more it is extended. For the scenario from Fig. 11.3(a), the extended facets are illustrated in Fig. 11.3(b). Unlike facets, an extended facet may obstruct the robot's view on other extended facets, moreover, these facets may intersect one another.

- *Generation of the control.* There may be two cases.
  - *The desired direction is not obstructed by the extended facets:* $\beta \notin \bigcup_k \hat{A}_k$. Then the robot is driven at the maximal speed in this direction

    $$\mathbf{v} := \overline{v}\mathbf{e}(\beta),$$

    where the vector $\mathbf{e}(\cdot)$ is defined in Fig. 11.2(b).
  - *The desired direction is obstructed by the extended facets:* $\beta \in \bigcup_k \hat{A}_k$. Then
    - A nearest obstructing facet k, i.e., that with the minimal value of $d_k(\beta)$, is chosen;
    - The finite set E_k is composed that consists of the following points:
      - The end-points of $\hat{A}_k$;
      - The end-points α_{end} of the angular intervals of other extended facets $\hat{A}_j, j \neq k$ provided that the end-point α_{end} first, lies in $\hat{A}_k$ and second, "obstructs" the robot's view on the kth facet, i.e., $d_j(\alpha_{\text{end}}) \leq d_k(\alpha_{\text{end}})$;
    - The points $\alpha_{\circlearrowleft}$ and $\alpha_{\circlearrowright}$ of E_k that are counter-clockwise and clockwise closest to the bearing β of the desired direction are determined;
    - Among the angles $\alpha_{\circlearrowleft}$ and $\alpha_{\circlearrowright}$, the minimizer α_0 of the discrepancy $|\beta - \alpha_0|$ is picked[2];
    - The robot is driven at the maximal speed in the direction given by the so found angle:

    $$\mathbf{v} := \overline{v}\mathbf{e}(\alpha_0). \tag{3.4}$$

For the scenario from Fig. 11.3(a), these rules of choosing the current direction of motion are illustrated in Fig. 11.4, where only two obstacles that directly affect this choice are displayed. There is only one extended facet that obstructs motion in the desired direction: the low-right facet of the large rectangle. The thin darkest blue (dark gray in print versions) cone corresponds to extension of this facet in the clockwise direction; let α_{rec} stand for its "clockwise" end. The thin lightest blue (light gray in print versions) cone corresponds to extension of the facet of a neighboring disk obstacle. The clockwise end-point α_{disk} of this extension participates into the set E_k since this end-point obstructs the view from the robot on the large rectangle. It follows that $\alpha_{\circlearrowleft} = \alpha_{\text{disk}}$, whereas $\alpha_{\circlearrowright} = \alpha_{\text{rec}}$. In the situation from Fig. 11.4(a) the robot moves in the direction given by α_{disk}, due to the orientation of $\mathbf{f}$. However in the scenario from Fig. 11.4(b) (as compared with Fig. 11.4(a), the vector $\mathbf{f}$ is turned clockwise so that it becomes closer to the right boundary ray than to the left one), the algorithm takes the direction of α_{disk}.

To elucidate principal features of the obstacle avoidance maneuver in the scenario from Fig. 11.4(a), we change spacing and sizes in Fig. 11.5 as compared with

[2] $\alpha_{\circlearrowleft}$ in the case of equal discrepancies.

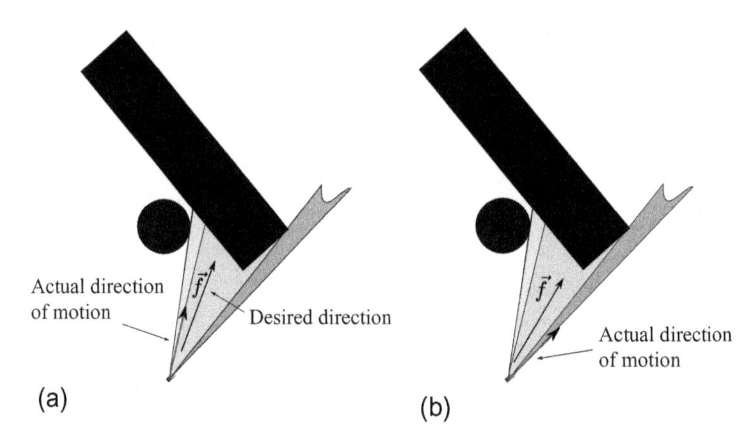

Figure 11.4 (a) The robot goes along the left boundary ray; (b) The robot goes along the right boundary ray.

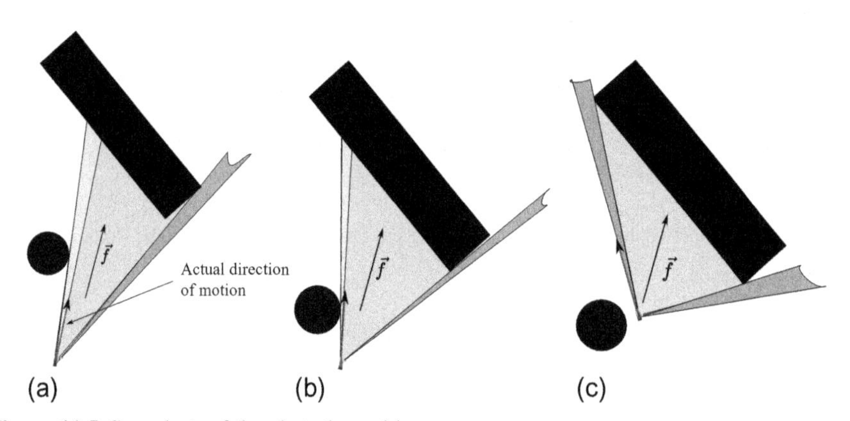

Figure 11.5 Snapshots of the obstacle avoidance maneuver.

Fig. 11.4(a). Figure 11.5 shows three characteristic snapshots of the maneuver. It should be stressed that the simple picture from Fig. 11.5(c) is not mandatory and is due to simplicity of the scene. By properly adding more convex obstacles, it is easy to maintain the logic of the algorithm operation identical to that from Fig. 11.4(a).

The use of a decaying function $\Delta_i(\cdot)$ in (3.2) allows the robot not to take overly precautionary measures against faraway obstacles, though constant functions $\Delta_i(\cdot)$ are always feasible. A reasonable modification results from putting (3.4) in use only if the distance to the nearest obstacle is less than a pre-specified threshold $d_* > 0$.

The proposed law does not estimate the velocity of the obstacle and does not attempt to predict its future positions. This is partly motivated by relevant troubles, up to infeasibility, possible low accuracy and essential extra computational burden. These problems are drastically enhanced if obstacles undergo general motions, including deformations, since then the velocity and future position are determined by infinitely many parameters. Another reason is that we are interested in disclosing a benchmark that can be reached without an aid of estimation or prediction.

Since the proposed navigation law is discontinuous, the closed-loop solutions are meant in the Filippov's sense; see Section 1.3. Given the initial state, such solution exists and does not blow up in a finite time since the controls are bounded. Its uniqueness is a more delicate problem. So we shall address all "Filippov's" solutions with only one exception. In the situation of repulsive discontinuity surface (the vector field points away from the surface on both sides), solutions that slide over the surface are dismissed. The theoretical reason is that they are unstable and so unviable in the face of unavoidable disturbances, even if they are small. The practical reason is that we tacitly assume sampled-data control by a digital device, which picks at random one of two control options on the above discontinuity surface and does not alter it during the sampling time $\tau > 0$. This causes an immediate escape from the surface and gives rise to no sliding solutions as $\tau \to 0$, which is the limit case that is caught by the model at hand.

11.4 Collision avoidance

The obstacles may not only arbitrarily move but also twist, skew, wriggle, or be otherwise deformed. We assume that all these are performed smoothly. To formally describe this, we use the Lagrangian formalism [424] by introducing a *reference configuration* $D_{j,*} \subset \mathbb{R}^2$ of the jth obstacle and the *configuration map* $\Phi_j(\cdot, t) : \mathbb{R}^2 \to \mathbb{R}^2$ that transforms $D_{j,*}$ into the current configuration $D_j(t) = \Phi_j[D_{j,*}, t]$.

Assumption 4.1. *The reference configuration $D_{j,*}$ of any obstacle is compact and bounded by a smooth Jordan curve. The configuration map $\Phi_j(\cdot, t)$ is defined on an open neighborhood of $D_{j,*}$ and is smooth and one-to-one, the determinant of its Jacobian matrix is everywhere nonzero.*

Let $\mathbf{V}_j(\boldsymbol{r}, t)$ stand for the velocity of the point $\boldsymbol{r} \in D_j(t)$, and $[\mathbf{T}_j(\boldsymbol{r}, t), \mathbf{N}_j(\boldsymbol{r}, t)]$ be the Frenet frame of $\partial D_j(t)$ at $\boldsymbol{r} \in \partial D_j(t)$. (The set $D_j(t)$ is to the left to the unit tangent vector $\mathbf{T}_j$, the unit normal vector $\mathbf{N}_j$ is directed inwards $D_j(t)$.) Finally, $W_{T,t,j}(\boldsymbol{r}, t) := \langle \mathbf{W}; \mathbf{T}_j(\boldsymbol{r}, t) \rangle$ and $W_{N,t,j}(\boldsymbol{r}, t) := \langle \mathbf{W}; \mathbf{N}_j(\boldsymbol{r}, t) \rangle$ are the tangential and normal components of a vector $\mathbf{W}$.

We start with conditions necessary for the robot to be capable of collision avoidance. The following definition tacitly assumes scenarios where the avoidance maneuver may start in an arbitrarily tight proximity of the obstacle.

Definition 4.1. The robot *is capable of avoiding collisions with the obstacle $D_j(t)$* if for any initial time t_0 and state of the robot $\boldsymbol{r}(t_0) \notin D_j(t_0)$, there exists an admissible velocity profile $\|\mathbf{v}(t)\| \leq \overline{v}, t \geq t_0$ under which the robot does not collide with this obstacle for $t \geq t_0$.

Let $[a]_- := \max\{0, -a\}$ stand for the negative part of $a \in \mathbb{R}$.

Lemma 4.1. *Suppose that the robot is capable of avoiding collisions with the obstacle $D_j(t)$. Then at any time t, the negative part of the normal velocity (i.e., the part associated with the outward normal to the boundary) of any boundary point $\boldsymbol{r} \in \partial D_j(t)$ does not exceed the robot's maximal speed $\overline{v}$:*

$$\left[\left(\mathbf{V}_j(\boldsymbol{r}, t) \right)_{N,t,j} (\boldsymbol{r}, t) \right]_- \leq \overline{v}. \tag{4.1}$$

Proof. Suppose to the contrary that

$$\left[\left(\mathbf{V}_j(\boldsymbol{r}_*, t_*) \right)_{N,t,j} (\boldsymbol{r}_*, t_*) \right]_- > \overline{v}$$

for some t_* and $\boldsymbol{r}_* \in \partial O_j(t_*)$. By continuity, there exist $\varepsilon > 0$ and $\varkappa > 0$ such that

$$\left[\left(\mathbf{V}_j(\boldsymbol{r}, t) \right)_{N,t,j} (\boldsymbol{r}, t) \right]_- \geq \overline{v} + \varepsilon$$
$$\text{whenever} \quad \boldsymbol{r} \in \partial D_j(t), \quad |t - t_*| < \varkappa, \quad |\boldsymbol{r} - \boldsymbol{r}_*| < \varkappa.$$

Let the robot start at time t_* at a point $\boldsymbol{r}(t_*) = \boldsymbol{r}_* - y\mathbf{N}(\boldsymbol{r}_*, t_*)$ from the outer normal, where $y > 0$ is so small that $\boldsymbol{r}(t_*) \notin D_j(t_*)$ and $|\boldsymbol{r}(t_*) - \boldsymbol{r}_*| < \varkappa/2$. By Definition 4.1, there exists an admissible velocity profile $\|\mathbf{v}(t)\| \leq \overline{v}, t \geq t_*$ under which

$$d(t) := \mathbf{dist}\left[\boldsymbol{r}(t); D_j(t) \right] > 0 \quad \forall t \geq t_*. \tag{4.2}$$

Let $\boldsymbol{r}_*(t)$ be the point of $\partial D_j(t)$ closest to $\boldsymbol{r}(t)$. By Lemma 4.1 for $t \in [t_*, t_* + \tau]$ and sufficiently small $y > 0$ and $\tau > 0$, we have $|\boldsymbol{r}_*(t) - \boldsymbol{r}_*| < \varkappa, |t - t_*| < \varkappa$, and so

$$V_j^N(t) := \left[\left(\mathbf{V}_j(\boldsymbol{r}_*(t), t) \right)_{N,t,j} (\boldsymbol{r}_*(t), t) \right]_- \geq \overline{v} + \varepsilon. \tag{4.3}$$

At the same time according to (4.7),

$$\dot{d}(t) = V_j^N(t) - \left\langle \mathbf{v}(t); N_j(\boldsymbol{r}_*(t), t) \right\rangle \leq -(v + \varepsilon) + v = -\varepsilon.$$

This and (4.2) yield $\varepsilon\tau \leq d(0) = y$, though y can be chosen independently of τ and as close to 0 as desired. This contradiction proves the lemma. □

By the following theorem, putting $<$ in place of $\leq$ makes the necessary condition (4.1) sufficient, and under this sufficient and "almost necessary" condition, collision avoidance is ensured by the proposed navigation law.

Theorem 4.1. *Suppose that the necessary condition* (4.1) *is true for any obstacle $D_j(t)$, at any time t, for any boundary point $\boldsymbol{r} \in \partial D_j(t)$, and with $<$ substituted for $\leq$. Moreover, suppose that this relationship does not degenerate as time goes to ∞:*

$$\left[\left(\mathbf{V}_j(\boldsymbol{r}, t) \right)_{N,t,j} (\boldsymbol{r}, t) \right]_- \leq \overline{v}_o^i < \overline{v} \quad \forall \boldsymbol{r} \in \partial D_j(t), j \in C_i, i, t. \tag{4.4}$$

Then the proposed navigation law keeps the robot in the obstacle-free part of the scene if $\Delta_i(\cdot)$ are chosen so that

$$\Delta_i(0) > \arcsin\left(\overline{v}_o^i / \overline{v} \right). \tag{4.5}$$

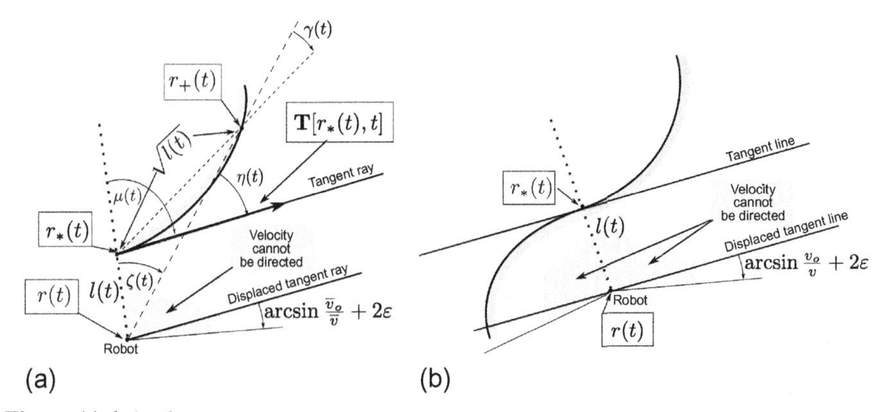

Figure 11.6 Angles.

Due to (4.4), this is possible. If the speed bound $\overline{v}_o^i$ is known, (4.5) can be directly used for controller tuning. Otherwise (4.5) provides general guidelines for tuning: by picking $\Delta_i(0) < \pi/2$ close enough to $\pi/2$, safety is guaranteed.

Proof. Suppose the contrary:

$$G := \{t \geq 0 : \boldsymbol{r}(t) \notin \bigcup_i D_i(t)\} \neq [0, \infty).$$

Since G is open in $[0, \infty)$, its leftmost connected component has the form $[0, \tau)$, where $\boldsymbol{r}(\tau) \in D_k(\tau)$ for some k. Let $\boldsymbol{r}_*(t)$ be the point of $\partial D_k(t)$ closest to $\boldsymbol{r}(t)$ for $t \leq \tau$, and $l(t) := \|\boldsymbol{r}(t) - \boldsymbol{r}_*(t)\|$. For $t \in (0, \tau)$, we introduce the following (see Fig. 11.6(a)):

- the point $\boldsymbol{r}_\pm(t) \in \partial D_k(t)$ that is counter-clockwise/clockwise ahead of $\boldsymbol{r}_*(t)$ and is separated by $\sqrt{l(t)}$ along the boundary $\partial D_k(t)$;
- the polar angle $\alpha(t)$ of $\overrightarrow{\boldsymbol{r}(t), \boldsymbol{r}_*(t)}$ in the robot's frame;
- the angle $\zeta(t)$ from $\overrightarrow{\boldsymbol{r}(t), \boldsymbol{r}_*(t)}$ to $\overrightarrow{\boldsymbol{r}(t), \boldsymbol{r}_+(t)}$;
- the angle $\gamma(t)$ from $\overrightarrow{\boldsymbol{r}(t), \boldsymbol{r}_+(t)}$ to $\overrightarrow{\boldsymbol{r}_*(t), \boldsymbol{r}_+(t)}$;
- the angle $\eta(t)$ from $\overrightarrow{\boldsymbol{r}_*(t), \boldsymbol{r}_+(t)}$ to $\mathbf{T}[\boldsymbol{r}_*(t), t]$;
- the angle $\mu(t)$ from $\overrightarrow{\boldsymbol{r}(t), \boldsymbol{r}_*(t)}$ to $\mathbf{T}[\boldsymbol{r}_*(t), t]$.

Here $\mu(t) = \zeta(t) + \gamma(t) + \eta(t)$ and $\eta(t) \to 0, \gamma(t) \to 0$ as $t \to \tau-$ since the vector $\mathbf{T}$ is tangential to the boundary, and

$$\frac{\|\boldsymbol{r}_*(t) - \boldsymbol{r}_+(t)\|}{\|\boldsymbol{r}(t) - \boldsymbol{r}_*(t)\|} \sim \sqrt{l(t)}/l(t) \to \infty \quad \text{as} \quad t \to \tau - .$$

Thus $\mu(t) - \zeta(t) \to 0 \quad$ as $\quad t \to \tau-$. By (4.5), there exist $\varepsilon > 0, \varkappa > 0$ such that

$$\Delta_i(d) - 2\varepsilon \geq \overline{\delta} := \arcsin \frac{\overline{v}_o^i}{\overline{v}} \quad \forall d \leq \varkappa,$$

where $C_i \ni k$. Now we focus on times $t < \tau$ so close to τ that

$$|\mu(t) - \zeta(t)| \leq \varepsilon \quad \text{and} \quad l(t) \leq \varkappa,$$

which implies that $d_k^{\min}(t) \leq \varkappa$ in (3.2).

The robot's view in the directions $\alpha \in [\alpha(t) - \zeta(t), \alpha(t)]$ is obstructed by $D_k(t)$. Hence the angular interval

$$I(t) := \left[\alpha(t) - \zeta(t) - \Delta_i[d_k^{\min}(t)], \alpha(t)\right]$$

is covered by the range $\hat{A}_k(t)$ of the extended facet $\hat{F}_k$. The respective part of $\hat{F}_k$ cannot be shadowed by other obstacles for $t \approx \tau$. Indeed as $t \to \tau-$, the locus of "shadowing" points collapses into a part of $\partial D_k(\tau)$ and so the distance from the locus to the other obstacles is lower limited by a non-zero value since the obstacles do not collide. Hence for the polar angle $\alpha_v(t)$ of the robot's velocity $\mathbf{v}(t)$, we have

$$\alpha_v(t) \notin I(t) \supset [\alpha(t) - \zeta(t) - \overline{\delta} - 2\varepsilon, \alpha(t)] \supset [\alpha(t) - \mu(t) - \overline{\delta} - \varepsilon, \alpha(t)].$$

By replacing $\boldsymbol{r}_+(t) \mapsto \boldsymbol{r}_-(t)$ in the foregoing, we also see that

$$\alpha_v(t) \notin [\alpha(t), \alpha(t) + \mu(t) + \overline{\delta} + \varepsilon].$$

Overall, $\mathbf{v}(t)$ is not directed to the shadowed sector from Fig. 11.6(b), i.e., the angle subtended by the velocity $\mathbf{v}(t)$ and the unit inner normal $\mathbf{N}[\boldsymbol{r}_*(t), t]$ is no less than $\pi/2 + \arcsin \frac{\overline{v}_o^i}{\overline{v}} + \varepsilon$. So the related normal component $v^N(t)$ of $\mathbf{v}(t)$ is such that

$$v^N(t) \leq -\overline{v} \sin\left(\arcsin \frac{\overline{v}_o^i}{\overline{v}} + \varepsilon\right) \leq -\overline{v}_o^i - \varepsilon_*, \quad \text{where}$$

$$\varepsilon_* := \overline{v}\left[\sin\left(\arcsin \frac{\overline{v}_o^i}{\overline{v}} + \varepsilon\right) - \sin \arcsin \frac{\overline{v}_o^i}{\overline{v}}\right] > 0.$$

At the same time, by employing the notation $V_k^N(t)$ from (4.3) and invoking (4.7), we see that

$$\dot{l}(t) = V_k^N(t) - v^N(t) \overset{(4.4)}{\geq} -\overline{v}_o^i + \overline{v}_o^i + \varepsilon_* = \varepsilon_* > 0,$$

which is incompatible with the relations $l(t) > 0 \; \forall t \in (0, \tau)$ and $l(t) \to 0$ as $t \to \tau-$. The contradiction obtained completes the proof. □

11.5 Achieving the main navigation objective

In general setting, this objective encompasses solving of dynamic mazes. Even for static mazes, this constitutes a challenging individual research topic, whereas deforming mazes lie in absolutely uncharted territory. To reduce the challenge to a reasonable level, this section is mainly aimed at convex obstacles. The obstacles still may undergo translations, rotations, and deformations. We show that the proposed navigation law ensures achieving the navigation objective if the inter-obstacle spacing or the robot's speed is large enough. The necessary spacing depends on the shapes and sizes of the obstacles and the speed excess ratios $e_i := \overline{v}/\overline{v}_o^i$ so that the smaller the

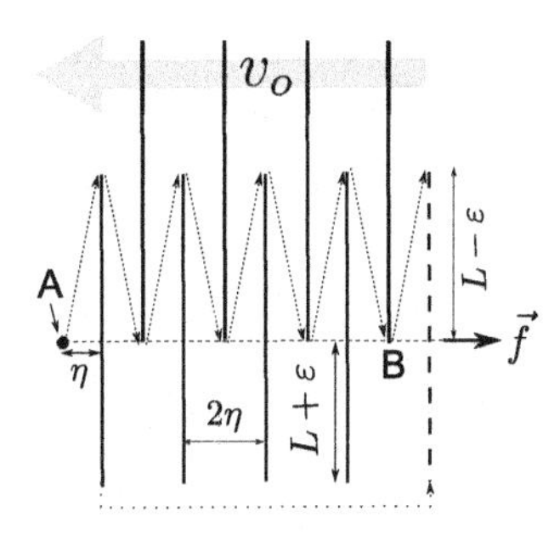

Figure 11.7 Counterexample.

size or e_i^{-1}, the smaller the spacing. In particular, the spacing requirement annihilates for convex obstacles as $e_i \to \infty$.

11.5.1 Counterexample

We first show that spacing and size are really relevant. Let the robot face $2N$ straight line segments of length $2L$ perpendicular to **f**; see Fig. 11.7, where $N = 4$. They are separated by the space 2η and drift in the direction of $-\mathbf{f}$ with the speed $v_o < v$. The odd segments are vertically aligned, the even ones are shifted upwards by $L+\varepsilon$, where $\varepsilon > 0, \varepsilon \approx 0$. The robot starts in the location A. The upper dotted line displays the robot's path under the proposed navigation law.

The robot needs time $T \geq 2N(L-\varepsilon)/v$ to reach point B. For this time, B moves a distance $\geq v_o T$ and displaces to the left of A if

$$v_o T > 2N\eta \Leftrightarrow e := v/v_o < (L-\varepsilon)/\eta.$$

So for any e, tight spacing η or large obstacle size L entail the robot's overall drift contrary to the desired direction. Moreover let just after being bypassed, the first segment S_1 move along the lower dotted path. That time is enough for S_1 to reach the dashed position if

$$(2L+2\eta N)/v_o < (2N-1)(L-\varepsilon)/v \Leftrightarrow e < (L-\varepsilon)[1-(2N)^{-1}](LN^{-1}+\eta)^{-1},$$

which holds for any e if $\eta \approx 0, N \approx \infty$. If each segment performs similar maneuver, the negative drift of the robot is never terminated.

Thus apart from the speed excess ratio, criteria for achieving the navigation objective should somehow impose requirements on the obstacles' sizes and spacing.

11.5.2 Main results

To flesh out this, we introduce geometric objects, called *hats*. Being the parts of the plane that should be free from the obstacles, they characterize the required inter-obstacle spacing. The hats are determined by an angle $\delta \in (0, \pi/2)$, which will depend on the speed ratio in the results to follow.

Consider the angle that is made of two rays obtained from the span of $\mathbf{f}$ via the counter clockwise and clockwise rotation, respectively, though the angle $\pi/2 - \delta$. Let us translate this angle into the unique position $\sphericalangle_j^\delta(t)$ in the plane where its both rays $R^\delta_{\circlearrowleft,j}(t)$ and $R^\delta_{\circlearrowright,j}(t)$ are tangential to the boundary of $D_j(t)$; see Fig. 11.8(a). For $\odot := \circlearrowleft, \circlearrowright$, let $S^\delta_{\odot,j}(t)$ be the segment of $R^\delta_{\odot,j}(t)$ between the vertex of the angle $\sphericalangle_j^\delta(t)$ and the furthest point of $R^\delta_{\odot,j}(t) \cap D_j(t)$. The *$\delta$-hat* $H_j^\delta(t)$ of the jth obstacle at time t is the part of the plane bounded by $S^\delta_{\circlearrowleft,j}(t), S^\delta_{\circlearrowright,j}(t)$ and the least part of the boundary $\partial D_j(t)$ that connects these segments and is closer to the vertex of the angle. The *height* $h_j^\delta(t)$ of this hat is the maximal distance from a point of $S^\delta_{\circlearrowleft,j}(t) \cup S^\delta_{\circlearrowright,j}(t)$ to the obstacle. The *δ_1-extended δ-hat* $\hat{H}_j^{\delta,\delta_1}(t)$ is the union of $H_j^\delta(t)$ with the sectors $D^{\delta+\delta_1}_{\circlearrowleft,j}(t), D^{\delta+\delta_1}_{\circlearrowright,j}(t)$; see Fig. 11.8(b). Here the sector $D^{\delta+\delta_1}_{\circlearrowleft,j}(t)$ is swept by the segment $S^\delta_{\circlearrowleft,j}(t)$ when rotating counter clockwise through the angle $\delta+\delta_1$; and the sector $D^{\delta+\delta_1}_{\circlearrowright,j}(t)$ is defined likewise.

Theorem 5.1. *Suppose that the following "two-sided" extension of* (4.4) *is true*

$$\left|\left(\mathbf{V}_j(\boldsymbol{r},t)\right)_{N,t,j}(\boldsymbol{r},t)\right| \le \overline{v}_o^i < \overline{v} \quad \forall \boldsymbol{r} \in \partial D_j(t), j \in C_i, i, t, \tag{5.1}$$

the obstacles are always convex and do not collide with each other, and for $\delta_i^\star := \arcsin \frac{\overline{v}_o^i}{\overline{v}}$, *the following claims hold:*

(a) *the $\delta_j^\star$-extended $\delta_i^\star$-hat of any obstacle of class C_i is always disjoint with the obstacles of class C_j and*

(b) *for $t = 0$ and any i, the location $\mathbf{r}(0)$ of the robot does not lie in the $\delta_i^\star$-hat of any obstacle of class C_i.*

Moreover let ***(a)*** *and* ***(b)*** *be true for a larger value of $\delta_i^\star$:*

$$\delta_i^\star \in \left(\overline{\delta}_i, \frac{\pi}{2}\right), \quad \textit{where} \quad \overline{\delta}_i := \arcsin \frac{\overline{v}_o^i}{\overline{v}}. \tag{5.2}$$

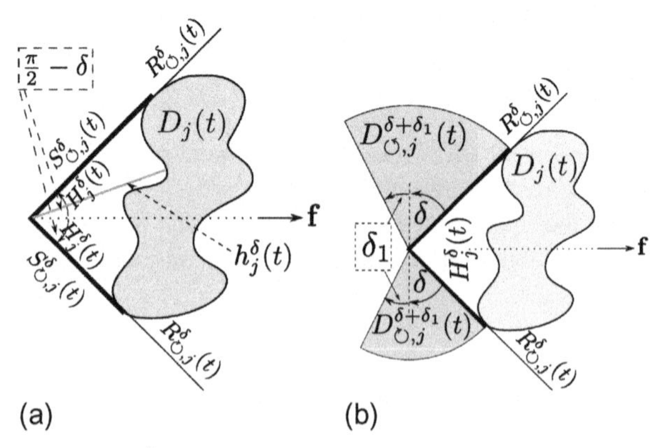

Figure 11.8 (a) The δ-hat $H_j^\delta(t)$; (b) The extended hat.

Then the proposed navigation controller drives the robot through the obstacle-free part of the environment. Moreover, it can be tuned so that the robot always drifts in the right direction: there exists $\eta > 0$ such that

$$\langle \mathbf{v}(t); \mathbf{f} \rangle \geq \eta > 0 \quad \forall t. \tag{5.3}$$

Specifically, this is true if in (3.1), *the functions $\Delta_i(\cdot)$ are chosen so that*

$$\Delta_i \left[h_j^{\delta_i^\star}(t) \right] = \Delta_i[0] \in \left(\overline{\delta}_i, \delta_i^\star \right) \quad \forall t, i, j \in C_i. \tag{5.4}$$

The proof of this theorem will be given in Section 11.5.3.

Remark 5.1.

- **(i)** If (b) is true for $\delta_i^\star := \overline{\delta}_i$, it necessarily holds for some $\delta_i^\star$ satisfying (5.2) (see (d) in Observation 5.1).
- **(ii)** Theorem 5.1 is true for arbitrary time horizon I, whether infinite $I = [0, \infty)$ or finite $I = [0, t_1], 0 < t_1 < \infty$. In any case, $t \in I$ in (5.1), (5.3), and (5.4).
- **(iii)** Picking $\Delta_i(\cdot) = \text{const} \in \left(\overline{\delta}_i, \delta_i^\star \right)$ shows that (5.4) can always be met. If an estimate of the hat height $h_j^{\delta_i^\star}(t) \leq h_i \ \forall j \in C_i, t$ is available, this choice can be altered for $d \geq h_i$ by any function continuously decaying from $\Delta_i(h_i)$.
- **(iv)** A guideline given by (5.4) is to pick $\Delta_i(0)$ close to $\overline{\delta}_i$, which may be used if estimation of $\delta_i^\star$ is troublesome.
- **(v)** For convex obstacles, all hats degenerate into parts of the obstacle boundary as $\overline{v}_o^i / \overline{v} \to 0$. Then (a) and (b) come to disjointness of the obstacles and the natural requirement that initially the robot is in the obstacle-free part of the plain: $\boldsymbol{r}(0) \notin D_j(0) \ \forall j$, whereas (5.4) comes to $\Delta_i(0) > 0$.
- **(vi)** Theorem 5.1 can be extended on some scenarios with non-convex obstacles by "replacement" of the obstacle by their convex hulls in the statement of the theorem. More precisely, this statement should be altered as follows:

- The requirement that the obstacles are always convex should be dropped;
- In **(a)**, not only the concerned hat but also the convex hull of the respective obstacle should be always disjoint with not only the obstacles of class C_j but also with their convex hulls;
- In **(b)**, the initial location of the robot should be outside not only the concerned hat but also the convex hull of the respective obstacle;
- In (5.4), $h_j^\delta(t)$ should be re-defined as the maximal distance from a point of $S_{\circlearrowleft,j}^\delta(t) \cup S_{\circlearrowright,j}^\delta(t)$ to not the obstacle itself by to its convex hull.

To estimate the hat's height in (iii), the following lemma is instructive for convex obstacles.

Lemma 5.1. *For a convex obstacle, the height $h_j^\delta(t)$ of the hat equals the distance from the vertex of $\sphericalangle_j^\delta(t)$ to the obstacle.*

Proof. Since $D_j(t)$ is convex, so is the function $\boldsymbol{r} \mapsto \textbf{dist}\left[\boldsymbol{r}; D_j(t)\right]$. So its maximum over both $S_{\circlearrowleft,j}^\delta(t)$ and $S_{\circlearrowright,j}^\delta(t)$ is attained at an end of the segment. This is not the end $\boldsymbol{p}$ different from the vertex of $\sphericalangle_j^\delta(t)$ since $\textbf{dist}\left[\boldsymbol{p}; D_j(t)\right] = 0$. $\square$

11.5.3 Proof of Theorem 5.1

We start with a deeper insight into the δ-hats of convex bodies, where $\delta \in (0, \pi/2)$.

We fix t and j, focus on $D := D_j(t)$, assume the polar angle of $\mathbf{f}$ to be zero, and add the adjective *free* to signal that the facet is calculated in the absence of the other obstacles. Let $\boldsymbol{r} = \rho(s)$ be a natural parametric representation of the boundary ∂D, where s is the arc length and D is to the left as s ascends, $\mathbf{T}(s) := \frac{d\rho}{ds}(s)$ and $\mathbf{N}(s)$ be the unit normal to ∂D directed inwards D. The ray

$$R_{\pm}(s) := \{\boldsymbol{r} = \boldsymbol{r}_{\pm}(x, s) := \rho(s) \pm x\mathbf{T}(s) : x \geq 0\}$$

is the locus of points from which the upper/lower edge of the free visible facet of D is given by $\rho(s)$. By the Frenet-Serrat equations $\boldsymbol{r}'_{\pm}(x, s) = \mathbf{T}(s) \pm x\varkappa(s)\mathbf{N}(s)$, where the signed curvature is non-negative $\varkappa(s) \geq 0$ since D is convex. Hence as s ascends, the ray $R_{\pm}(s)$ (nonstrictly) displaces to the left with respect to its current position observed from the ray origin. By the same argument, $\mathbf{T}(s)$ rotates counter-clockwise, and makes one full turn as s runs the entire perimeter of ∂D.

As a result, we arrive at the following.

Observation 5.1.

(a) *Let s_- and s_+ be values of s for which the polar angle $\alpha(s)$ of $\mathbf{T}(s)$ equals δ and $\pi - \delta$, respectively. The locus $L_j(t)$ of points $\boldsymbol{r} \notin D_j(t)$ whose view in the direction of $\mathbf{f}$ is obstructed by the free δ-facet of $D_j(t)$ is bounded by $R_+(s_+)$, $R_-(s_-)$, and the respective part of $\partial D_j(t)$; see Fig. 11.9(a).*

(b) *Let $s_{\pm}^{\delta}$ be the largest/smallest value of s for which $\alpha(s) = 3/2\pi \pm \delta$, respectively. The δ-hat $H_j^{\delta}(t)$ of $D_j(t)$ is bounded by $R_+(s_-^{\delta})$, $R_-(s_+^{\delta})$, and the least part of $\partial D_j(t)$ that connects these rays and excludes $D_j(t)$ from the hat.*

(c) *$H_j^{\delta_1}(t) \subset H_j^{\delta_2}(t), s_-^{\delta_1} \geq s_-^{\delta_2}, s_+^{\delta_2} \geq s_+^{\delta_1}$ if $\delta_1 \leq \delta_2$.*

(d) *$s_-^{\delta} \to s_-^{\delta_\star}, s_+^{\delta} \to s_+^{\delta_\star}$ as $\delta \to \delta_\star +$, and the ends of the segments $S_{\circlearrowright,j}^{\delta}(t), S_{\circlearrowleft,j}^{\delta}(t)$ from Fig. 11.8 converge to the respective ends of $S_{\circlearrowright,j}^{\delta_\star}(t), S_{\circlearrowleft,j}^{\delta_\star}(t)$ uniformly over t.*

Corollary 5.1. *The interiors of the infinite sectors shadowed in Fig. 11.9(b) do not contain points for which the two directions to the obstacle's edges have equal angular discrepancy with respect to* **f**.

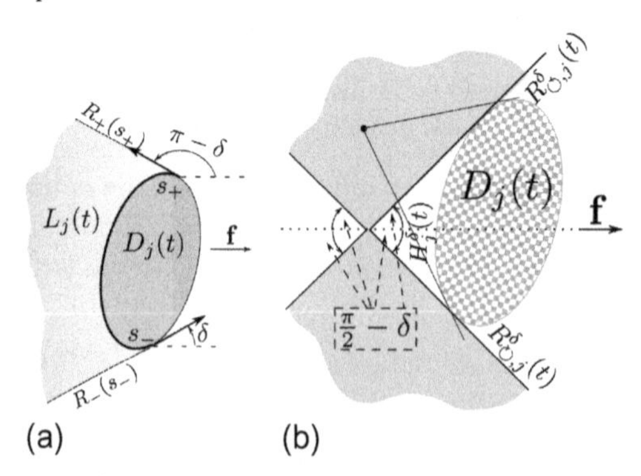

Figure 11.9 (a) Where the direction of **f** is shadowed by the extended facet? (b) Infinite sectors.

This holds by (c) in Observation 5.1 since such points are vertices of hats. Exactly at these points the navigation law is discontinuous at time t if the effect of the other obstacles can be neglected. The ray $R^{\delta}_{\aleph,j}(t), \aleph = \circlearrowleft, \circlearrowright$ evidently moves without rotation. Its *normal velocity* $V_{\aleph}(t)$ is that with respect to itself; with the positive direction being to the left when observing from the ray origin.

Lemma 5.2. *Putting* $\sigma_{\circlearrowleft} := -1, \sigma_{\circlearrowright} := 1$, *we have for* $\aleph = \circlearrowleft, \circlearrowright$

$$V_{\aleph}(t + /-) = \sigma_{\aleph} \min_{\boldsymbol{r} \in R^{\delta}_{\aleph,j}(t) \cap D_j(t)} / \max \; V_j^N(\boldsymbol{r}, t).$$

Proof. Let $\aleph = \circlearrowleft$ and $\rho(s)$ be the natural parametric representation of the reference configuration boundary $\partial D_{*,j}$. We put

$$\Phi_j[s,t] := \Phi_j[\rho(s), t], \quad \mathbf{V}_j(s,t) := \mathbf{V}_j[\rho(s), t],$$

etc. There are s_1, s_2 such that $\Phi_j[s,t] \in R^{\delta}_{\circlearrowleft,j}(t) \cap D_j(t) \Leftrightarrow s \in [s_1, s_2]$. Let $\mathbf{n}_{\circlearrowleft}$ be the positively oriented unit vector normal to $R^{\delta}_{\circlearrowleft,j}(t)$. The line

$$\langle \boldsymbol{r}; \mathbf{n}_{\circlearrowleft} \rangle = \zeta(t) := \max_{\boldsymbol{r}_* \in D_j(t)} \langle \boldsymbol{r}; \mathbf{n}_{\circlearrowleft} \rangle = \max_s \langle \mathbf{n}_{\circlearrowleft}; \Phi_j(s,t) \rangle$$

contains $R^{\delta}_{\circlearrowleft,j}(t)$ and so $V_{\circlearrowleft}(t\pm) = \dot{\zeta}(t\pm)$. By Corollary 2 in [435, Section 2.8] combined with Definition 2.3.4 and Proposition 2.1.2 [435], the one-sided derivatives $\dot{\zeta}(t\pm)$ do exist and

$$\dot{\zeta}(t + /-) = \max_{\mu} / \min_{\mu} \int_{s_1}^{s_2} \frac{\partial \langle \mathbf{n}_{\circlearrowleft}; \Phi_j(s,t) \rangle}{\partial t} \mu(\mathrm{d}s),$$

where μ ranges over all probability measures on $[s_1, s_2]$. So this max / min is equal to max / $\min_{s \in [s_1, s_2]}$ of the integrand

$$\frac{\partial \langle \mathbf{n}_{\circlearrowleft}; \Phi_j(s,t) \rangle}{\partial t} = \langle \mathbf{n}_{\circlearrowleft}; \mathbf{V}_j(s,t) \rangle \overset{\text{(a)}}{=} -V_j^N(s,t),$$

where (a) holds since $\mathbf{n}_{\circlearrowleft} = -\mathbf{N}_j(s,t) \; \forall s \in [s_1, s_2]$. This completes the proof for $\aleph = \circlearrowleft$. The case $\aleph = \circlearrowright$ is considered likewise. □

Now we start to examine the robot driven by the proposed navigation law. The δ-facet of the obstacle is given by (3.2) and (3.3), where δ_k is replaced by δ.

Lemma 5.3.

(a) *If the robot is on* $S^{\delta_i^\star}_{\circlearrowleft,j}(t) \setminus \{p_j^{\delta_i^\star}(t)\}, j \in C_i$, *the normal component of its relative velocity with respect to each of the rays* $R^{\delta_i^\star}_{\circlearrowleft,j}(t), R^{\delta_i^\star}_{\circlearrowright,j}(t)$ *is no less than* $\overline{v} \sin \delta^i - \overline{v}^i_o$, *where* $\delta^i := \Delta_i[0]$.

(b) *If the robot is on* $S^{\delta_i^\star}_{\circlearrowright,j}(t) \setminus \{p_j^{\delta_i^\star}(t)\}, j \in C_i$, *the above normal component does not exceed* $-(\overline{v} \sin \delta^i - \overline{v}^i_o)$.

Proof. We focus on (a), the claim (b) is proved likewise.

Since $\boldsymbol{r}(t) \in S^{\delta_i^\star}_{\circlearrowleft,j}(t)$ and $D_j(t)$ is convex, $D_j(t)$ obstructs the view in the direction of $\mathbf{f}$, lies in the angle $\sphericalangle$ subtended by $R^{\delta_i^\star}_{\circlearrowleft,j}(t)$ and the tangent ray T from Fig. 11.10(a), and its free facet corresponds to an angular interval of the form $\mathcal{A}_j = \left(-\alpha, \frac{\pi}{2} - \delta_i^\star\right)$.

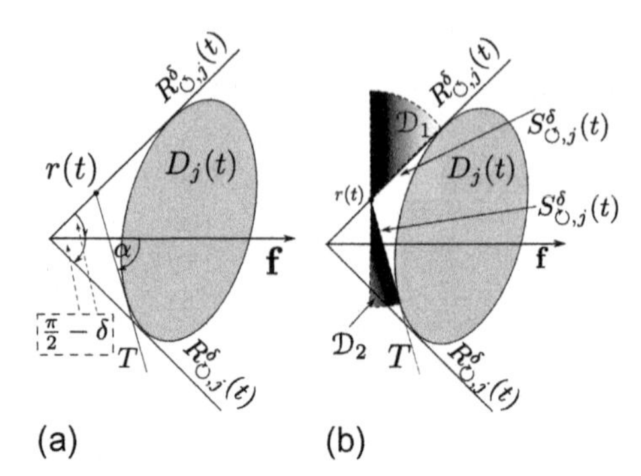

Figure 11.10 (a) The tangent ray T; (b) Disk sectors that should be free.

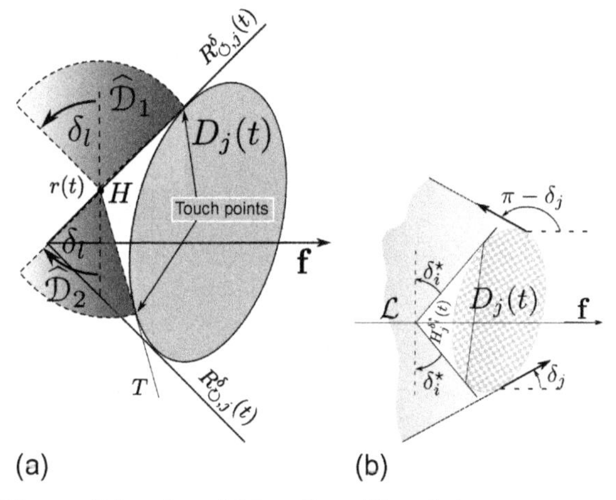

Figure 11.11 (a) Larger disk sectors; (b) Location of the robot.

So $\alpha > \frac{\pi}{2} - \delta_i^\star$, whereas $\alpha \leq \frac{\pi}{2} + \delta_i^\star$ since $\sphericalangle$ does not exceed π radian. By (3.1), (3.2), (5.4), and Lemma 5.1, the range of the free extended facet $\hat{\mathscr{A}}_j = \left(-\alpha - \delta^i, \frac{\pi}{2} - \delta_i^\star + \delta^i\right)$ does not cover the unit circle since its angular span $\frac{\pi}{2} - \delta_i^\star + \delta^i + [\alpha + \delta^i] \leq \pi + 2\delta^i < 2\pi$ thanks to $\delta^i < \frac{\pi}{2}$. So by (5.4), the robot's absolute velocity $\mathbf{v}$ has the polar angle $\frac{\pi}{2} - \delta_i^\star + \delta^i$ unless some of the sectors $\mathfrak{D}_1, \mathfrak{D}_2$ from Fig. 11.10(b) is partly shadowed by an extended facet of another obstacle $D_k(t)$. Now we are going to show that this does not hold.

Suppose the contrary. Then an obstacle $D_k(t)$ (of class C_l) contains a point p whose rotation through an angle $\leq \delta_l$ about $\boldsymbol{r}(t)$ goes through $\mathbb{R}^2 \setminus D_j(t)$ and ends on $\mathfrak{D}_1 \cup \mathfrak{D}_2$. The locus of such points $p \in \mathbb{R}^2$ lies in the union of the domain H from Fig. 11.11(a) and two larger sectors $\hat{\mathfrak{D}}_1$ and $\hat{\mathfrak{D}}_2$ that are obtained from $\mathfrak{D}_1$ and $\mathfrak{D}_2$, respectively, by extra rotation of the radius through angle δ_l. However by (5.4), this union is a subset of the $\delta_l^\star$-extended $\delta_i^\star$-hat of $D_j(t)$. Thus $D_k(t)$ intersects this hat, in violation of (a) in Theorem 5.1. Hence $\mathbf{v}$ has the polar angle $\frac{\pi}{2} - \delta_i^\star + \delta^i$ indeed.

So $\mathbf{v}$ subtends an angle of δ^i with the ray $R_{\circlearrowleft,j}(t)$ and an angle of $2\delta_i^\star - \delta^i$ with $-R_{\circlearrowright,j}(t)$. So the normal components of $\mathbf{v}$ are

$$\langle \mathbf{v}; \mathbf{n}_{\circlearrowleft} \rangle = \overline{v} \sin \delta^i \quad \text{and} \quad \langle \mathbf{v}; \mathbf{n}_{\circlearrowright} \rangle = \overline{v} \sin(2\delta_i^\star - \delta^i),$$

where

$$\sin(2\delta_i^\star - \delta^i) - \sin \delta^i = 2 \sin(\delta_i^\star - \delta^i) \cos \delta_j^\star > 0,$$

and so no less than $\overline{v} \sin \delta^i$ in both cases. It remains to note that the both rays are translated at the speed $\leq \overline{v}_o^i$ by (5.1) and Lemma 5.2. □

By (5.2), (b) in Theorem 5.1, and Corollary 5.1, we get the following.

Corollary 5.2. *The robot is always outside the $\delta_i^\star$-hats of all obstacles of class C_i.*

Proof. If at time t the ray R emitted from the robot in the direction of $\mathbf{f}$ hits no extended facet, the robot's velocity $\mathbf{v} = \overline{v}\mathbf{f}$ and so (5.3) does hold with $\varepsilon := \overline{v}$. Suppose that R hits an extended facet; let $j \in C_i$ be the index of the first of them and δ_j be defined from (3.2). By Corollary 5.2, the robot is not in the $\delta_i^\star$-hat of the jth obstacle. So in view of (a) in Observation 5.1, the robot is located in the shadowed domain from Fig. 11.11(b).

Let the robot is above the line $\mathcal{L}$ from Fig. 11.11(b). Let α_j^+ be the polar angle of the upper edge of the free facet of $D_j(t)$. Then

$$-\delta_j \leq \alpha_j^+ \leq \pi/2 - \delta_i^\star \Rightarrow 0 \leq \alpha_j^+ + \delta_j \leq \pi/2 - \delta_i^\star + \delta_j,$$

where $\alpha_j^+ + \delta_j$ is the upper end of the extended facet. Since $\delta_i^\star > \delta^i \geq \delta_j$ due to (5.4), its deviation from the angle 0 of $\mathbf{f}$ does not exceed $\beta \leq \pi/2 - \delta_i^\star + \delta^i < \pi/2$. So were the robot's velocity $\mathbf{v}$ directed to that upper end, the angle γ between $\mathbf{v}$ and $\mathbf{f}$ would be acute $\gamma \leq \beta$. That direction can be dismissed by either the lower edge of the same obstacle or by another obstacle. However this may result only in decrease of γ. Thus in any case, $\gamma \leq \beta$ and so

$$\langle \mathbf{v}; \mathbf{f} \rangle = \overline{v} \cos \gamma \geq \overline{v} \cos\left(\frac{\pi}{2} - \delta_i^\star + \delta^i\right) = \overline{v} \sin\left(\delta_i^\star - \delta^i\right) > 0,$$

i.e., (5.3) does hold with $\varepsilon := \overline{v} \min_i \sin\left(\delta_i^\star - \delta^i\right)$. The case where the robot is below $\mathcal{L}$ is considered likewise. □

11.6 Illustrations of the main results for special scenarios

For each of them, we start with computation of the hats and related specifications of Theorem 5.1. In some cases, these hats appear to be rather complex. So we finish with easier comprehensible corollaries that result from replacement of the hats with their upper estimates by simple shapes.

11.6.1 Disk-shaped obstacles

In robotics research, obstacles are often represented by disks. Let the jth obstacle be the disk of the constant radius R_j with the center $\boldsymbol{r}_j(t)$ whose velocity is $\mathbf{v}_j(t)$. The "collision avoidance" condition (4.4) is equivalent to (5.1) and shapes into

$$\|\mathbf{v}_j(t)\| \leq \overline{v}_o^i < \overline{v} \quad \forall j \in C_i. \tag{6.1}$$

Elementary geometrical considerations show that the δ_i-hat of the jth obstacle is the set shadowed in dark in Fig. 11.12(a). The δ_l-extended δ_i-hat is obtained via union of the hat with two slightly shadowed disk sectors. In the basic case from Theorem 5.1, where $\delta_r = \overline{\delta}_r$ and $\overline{\delta}_r$ is defined in (5.2), Fig. 11.12(a) shapes into Fig. 11.12(b), where

$$\xi_r := \frac{\overline{v}_o^r}{\overline{v}}.$$

By bringing the pieces together, we arrive at the following.

Proposition 6.1. *If* (6.1) *and* (4.5) *are true, the robot is driven through the obstacle-free part of the environment. Suppose also that for any obstacle $j \in C_i$, its extended hat from Fig. 11.12(b) is always disjoint with the obstacles of class C_l, and the robot is initially outside its hat from Fig. 11.12(b). Moreover let these claims be true for the hats from Fig. 11.12(a) with some $\delta_r > \overline{\delta}_r \; \forall r$. Then the robot always drifts in the right direction, i.e.,* (5.3) *is true, if $\Delta_i(\cdot)$ are chosen so that*

$$\Delta_i\left[R_j\frac{1-\cos\delta_i}{\cos\delta_i}\right] = \Delta_i[0] \in (\overline{\delta}_i, \delta_i) \quad \forall i. \tag{6.2}$$

In (6.2), R_j can be replaced by any its known upper bound. If this bound is not available, $\Delta_i(\cdot)$ can be picked constant.

Now we apply Proposition 6.1 to more special scenarios and, by sacrificing a part of its content, provide convergence criteria in terms of distances instead of hats.

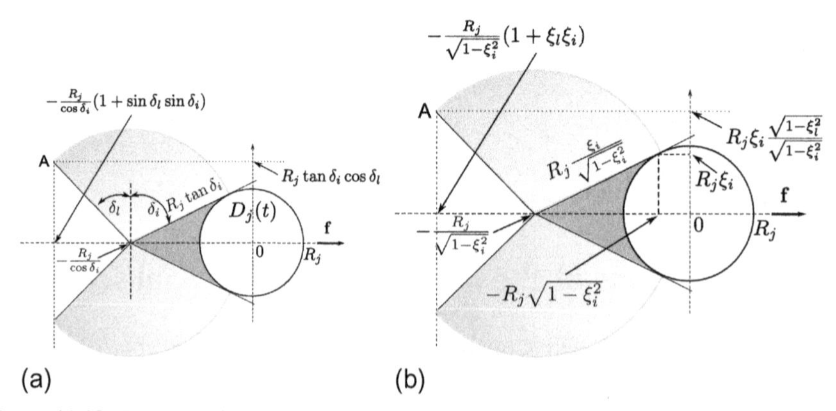

Figure 11.12 The hat and extended hat of the disk.

11.6.1.1 Irregularly and unpredictably moving disk-shaped obstacles with a common speed bound

Now $\overline{v}_o^i = \overline{v}_o < \overline{v}\ \forall i$. Let $L_{j,k} = L_{k,j}$ be a lower estimate of the spacing between the jth and kth obstacles. Since δ-hat is covered by the disk passing through its vertex (see Lemma 5.1) and the extended hat is covered by the disk centered at the origin and passing through point A in Fig. 11.12(b), we arrive at the following.

Corollary 6.1. *The proposed navigation controller guarantees no collisions with the obstacles and constant drift in the right direction if*

$$\frac{\mathbf{dist}\left[\boldsymbol{r}(0); D_j(0)\right]}{R_j} > \Omega(\xi) := \left[\frac{1}{\sqrt{1-\xi^2}} - 1\right], \quad \xi := \frac{\overline{v}_o}{\overline{v}};$$

$$\frac{L_{j,k}}{\max\{R_k, R_j\}} > \Upsilon(\xi) := \frac{\sqrt{1+3\xi^2} - \sqrt{1-\xi^2}}{\sqrt{1-\xi^2}}, \tag{6.3}$$

and the "parameter" $\Delta(\cdot)$ of the controller is chosen so that

$$\Delta\left[\Omega(\xi)R_j + \varepsilon\right] = \Delta[0] \in \left(\arcsin\xi, \arcsin\xi^\star\right)\ \forall j. \tag{6.4}$$

Here $\varepsilon > 0$ may be arbitrarily small and $\xi^\star > \xi$ is picked so that (6.3) *remains true with ξ replaced by $\xi^\star$.*

For common bounds $R_j \le R$, $L_{j,k} = L$, (6.3) takes the form

$$R^{-1}\mathbf{dist}\left[\boldsymbol{r}(0); D_j(0)\right] > \Omega(\xi), \quad L/R > \Upsilon(\xi).$$

Figure 11.13 illustrates the function $\Upsilon(\cdot)$. For example, if the robot is four times faster than the obstacles, the inter-obstacle spacing should exceed only nearly negligible 6% of the obstacle diameter = size. In Fig. 11.14, the requirements to the initial location of the robot are even more liberal.

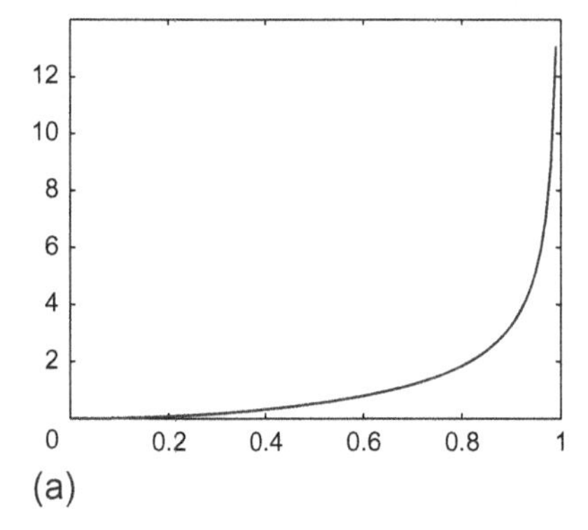

ξ	0	1/8	1/7	1/6
$\Upsilon(\xi)$	0	3%	4%	6%
ξ	1/4	1/3	1/2	$1/\sqrt{2}$
$\Upsilon(\xi)$	12%	22%	52%	124%

(b)

Figure 11.13 (a) Graph of $\Upsilon(\cdot)$; (b) Table of values.

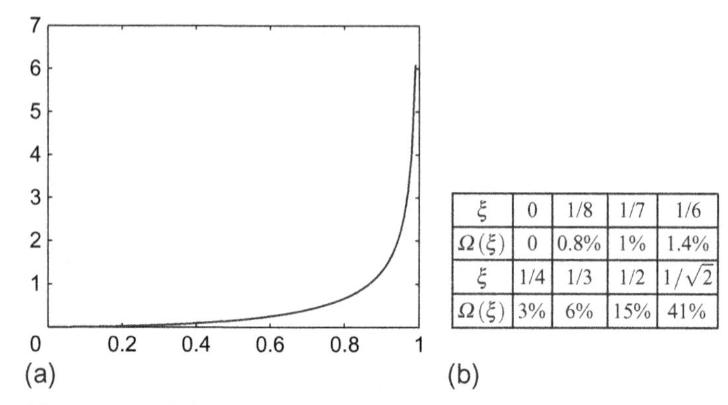

ξ	0	1/8	1/7	1/6
$\Omega(\xi)$	0	0.8%	1%	1.4%
ξ	1/4	1/3	1/2	$1/\sqrt{2}$
$\Omega(\xi)$	3%	6%	15%	41%

Figure 11.14 (a) Graph of $\Omega(\cdot)$; (b) Table of values.

11.6.1.2 Motion along a corridor obstructed with irregularly and unpredictably moving disk-shaped obstacles

We assume that the obstacles have a common velocity bound $\overline{v}_o$ and radius R; see Fig. 11.15(a). They are partitioned into groups so that the disk centers are "horizontally" (in Fig. 11.15(a)) aligned within any group. The "horizontal" projection of the centers moves within a steady interval, which are disjoint for distinct groups. The walls of corridor are rectangles parallel to **f**, which are smoothly concatenated with two end-disks of proper radii. The obstacles move without collisions. Initially the robot is inside the "rectangular" part of the corridor, to the left of all intervals, and in touch with no obstacle. The velocity ratio $\xi := \overline{v}_o/\overline{v} < 1$ guarantees collision avoidance by Theorem 4.1. Starting with the case of one group, we are interested in the value of ξ for which

(p) *the robot driven by the properly tuned navigation controller does pass the corridor in the desired direction, no matter how small the minimal "vertical" spacing between the obstacles is.*

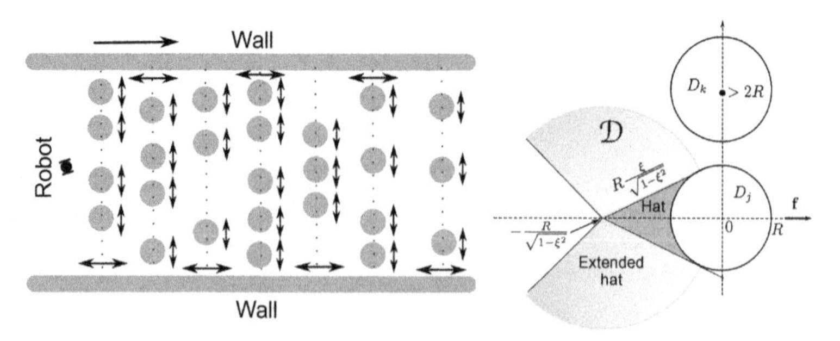

Figure 11.15 (a) The corridor; (b) Two close obstacles.

The hats of the walls lie to the left of them. So (a) and (b) in Theorem 5.1 are true for the walls of the corridor. For the disks, (a) holds if the extended hat from Fig. 11.12(b) lies in the horizontal strip spanned by the disk. This holds if and only if

$$\frac{\xi}{\sqrt{1-\xi^2}} < 1 \Leftrightarrow \xi < 1/\sqrt{2} \Leftrightarrow \overline{v} > \sqrt{2}\overline{v}_o \approx 1.41\overline{v}_o,$$

i.e., the robot is $\approx$41% faster than the obstacles. Then **(p)** holds if initially the robot is at a distance $d_{\text{in}} > R\Omega(1/\sqrt{2}) = \sqrt{2}R \approx 1.41R$ from the leftmost interval. This extends on many groups if the distance d_i between the ith and $(i+1)$th intervals exceeds $R\Omega(1/\sqrt{2})$ for any i. We remark that controller tuning depends only on d_{in} and d_i: it suffices to choose $\Delta(\cdot)$ so that

$$\Delta\left[\sqrt{2}R\right] = \Delta[0] \in \left(\pi/4, \arcsin \xi^\star\right),$$

where $\xi^\star > 1/\sqrt{2}$ is in turn, chosen so that $d_{\text{in}}, d_i > R\Omega(\xi^\star)\ \forall i$.

In this example, we omitted specifications of the controller tuning (6.2) for the sake of brevity. They are technical, elementary, and basically in the vein of (6.4).

11.6.2 Thin cigar-shaped obstacles

Cigar is the body obtained by smooth concatenation of a rectangle with two end-disks of proper radii; see Fig. 11.16(a). Let $\varphi \in (-\pi, \pi)$ stand for the angle from the desired direction **f** to the cigar centerline. If $|\varphi| \leq \frac{\pi}{2} - \delta$ or $|\pi - \varphi| \leq \frac{\pi}{2} - \delta$, the δ-hat and extended hat of the cigar are identical to those of the respective end-disk; see Figs. 11.16(b) and 11.12. For $\varphi \in (\pi/2 - \delta, \pi/2 + \delta)$, the δ-hat can be found via elementary geometrical considerations and is shown in Fig. 11.16(a) in lighter dark. The δ_1-extended δ-hat is obtained by adding the disk sectors swept by the segments $S_{\circlearrowleft}$ and $S_{\circlearrowright}$ from Fig. 11.16(a) when rotating counter clockwise and clockwise, respectively, around the hat's vertex p through the angle $\delta + \delta_1$. If $-\varphi \in (\pi/2 - \delta, \pi/2 + \delta)$, the picture is symmetric.

To simplify formulas, we confine ourselves to asymptotic analysis as $R \to 0$ and consider thin cigars with $R \approx 0$, called *segments*. The subsequent characterization of segment hats is true with as high precision as desired and the final claims are valid if R is small enough. Transition to "non-asymptotic" exact claims results from elementary substitution of the true hats in place of their asymptotic counterparts.

For $|\varphi - \pi/2| < \delta$, the hats of the segment are depicted in Fig. 11.17 and obtained via elementary geometrical considerations; the extended hat is composed of the hat and two slightly shadowed disk sectors. For $|\varphi + \pi/2| < \delta$, the picture is symmetric. For the other φ, the hats are empty.

Now we consider a scene cluttered with moving disjoint segments; see Fig. 11.18(a). The jth of them has time-varying half-length $L_j(t)$ and rotates with

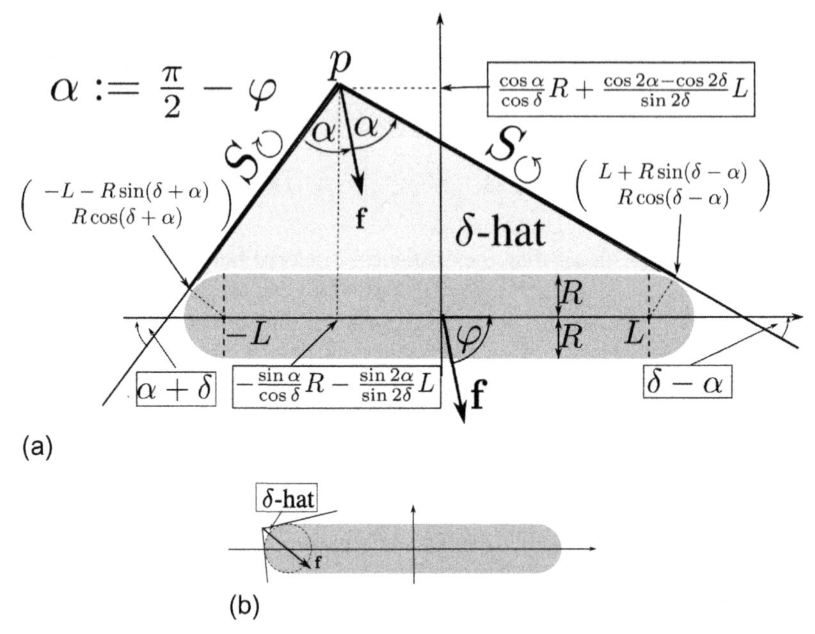

Figure 11.16 Cigar and its hat (a) $\alpha \in [-\delta, +\delta]$; (b) $\alpha \in (\delta, \pi/2]$.

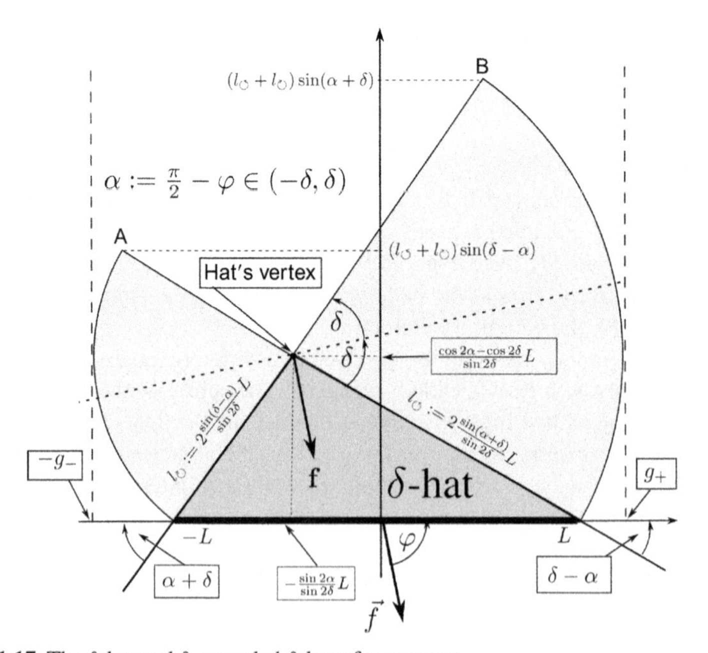

Figure 11.17 The δ-hat and δ-extended δ-hat of a segment.

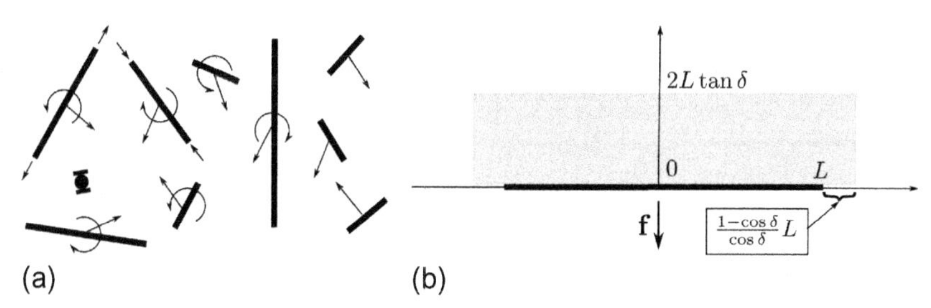

Figure 11.18 (a) Scene cluttered with dynamic segments; (b) Rectangle containing the extended hat.

the angular velocity $\omega_j(t)$; the velocity of its center is $\mathbf{v}_j(t)$. For the sake of brevity, we assume common bounds for the parameters of all segments:

$$L_j(t) \le L, \quad |\dot{L}_j(t)| \le L_1, \quad |\omega_j(t)| \le \omega, \quad \|\mathbf{v}_j(t)\| \le \overline{v}_o \quad \forall t,$$

and consider controller with only one class: $\Delta_i(\cdot) = \Delta(\cdot)$.

The speed of any segment's point does not exceed $v_0 + L_1 + L\omega$. So Theorems 4.1 and 5.1 imply the following.

Corollary 6.2. *Let the robot be faster than the obstacles:*

$$\overline{v}_\Sigma := \overline{v}_o + L_1 + L\omega < \overline{v}. \tag{6.5}$$

If $\Delta(\cdot)$ is chosen so that

$$\Delta(0) > \overline{\delta} := \arcsin \frac{\overline{v}_\Sigma}{\overline{v}},$$

no collision with the obstacles occurs. Let for some $\delta := \delta^\star > \overline{\delta}$, the δ-extended δ-hat of any segment j (see Fig. 11.17) be always disjoint with the other segments, and the robot be initially outside their δ-hats. Then the navigation controller can be tuned so that the robot always drifts in the right direction, i.e., (5.3) is true. For this to hold, it suffices to pick $\Delta(\cdot)$ so that

$$\Delta_i \left[\frac{1 - \cos 2\delta^\star}{\sin 2\delta^\star} \right] = \Delta_i[0] \in (\overline{\delta}, \delta^\star). \tag{6.6}$$

Now we apply this corollary to more special scenarios.

11.6.2.1 Steady-size segments irregularly and unpredictably moving so that they remain perpendicular to the desired direction f

The directions of their velocities and so the paths of the centers are not anyhow restricted. In this case, $L_1 = 0, \omega = 0$ and (6.5) shapes into $\xi := \overline{v}_o/\overline{v} < 1$. It is easy to see that the δ-extended δ-hat is contained in the rectangle from Fig. 11.18(b), where

$$\tan\delta = \Gamma(\xi) := \frac{\xi}{\sqrt{1-\xi^2}}$$

and

$$\frac{1}{2}\frac{1-\cos\delta}{\cos\delta} = \Xi(\xi) := \frac{1-\sqrt{1-\xi^2}}{2\sqrt{1-\xi^2}}$$

for $\delta = \arcsin\xi$. Let the spacing between any two obstacles be always no less than a constant d_f in the direction of $\mathbf{f}$ and no less than $d_f^{\perp}$ in the perpendicular direction. Then we arrive at the following.

Corollary 6.3. *Obstacle avoidance and constant drift in the right direction* (5.3) *are ensured if there is enough space between the obstacles*

$$\frac{d_f}{2L} > \Gamma(\xi), \quad \frac{d_f^{\perp}}{2L} > \Xi(\xi),$$

the initial distance from the robot to any of them is no less than $\Gamma(\xi)L$, and $\Delta(\cdot)$ is chosen so that

$$\Delta[1/2\Gamma(\xi)L+\varepsilon] = \Delta[0] \in \left(\arcsin\xi, \arcsin\xi^{\star}\right)$$

for some $\varepsilon > 0$. Here $\xi^{\star} > \xi$ is picked so that

$$\frac{d_f}{2L} > \Gamma(\xi^{\star}), \quad \frac{d_f^{\perp}}{2L} > \Xi(\xi^{\star}).$$

The function $\Gamma(\cdot)$ is illustrated in Fig. 11.19. For example, if the robot is at least eight times faster than the obstacles, spacing between them in the direction of $\mathbf{f}$ should exceed only 13% of the obstacle length for the controller to succeed. If the robot is twice faster, this percentage increases to $\approx$ 58% of the length. As follows from Fig. 11.20, the requirements to spacing in the perpendicular direction are much more liberal.

It is instructive to compare Corollary 6.3 with the counterexample from Section 11.5.1.

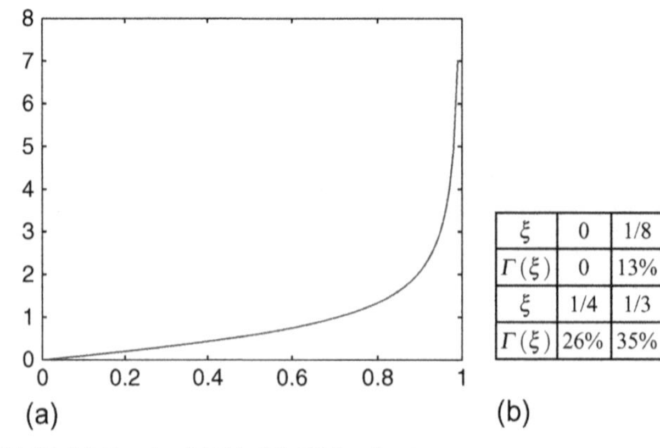

(b)

ξ	0	1/8	1/7	1/6
$\Gamma(\xi)$	0	13%	14%	17%
ξ	1/4	1/3	1/2	$1/\sqrt{2}$
$\Gamma(\xi)$	26%	35%	58%	100%

Figure 11.19 (a) Graph of $\Gamma(\cdot)$; (b) Table of values.

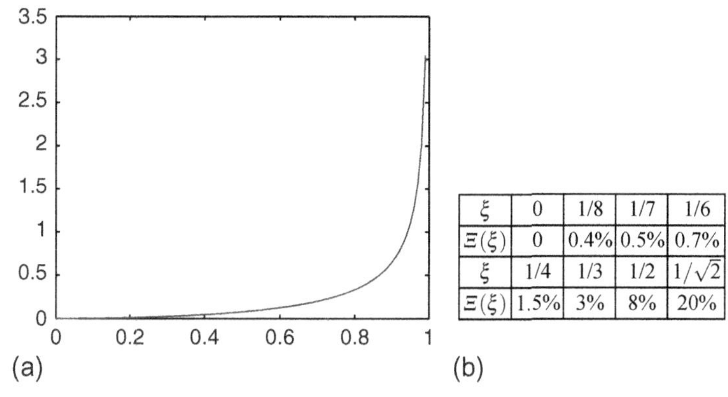

ξ	0	1/8	1/7	1/6
$\Xi(\xi)$	0	0.4%	0.5%	0.7%
ξ	1/4	1/3	1/2	$1/\sqrt{2}$
$\Xi(\xi)$	1.5%	3%	8%	20%

Figure 11.20 (a) Graph of $\Xi(\cdot)$; (b) Table of values.

11.6.3 Navigation in the field of rotating segments

Now we revert to the example concerned by the introduction to this chapter. We consider an infinite planar uniform quadrangular grid with the cell dimension $d_{\text{cell}} \times d_{\text{cell}}$; see Fig. 11.21(a). It is not limited along the abscissa axis and in the negative direction of the ordinate axis, which is identical to the desired direction **f**. At the same time, the grid has an "upper" row.

Every vertex of the grid is a pivot point for a segment with length $2L$; see Fig. 11.21(b). The segments rotate about their centers with a common angular speed $\omega > 0$ in the directions and from initial orientations shown in Fig. 11.21(b). This is possible if the segments at the ends of the cell's diagonal do not collide: $d_{\text{cell}} > \sqrt{2}L$. To make the things interesting, we focus on the case where the disks swept by the segments overlap $d_{\text{cell}} < 2L$. Then the part of the plane with no collision threat consists of infinitely many compact disconnected areas, shown in Fig. 11.22 in white.

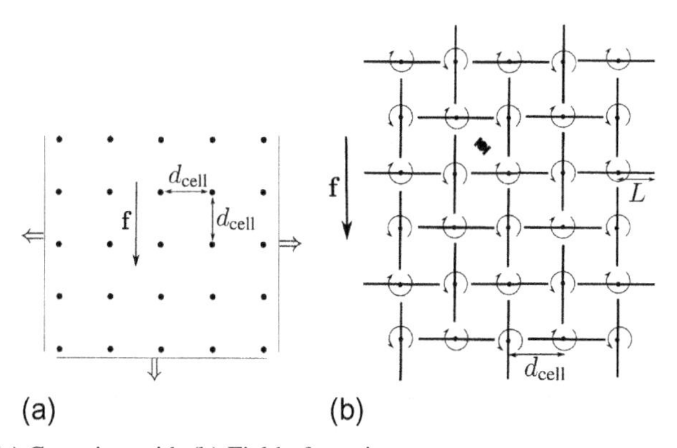

Figure 11.21 (a) Cartesian grid; (b) Field of rotating segments.

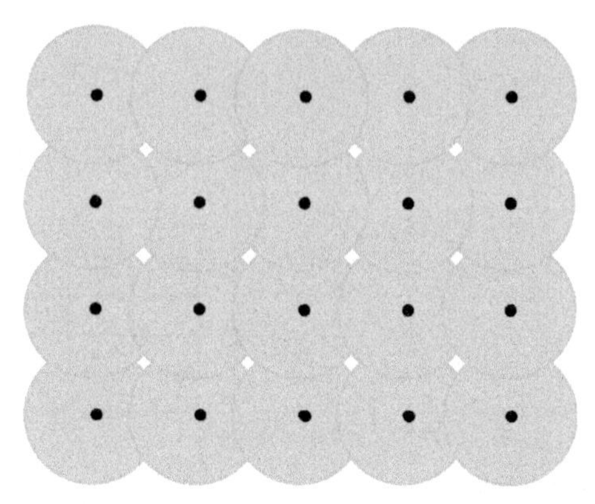

Figure 11.22 White areas where collisions with the obstacles are impossible.

Motion through this part with constant drift in the direction of **f** is not feasible. Initially the robot is in touch with no segment and above the upper row of the grid.

We assume that the robot is able to overtake any point of any obstacle: $\overline{v} > \omega L$. Then by Theorem 4.1, the proposed navigation controller ensures collision avoidance if

$$\Delta(0) > \arcsin \frac{L\omega}{\overline{v}}.$$

Corollary 6.4. *The proposed navigation controller provides not only collision avoidance but also constant drift in the right direction* (5.3) *if*

$$\frac{d_{cell}}{L} > 2\sin\delta + \frac{1}{\cos\delta}, \quad \mathbf{dist}\,[\boldsymbol{r}(0);\mathcal{L}] > \frac{1-\cos 2\delta}{\sin 2\delta}L \tag{6.7}$$

for $\delta = \overline{\delta} := \arcsin(L\omega/\overline{v})$*, where* $\mathcal{L}$ *is the line spanned by the upper row of the grid. For these to hold,* $\Delta(\cdot)$ *should be chosen so that*

$$\Delta\left[\frac{1-\cos 2\delta^\star}{\sin 2\delta^\star}L + \varepsilon\right] = \Delta[0] \in \left(\overline{\delta}, \delta^\star\right)$$

for some $\varepsilon > 0$ *and* $\delta^\star > \overline{\delta}$ *such that* (6.7) *is true with* $\delta := \delta^\star$.

If $d_{\text{cell}} < 2L$, (6.7) $\Rightarrow$ $\delta < 26.26°$ and $\overline{v}/(L\omega) > 2.26$, i.e., the robot should be $\approx 126\%$ faster than the obstacles for the constant drift in the right direction.

The right-hand side of the first inequality from (6.7) treated as a function of $\xi = (\omega L)/\overline{v}$ is illustrated in Fig. 11.23. Since $d_{\text{cell}}/L > \sqrt{2} \approx 1.4142$ (excess $\approx 41\%$), that inequality is fulfilled for robots that are at least 5.08 times faster than the obstacles. For them, the controller ensures constant drift in the right direction irrespective of how small the white cells in Fig. 11.22(a) are. For slower robots, the ratio d_{cell}/L should exceed $\approx 100 + 41\%$ as is shown in Fig. 11.23(b).

Proof of Corollary 6.4. Since the right-hand side of the second inequality from (6.7) estimates the height of the δ-hat from above in Fig. 11.17, the robot is initially outside the $\delta^\star$-hats of all horizontal segments. The same is true for the vertical segments since their hats are empty.

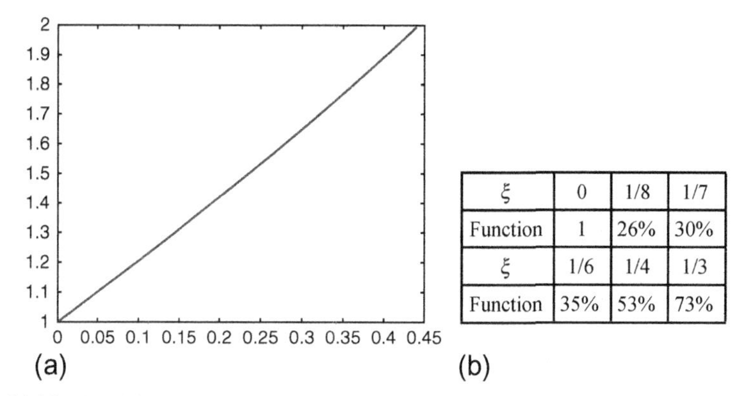

ξ	0	1/8	1/7
Function	1	26%	30%
ξ	1/6	1/4	1/3
Function	35%	53%	73%

Figure 11.23 The right-hand side of (6.7) (a) graph, (b) table of values.

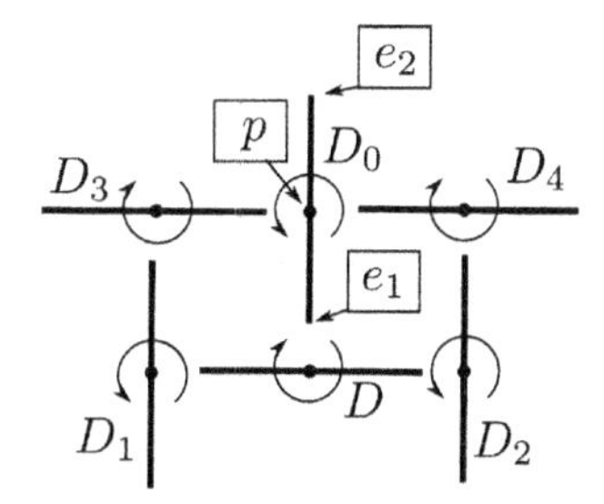

Figure 11.24 Five neighboring obstacles.

It remains to check that the extended hat of any obstacle D is always disjoint with the other obstacles. In doing so, it suffices to examine only five neighbors $D_0, \dots, D_4$ of D from Fig. 11.24 for $|\alpha| < \delta$ since otherwise the hat is empty.

D₀. The δ-extended δ-hat is disjoint with D_0 if in Fig. 11.17, the ends e_i of D_0 lie above the lines spanned by the segments of the lengths $l_{\circlearrowright}$ and $l_{\circlearrowleft}$. Analytically this means that

$$\begin{array}{rl} -x_i \sin(\delta+\alpha) + y_i \cos(\delta+\alpha) \geq L \sin(\delta+\alpha) & \forall \alpha \in (-\delta, \delta) \\ x_i \sin(\delta-\alpha) + y_i \cos(\delta-\alpha) \geq L \sin(\delta-\alpha) & \forall i = 1, 2 \end{array}.$$

Here

$$x_i = -d_{\text{cell}} \sin\alpha - (-1)^i L \sin 2\alpha, \quad y_i = d_{\text{cell}} \cos\alpha + (-1)^i L \cos 2\alpha$$

are the coordinates of e_i in the local frame of D; see Fig. 11.17. Via elementary trigonometrical identities, these inequalities are shaped into

$$d_{\text{cell}} \cos\delta \geq L \max_{\alpha \in [0,\delta]} \max \left\{ \begin{array}{l} \sin(\delta+\alpha) + |\cos(\delta-\alpha)| \\ \sin(\delta-\alpha) + |\cos(\delta+\alpha)| \end{array} \right\}.$$

This is true by (6.7) since $\max_\alpha \dots$ does not exceed $\sin 2\delta + 1$.

D_1, D_2. We focus on D_1, whereas D_2 is considered likewise. The pivot of D_1 has the coordinates

$$x = -d_{\text{cell}} \cos \alpha, \quad y = -d_{\text{cell}} \sin \alpha;$$

the unit vector normal to D_1 is $(\cos 2\alpha, \sin 2\alpha)$. So the line $\mathcal{L}_1$ spanned by D_1 is given by $x \cos 2\alpha + y \sin 2\alpha = -d_{\text{cell}} \cos \alpha$. The δ-extended δ-hat from Fig. 11.17 is disjoint with D_1 if the distance from the vertex V of the δ-hat from Fig. 11.17 to $\mathcal{L}_1$ exceeds the radius $l_{\circlearrowright}$ of the left shadowed disk segment. With regard to the coordinates of V given in Fig. 11.17, this means that for all $\alpha \in [-\delta, \delta]$,

$$-L\tfrac{\sin 2\alpha}{\sin 2\delta} \cos 2\alpha + L\tfrac{\cos 2\alpha - \cos 2\delta}{\sin 2\delta} \sin 2\alpha > -d_{\text{cell}} \cos \alpha + 2L\tfrac{\sin(\delta-\alpha)}{\sin 2\delta}$$
$$\Updownarrow$$
$$\tfrac{d_{\text{cell}}}{L} > \Theta(\alpha) := \tfrac{2}{\sin 2\delta} \left[\cos 2\delta \sin \alpha + \sin \delta - \cos \delta \tan \alpha\right].$$

By elementary calculus, $\Theta'(\alpha) \le 0$ and so $\max_{\alpha \in [-\delta,\delta]} \Theta(\alpha) = \Theta(-\delta)$, which shapes the condition into

$$\frac{d_{\text{cell}}}{L} > \frac{2}{\sin 2\delta}\left[-\cos 2\delta \sin \delta + 2 \sin \delta\right] = \frac{3 - 2\cos^2 \delta}{\cos \delta}.$$

The difference between the expression on the right and the right-hand side of the first inequality from (6.7) equals

$$2 \sin \delta + \frac{1}{\cos \delta} - \frac{3 - 2\cos^2 \delta}{\cos \delta} = \frac{\sin 2\delta + \cos 2\delta - 1}{\cos \delta} =$$
$$\sqrt{2}\frac{\sin(2\delta + 45^\circ) - \sin 45^\circ}{\cos \delta} > 0 \quad \text{since } 0 < \delta < 26.26^\circ.$$

Thus D_1 is disjoint with the δ-extended δ-hat if (6.7) holds.

D_3, D_4. We focus on D_3, whereas D_4 is considered likewise. Since D_3 is always parallel to D_4, it suffices to show that the relative ordinate $y_p = d_{\text{cell}}(\cos \alpha - \sin \alpha)$ of its pivot exceeds those of A and B in Fig. 11.17:

$$d_{\text{cell}}(\cos \alpha - \sin \alpha) > L\frac{\cos \alpha}{\cos \delta} \max\{\sin(\delta - \alpha); \sin(\delta + \alpha)\}$$

for all $\alpha \in [-\delta, \delta]$. Only $\alpha \in [0, \delta]$ can be examined since for them, $\alpha := -\alpha$ keeps the inequality true. Then it shapes into

$$d_{\text{cell}}(\cos \alpha - \sin \alpha) > L\frac{\cos \alpha}{\cos \delta} \sin(\delta + \alpha).$$

Since the left- and right-hand sides descend and ascend, respectively, as $\alpha \in [0, \delta]$ grows, the condition is equivalent to

$$\frac{d_{\text{cell}}}{L} > \frac{\sin 2\delta}{\cos \delta - \sin \delta} = 2 \sin \delta + \frac{1}{\cos \delta} - \frac{\cos \delta - \sin \delta(1 + \sin 2\delta)}{\cos \delta(\cos \delta - \sin \delta)}.$$

The numerator of the last ratio decays on the interval $0 < \delta < 26.26^\circ$; its value at the right end $> 0.1 > 0$ by and elementary estimation. So this ratio is positive. Thus the inequality holds by (6.7) and D_3 is disjoint with the δ-extended δ-hat of D. $\square$

11.7 Simulations

For the velocity controlled mobile robot, the control was updated at a period of 0.1 s, and 40 rays in Fig. 11.7(a) were used to detect obstacles. Their edges were associated with the difference ≥ 2 m in two sequentially detected distances. Following a given azimuth was modeled as reaching a faraway target point. The function $\Delta_i(\cdot)$ in (3.2) was common for all obstacles and specified as a continuous piecewise linear function, whose fractures are as follows:

d (m)	0	0.5	1.0	1.5	2.0	2.5	3.0	100.0
Δ (rad)	1.52	1.27	1.21	0.43	0.2	0.02	0.01	0.003

They result from optimization by a genetic algorithm (MATLAB 2013a, Optimization Toolbox v6.3), minimizing the time of target reaching while respecting the safety margin 1 m to the obstacles in the scenario from Fig. 11.25.

In the simulation from Fig. 11.25, the robot efficiently converges to the target in a complex scene with many translating and rotating obstacles. In the simulation from Fig. 11.26, the mission was troubled by robot's dynamics and disturbances. Unicycle dynamics were used with the turning rate limit of 1 rad/s. A random Gaussian disturbance with the standard deviation of 0.1 rad/s was added to the control input. These entailed no problems for the method, displaying a promising potential to cope with noise and dynamic constraints.

In the scenario from Fig. 11.25, the proposed guidance law (PGL) was compared with the popular Velocity Obstacle Approach (VOA). The latter was given advantage over PGL by access to the obstacle's full velocity and knowledge of its future movement within the next 10 s. VOA also took into account the desired offset from the obstacle and a weighted cost to compromise progression to the target and separation

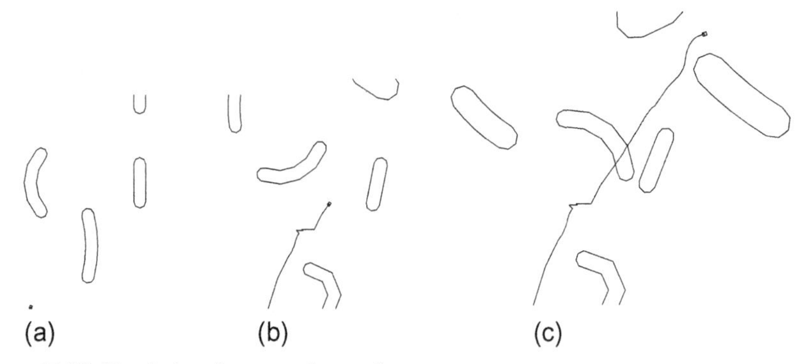

Figure 11.25 Simulations in a complex environment.

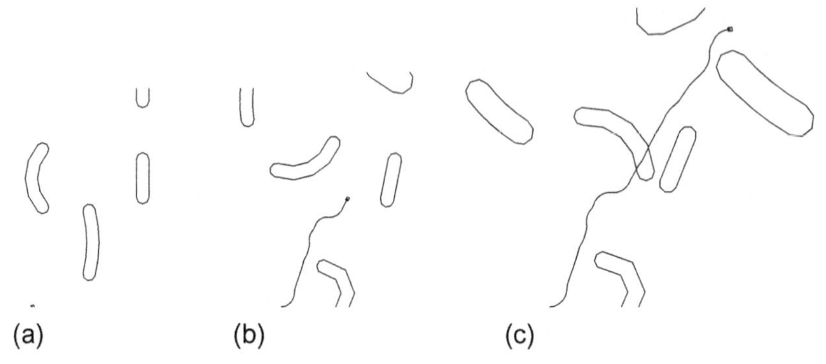

Figure 11.26 Simulation with vehicle dynamics and noise.

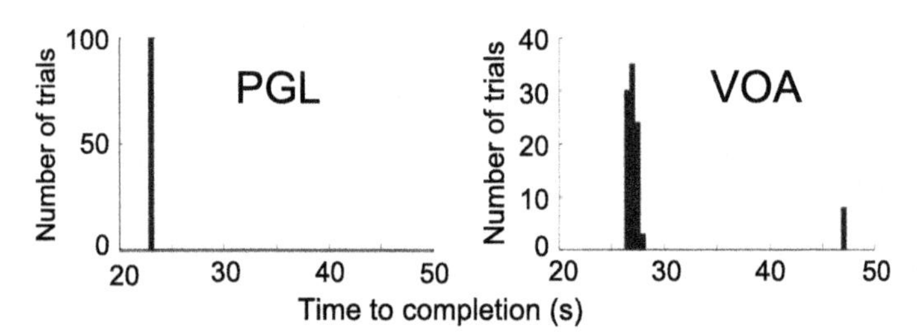

Figure 11.27 Comparison between the PGL and VOA.

from obstacles. These parameters were optimized, like for PGL. A Gaussian noise with the standard deviation of 0.1 rad/s was added to the heading direction. The performance measure was the time taken to reach the target. In Fig. 11.27, PGL on average outperforms VOA: VOA took a longer (occasionally, a much longer) path around the obstacles.

Safe cooperative navigation of multiple wheeled robots in unknown steady environments with obstacles

12

12.1 Introduction

As autonomous wheeled robots are used in greater concentrations, the probability of multiple vehicle encounters correspondingly increases. Techniques suitable for navigating a single robot through an obstacle-filled area would not always be successful in such situations. In this chapter, we present a solution to the problem of reactively controlling multiple autonomous wheeled robots in a cluttered environment based on model predictive control. Unlike other approaches in this area, we do not require explicit ordering of the robots, each robot can plan trajectories simultaneously and for the most part, independently of the others, and communication exchange only occurs once for each update of the control.

This type of problem is classed as a networked control problem, which is an active field of research. For examples of more generalized work in this area, we refer the reader to [309, 310].

Control laws have been proposed in [322, 354] to avoid collisions between two or more robots traveling at constant speeds. For acceleration-constrained holonomic mobile robots, it can be shown that collision avoidance is feasible for up to three agents by using a potential field type approach [335]. Some approaches ensure collision avoidance for unicycle robots with speed actuation, but this requires global information on the other robots [346]. Collision avoidance can be achieved for an arbitrarily large group of agents if the control inputs are allowed to be unbounded [327]. It can also be shown that an arbitrarily large group of agents can avoid one another if the robot's velocity is the control input [339]. However, many of the proposed solutions were not supported by rigorous justification of the collision-avoidance capability, which could possibly lead to failure in special and a priori unknown circumstances [280]. A decentralized collision-free navigation strategy for target capturing by a team of non-holonomic Dubins-car-like vehicles is proposed and rigorously justified in [436].

Off-line path planning can be used to find the optimal trajectories for a set of robots, and these trajectories can then be tracked to avoid collision [190]. Alternatively, if the paths to be followed are given, an appropriate velocity profile can be found to avoid collision [320]. This problem can be treated in the vein of resource allocation in

Safe Robot Navigation Among Moving and Steady Obstacles. http://dx.doi.org/10.1016/B978-0-12-803730-0.00012-3

discrete event systems and methods borrowed from this area may be employed to solve it [350, 437]; an example with the use of improved movement patterns can be found in [438]. The main drawback of these approaches is that they are based on segmentation of the environment into a discrete graph, which assumes a known and structured workspace. Furthermore, they typically give rise to control architectures that are essentially equivalent to centralized decision making, with no clear and easy way to achieve decentralization. On the positive side, these approaches are able to ensure avoidance of deadlock situations. This task is particularly difficult in general when the mobile robots operate in ad-hoc environments under truly decentralized control laws [356, 370].

Robust model predictive control (MPC) is increasingly being used in the context of vehicle navigation as it is resistant to disturbance, and decentralized MPC can produce a near-optimal solution for a multi-agent system by using dual decomposition to find a set of trajectories for the platoon of mobile robots [360, 361]. While this is more efficient than centralized optimization, it requires many iterations of communication exchange between the robots per every control update in order to converge to a solution. Another approach based on multiplexed MPC has been used to control mobile robots in [356]. The robust control for each robot is computed by updating a finite trajectory for each robot sequentially. While multiplexed MPC is suited for real-time implementation, a possible disadvantage is that path planning cannot be carried out simultaneously in two adjacent robots. Approaches also have been proposed that permit single communication exchange per control update [371]. This is done by explicitly injecting coherence between the robots' paths into the control objective to prevent the robots from changing their planned trajectories significantly after transmitting them to the companions. Other approaches with similar structures have been proposed that have the added benefit of not requiring synchronization between the robots [370]. However, this comes at the expense of an increase in communication overhead.

In this chapter, we propose a MPC type method that requires only a single communication exchange per control update and addresses the issue of communication delay. The proposed approach does not employ imposing an artificial and auxiliary coherence objective and may be suited to real-time implementation, while retaining robustness properties. We do require that the sampling times of the robots are synchronized, but communication-based algorithms are available for the task of clock synchronization [5, 439, 440].[1]

The main results of this chapter were originally published in [145]. A preliminary version of some of these results was also reported in the conference paper [441].

The body of the chapter is organized as follows. In Section 12.2, the problem statement is explicitly defined and the model of the mobile robot is given. In Section 12.3, we present the structure of the navigation system. Section 12.4 offers simulated results with a perfect unicycle model, while in Section 12.5, experiments with real-world wheeled robots are discussed.

[1] In the experiments described in this chapter, the robots were synchronized by a base station.

12.2 Problem statement

We consider N_h mobile robots traveling in a plane, each of which is associated with a steady point target $\mathcal{T}_i, i \in [1 : N_h]$. The plane contains a set of unknown, untraversable, static, and closed obstacles $D_j \not\ni \mathcal{T}_i, j \in [1 : n]$; see Fig. 12.1. The objective is to design a navigation law that drives every robot toward the assigned target through the obstacle-free part of the plane $\mathbb{R}^2 \setminus D, D := D_1 \cup \cdots \cup D_n$. Moreover, the distance from the robot to every obstacle and other robots should constantly exceed the given safety margin d_{safe}.

We use the discrete-time point-mass unicycle-like model of wheeled robots discussed in Section 3.2. For simplicity, the time step is normalized to unity.

Specifically, for $k < t \leq (k+1)$ and any robot:

$$\boldsymbol{r}(t) = \boldsymbol{r}(k) + \int_k^t \begin{bmatrix} \cos(\theta(t')) \\ \sin(\theta(t')) \end{bmatrix} \cdot v(t')\, dt',$$
$$v(t) = \text{sat}_0^{\overline{v}}\left[v(k) + u_v(k) \cdot (t-k)\right], \quad |u_v(k)| \leq \overline{u}_v,$$
$$\theta(t) = \theta(k) + u_\theta(k) \cdot (t-k), \quad |u_\theta(k)| \leq \overline{u}_\theta. \tag{2.1}$$

Here k is the time index, $\boldsymbol{r} = \mathbf{col}\,(x, y)$ is the vector of the robot's Cartesian coordinates in the world frame, v is its speed, θ is the orientation angle, and $\overline{v}$ is the maximal achievable speed. The symbol $\text{sat}_0^{\overline{v}}(\cdot)$ stands for saturation at the lower 0 and upper $\overline{v}$ thresholds, respectively, i.e.,

$$\text{sat}_a^b(x) := \min\{\max\{x, a\}, b\},$$

where $\max\{a, b\}$ represents the greater of a or b; similarly for $\min\{a, b\}$. Saturation of the speed v at 0 excludes meaningless negative speed values, whereas saturation at the upper threshold $\overline{v}$ models a physical limitation on the robot: its inability to achieve arbitrarily high speeds due to friction, limited engine power, etc.

At time k, every robot has knowledge of its speed $v(k)$, orientation $\theta(k)$, and position $\boldsymbol{r}(k)$. To accomplish its mission, the robot also has access to the position of its target. The sensor system also scans the region within the sensor range L by rotating

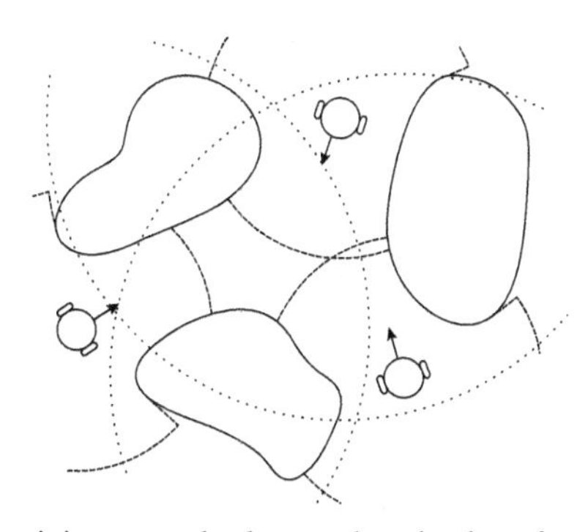

Figure 12.1 Scenario containing several robots and static obstacles, together with the obstacle-detection radius and a larger communication radius.

the ray-of-sight and giving the distance ρ to the nearest obstacle in the current sight direction (if $\rho \leq L$). We assume this rotation is fast enough as compared with the robot motion and neglect effects caused by conventional time sampling of observations. Modulo this, the robot has access to the visible facet of the obstacle part lying within the sensor range.

Furthermore, the robots have capability of communication with companions: every robot is able to broadcast to other robots within a given radius C. Any communication delay is assumed to be less than the time step, so that any data transmitted at time step k from one robot is available at time $k+1$ at all robots within the communication radius.

The robots need not be identical, i.e., not only the variables but also the parameters in (2.1) and the sensing radius L depend on the robot index i. However, we do assume that the communication radius C is the same for all robots. (In other words, the communication graph is indirected.) Furthermore, we assume that each robot has knowledge of the fixed parameters of all other robots inside the communication radius, by broadcast or otherwise.

Remark 2.1. There is a limit to the number of agents that lie within the communication range of a particular robot, due to the constraint of the minimum distance between the robots. This entails a bound on the maximum communication burden on each robot caused by scaling the number of robots.

Finally, we tacitly assume the robots are subjected to external disturbances. Therefore, (2.1) is interpreted as a *nominal* model. The *nominal trajectories* generated by this model may deviate from the actual ones. We also assume the disturbance can be modeled by bounded addends to two control signals, and also by a bounded side-slip angle between the robot's heading and the velocity vector. This causes bounded translational, rotational, and speed alterations of the expected state of the nominal model of the mobile robot. To highlight the major points by dropping secondary and standard technical details, we assume perfect measurement of the robot's state. At the same time, bounded state estimation/measurement errors can be taken into account in a similar way by, e.g., interpreting (2.1) as a nominal model for evolution of the state estimation/measurement, which point of view transforms in effect the estimation error into disturbance.

12.3 Proposed navigation system

12.3.1 Architecture of navigation system

Navigation is based on generating probational trajectories over the relatively short planning horizon $[k, (k + \tau)]$ at every time step k. Probational trajectory is nominal and is specified by a finite sequence of way points, which satisfy the constraints of the nominal model (2.1). Each way-point is attributed to a particular time step and has an associated position, speed, heading, and control input. Therefore, the entire nominal trajectory between the way-points can easily be computed. Probational trajectories necessarily halt at the terminal planning time. When necessary, we interpret them as being prolonged by the "stay still" maneuver.

Our proposed navigation system consists of a high level trajectory planning module (TPM) and a lower level trajectory tracking module (TTM). TPM updates *probational* trajectories basically at every time step k; TTM updates the control signals at a more frequent rate. The goal of TTM is to drive the robot along the current probational trajectory in the face of disturbances. TTM is assembled of the path tracking module (PTM) and longitudinal tracking module (LTM), responsible to compensate for the deviations in space and time, respectively.

TPM first attempts to select the probational trajectory from the set of *planned* (nominal) trajectories. They are generated by TPM at every time step k and start at the current (actual) state of the robot. However, selection may give a void result. Then the probational trajectory from the previous time step remains in use. (Formally, the current probational trajectory is set to be the previous one shifted by the time step.) In this case, the robot continues to follow a previously adopted trajectory. To distinguish between these two types of probational trajectories, they are said to be *updated* and *followed*, respectively.

Remark 3.1. As a result, the mobile robot evolves over consecutive time intervals during each of which the probational trajectory is not updated. For every of them, the robot is in fact driven by TTM from the initial state identical to that of the tracked probational trajectory.

At any sampling time, every robot broadcasts its current state (position, orientation, and velocity) and probational trajectory, which data arrive at the robots in the communication range one time step later. To avoid collisions with the companions, the robot also attempts to reconstruct their current probational trajectories, starting with reconstruction of the planned ones. However, the latter start at the current states of the companions, which are unknown to the robot at hand. So the robot first estimates these states on the basis of the nominal model (2.1) and just received states at the previous time step. (The estimate may differ from the actual state because of the disturbances.) After this the robot copycats generation of the planned trajectories for every robot in the communication range, with substituting the estimate in place of the true state in doing so. This gives rise to the set of *presumable planned trajectories*.

The upper index * is used to mark variables associated with the planned trajectories; this index is discarded to emphasize that the concerned trajectory is in fact probational; the hat $\hat{\{\}}^*$ is added to signal that a presumable planned trajectory is concerned. These variables depend on two arguments $(j|k)$, where k is the time instant when the trajectory is generated and $j \geq 0$ is the number of time steps into the future: the related value concerns the state at time $k+j$. Whenever a certain robot is considered, the lower index $\{\}_i$ refers to other robots within the communication range C, whereas its absence indicates that we are referring to the robot at hand.

12.3.2 Enhanced safety margins adopted by TPM

To ensure collision avoidance, TPM respects more conservative safety margins than d_{safe}. They take into account deviations from the probational trajectory caused by disturbances and are based on estimation of these mismatches. The first such estimate addresses the performance of TTM.

(p.1) $\Delta_{\text{track}}\boldsymbol{r}$—an upper bound on the translational deviation over the planning horizon between the probational trajectory and the real motion of the robot driven by the TPM along this trajectory in the face of disturbances.

Computation of $\Delta_{\text{track}}\boldsymbol{r}$ takes into account the particular design of TPM and is discussed in Section 12.3.4. To decouple designs of TPM and TTM, the control capacity is a priori distributed between TPM and TTM. Specifically, TPM must generate the controls only within the reduced bounds determined by

$$u_{v,nom} := \mu_{u_v} \cdot \overline{u}_v, \quad \dot{\theta}_{nom} := \mu_{\dot{\theta}} \cdot \overline{u}_\theta \tag{3.1}$$

and so that the resultant speed does not exceed the reduced threshold

$$v_{nom} := \mu_v \cdot \overline{v}. \tag{3.2}$$

The remaining control capacity is allotted to TTM. Here $\mu_{u_v}, \mu_{\dot{\theta}}, \mu_v \in (0,1)$ are tunable design parameters. Along with the system parameters, they uniquely determine the estimate $\Delta_{\text{track}}\boldsymbol{r}$ (for the adopted design of TPM), as will be shown in Section 12.3.4. In particular, $\Delta_{\text{track}}\boldsymbol{r}$ does not depend on the choice of the probational trajectory.

To simplify notations, we assume that $\Delta_{\text{track}}\boldsymbol{r}$ is the same for all robots; however the control scheme is still feasible if this quantity varies between them.

The following more conservative safety margins account for not only disturbances but also the use of time sampling in measurements of relative distances:

$$d_{\text{tar}} := d_{\text{safe}} + \frac{\overline{v}}{2} + \Delta_{\text{track}}\boldsymbol{r}, \quad d_{\text{mut}} := d_{\text{safe}} + 2\left[\frac{\overline{v}}{2} + \Delta_{\text{track}}\boldsymbol{r}\right]. \tag{3.3}$$

To illustrate their role, we introduce the following.

Definition 3.1. An ensemble of probational trajectories generated by all robots at a given time is said to be *mutually feasible* if the following claims hold:

(i) The distance from any way-point to any static obstacle exceeds d_{tar} for any trajectory;
(ii) For any two trajectories, the distance between any two matching way-points exceeds d_{mut}.

The motivation behind this definition is illuminated by the following.

Lemma 3.1. *Let an ensemble of probational trajectories adopted for use at time step $k-1$ be mutually feasible. Then despite of the external disturbances, the robots do not collide with the static obstacles and one another on the time interval $[k-1,k]$ and, moreover, respect the required safety margin d_{safe}.*

Proof. For any $t \in [k-1,k]$, Remark 3.1 implies that the robot deviates from the related position on the probational trajectory by no more than $\Delta_{\text{track}}\boldsymbol{r}$. Consider the end k_* of the interval$[k-1,k]$ nearest to t; the times t and k_* are separated by no more than half time-step. So thanks to the upper bound on the speed from (2.1), the nominal positions at times t and k_* differ by no more than $\frac{\overline{v}}{2}$. Overall, the distance between the way-point on the probational trajectory that is related to time step k_* and the actual position at time t does not exceed $\frac{\overline{v}}{2} + \Delta_{\text{track}}\boldsymbol{r}$. Since the distance from this way-point to any static obstacle exceeds d_{tar}, the distance from the real position at time t is no less than $d_{\text{tar}} - \frac{\overline{v}}{2} - \Delta_{\text{track}}\boldsymbol{r} = d_{\text{safe}}$. For any couple of robots, the distance between the concerned way-points exceeds d_{mut}. So the distance between the real positions at time t is no less than $d_{\text{mut}} - 2[\frac{\overline{v}}{2} + \Delta_{\text{track}}\boldsymbol{r}] = d_{\text{safe}}$. □

Remark 3.2. For arbitrary (not unit) sampling time Δt, the addend $\frac{\overline{v}}{2}$ in (3.3) should be replaced by $\frac{\overline{v}}{2}\Delta t$. This addend asymptotically vanishes as $\Delta t \to 0$; so typically does $\Delta_{\text{track}}\boldsymbol{r}$ provided that the discrete planning horizon τ is upper bounded. Thus the conservatism imposed by the extra addends in (3.3) may be attenuated by reducing the time step employed during the planning process. As the time-step is conversely increased, $\Delta_{\text{track}}\boldsymbol{r}$ grows but remains bounded even if $\Delta t \to \infty$, as will be discussed in Section 12.3.4. However the other addend $\frac{\overline{v}}{2}\Delta t$ grows without limits as $\Delta t \to \infty$.

Overly enhanced margins (3.3) may make the control objective unachievable; for example, the corridors amidst the obstacles may be too narrow to accommodate a distance of no less than d_{tar} to the both sides of the corridor, whereas the target locations and the initial positions of the robots may fail to satisfy the enhanced inter-robot margin d_{mut}. In fact, feasibility of the enhanced margins is the basic requirement regulating the practical choice of Δt. In the case of heavy uncertainty about the scene, a reasonable option is to pick Δt so that the quantization error $\frac{\overline{v}}{2}\Delta t$ is comparable with the limit bound on the disturbance-induced error $\Delta_{\text{track}}\boldsymbol{r}$.

12.3.3 Trajectory planning module

This module uses an additional tunable parameter, $\Delta\Lambda \in (0, 1)$. This determines the sampling period, or "mesh size" employed for generation of trajectories.

TPM of every robot iteratively executes the following steps:

S.1 Generation of a finite set $\mathcal{P}$ of planned trajectories, each starting at the current robot state $\boldsymbol{r}(k), v(k), \theta(k)$; see Section 12.3.3.1 for details;

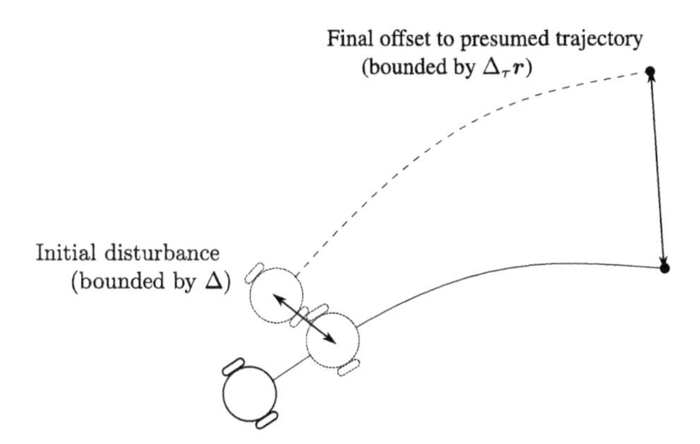

Figure 12.2 The disturbance over a single time-step determines a larger bound for the deviation to the re-planned trajectory.

S.2 Refinement of $\mathcal{P}$ to only mutually feasible trajectories; see Section 12.3.3.2 for details;
S.3 If $\mathcal{P}$ is empty, the probational trajectory is inherited from the previous time step (modulo a proper time shift);
S.4 Selection of the probational trajectory from $\mathcal{P}$ if $\mathcal{P}$ is not empty; see Section 12.3.3.3 for details;
S.5 Application of the first control $[u_v^*(0|k), u_\theta^*(0|k)]$ related to the selected probational trajectory to the robot;
S.6 Transmission of the chosen probational trajectory and the current state $\boldsymbol{r}(k)$, $v(k)$, $\theta(k)$ to all robots within the communication radius C;
S.7 $k := k+1$ and go to S.12.3.3.

This control process is executed co-currently in each robot. Step S.2 is solely based on the currently observed part of the environment and data received from and sent to other robots in communication range since time step $k-1$. This step involves construction of the set of presumable planned trajectories for all robots in the communication range.

12.3.3.1 Generation of planned trajectories (Step S.1)

Trajectory generation could potentially use many forms. However we use a simplified version partly to improve tractability and partly since this version displayed a good performance in the case of a single mobile robot [264]. Specifically, the generated set of trajectories $\mathcal{P}$ consists of two parts $\mathcal{P}_+$ (*cruising* trajectories) and $\mathcal{P}_-$ (*slowing* trajectories), with each part being composed of trajectories matching a given pattern of speed evolution over the planning horizon:

- The pattern $p_+ = (+ - - \cdots)$ means that the speed $v^*(0|k) := v(k)$ is first increased:

$$v^*(1|k) = \min\left\{v^*(0|k) + \Delta\overline{v}; v_{nom}\right\} \tag{3.4}$$

  by adding the constant $\Delta\overline{v} := u_{v,nom} \times (1 \text{ time unit}) > 0$, where $u_{v,nom}$ is taken from (3.1), (with subduing the result to the reduced speed upper threshold from (3.2) if necessary). The planned speed is then constantly decreased by subtracting $\Delta\overline{v}$ at all subsequent time steps (with not allowing the speed to take the meaningless negative value): for $j \geq 1$,

$$v^*(j+1|k) = \min\left\{\max\left\{v^*(j|k) - \Delta\overline{v}; 0\right\}; v_{nom}\right\}; \tag{3.5}$$

- The pattern $p_- = (- - - \cdots)$ means the speed is constantly decreased in accordance with (3.5).

The planning horizon τ is selected from the requirement that when following any generated trajectory, the robot halts within this horizon:

$$\tau := \begin{cases} \tau_- = \left\lceil \dfrac{v(k)}{u_{v,nom}} \right\rceil & \text{for slowing trajectories,} \\ \tau_+ = \left\lceil \dfrac{v(k)}{u_{v,nom}} \right\rceil + 2 & \text{for cruising trajectories,} \end{cases} \tag{3.6}$$

where $\lceil a \rceil$ is the integer ceiling of the real number a. This horizon is determined by the current robot speed $v(k)$ and may vary over time. To ensure obstacle avoidance, we limit attention to trajectories for which the robot halts within the planning time horizon and observed part of the environment.

Remark 3.3. The length of any planned path does not exceed $l = \overline{v} + \frac{\overline{v}^2}{2 \cdot u_{v,nom}}$. In order that the trajectory not only remains in the observed part of the environment but also its possible collisions with the static obstacles are detectable, we require that $(l + d_{\text{tar}})$ is less than the sensor range L. We also assume that the communication range C is large enough

$$C > 2\overline{v} + 2\Delta_{\text{track}}\boldsymbol{r} + 2l + d_{\text{mut}}.$$

The trajectories are constrained to invariably consist of a sharp turn, followed by a reduced turn, followed by a straight section. This concept is similar to the dynamic window approach and its modern adaptations, and provide s good performance in many situations [257]. The turns should be planned so that the resultant path be trackable in the face of external disturbances. This means that the curvature of the path must fall below the level accessible by the robot. However, this level depends on the actual speed v, which is influenced by unknown disturbances. In the face of this uncertainty, we require that the curvature of the planned path must fit the worst case scenario. In other words, it should not exceed

$$\varkappa_{nom} := \frac{\overline{u}_\theta}{\overline{v}}. \tag{3.7}$$

In fact, we require a bit more: it should not exceed $\mu_\varkappa \cdot \varkappa_{nom}$. Here $\mu_\varkappa \in (0, 1)$ is a design parameter similar to those from (3.1) and (3.2) and introduced for the same reason as they. The turning pattern of the planned trajectory can thus be described by a unique scalar Λ:

$$u_\theta^*(j|k) = \min\left\{\dot{\theta}_{nom}, \mu_\varkappa \varkappa_{nom} \min\left\{v^*(j|k), v^*(j+1|k)\right\}\right\} \operatorname{sgn}(\Lambda) \cdot \operatorname{sat}_0^1(|\Lambda| - j). \tag{3.8}$$

(We recall that $\operatorname{sat}_0^1(x) := \min\{\max\{x, 0\}, 1\}$.) At every time step k, several turning patterns are generated. They are enumerated by an integer m and correspond to the following values of Λ:

$$\Lambda = m \cdot \Delta\Lambda, \quad m = 0, 1, \ldots, \left\lceil \frac{\tau}{\Delta\Lambda} \right\rceil. \tag{3.9}$$

For each m, the planned trajectory is thus uniquely determined by the initial state, nominal equations (2.1), and the above generation rule.

Generation of the planned trajectories is summarized in Fig. 12.3.

12.3.3.2 Refinement of the planned trajectories (Step S.2)

Refinement is prefaced by the following computations, which are carried out at the robot at hand for every robot i that was within the communication range at the previous time step $k - 1$:

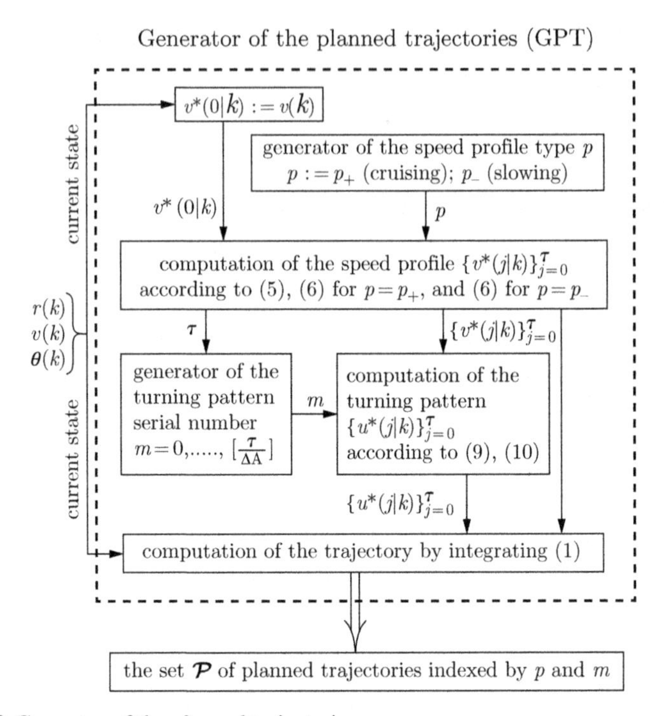

Figure 12.3 Generator of the planned trajectories.

- The nominal motion equations (2.1) are integrated for robot i from $k-1$ to k with the initial data $\boldsymbol{r}_i(k-1), \theta_i(k-1), v_i(k-1)$ and controls $[u^*_{v,i}(0|k-1), u^*_{\theta,i}(0|k-1)]$ to acquire the estimated current state $\hat{\boldsymbol{r}}_i(k), \hat{\theta}_i(k), \hat{v}_i(k)$[2];
- The procedure from Section 12.3.3.1 is carried out for the ith robot, with the true current state $\boldsymbol{r}_i(k), \theta_i(k), v_i(k)$ replaced by the estimated one $\hat{\boldsymbol{r}}_i(k), \hat{\theta}_i(k), \hat{v}_i(k)$.

This procedure is summarized in Fig. 12.4 and results in the set $\hat{\mathscr{P}}_i$ of presumable planned trajectories of the ith robot.

To proceed, we introduce the following estimates:

(p.1) Δ—an upper bound on the deviation over the single time step between the nominal and real full states of the robot driven by the TTM along the probational trajectory in the face of disturbances;

(p.2) $\Delta_\tau \boldsymbol{r}$— an upper bound on the translational deviation between two nominal trajectories over the entire planning horizon of duration τ, provided that the difference between their initial states does not exceed Δ and the both follow common speed and turning patterns selected by the trajectory generation rule from Section 12.3.3.1. This is illustrated by Fig.12.2.

[2] We recall that these data and controls were broad-casted by the ith robot at time $k-1$. So they are received by the robot at hand at time k.

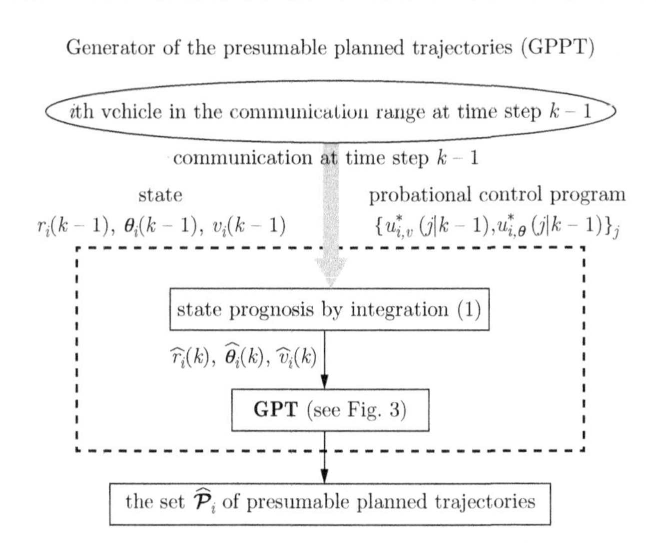

Figure 12.4 Generator of the presumable planned trajectories.

The constant Δ is similar to $\Delta_{\text{track}}\boldsymbol{r}$; its computation is discussed in Section 12.3.4. Given Δ, the bound $\Delta_\tau \boldsymbol{r}$ can be easily computed based on (2.1). More or less conservative bounds can be used, up to the tight bounds based on computer-aided calculation of the reachable sets. However our experimental results showed that even conservative estimates are able to entail good performance. Partly in view of this and partly due to the fact that computation of conservative bounds is elementary, we do not come into the related details.

We note that Remark 3.2 is evidently extended on Δ and $\Delta_\tau \boldsymbol{r}$. The following remark is immediate from the foregoing and partly elucidates the role of $\Delta_\tau s$.

Remark 3.4. The mismatch $\max_{j=0,\dots,\tau} \|\boldsymbol{r}_i(j|k) - \hat{\boldsymbol{r}}_i(j|k)\|$ between the presumable planned trajectory and the true planned trajectory does not exceed $\Delta_\tau \boldsymbol{r}$.

Here and throughout, $\|\cdot\|$ is the standard Euclidean norm.

In order to ensure that the probational trajectories are mutually feasible, the set of planned trajectories is subjected to a series of refinements. First, the robot at hand refines the generated set of presumable planned trajectories $\{\hat{\boldsymbol{r}}_i^*(j|k), \dots\}_j$ of any other robot i that was in the communication range at the previous time step $k-1$. Specifically, it discounts any presumable trajectory that interferes with the true probational trajectory of the robot at hand $\{\boldsymbol{r}(j|k-1), \dots\}_j$ selected at the previous time step $k-1$, i.e., such that

$$\|\hat{\boldsymbol{r}}_i^*(j|k) - \boldsymbol{r}(j+1|k-1)\| \leq d_{\text{mut}} - \Delta_\tau \boldsymbol{r} \quad \text{for some} \quad j. \tag{3.10}$$

This operation is summarized in Fig. 12.5 and gives rise to the set of the refined presumable planned trajectories $\tilde{\mathscr{P}}_i$ of the ith robot.

After this, the robot at hand refines its own set of planned trajectories by successively discounting those interfering with:

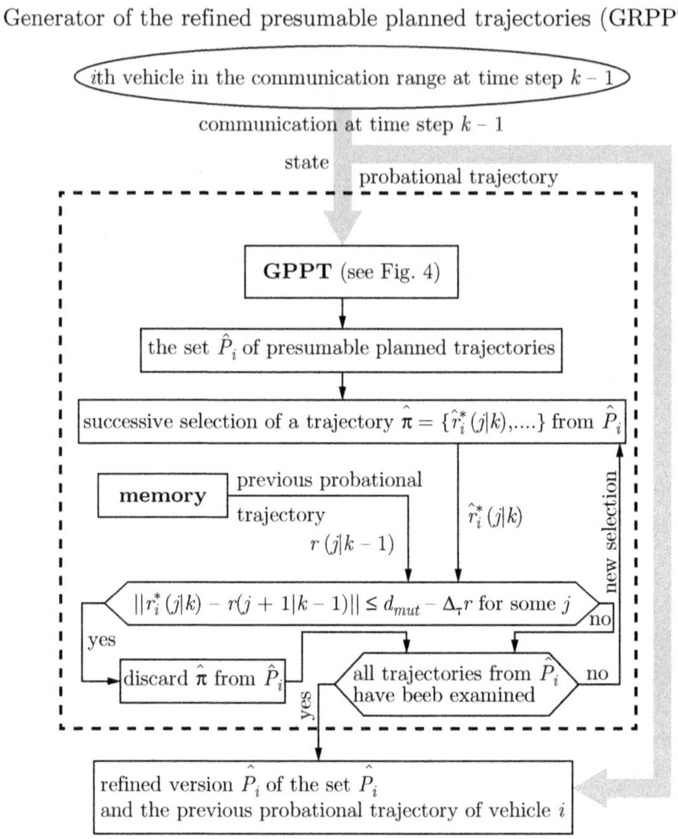

Figure 12.5 Generator of the presumable planned trajectories.

(r.1) Static obstacles, such that

$$\|\boldsymbol{r}^*(j|k) - x\| < d_{\text{tar}} \quad \text{for some} \quad x \in D \quad \text{and} \quad j \tag{3.11}$$

(r.2) Probational trajectories of other robots received from them within the time interval $[k-1, k]$, such that

$$\|\boldsymbol{r}^*(j|k) - \boldsymbol{r}_i(j+1|k-1)\| < d_{\text{mut}} \quad \text{for some } j \tag{3.12}$$

(r.3) Any refined presumable planned trajectory of any other robot i that was in the communication range of the robot at hand at the previous time step $k-1$:

$$\left\|\boldsymbol{r}^*(j|k) - \hat{\boldsymbol{r}}_i^*(j|k)\right\| < d_{\text{mut}} + \Delta_\tau \boldsymbol{r} \text{ for some } j \text{ and } \{\hat{\boldsymbol{r}}_i^*(j|k), \ldots\} \in \tilde{\mathscr{P}}_i. \tag{3.13}$$

As stated previously, if this refinement sweeps away all trajectories, we reuse the probational trajectory from the previous time step.

Lemma 3.2. *The proposed refinement procedure does ensure that the probational trajectories generated at time step k are mutually feasible provided that the probational trajectories were mutually feasible at the previous time step $k-1$.*

Proof. Collisions with the static obstacles are excluded (more precisely, (i) in Definition 3.1 is ensured) by (r.1). As for the robot to robot collisions, we consider two robots i and j and examine separately the following cases:

- *At the previous time step, the robots were not in the communication range of each other.* For the unit time, each of them covers a distance $d \leq \overline{v}$ due to (2.1). So their current positions are separated by no less than $C - 2\overline{v}$. Each of them differs from the matching "probational" position by no more than $\Delta_{\text{track}}\boldsymbol{r}$ thanks to (p.12.3.2) in Section 12.3.2. Thus the latter positions are separated by no less than $C - 2\overline{v} - 2\Delta_{\text{track}}\boldsymbol{r}$. The lengths of the both probational trajectories departing from these states do not exceed l. So these trajectories are separated by no less than $C - 2\overline{v} - 2\Delta_{\text{track}}\boldsymbol{r} - 2l > d_{\text{mut}}$, where the last inequality follows from Remark 3.3.
- *At the previous time step, the robots communicated with each other.*
  - *Both robots engage followed probational trajectories.* These trajectories were in a mutually feasible ensemble at the previous time step, thus they will not interfere.
  - *One of the robots (say i) engages the followed trajectory, whereas the other uses an updated one.* Since the other robot has knowledge of the first trajectory due to communication, its updated probational trajectory does not collide with that of the ith robot thanks to (r.2).
  - *Both robots employ updated trajectories.* Let us interpret robots j and i as the "robot at hand" and the "other robot", respectively; according to our conventions, this means dropping the index j everywhere. The probational trajectories were selected from the set of planned trajectories

$$\boldsymbol{r}(\cdot|k) = \boldsymbol{r}^*(\cdot|k), \quad \boldsymbol{r}_i(\cdot|k) = \boldsymbol{r}_i^*(\cdot|k).$$

The trajectory $\boldsymbol{r}_i^*(\cdot|k)$ has passed the test from (r.2) at robot i and so does not interfere with the trajectory transmitted from the robot at hand at time step $k-1$, i.e.,

$$\|\boldsymbol{r}_i^*(q|k) - \boldsymbol{r}(q+1|k-1)\| > d_{\text{mut}} \quad \forall q.$$

For the related presumable planned trajectory, we have

$$\|\hat{\boldsymbol{r}}_i^*(q|k) - \boldsymbol{r}(q+1|k-1)\| \geq \|\boldsymbol{r}_i^*(q|k) - \boldsymbol{r}(q+1|k-1)\| - \underbrace{\left\|\hat{\boldsymbol{r}}_i^*(q|k) - \boldsymbol{r}_i^*(q|k)\right\|}_{\leq \Delta_\tau \boldsymbol{r} \text{ by Remark 3.4}} \geq d_{\text{mut}} - \Delta_\tau \boldsymbol{r}. \tag{3.14}$$

So this trajectory was among those against which $\boldsymbol{r}^*(\cdot|k)$ was tested in correspondence with (3.13). Since $\boldsymbol{r}^*(\cdot|k)$ has survived the refinement procedure, this test was passed:

$$\left\|\boldsymbol{r}^*(q|k) - \hat{\boldsymbol{r}}_i{}^*(q|k)\right\| > d_{\text{mut}} + \Delta_\tau \boldsymbol{r} \quad \forall q.$$

The proof is completed by invoking the under-braced inequality from (3.14) :

$$\|\boldsymbol{r}(q|k) - \boldsymbol{r}_i(q|k)\| = \left\|\boldsymbol{r}^*(q|k) - \boldsymbol{r}_i^*(q|k)\right\| \geq \left\|\boldsymbol{r}^*(q|k) - \hat{\boldsymbol{r}}_i^*(q|k)\right\| - \left\|\boldsymbol{r}_i^*(q|k) - \hat{\boldsymbol{r}}_i{}^*(q|k)\right\| \geq d_{\text{mut}} + \Delta_\tau \boldsymbol{r} - \Delta_\tau \boldsymbol{r} = d_{\text{mut}}.$$

□

By retracing the arguments of the first part of the proof and invoking Remark 3.5, we arrive at the following.

Remark 3.5. We assume that initially there are no robots in communication range of each other and every robot is far enough from the static obstacles so that at least one planned trajectory survives (r.1). Then the generated set of probational trajectories is well-defined and mutually feasible, so it also follows that this set remains mutually feasible in subsequent time steps.

12.3.3.3 Selection of the probational trajectory

From the set of the planned trajectories surviving the refinement procedure, the final trajectory is selected to furnish the minimum of the cost functional:

$$J = \|\boldsymbol{r}^*(\tau|k) - \mathcal{T}\| - c \cdot v^*(1|k), \tag{3.15}$$

where $\mathcal{T}$ is the target for this robot and $c > 0$ is a given weighting coefficient, which may depend on the robot. This minimization aims to align the robot with the target while preferring trajectories with faster initial planned speeds.

12.3.3.4 Entire operation of the trajectory planning module

This operation is summarized by the pseudo-code Algorithm 1 and in Fig. 12.6.

An upper bound on the total computation load is proportional to:

$$O(N_v, N_t, N_p) = N_v \cdot N_t^2 \cdot \tau, \quad N_t = 4 \cdot \left\lceil \frac{\tau}{\Delta\Lambda} \right\rceil + 2, \tag{3.16}$$

where N_v is the number of the robots in the communication range and N_t is the number of planned trajectories. As for the communication load, only four reals are required to encode the state for communication, with extra two integers being needed to encode the index of the trajectory inside $\mathcal{P}$ and the progression along the trajectory.

12.3.4 Trajectory tracking module

This module is used to compensate for deviation from the chosen trajectory caused by external disturbances. To compensate for space errors, we adopt the sliding mode path tracking approach presented in [397] for pure steering control of the single four-wheeled vehicle traveling at the constant nominal longitudinal speed. Chosen for its provable bounds on tracking error in the face of a bounded wheels slip, this control law is adapted for trajectory tracking purposes in this paper. In particular, this requires to add a longitudinal tracking module (LTM) to the path tracking module (PTM) to maintain the temporal difference between the planned and actual positions. Other examples of control approaches suitable for this task include MPC path tracking and combined “sliding mode”–MPC, see, e.g., [442] and [176], respectively, for representative samples.

Input: The current state of the robot: $\boldsymbol{r}(k)$, $\theta(k)$, $v(k)$
The probational trajectory that is the previous output of the algorithm:
$\{\boldsymbol{r}^*(j|k-1), \ldots\}$
The trajectories received from the other robots: $\{\boldsymbol{r}_i^*(j|k-1), \ldots\}$
The currently sensed obstacle set
Output: Current probational trajectory: $\{\boldsymbol{r}^*(j|k), \ldots\}$
Generate the set of planned trajectories $\mathscr{P}$
Remove from $\mathscr{P}$ the trajectories that are closer than d_{tar} to the currently sensed obstacle set
foreach *Received trajectory* $\{\boldsymbol{r}_i^*(j|k-1), \ldots\}$ **do**
  Remove from $\mathscr{P}$ the trajectories that are closer than d_{mut} to $\boldsymbol{r}_i^*(j|k-1)$
  Generate the set of presumable planned trajectories $\hat{\mathscr{P}}_i$ based on $\boldsymbol{r}_i^*(1|k-1), \theta_i^*(1|k-1), v_i^*(1|k-1)$
  Remove from $\hat{\mathscr{P}}_i$ the trajectories that are closer than $d_{\text{mut}} - \Delta_\tau \boldsymbol{r}$ to $\boldsymbol{r}^*(j|k-1)$
  Remove from $\mathscr{P}$ the trajectories that are closer than $d_{\text{mut}} + \Delta_\tau s$ to some trajectory in the remaining set $\hat{\mathscr{P}}_i$
end
if $\mathscr{P} = \emptyset$ **then**
  Reuse the trajectory from the previous time-step
end
else
  Select a minimizer of (3.15) over $\mathscr{P}$
end

Algorithm 1: Operation of TPM.

12.3.4.1 Path tracking module

We consider three types of slip/disturbance—bias in the longitudinal acceleration, bias in the turning rate, and a side slip causing an angular difference between the robot orientation and the velocity direction.

Since PTM updates controls at much higher rate than TPM, now we employ a continuous-time model of the robot's motion. Furthermore, since disturbances are our main concern now, we explicitly model them, assuming that the maximal feasible rotation rate bias due to disturbance is proportional to the robot speed. To come into details, we introduce the following variables (see Fig. 12.7):

- η — the curvilinear abscissa of the point on the reference path that is closest to the robot;
- $\boldsymbol{\rho}(\eta) \in \mathbb{R}^2$ — a regular parametric representation of the reference path in the world frame;
- $\varkappa(\eta)$ — the signed curvature of the reference path at the point $\boldsymbol{\rho}(\eta)$;
- z — the distance from the robot to the reference path;
- φ — the angular difference between the robot orientation and the velocity direction caused by the wheels slip;
- $\hat{\theta}$ — the angle between the robot's orientation and the orientation of a virtual robot perfectly tracking the reference path, positioned on $\boldsymbol{\rho}(\eta)$;
- w_θ — the coefficient of the rotation rate bias due to disturbance;
- w_v — the longitudinal acceleration bias due to disturbance.

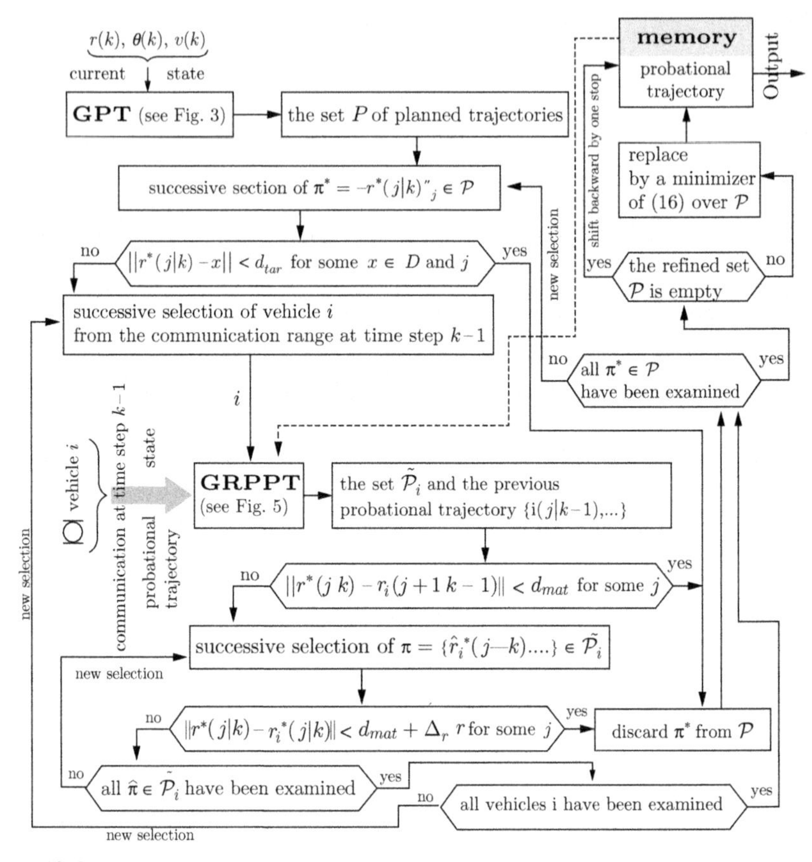

Figure 12.6 Trajectory planning module.

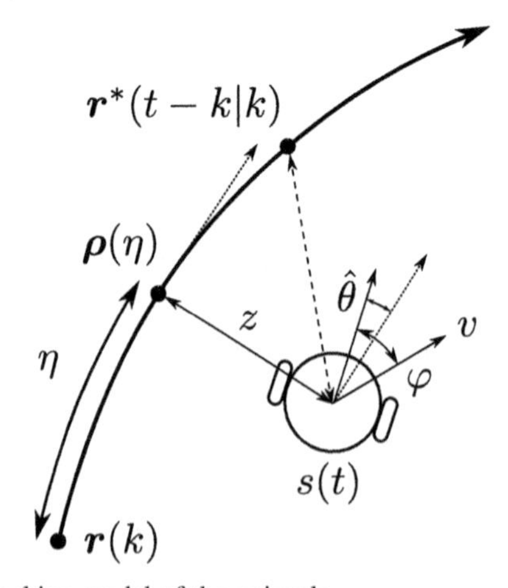

Figure 12.7 The path tracking model of the unicycle.

Assumption 3.1. *The error and slip parameters are unknown, bounded quantities:*

$$|\varphi| \leq \overline{\varphi}, \quad |w_\theta| \leq \overline{w}_\theta, \quad |w_v| \leq \overline{w}_v. \tag{3.17}$$

Lemma 3.3. *The trajectory tracking kinematic model is as follows:*

$$\dot{\eta} = \frac{v\cos(\hat{\theta} - \varphi)}{1 - \varkappa(\eta)z}, \quad \dot{z} = -v\sin(\hat{\theta} - \varphi), \tag{3.18}$$

$$\dot{v} = \begin{cases} u_v + w_v & \text{if } 0 < v < \overline{v} \text{ or } \left| \begin{array}{l} v = 0 \\ u_v + w_v > 0 \end{array} \right. \text{ or } \left| \begin{array}{l} v = \overline{v} \\ u_v + w_v < 0 \end{array} \right., \\ 0 & \text{otherwise}, \end{cases} \tag{3.19}$$

$$\dot{\hat{\theta}} = \varkappa(\eta)\frac{v\cos(\hat{\theta} - \varphi)}{1 - \varkappa(\eta)z} + u_\theta + v w_\theta. \tag{3.20}$$

The proof of this lemma is similar to, e.g., that of Lemma 3.1 from [119], where the bicycle error tracking model is presented. To adapt this model to the case at hand, it suffices to put $u_\theta + v w_\theta$ in place of $\frac{v}{L}[\tan(\delta + \beta) - \tan\varphi]$. This takes into account the differences between a unicycle and bicycle (i.e. that the angular velocity is controlled directly), as well as an angular disturbance (which is proportional to the velocity so that there is no disturbance when the vehicle is stationary).

To generate the steering control signal, we employ the law from [397], slightly adopted to the current context:

$$u_\theta(t) = \mathbf{sat}_{-\overline{u}_\theta}^{\overline{u}_\theta}\left[v\frac{\varkappa_v - p(|S|, z)}{\varkappa_v - \underline{\varkappa}(z)}\Xi + vp(|S|, z)\mathrm{sgn}S\right], \tag{3.21}$$

$$\text{where} \quad \Xi := \frac{\varkappa(\eta)\cos(\hat{\theta})}{1 - \varkappa(\eta)z} + \frac{d\chi}{dz}(z)\sin(\hat{\theta}), \quad S := \hat{\theta} - \chi(z).$$

Here the lower bound $\varkappa_{nom}$ on the reference path curvature is given by (3.7). The functions $\chi(z)$, $p(|S|, z)$ and $\underline{\varkappa}(z)$ are user selectable, subject to some restrictions described in [397].

Assumption 3.2. *There exists a choice of the coefficients μ's (with indices) from* (3.1), (3.2), *and* (3.8) *and functions $\chi(z)$, $p(|S|, z)$, and $\underline{\varkappa}(z)$, such that all assumptions from [397] hold.*

These assumptions employ some constants, which are as follows in the context of this paper $\overline{\varphi} := \varphi_{\text{est}} := \overline{\varphi}, \varkappa_v := \varkappa_{nom}, \overline{\varkappa} := \mu_\varkappa \varkappa_{nom}, \varkappa_{v,u} := \varkappa_{nom} - \overline{w}_\theta$.

Constructive sufficient conditions for Assumption 3.2 to hold, as well as details of the required functions are presented in [397]. We note that these conditions can be satisfied by a proper choice of the coefficients $\mu_{u_v}, \mu_a, \mu_{\dot{\theta}}, \mu_v$ in (3.1), (3.2) provided that minor and partly unavoidable requirements are met. According to [397], the resultant tracking errors $\overline{z}_{err} \geq |z|$ ($\overline{z}_{est}$ in notations of [397]) and $\overline{\theta}_{err} \geq |\hat{\theta}|$ are explicitly determined by these coefficients (for given parameters of the robots) and are not influenced by both the path selected by TPM and the speed profile $v(\cdot)$.[3] It is worth noting that typically $\overline{z}_{err}\mu_\varkappa \varkappa_{nom} \ll 1$ and $\overline{\theta}_{err} + \varphi_{max} \ll \pi/2$ [397].

12.3.4.2 Longitudinal tracking module

Though the trajectory to be tracked is given by the sequence of way-points, it can be easily interpolated on any interval between sampling times to give rise to its continuous-time counterpart. The related variables will be marked by *. Whereas PTM makes the curvilinear ordinate z close to that $z^* = 0$ of the traced trajectory, the goal of LTM is to equalize the abscissas η and η^*. To this end, we employ the following longitudinal tracking controller

$$u_v(t) = u_v^*(t) - u_{exc}\, \mathbf{sgn}\left\{k_0\left[\eta - \eta^*(t)\right] + k_1\left[v - v^*(t)\right]\right\}, \tag{3.22}$$

$$\text{where} \quad u_{exc} := \overline{u}_v - u_{v,nom} \overset{(3.1)}{=} (1-\mu_{u_v})\overline{u}_v > 0$$

and $k_i > 0$ are tunable controller parameters.

In order that the longitudinal control be realistic, the robot should be controllable in the longitudinal direction, i.e., the available control range exceeds that of disturbances: $\overline{u}_v > \overline{w}_v$.

Proposition 3.1. *Suppose that the controller parameters $\mu_{u_v} \in (0,1), k_0 > 0$, $k_1 > 0$ are chosen so that*

$$(1-\mu_{u_v})\overline{u}_v > \overline{w}_v, \quad (1-\mu_{u_v})\overline{u}_v - \overline{w}_v > \frac{k_0}{k_1}\left[2\overline{v} + \overline{w}_\eta\right], \tag{3.23}$$

where

$$\overline{w}_\eta := \frac{\overline{z}_{err}\mu_\varkappa \varkappa_{nom} + 1/2\left(\overline{\theta}_{err} + \overline{\varphi}\right)^2}{1 - \overline{z}_{err}\mu_\varkappa \varkappa_{nom}}. \tag{3.24}$$

Then the maximum longitudinal deviation $|\eta - \eta^*|$ along the trajectory is bounded by $\frac{k_1 \overline{w}_\eta}{k_0}$.

Proof. We first put

$$\eta_\Delta(t) := \eta(t) - \eta^*(t), \quad v_\Delta(t) := v(t) - v^*(t), \quad S := k_0\eta_\Delta + k_1 v_\Delta$$

and note that due to (4.9) and (3.19),

[3] To reduce the case at hand to the constant-speed robot considered in [397], we employ the change of the independent variable: the time t is replaced by the covered path $\int v \, dt$. Due to the multiplier v in the right-hand sides of (4.9), (3.20), (3.21), this hint sweeps v away and reduces the equations to the form considered in [397], with the only exception in the saturation thresholds in (3.21). They become $\pm\frac{\overline{u}_\theta}{v}$ and may vary over time. It is easy to see by inspection that all arguments from [397] concerning the robot with constant thresholds $\pm\varkappa_{nom}$ remain valid for the case at hand since $\frac{\overline{u}_\theta}{v} \geq \varkappa_{nom}$ ($\Leftarrow v \leq \overline{v}$).

$$\dot{\eta}_\Delta = \frac{v\cos(\hat{\theta}-\varphi)}{1-\varkappa(\eta)z} - v^* = v_\Delta + w_\eta, \tag{3.25}$$

$$\dot{v}_\Delta = \begin{cases} \sigma := -u_{exc}\,\mathbf{sgn}S + w_v & \text{if } 0 < v < \overline{v} \text{ or } \left| \begin{array}{l} v = 0 \\ \sigma > 0 \end{array} \right. \text{ or } \left| \begin{array}{l} v = \overline{v} \\ \sigma < 0 \end{array} \right., \\ -u_v^*(t) & \text{otherwise.} \end{cases}$$

In (3.25),

$$w_\eta := \frac{v\cos(\hat{\theta}-\varphi)}{1-\varkappa(\eta)z} - v \Rightarrow |w_\eta| \overset{(3.24)}{\leq} \overline{w}_\eta.$$

We are going to show that in the domain $0 < v < \overline{v}$, the discontinuity surface $S = 0$ is sliding. Indeed, in a close vicinity of any concerned point, we have

$$\begin{aligned} \dot{S}\,\mathbf{sgn}S &= [k_0\dot{\eta}_\Delta + k_1\dot{v}_\Delta]\,\mathbf{sgn}S = k_1(-u_{exc} + w_v\,\mathbf{sgn}S) + k_0(v_\Delta + w_\eta)\,\mathbf{sgn}S \\ &\overset{(3.17)}{\leq} -k_1(u_{exc} - \overline{w}_v) + k_0(|v_\Delta| + |w_\eta|) \leq -k_1(u_{exc} - \overline{w}_v) + k_0(2\overline{v} + \overline{w}_\eta) \\ &\leq -k_1\left[(1-\mu_{u_v})\overline{u}_v - \overline{w}_v\right] + k_0(2\overline{v} + \overline{w}_\eta) \overset{(3.23)}{<} 0. \end{aligned}$$

According to the architecture of our control system, the reference trajectory always starts at the real state. So tracking is commenced with zero error $\eta_\Delta = 0, v_\Delta = 0$, i.e., on the sliding surface. Hence tracking proceeds as sliding motion over the surface $S = 0$ while $0 < v < \overline{v}$. During this motion,

$$\begin{aligned} S = k_0\eta_\Delta + k_1 v_\Delta = 0 \Rightarrow v_\Delta = -\frac{k_0}{k_1}\eta_\Delta \overset{(3.25)}{\Rightarrow} \dot{\eta}_\Delta = v_\Delta + w_\eta = -\frac{k_0}{k_1}\eta_\Delta + w_\eta \\ \Rightarrow \eta_\Delta(t) = \int_0^t e^{-\frac{k_0}{k_1}(t-\tau)} w_\eta(\tau)\,d\tau \Rightarrow |\eta_\Delta(t)| \leq \int_0^t e^{-\frac{k_0}{k_1}(t-\tau)}\overline{w}_\eta\,d\tau \leq \frac{k_1\overline{w}_\eta}{k_0}. \end{aligned}$$

Now we analyze the marginal cases where $v = 0$ or $v = \overline{v}$. If a state where $v = 0$ is reached with $\eta < \eta^*$, we have

$$\sigma = -u_{exc}\,\mathbf{sgn}\left[k_0(\eta - \eta^*) - k_1 v_*\right] + w_v > 0.$$

Then the formula for $\dot{v}_\Delta$ is identical to that in the domain $0 < v < \overline{v}$, and the above analysis remains valid. If conversely, $\eta \geq \eta^*$, the robot is steady $\dot{\eta} = 0$, whereas $\dot{\eta}_* \geq 0$, so the error $\eta - \eta^*$ does not increase. If a state where $v = \overline{v}$ is reached with $\eta > \eta^*$, we have

$$\sigma = -u_{exc}\,\mathbf{sgn}\left[k_0(\eta - \eta^*) + k_1(\overline{v} - v_*)\right] + w_v < 0,$$

and the arguments concerning the domain $0 < v < \overline{v}$ remain valid. If conversely, $\eta \leq \eta^*$, the robot moves at the maximal speed $\dot{\eta} = \overline{v}$, whereas $\dot{\eta}_* \leq \mu_v\overline{v} < \overline{v}$, so the error $\eta - \eta^*$ decrease.

Thus the longitudinal deviation does not exceed $\frac{k_1\overline{w}_\eta}{k_0}$ in any case. □

It should be noted that conditions (3.23) can always be satisfied by proper choice of the controller parameters μ_v, k_0, k_1.

In conclusion, we note that the overall positional deviation $\|\boldsymbol{r}(t) - \boldsymbol{r}^*(t)\|$ between the actual and reference trajectories does not exceed $|z| + |\eta - \eta^*|$. So the estimate introduced by (p.12.3.2) in Section 12.3.2 can be taken in the form

$$\Delta_{\text{track}}\boldsymbol{r} = \overline{z}_{est} + \frac{k_1 \overline{w}_\eta}{k_0}.$$

The overall architecture of the control system is illustrated in Fig. 12.8.

12.3.5 Collision avoidance

The performance of the entire control system is addressed in the following.

Proposition 3.2. *Let every robot be driven by the proposed control law. Then the robots do not collide with the static obstacles and one another and, moreover, respect the required safety margin* d_{safe}.

Proof. By Lemma 3.2 and Remark 3.5, the ensemble of probational trajectories adopted for use at any step is mutually feasible. Lemma 3.1 completes the proof. □

12.3.6 Concluding remarks

The proposed trajectory tracking strategy belongs to the class of discontinuous sliding mode control laws discussed in Section 1.3. In our simulations and experiments, the discontinuous signum function employed by the control law was replaced with the saturated linear function sat_{-1}^{+1}. This was found to give more than satisfactory performance during trajectory tracking.

Modulo slight extension, the proposed algorithm also displayed good resistance to packet dropouts, which are common in wireless communications [443]. The only trouble caused by a packet dropout is that the robot losses access to the state and probational trajectory of some "other" robot at the previous time step and so is unable to carry out the proposed refinement procedure; see Section 12.3.3.2. This trouble can be easily overcome by employing the latest known data about the "other" robot, which is brought by the last successful transmission. This however gives more conservative approximations of the true planned trajectories If this conservatism appears to be too large, path following would be engaged. Thus it can be reasonably inferred that packet loss will reduce the system performance, though can hardly violate robustness.

The main advantage of the proposed approach is its "reactiveness": collision avoidance only requires a single time step of latency for communication between robots before mutual action is taken. This also motivates the simplified planning approach that was proposed, as opposed to other types of path planners; see, e.g., [356, 444]. While some aspects may seem restrictive, simulations and experiments (see Sections 12.4 and 12.5, respectively) have shown that the proposed navigation strategy does allow every robot to efficiently converge to its assigned target in most circumstances. The only caveat is that in simulation and testing the robot was instructed to rotate in place when stationary in order to present different initial states to the planning algorithm, which is needed mainly as a consequence of the upper bound on path curvature.

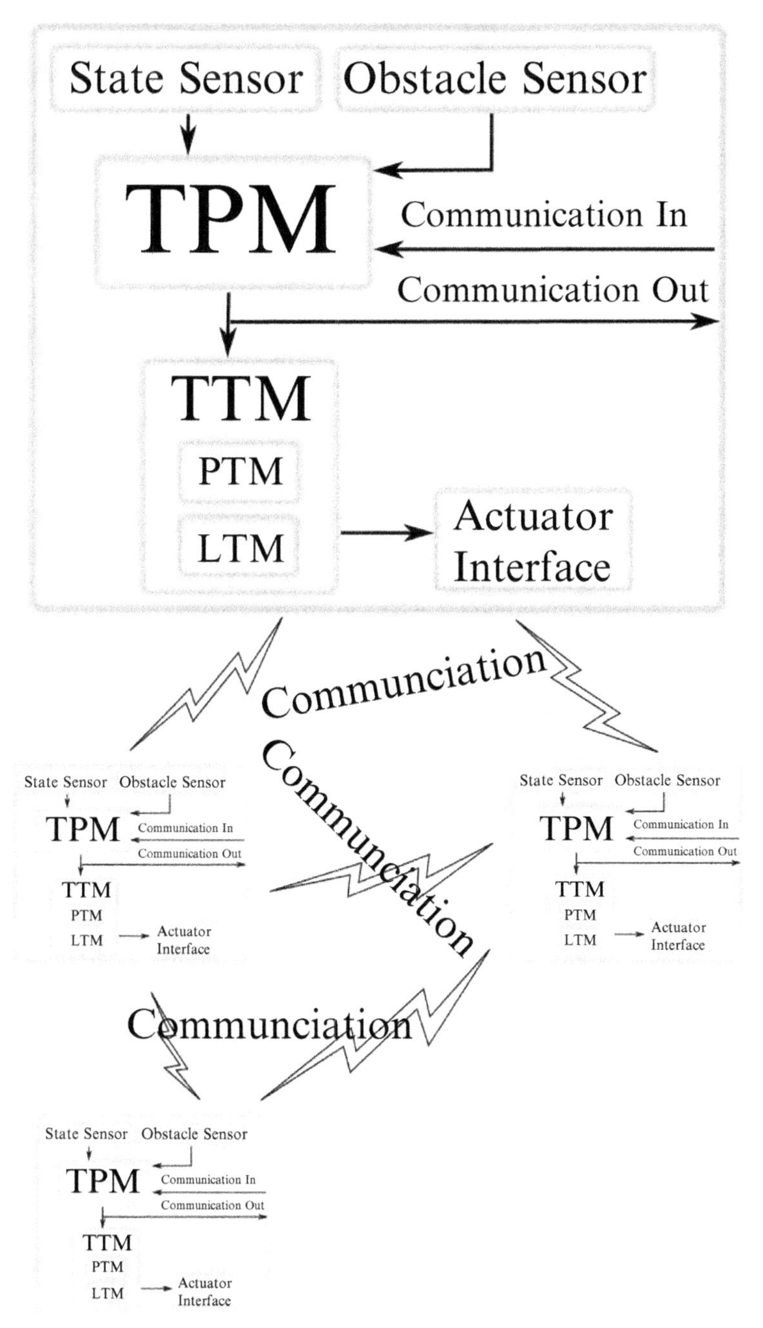

Figure 12.8 The overall architecture of the control system.

12.4 Simulation results

Simulations were carried out in MATLAB with a perfect unicycle model and parameters given in Table 12.1. TPM and TTM updates occurred at 1 Hz and 10 Hz, respectively. To examine the sensor noise implications, state measurements were corrupted by bounded, random, uniformly distributed zero mean errors. The magnitude of these was 0.1 m for the translational position measurement and 0.02 rad for the heading measurement. Acquisition of data about obstacles was modeled as rotation of the detection ray with the step of 0.157 rad, finding the distance to the obstacle reflection point, and interpolation of the reflection points by straight line segments. In the cases where the probational trajectory was exhausted, the following control was used to prevent the robot stalling: $u_v = -1.0 \cdot v(k)$, and $u_\theta = -0.5$. Also once the robots were within 0.5 m of their target positions, the speed and control signals were artificially set to zero.

The obtained closed loop trajectories are displayed in Figs. 12.9–12.11. In all cases when the noise was absent, the inter robot distance never dropped below 0.995 m and the robot-obstacle distance never dropped below 0.495 m, in harmony with the chosen value of d_{tar}. When the sensor noise was added, as in the simulations shown here, the minimum observed inter-robots distance was 0.6547 m. In every case, the robots successfully converged to the assigned targets.

Figure 12.9 shows a collision avoidance maneuver for four robots. This scenario may be encountered in an office, warehouse, factory or urban environment. The robots smoothly move around each other and adjust their speeds to prevent collisions. This is best seen in the third image, where two pairs of robots are avoiding head-on collisions.

Figure 12.10 represents an obstacle free environment where thirty robots are initially aligned around the edge of a circle; every robot must exchange positions with the robot on the opposite side of the circle. Similar scenarios may be found in [339]. While the motion looks like relatively chaotic, all robots still converged to the desired locations. At some points, some robots were stationary, and performing the rotation maneuver described in Section 12.3.6 that allowed them to find a trajectory after a short delay.

Table 12.1 Parameters used for simulations

L	4.0 m	$u_{\theta,nom}$	1.0 rad/s	$\chi(z) = \mathbf{sgn}(z) \cdot \min\{0.50 \cdot \lvert z\rvert, 0.50\}$
C	8.5 m	$u_{v,nom}$	0.20 ms^{-2}	$\underline{\varkappa}(z) = 0.05 + \min\{1.0 \cdot \lvert z\rvert, 0.70\varkappa_v\}$
d_{tar}	0.50 m	$\Delta\Lambda$	0.25	$p(S, z) = \min\{\underline{\varkappa}(z) + 1.0S, \varkappa_v\}$
$\overline{v}$	1.1 ms^{-1}	k_0	1	
$\overline{u}_\theta$	3.0 rad/s	k_1	1	
$\overline{u}_v$	0.30 ms^{-2}	c	1	
v_{nom}	1.0 ms^{-1}	$\mu_\varkappa$	0.9	

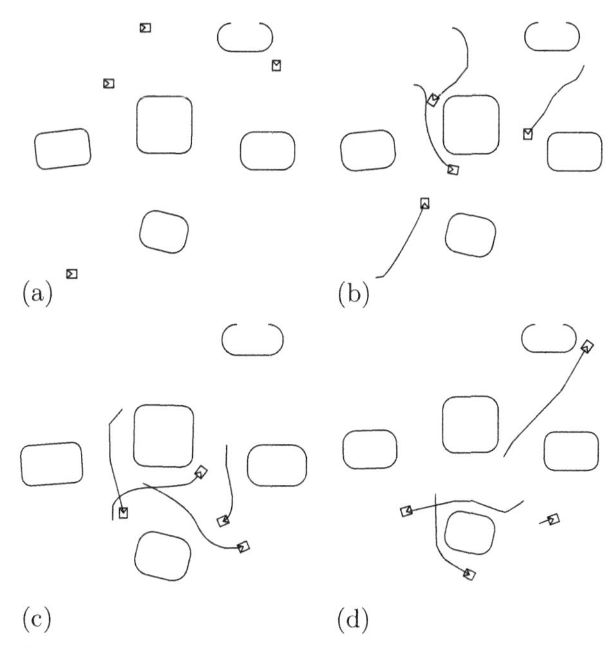

Figure 12.9 Trajectory using the MPC navigation approach with four robots.

Figure 12.11 shows the results with twelve robots in a constrained environment. This experiment highlights that in complex situations higher order planning is useful to generate efficient solutions. However even in this situation, the robots were all able to progress toward their respective targets. Most importantly, collisions were always avoided.

Figure 12.12 shows the computation time of each execution of the algorithm on each robot on a 3 GHz Desktop PC. The total computation time was under 50 ms even for thirty robots, which implies that the proposed approach is suitable for real time implementation.

In order to compare the performance of the proposed algorithm with other types of collision avoidance controllers, simulations were also carried out with a potential field type method [280, 326]. Different elements from these papers were combined to give a system which solves a comparable problem to ours, resulting in a system model which is the continuous time equivalent of (2.1).

The nearest point on the static obstacle was treated as an additional virtual robot for the purposes of collision avoidance, as in [280]. The control law is given by the following expression, loosely based on [326]:

$$\begin{bmatrix} v_t(t) \\ w_t(t) \end{bmatrix} = A^{-1}(\theta) \cdot \left(-\frac{1}{2}\nabla\gamma - \sum_i \nabla V_i \right)$$

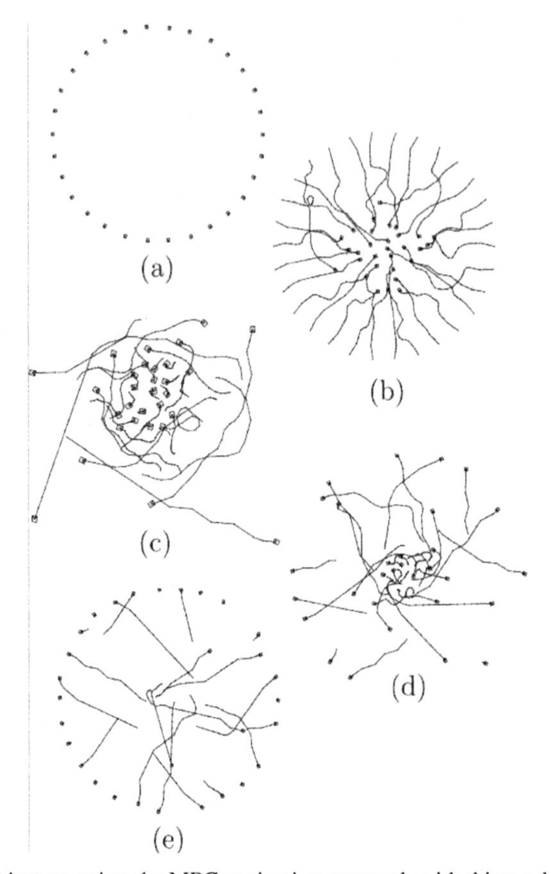

Figure 12.10 Trajectory using the MPC navigation approach with thirty robots.

Here γ is a navigation function representing the distance to the robot's target, V_i is a repulsion potential to each other robot

$$V_i = \begin{cases} \left(\frac{1}{\beta_i^2} - \frac{1}{d_{\text{sep}}^2}\right)^2 & \text{if } \beta_i \geq d_{\text{sep}}, \\ 0 & \text{if } \beta_i < d_{\text{sep}}, \end{cases}$$

β_i is the distance to the corresponding robot, and the matrix $A(\theta)$ rotates the control signal between the global and robot's reference frames:

$$A(\theta) = \begin{bmatrix} \cos(\theta) & -l\sin(\theta) \\ \sin(\theta) & l\cos(\theta) \end{bmatrix}$$

The control signals were adapted from the calculation above. The signal $u_\theta(t)$ was set to $w_t(t)$ saturated to $u_{\theta,nom}$, and the longitudinal acceleration was set as:

$$u_v(t) = u_{v,nom} \cdot \mathbf{sgn}\left(v_t(t) - v(t)\right).$$

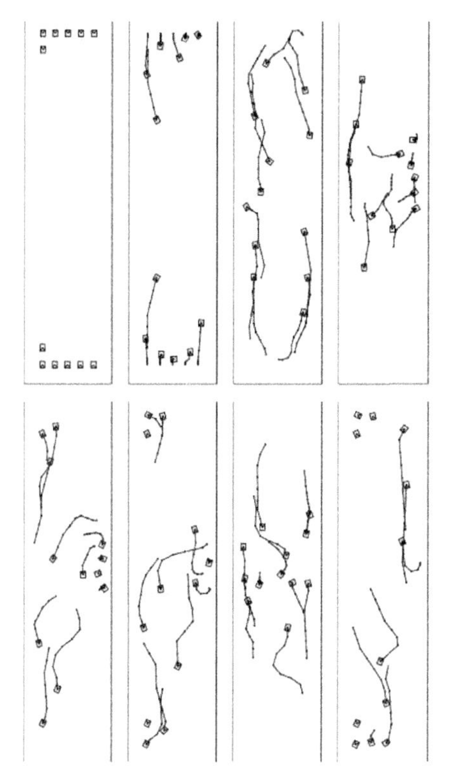

Figure 12.11 Trajectory using the MPC navigation approach with twelve robots in a constrained environment.

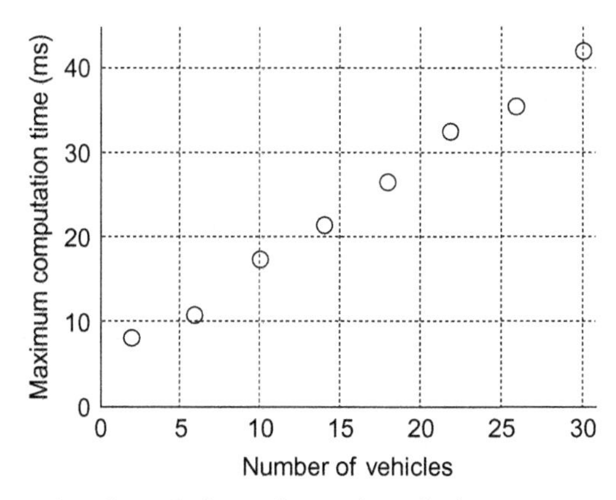

Figure 12.12 Computation time relative to the number of robots present.

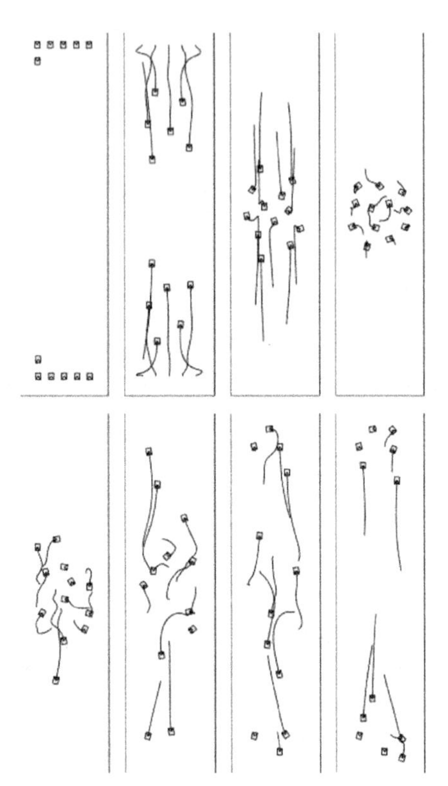

Figure 12.13 Trajectory using the potential field approach to the same environment as in Fig. 12.11.

For our simulations, the control law was updated with a sampling time of 0.1 s, and the parameters $d_{\text{sep}} = 4\,\text{m}, l = 1$.

The results of the potential field method (PFM) are shown in Fig. 12.13. Qualitatively, the motion seems comparable to our approach. However our method affords some advantages. To quantitatively compare the results, we measure the portion of the total distance traveled by each robot while they are more than 5 m from their target point, and find the total for all robots. For the PFM, this was 280.52 m, while the MPC approach (without sensor noise) had an aggregate distance of 282.40 m. For comparison the shortest distance that may be achieved without any collision avoidance is 243 m. This difference may be accounted by the fact the minimum distance between robots was 0.951 m for the PFM, whereas it was 1.001 m for the MPC approach. More thorough parameter tuning may allow the same separation as the MPC approach, but this highlights the difficulty of specifying arbitrary robot separations with the PFM when acceleration constraints must be respected. The main difference is that the PFM took longer to perform the maneuver—it took 96.7 s as opposed to 70 s for the MPC approach for all robots to become within 5 m of their targets.

12.5 Experimental results with wheeled robots

To show real-time applicability of the proposed navigation system, experiments were carried out with two Pioneer P3-DX mobile robots described in Section 1.4.1. In order to localize the robots, they were launched from known starting positions and headings, and odometry feedback was used during the maneuver. In our experiments, this method resulted in a good enough performance—a previous report has shown the systematic error relative to distance traveled is under 1.5% for certain maneuvers [445].[4] At each control update, the target translational acceleration and turning rate of the low level wheel controllers were set to match the results of the navigation algorithm. The laser detections in the circle of radius d_{tar} around the estimated position of the other robot were excluded when computing the navigation approach to prevent detections of the other robot affecting the path computations. The values of the parameters used in the experiments are listed in Table 12.2.[5]

The navigation scheme proposed in this chapter assumes decentralized communication. However for purely technical reasons, this decentralization was emulated on a common base station in the experiments so that the robots communicated with each other not directly but through this station. Communication was established via sending UDP packets over a 802.11g WLAN, with a checksum error detection scheme being employed [443]. A substantial rate of packet dropouts was observed, which according to Section 12.3.6, does not affect robustness but does affect performance. The event of packet dropout was handled similar to the event where refinement of the set of planned trajectories results in the empty set: the previous probational trajectory remains in use. TTM and TPM were run at a common rate, the communication range covered the entire operational zone of the robots.

In the first scenario demonstrated in Figs. 12.14 and 12.15, the robots are both trying to move through a gap that is only wide enough for one robot. The "top" robot slows down and allows the "bottom" robot to pass before moving into the gap itself.

In the second scenario demonstrated in Figs. 12.16 and 12.17, the robots must pass each other while a static obstacle is present. Robot 2 (which begins on the left) takes a longer path to prevent collision occurring.

Table 12.2 Parameters used for experiments

L	4.0 m	$u_{v,nom}$	0.10 ms^{-2}
d_{tar}	0.30 m	$\Delta\Lambda$	0.50
$\overline{v}$	0.50 ms^{-1}	k_0	1
$\overline{u}_\theta$	0.80 rad/s	k_1	1
$\overline{u}_v$	0.30 ms^{-2}	c	1
v_{nom}	0.40 ms^{-1}	$\mu_\varkappa$	0.9
$u_{\theta,nom}$	0.60 rad/s	C	∞

Note: The communication range C is shown as ∞ since the robots were always communicating.

[4] For longer experiments, accumulation of odometry errors over time is probable. To compensate for this, absolute positioning systems such as Indoor GPS or camera tracking may be employed.

[5] The physical parameters of the robots result from conservative estimates, whereas exact identification was not carried out.

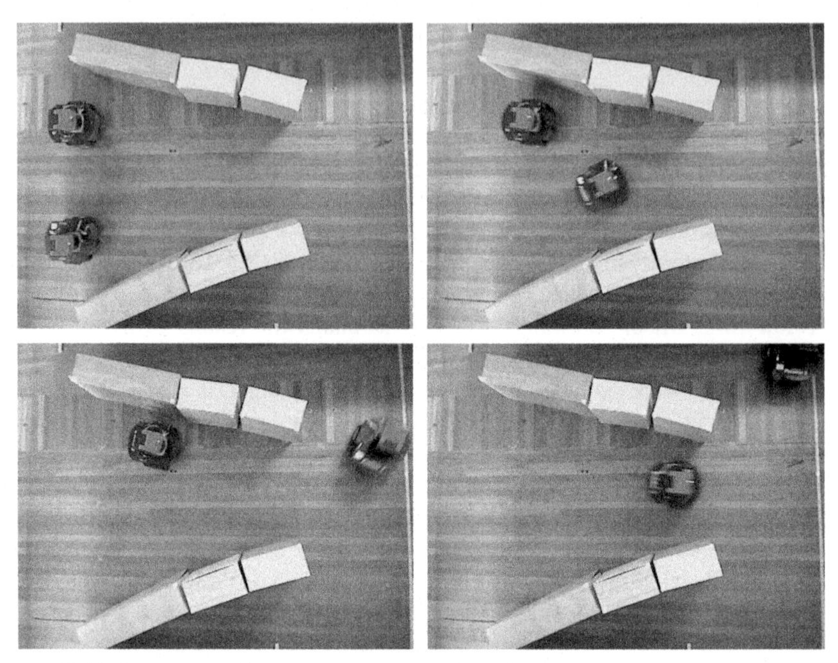

Figure 12.14 Sequence of images showing the real experiment for a parallel encounter.

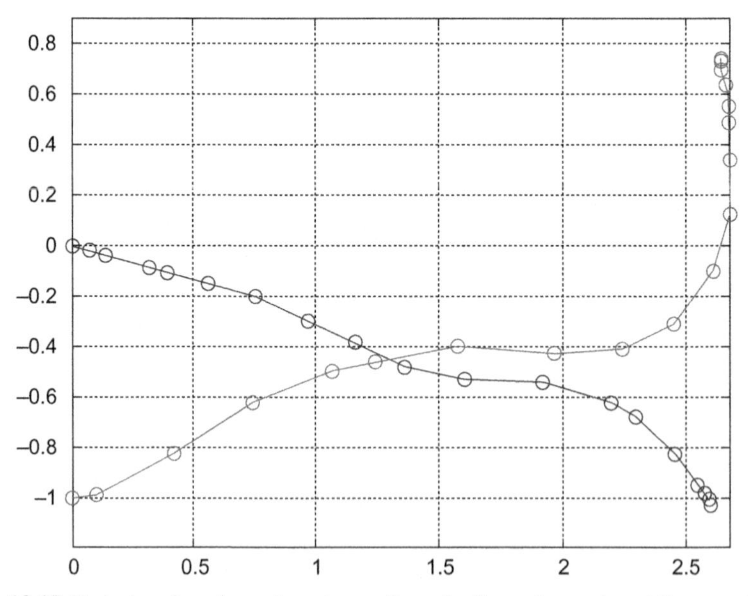

Figure 12.15 Trajectory based on odometry readings for the real experiment for a parallel encounter.

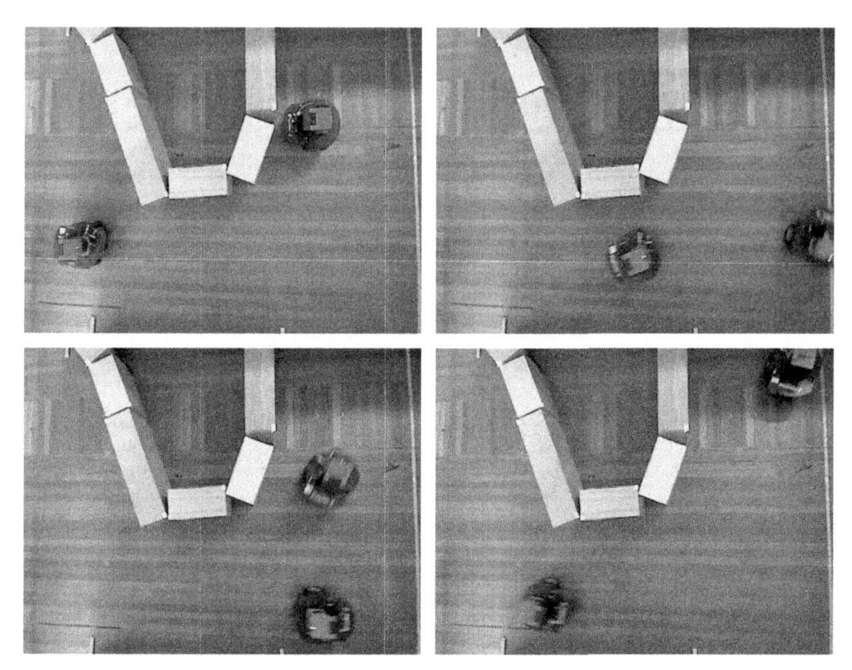

Figure 12.16 Sequence of images showing the real experiment for a head-on encounter.

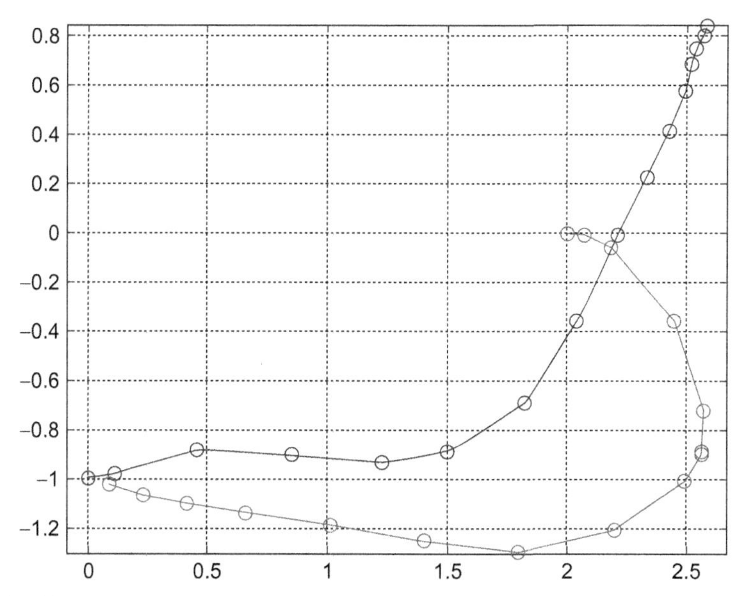

Figure 12.17 Trajectory based on odometry readings for the real experiment for a head-on encounter.

Bibliography

[1] USDoD, Unmanned Aircraft Systems Roadmap, 2005-2030, tech. rep., Office of the Secretary of Defense, Washington, 2005.

[2] D.A. Shoenwald, AUVs: in space, air, water, and on the ground, IEEE Control Syst. Mag. 20 (6) (2000) 15-19.

[3] M. Quigley, B. Barber, S. Griffiths, M. Goodrich, Towards real-world searching with fixed-wing mini-UAV's, in: Proceedings of the IEEE/RSJ International Conference on Intelligent Robots and Systems, Alberta, Canada, 2005, pp. 3028-3033.

[4] A. Ahmadzadeh, J. Keller, G. Pappas, A. Jadbabaie, V. Kumar, Multi-UAV cooperative surveillance with spatio-temporal specifications, in: Proceedings of the 45th IEEE Conference on Decision and Control, San Diego, CA, 2006, pp. 5293-5298.

[5] A. Girard, A.S. Howell, J.K. Hedrick, Border patrol and surveillance missions using multiple unmanned air vehicles, in: Proceedings of the 43th IEEE Conference on Decision and Control, Paradise Island, Bahamas, 2004, pp. 620-625.

[6] J. Wang, J. Steiber, B. Surampudi, Autonomous ground vehicle control system for high-speed and safe operation, in: Proceedings of the American Control Conference, Seattle, Washington, USA, 2008, pp. 218-223.

[7] J.C. Latombe, Robot Motion Planning, Kluwer Academic Publishers, London, 1991.

[8] J. Minguez, L. Montano, Robot navigation in very complex, dense and cluttered indoor/outdoor environments, in: Proceedings of the 15th IFAC World Congress, Barcelona, Spain, 2002, pp. 218-223.

[9] L. Lapierre, R. Zapata, P. Lepinay, Combined path-following and obstacle avoidance control of a wheeled robot, Int. J. Robot. Res. 26 (4) (2007) 361-375.

[10] Z. Shiller, Online suboptimal obstacle avoidance, Int. J. Robot. Res. 19 (5) (2000) 480-497.

[11] I. Kamon, E. Rivlin, Sensory-based motion planning with global proofs, IEEE Trans. Robot. Autom. 13 (6) (1997) 814-822.

[12] I. Kamon, E. Rimon, E. Rivlin, TangentBug: a range-sensor-based navigation algorithm, Int. J. Robot. Res. 17 (9) (1998) 934-953.

[13] Y.-H. Liu, S. Arimoto, Path planning using a tangent graph for mobile robots among polygonal and curved obstacles, Int. J. Robot. Res. 11 (4) (1992) 376-382.

[14] N.A. Vlassis, N.M. Sgouros, G. Efthivolidis, G. Papakonstantinou, Global path planning for autonomous qualitative navigation, in: Proceedings of the IEEE Conference on Tools with Artificial Intelligence, Toulouse, France, 1996, pp. 354-359.

[15] S. Belkhous, A. Azzouz, M. Saad, C. Nerguizian, V. Nerguizian, A novel approach for mobile robot navigation with dynamic obstacles avoidance, J. Intell. Robot. Syst. 44 (3) (2005) 187-201.

[16] A. Savkin, M. Hoy, Reactive and the shortest path navigation of a wheeled mobile robot in cluttered environments, Robotica 31 (2) (2013) 323-330.

[17] F. Large, C. Lauger, Z. Shiller, Navigation among moving obstacles using the NLVO: principles and applications to intelligent vehicles, Auton. Robots 19 (2) (2005) 159-171.

[18] R. Kulić, Z. Vukić, Methodology of concept control synthesis to avoid unmoving and moving obstacles, J. Intell. Robot. Syst. 45 (1) (2006) 267-294.
[19] P. Fiorini, Z. Shiller, Motion planning in dynamic environments using velocity obstacles, Int. J. Robot. Res. 17 (7) (1998) 760-772.
[20] Z. Qu, J. Wang, C.E. Plaisted, A new analytical solution to mobile robot trajectory generation in the presence of moving obstacles, IEEE Trans. Robot. 20 (6) (2004) 978-993.
[21] M. Erdmann, T. Lozano-Perez, On multiple moving objects, Algorithmica 2 (4) (1987) 477-521.
[22] T. Fraichard, Trajectory planning in a dynamic workspace: a'state-time space'approach, Adv. Robot. 13 (1) (1998) 75-94.
[23] J. Reif, M. Sharir, Motion planning in the presence of moving obstacles, J. ACM 41 (4) (1994) 764-790.
[24] J. Canny, The Complexity of Robot Motion Planning, MIT Press, Cambridge, MA, 1988.
[25] D. Hsu, R. Kindel, J.-C. Latombe, S. Rock, Randomized kinodynamic motion planning with moving obstacles, Int. J. Robot. Res. 21 (3) (2002) 233-255.
[26] E. Frazzoli, M. Dahleh, E. Feron, Real-time motion planning for agile autonomous vehicles, J. Guid. Control Dyn. 25 (1) (2002) 116-129.
[27] F. Lamiraux, D. Bonnafous, O. Lefebvre, Reactive path deformation for nonholonomic mobile robots, IEEE Trans. Robot. 20 (6) (2004) 967-977.
[28] J. Minguez, L. Montano, Sensor-based robot motion generation in unknown, dynamic and troublesome scenarios, Robot. Auton. Syst. 52 (4) (2005) 290-311.
[29] F. Belkhouche, B. Belkhouche, A method for robot navigation toward a moving goal with unknown maneuvers, Robotica 23 (6) (2005) 709-720.
[30] Y. Zhu, T. Zhang, J. Song, X. Li, A new hybrid navigation algorithm for mobile robots in environments with incomplete knowledge, Knowl Based Syst 27 (2012) 302-313.
[31] M. Deng, A. Inoue, Y. Shibata, K. Sekiguchi, N. Ueki, An obstacle avoidance method for two wheeled mobile robot, in: Proceedings of the 2007 IEEE International Conference on Networking, Sensing and Control, London, UK, 2007, pp. 689-692.
[32] H. Teimoori, A. Savkin, A biologically inspired method for robot navigation in a cluttered environment, Robotica 28 (5) (2010) 637-648.
[33] A. Matveev, H. Teimoori, A. Savkin, A method for guidance and control of an autonomous vehicle in problems of border patrolling and obstacle avoidance, Automatica 47 (2011) 515-524.
[34] M. Seder, K. Macek, I. Petrovic, An integrated approach to real-time mobile robot control in partially known indoor environments, in: Proceedings of the 31st Annual Conference of the IEEE Industrial Electronics Society, Raleigh, NC, USA, 2005, pp. 1785-1790.
[35] D. Fox, W. Burgard, S. Thrun, The dynamic window approach to collision avoidance, IEEE Robot. Autom. Mag. 4 (1) (1997) 23-33.
[36] R.C. Simmons, The curvature-velocity method for local obstacle avoidance, in: IEEE International Conference on Robotics and Automation, vol. 4, Minneapolis, MI, USA, 1996, pp. 3375-3382.
[37] Y.K. Nak, R. Simmons, The lane-curvature method for local obstacle avoidance, in: IEEE International Conference on Robotics and Automation, vol. 3, Lueven, Belgium, 1998, pp. 1615-1621.
[38] A. Chakravarthy, D. Ghose, Obstacle avoidance in a dynamic environment: a collision cone approach, IEEE Trans. Syst. Man Cybern. 28 (5) (1998) 562-574.

[39] T. Fraichard, H. Asama, Inevitable collision states. A step towards safer robots? in: IEEE International Conference on Intelligent Robots and Systems, vol. 1, Las Vegas, NV, USA, 2003, pp. 388-393.
[40] E. Owen, L. Montano, A robocentric motion planner for dynamic environments using the velocity space, in: IEEE International Conference on Intelligent Robots and Systems, vol. 1, Beijing, China, 2006, pp. 2833-2838.
[41] S.M. LaValle, Planning Algorithms, Cambridge University Press, Cambridge, NY, 2006.
[42] A. Wu, J. How, Guaranteed infinite horizon avoidance of unpredictable, dynamically constrained obstacles, Auton. Robots 32 (3) (2012) 227-242.
[43] J. Borenstein, Y. Koren, Real-time obstacle avoidance for fast mobile robots, IEEE Trans. Syst. Man Cybern. 19 (5) (1989) 1179-1187.
[44] O. Khatib, Real-time obstacle avoidance for manipulators and mobile robots, Int. J. Robot. Res. 5 (1) (1986) 90-98.
[45] M. Rubagotti, M.D. Vedova, A. Ferrara, Time-optimal sliding mode control of a mobile robot in a dynamic environment, IET Control Theory Appl. 5 (16) (2011) 1916-1924.
[46] J. Borenstein, Y. Koren, The vector field histogram -fast obstacle avoidance for mobile robots, IEEE Trans. Robot. Autom. 7 (3) (1991) 278-288.
[47] A. Elfes, Sonar-based real-world mapping and navigation, IEEE J. Robot. Autom. 3 (3) (1987) 249-265.
[48] J. Minguez, L. Montano, Nearness diagram navigation: collision avoidance in troublesome scenarios, IEEE Trans. Robot. Autom. 20 (1) (2004) 45-59.
[49] D. Nakhaeinia, S. Tang, S.M. Noor, O. Motlagh, A review of control architectures for autonomous navigation of mobile robots, Int. J. Phys. Sci. 6 (2011) 169-174.
[50] A. Ferrara, M. Rubagotti, Sliding mode control of a mobile robot for dynamic obstacle avoidance based on a time-varying harmonic potential field, in: ICRA 2007 Workshop: Planning, Perception and Navigation for Intelligent Vehicles, Rome, Italy, 2007.
[51] J. Chunyu, Z. Qu, E. Pollak, M. Falash, Reactive target-tracking control with obstacle avoidance of unicycle-type mobile robots in a dynamic environment, in: American Control Conference, Baltimore, MD, USA, 2010, pp. 1190-1196.
[52] A. Savkin, C. Wang, A simple biologically-inspired algorithm for collision free navigation of a unicycle-like robot in dynamic environments with moving obstacles, Robotica 31 (6) (2013) 993-1001.
[53] S.R. Lindemann, I.I. Hussein, S.M. LaValle, Real time feedback control for nonholonomic mobile robots with obstacles, in: Proceedings of the 45th IEEE Conference on Decision and Control, San Diego, CA, USA, 2006, pp. 2406-2411.
[54] E. Masehian, Y. Katebi, Robot motion planning in dynamic environments with moving obstacles and target, Int. J. Mech. Syst. Sci. Eng. 1 (1) (2007) 20-25.
[55] R. Kuc, B. Barshan, Navigating vehicles through an unstructured environment with sonar, in: Proceedings of the 1989 IEEE International Conference on Robotics and Automation, vol. 3, Scottsdale, AZ, 1989, pp. 1422-1426.
[56] S. Yang, M. Meng, Neural network approaches to dynamic collision-free trajectory generation, IEEE Trans. Syst. Man Cybern. B Cybern. 31 (3) (2001) 302-318.
[57] Y. Yagi, H. Nagai, K. Yamazawa, M. Yachida, Reactive visual navigation based on omnidirectional sensing-path following and collision avoidance, J. Intell. Robot. Syst. 31 (4) (2001) 379-395.
[58] A. Ferreira, F. Pereira, R. Vassallo, T. Filho, M. Filho, An approach to avoid obstacles in mobile robot navigation: the tangential escape, SBA: Controle Autom. 19 (4) (2008) 395-405.

[59] S. Tang, S. Ang, D. Nakhaeinia, B. Karasfi, O. Motlagh, A reactive collision avoidance approach for mobile robot in dynamic environments, J. Autom. Control Eng. 1 (1) (2013) 16-20.

[60] A. Matveev, C. Wang, A. Savkin, Real-time navigation of mobile robots in problems of border patrolling and avoiding collisions with moving and deforming obstacles, Robot. Auton. Syst. 60 (6) (2012) 769-788.

[61] A. Matveev, M. Hoy, A. Savkin, A method for reactive navigation of nonholonomic robots in the presence of obstacles, in: Proceedings of the 18th IFAC World Congress, Milano, Italy, 2011, pp. 11894-11899.

[62] J. Minguez, L. Montano, Nearness diagram (ND) navigation: collision avoidance in troublesome scenarios, IEEE Trans. Robot. Autom. 20 (1) (2004) 45-59.

[63] V.J. Lumelsky, T. Skewis, Incorporating range sensing in the robot navigation function, IEEE Trans. Syst. Man Cybern. 20 (5) (1990) 1058-1069.

[64] V.J. Lumelsky, A.A. Stepanov, Path-planning strategies for a point mobile automaton amidst unknown obstacles of arbitrary shape, in: I.J. Cox, G.T. Wilfong (Eds.), Autonomous Robots Vehicles, Springer, New York, (1990) 1058-1068.

[65] A.S. Matveev, A.V. Savkin, Qualitative Theory of Hybrid Dynamical Systems, Birkhäuser, Boston, 2000.

[66] A. van der Schaft, H. Schumacher, An Introduction to Hybrid Dynamical Systems, Lecture Notes in Control and Information Sciences, vol. 251, Springer, London, 2000.

[67] A.V. Savkin, R.J. Evans, Hybrid Dynamical Systems. Controller and Sensor Switching Problems, Birkhäuser, Boston, 2002.

[68] D. Liberson, Switching in Systems and Control, Birkhäuser, Boston, 2003.

[69] R. Goebel, R. Sanfelice, A. Teel, Hybrid Dynamical Systems: Modeling, Stability, and Robustness, Princeton University Press, Princeton, NJ, 2012.

[70] V. Utkin, Sliding Modes in Control Optimization, Springer-Verlag, Berlin, 1992.

[71] A. Sabanovic, Variable structure systems with sliding modes in motion control—a survey, IEEE Trans. Ind. Inform. 7 (2) (2011) 212-223.

[72] A. Pamosoajia, P. Cat, K. Hong, Sliding-mode and proportional-derivative-type motion control with radial basis function neural network based estimators for wheeled vehicles, Int. J. Syst. Sci. 45 (12) (2014) 2515-2528.

[73] V. Utkin, S. Drakunov, H. Hashimoto, F. Harashima, Robot path obstacle avoidance control via sliding mode approach, in: IEEE/RSJ International Workshop on Intelligent Robots and Systems '91, vol. 3, San Antonio, TX, 1991, pp. 1287-1290, doi:10.1109/IROS.1991.174678.

[74] J. Guldner, V. Utkin, Sliding mode control for an obstacle avoidance strategy based on an harmonic potential field, in: 32nd IEEE Conference on Decision and Control, vol. 1, San Antonio, TX, 1993, pp. 424-429, doi:10.1109/CDC.1993.325112.

[75] C. Edwards, S. Spurgeon, Sliding Mode Control, Taylor and Francis Ltd, London, 1988.

[76] X. Yu, J. Xu, Variable Structure Systems: Towards 21st Century, Springer-Verlag, Berlin/Heidelberg, 2002.

[77] B. Bandyopadhyay, S. Janardhanan, Discrete-time Sliding Mode Control: A Multirate Output Feedback Approach, Springer, Berlin/Heidelberg, 2005.

[78] G. Bartolini, L. Fridman, A. Pisano, E. Usai, Modern Sliding Mode Control Theory: New Perspectives and Applications, Lecture Notes in Control and Information Sciences, vol. 375, Springer, Berlin/Heidelberg, 2008.

[79] J. Liu, X. Wang, Advanced Sliding Mode Control for Mechanical Systems, Tsinghua University Press/Springer, Beijing/Heidelberg, 2011.

[80] Y. Shtessel, C. Edwards, L. Fridman, A. Levant, Sliding Mode Control and Observation, Birkhäuzer, Boston, 2013.
[81] G. Bartolini, A. Ferrara, E. Usai, Chattering avoidance by second order sliding-mode control, IEEE Trans. Autom. Control 43 (2) (1998) 241-246.
[82] K. Young, V. Utkin, U. Ozguner, Control engineer's guide to sliding mode control, IEEE Trans. Control Syst. Technol. 7 (3) (1999) 324-342.
[83] G. Bartolini, A. Ferrara, E. Usai, V. Utkin, On multi-input chattering-free second order sliding mode control, IEEE Trans. Autom. Control 45 (9) (2000) 1711-1717.
[84] L. Fridman, An averaging approach to chattering, IEEE Trans. Autom. Control 46 (8) (2001) 1260-1264.
[85] L. Fridman, Chattering analysis in sliding mode systems with inertial sensors, Int. J. Control 76 (9/10) (2003) 906-912.
[86] C. Edwards, C.E. Fossas, L. Fridman (Eds.), Advances in Variable Structure and Sliding Mode Control, Springer, Berlin, 2006.
[87] H. Lee, V.I. Utkin, A. Malinin, Chattering reduction using multiphase sliding mode control, Int. J. Control 82 (9) (2009) 1720-1737.
[88] A. Levant, Chattering analysis, IEEE Trans. Autom. Control 55 (6) (2010) 1380-1389.
[89] M. Chen, M. Tseng, A New Design for Noise-Induced Chattering Reduction in Sliding Mode Control, INTECH Open Access Publisher, 2011, ISBN 9789533071626, URL http://books.google.com.au/books?id=ydzNoAEACAAJ.
[90] L. Wu, P. Shi, X. Su, Sliding Mode Control of Uncertain Parameter-Switching Hybrid Systems, John Wiley and Sons, Chichester, UK, 2014.
[91] E.P. Company, Data sheet for encoder model 775, 2013, URL http://www.encoder.com/model775.html.
[92] V.A. Yakubovich, G.A. Leonov, A.K. Gelig, Stability of Stationary Sets in Control Systems with Discontinuous Nonlinearities, World Scientific, Singapore, 2004.
[93] A. Filippov, Differential Equations with Discontinuous Right-hand Sides, Kluwer Academic Publishers, Dordrecht, the Netherlands, 1988.
[94] R.T. Rockafellar, Convex Analysis, Princeton University Press, Princeton, NJ, 1970.
[95] G. Smirnov, Introduction to the Theory of Differential Inclusions, American Mathematical Society, Providence, RI, 2002.
[96] N. Dadkhah, B. Mettler, Survey of motion planning literature in the presence of uncertainty: considerations for UAV guidance, J. Intell. Robot. Syst. 65 (1) (2012) 233-246.
[97] M. Hoy, A. Matveev, A. Savkin, Algorithms for collision-free navigation of mobile robots in complex cluttered environments: a survey, Robotica 33 (3) (2015) 463-497.
[98] H. Durrant-Whyte, T. Bailey, Simultaneous localization and mapping: part I, IEEE Robot. Autom. Mag. 13 (2) (2006) 99-110.
[99] M.E. Jefferies, W. Yeap (Eds.), Robotics and Cognitive Approaches to Spatial Mapping, vol. 38, Springer, Berlin/Heidelberg, 2008.
[100] S. Thrun, Learning occupancy grid maps with forward sensor models, Auton. Robots 15 (2) (2003) 111-127.
[101] J. Ng, T. Braunl, Performance comparison of bug navigation algorithms, J. Intell. Robot. Syst. 50 (1) (2007) 73-84.
[102] Y. Gabriely, E. Rimon, CBUG: a quadratically competitive mobile robot navigation algorithm, IEEE Trans. Robot. 24 (6) (2008) 1451-1457.
[103] M. Katsev, A. Yershova, B. Tovar, R. Ghrist, S.M. LaValle, Mapping and pursuit-evasion strategies for a simple wall-following robot, IEEE Trans. Robot. 27 (1) (2011) 113-128.
[104] S. Suri, E. Vicari, P. Widmayer, Simple robots with minimal sensing: from local visibility to global geometry, Int. J. Robot. Res. 27 (9) (2008) 1055-1067.

[105] L. Consolini, M. Tosques, A path following problem for a class of non-holonomic control systems with noise, Automatica 41 (6) (2005) 1009-1016.
[106] E. Ostertag, An improved path-following method for mixed H-2/H-infinity controller design, IEEE Trans. Autom. Control 53 (8) (2008) 1967-1971.
[107] R.C. Arkin, Motor schemas-based mobile robot navigation, Int. J. Robot. Res. 8 (4) (1989) 92-112.
[108] R.C. Arkin, Behavior-based robot navigation for extended domains, Adapt. Behav. 1 (2) (1992) 201-225.
[109] P. Maes, R.A. Brooks, Learning to coordinate behaviors, in: Proceedings of the AAAI, 1990, pp. 796-802.
[110] M.J. Mataric, Behavior-based control: main properties and implications, in: Proceedings of the IEEE International Conference on Robotics and Automation, 1992, pp. 46-54.
[111] C. Goerzen, Z. Kong, B. Mettler, A survey of motion planning algorithms from the perspective of autonomous UAV guidance, J. Intell. Robot. Syst. 57 (1-4) (2009) 65-100.
[112] A.S. Lopez, R. Zapata, M.A. Osorio-Lama, Sampling-based motion planning: a survey, Comput. Syst. 12 (1) (2008) 5-24.
[113] M. Innocenti, L. Pollini, D. Turra, A fuzzy approach to the guidance of unmanned air vehicles tracking moving targets, IEEE Trans. Control Syst. Technol. 16 (6) (2008) 1125-1137.
[114] M. Gomez, R.V. Gonzalez, T. Martinez-Marin, D. Meziat, S. Sanchez, Optimal motion planning by reinforcement learning in autonomous mobile vehicles, Robotica 30 (2) (2012) 159-170.
[115] B. Douillard, D. Fox, F. Ramos, H. Durrant-Whyte, Classification and semantic mapping of urban environments, Int. J. Robot. Res. 30 (1) (2011) 5-32.
[116] L. Gracia, J. Tornero, Kinematic modeling and singularity of wheeled mobile robots, Adv. Robot. 21 (7) (2007) 793-816.
[117] L. Gracia, J. Tornero, Kinematic modeling of wheeled mobile robots with slip, Adv. Robot. 21 (11) (2007) 1253-1279.
[118] K. Kozlowski, Robot Motion and Control, Springer, Berlin/Heidelberg, 2009.
[119] A. Micaelli, C. Samson, Trajectory tracking for unicycle-type and two-steering-wheels mobile robots, Tech. Rep. 2097, INRIA, 1993.
[120] T. Fossen, Guidance and Control of Ocean Vehicles, Wiley, NY, 1994.
[121] J. Ben-Asher, I. Yaesh, Advances in Missile Guidance Theory, AIAA, Inc., VA, 1998.
[122] E. Low, I. Manchester, A. Savkin, A biologically inspired method for vision-based docking of wheeled mobile robots, Robot. Auton. Syst. 55 (10) (2007) 769-784.
[123] I. Manchester, A. Savkin, Circular navigation missile guidance with incomplete information and uncertain autopilot model, J. Guidance Control Dyn. 27 (6) (2004) 1076-1083.
[124] I. Manchester, A. Savkin, Circular navigation guidance law for precision missile/target engagement, J. Guidance Control Dyn. 29 (2) (2006) 1287-1292.
[125] A. Savkin, H. Teimoori, Bearings-only guidance of a unicycle-like vehicle following a moving target with a smaller minimum turning radius, IEEE Trans. Autom. Control 55 (10) (2010) 2390-2395.
[126] H. Teimoori, A. Savkin, Equiangular navigation and guidance of a wheeled mobile robot based on range-only measurements, Robot. Auton. Syst. 58 (2) (2010) 203-215.
[127] A. Matveev, H. Teimoori, A. Savkin, Navigation of a unicycle-like mobile robot for environmental extremum seeking, Automatica 47 (1) (2011) 85-91.
[128] A. Matveev, H. Teimoori, A. Savkin, Range-only measurements based target following for wheeled mobile robots, Automatica 47 (1) (2011) 177-184.

[129] A. Richards, J.P. How, Robust variable horizon model predictive control for vehicle maneuvering, Int. J. Robust Nonlinear Control 16 (7) (2006) 333-351.

[130] L. Lapierre, B. Jouvencel, Robust nonlinear path-following control of an AUV, IEEE J. Oceanic Eng. 33 (2) (2008) 89-102.

[131] A. Albagul, Wahyudi, Dynamic modeling and adaptive traction control for mobile robots, in: Proceedings of the 30th Annual Conference of IEEE Industrial Electronics Society, vol. 1, Busan, Korea, 2004, pp. 614-620.

[132] R. Balakrishna, A. Ghosal, Modeling of slip for wheeled mobile robots, IEEE Trans. Robot. Autom. 11 (1) (1995) 126-132.

[133] A. Matveev, M. Hoy, J. Katupitiya, A. Savkin, Nonlinear sliding mode control of an unmanned agricultural tractor in the presence of sliding and control saturation, Robot. Auton. Syst. 61 (9) (2013) 973-987.

[134] G. Bevan, H. Gollee, J. O'reilly, Automatic lateral emergency collision avoidance for a passenger car, Int. J. Control 80 (11) (2007) 1751-1762.

[135] Y. Yoon, J. Shin, H.J. Kim, Y. Park, S. Sastry, Model-predictive active steering and obstacle avoidance for autonomous ground vehicles, Control Eng. Pract. 17 (7) (2009) 741-750.

[136] S. Bereg, D. Kirkpatrick, Curvature-bounded traversals of narrow corridors, in: Proceedings of the 21st Annual Symposium on Computational Geometry, Pisa, Italy, 2005, pp. 278-287.

[137] A. Bicchi, G. Casalino, C. Santilli, Planning shortest bounded-curvature paths for a class of nonholonomic vehicles among obstacles, J. Intell. Robot. Syst. 16 (4) (1996) 387-405.

[138] W. Travis, A.T. Simmons, D.M. Bevly, Corridor navigation with a LiDAR/INS Kalman filter solution, in: Proceedings of the IEEE Intelligent Vehicles Symposium, Tokyo, Japan, 2005.

[139] P. Moghadam, W.S. Wijesoma, J.F. Dong, Improving path planning and mapping based on stereo vision and lidar, in: Proceedings of the International Conference on Control, Automation, Robotics and Vision, Hanoi, Vietnam, 2008.

[140] C. Shi, Y. Wang, J. Yang, A local obstacle avoidance method for mobile robots in partially known environment, Robot. Auton. Syst. 58 (5) (2010) 425-434.

[141] W.H. Huang, B.R. Fajen, J.R. Fink, W.H. Warren, Visual navigation and obstacle avoidance using a steering potential function, Robot. Auton. Syst. 54 (4) (2006) 288-299.

[142] F. Bonin-Font, A. Ortiz, G. Oliver, Visual navigation for mobile robots: a survey, J. Intell. Robot. Syst. 53 (3) (2008) 263-296.

[143] W.E. Green, P.Y. Oh, Optic-flow-based collision avoidance, IEEE Robot. Autom. Mag. 15 (1) (2008) 96-103.

[144] J. van den Berg, D. Wilkie, S.J. Guy, M. Niethammer, D. Manocha, LQG-obstacles: feedback control with collision avoidance for mobile robots with motion and sensing uncertainty, in: Proceedings of the IEEE International Conference on Robotics and Automation, St. Paul, MN, USA, 2012, pp. 346-353.

[145] M. Hoy, A. Matveev, A. Savkin, Collision free cooperative navigation of multiple wheeled robots in unknown cluttered environments, Robot. Auton. Syst. 60 (10) (2012) 1253-1266.

[146] L.E. Dubins, On curves of minimal length with a constraint on average curvature and with prescribed initial and terminal positions and tangents, Am. J. Math. 79 (3) (1957) 497-516.

[147] J.A. Reeds, L.A. Shepp, Optimal paths for a car that goes both forwards and backwards, Pac. J. Math. 145 (2) (1990) 367-393.

[148] D.J. Balkcom, P.A. Kavathekar, M.T. Mason, Time-optimal trajectories for an omni-directional vehicle, Int. J. Robot. Res. 25 (10) (2006) 985-999.

[149] H. Chitsaz, S.M. LaValle, D.J. Balkcom, M.T. Mason, Minimum wheel-rotation paths for differential-drive mobile robots, Int. J. Robot. Res. 28 (1) (2009) 66-80.

[150] C. Trevai, J. Ota, T. Arai, Multiple mobile robot surveillance in unknown environments, Adv. Robot. 21 (7) (2007) 729-749.

[151] L. Armesto, V. Girbes, M. Vincze, S. Olufs, P. Munoz-Benavent, Mobile robot obstacle avoidance based on quasi-holonomic smooth paths, in: Advances in Autonomous Robotics, Lecture Notes in Computer Science, vol. 7429, Springer, Berlin/Heidelberg, 2012, pp. 244-255.

[152] V.A. Tucker, The deep fovea, sideways vision and spiral flight paths in raptors, J. Exp. Biol. 203 (24) (2001) 3745-3754.

[153] D.N. Lee, Guiding movements by coupling taus, Ecol. Psychol. 10 (3-4) (1998) 221-250.

[154] J.M. Camhi, E.N. Johnson, High-frequency steering maneuvers mediated by tactile cues: antennal wall-following in the cockroach, J. Exp. Biol. 202 (5) (1999) 631-643.

[155] M.V. Srinivasan, S.W. Zhang, J.S. Chahl, E. Barth, S. Venkatesh, How honeybees make grazing landings on flat surfaces, Biol. Cybern. 83 (3) (2000) 171-183.

[156] G. Flierl, D. Grunbaum, S. Levin, D. Olson, From individuals to aggregations: the interplay between behavior and physics, J. Theor. Biol. 196 (4) (1999) 397-454.

[157] N.W. Bode, A.J. Wood, D.W. Franks, Social networks and models for collective motion in animals, Behav. Ecol. Sociobiol. 65 (2) (2011) 117-130.

[158] M.A. Ahmadi-Pajouh, F. Towhidkhah, S. Gharibzadeh, M. Mashhadimalek, Path planning in the hippocampo-prefrontal cortex pathway: an adaptive model based receding horizon planner, Med. Hypotheses 68 (6) (2007) 1411-1415.

[159] F. Kendoul, Survey of advances in guidance, navigation, and control of unmanned rotorcraft systems, J. Field Robot. 29 (2) (2012) 315-378.

[160] V. Boquete, R. Garcia, R. Barea, M. Mazo, Neural control of the movements of a wheelchair, J. Intell. Robot. Syst. 25 (3) (1999) 213-226.

[161] C. Wang, A. Savkin, T. Nguyen, H.T. Nguyen, An algorithm for collision free navigation of an intelligent powered wheelchair in dynamic environments, in: 12th International Conference on Control, Automation, Robotics and Vision, Guangzhou, China, 2012, pp. 1571-1575.

[162] C. Wang, A. Matveev, A. Savkin, R. Clout, H. Nguyen, A real-time obstacle avoidance strategy for safe autonomous navigation of intelligent hospital beds in dynamic uncertain environments, in: Proceedings of Australasian Conference on Robotics and Automation, Sydney, Australia, 2013.

[163] A. Richards, J.P. How, Robust distributed model predictive control, Int. J. Control 80 (9) (2007) 1517-1531.

[164] D.Q. Mayne, S. Rakovic, Model predictive control of constrained piecewise affine discrete-time systems, Int. J. Robust Nonlinear Control 13 (3-4) (2003) 261-279.

[165] P.O. Scokaert, D.Q. Mayne, Min-max feedback model predictive control for constrained linear systems, IEEE Trans. Autom. Control 43 (8) (1998) 1136-1142.

[166] Y. Kuwata, A. Richards, T. Schouwenaars, J.P. How, Distributed robust receding horizon control for multivehicle guidance, IEEE Trans. Control Syst. Technol. 15 (4) (2007) 627-641.

[167] A. Richards, J.P. How, Robust stable model predictive control with constraint tightening, in: Proceedings of the American Control Conference, Minneapolis, MN, USA, 2006.

[168] W. Langson, I. Chryssochoos, S.V. Rakovic, D.Q. Mayne, Robust model predictive control using tubes, Automatica 40 (1) (2004) 125-133.

[169] D.Q. Mayne, E.C. Kerrigan, E.J. van Wyk, P. Falugi, Tube-based robust nonlinear model predictive control, Int. J. Robust Nonlinear Control 21 (11) (2011) 1341-1353.

[170] E. Scholte, M.E. Campbell, Robust nonlinear model predictive control with partial state information, IEEE Trans. Control Syst. Technol. 16 (4) (2008) 636-651.
[171] M. Defoort, J. Palos, A. Kokosy, T. Floquet, W. Perruquetti, Performance-based reactive navigation for non-holonomic mobile robots, Robotica 27 (2) (2009) 281-290.
[172] S.V. Rakovic, E.C. Kerrigan, K.I. Kouramas, D.Q. Mayne, Invariant approximations of the minimal robust positively invariant set, IEEE Trans. Autom. Control 50 (3) (2005) 406-410.
[173] L. Blackmore, M. Ono, B.C. Williams, Chance-constrained optimal path planning with obstacles, IEEE Trans. Robot. 27 (6) (2011) 1080-1094.
[174] N.E. Du Toit, J.W. Burdick, Robot motion planning in dynamic, uncertain environments, IEEE Trans. Robot. 28 (1) (2012) 101-115.
[175] L. Magni, D. Raimondo, F. Allgower, Nonlinear Model Predictive Control: Towards New Challenging Applications, Springer, Berlin, 2009.
[176] M. Rubagotti, D.M. Raimondo, A. Ferrara, L. Magni, Robust model predictive control with integral sliding mode in continuous-time sampled-data nonlinear systems, IEEE Trans. Autom. Control 56 (3) (2010) 556-570.
[177] M. Defoort, A. Kokosy, T. Floquet, W. Perruquetti, J. Palos, Motion planning for cooperative unicycle-type mobile robots with limited sensing ranges: a distributed receding horizon approach, Robot. Auton. Syst. 57 (11) (2009) 1094-1106.
[178] Y. Zhu, U. Ozguner, Constrained model predictive control for nonholonomic vehicle regulation problem, in: Proceedings of the 17th IFAC World Congress, Seoul, South Korea, 2008, pp. 9552-9557.
[179] A. Tahirovic, G. Magnani, PB/MPC navigation planner, in: Passivity-Based Model Predictive Control for Mobile Vehicle Motion Planning, Springer, Berlin, 2013, pp. 11-24.
[180] J.M. Park, D.W. Kim, Y.S. Yoon, H.J. Kim, K.S. Yi, Obstacle avoidance of autonomous vehicles based on model predictive control, Proc. Inst. Mech. Eng. D J. Autom. Eng. 223 (12) (2009) 1499-1516.
[181] R. Gonzalez, M. Fiacchini, J.L. Guzman, T. Alamo, F. Rodriguez, Robust tube-based predictive control for mobile robots in off-road conditions, Robot. Auton. Syst. 59 (10) (2011) 711-726.
[182] D.H. Shim, H. Chung, S.S. Sastry, Conflict-free navigation in unknown urban environments, IEEE Robot. Autom. Mag. 13 (3) (2006) 27-33.
[183] R. Diankov, J. Kuffner, Randomized statistical path planning, in: Proceedings of the 2007 IEEE/RSJ International Conference on Robots and Systems, San Diego, CA, USA, 2007.
[184] S. Karaman, E. Frazzoli, Sampling-based algorithms for optimal motion planning, Int. J. Robot. Res. 30 (7) (2011) 846-894.
[185] B.M. Sathyaraj, L.C. Jain, A. Finn, S. Drake, Multiple UAVs path planning algorithms: a comparative study, Fuzzy Optim. Decis. Making 7 (3) (2008) 257-267.
[186] S. Koenig, M. Likhachev, Fast replanning for navigation in unknown terrain, IEEE Trans. Robot. 21 (3) (2005) 354-363.
[187] S. Garrido, L. Moreno, D. Blanco, P. Jurewicz, Path planning for mobile robot navigation using Voronoi diagram and fast marching, Int. J. Robot. Autom. 2 (1) (2011) 42-64.
[188] V. Kallem, A.T. Komoroski, V. Kumar, Sequential composition for navigating a nonholonomic cart in the presence of obstacles, IEEE Trans. Robot. 27 (6) (2011) 1152-1159.
[189] M.P. Vitus, V. Pradeep, G.M. Hoffmann, S.L. Waslander, C.J. Tomlin, Tunnel-MILP: path planning with sequential convex polytopes, in: Proceedings of the AIAA Guidance, Navigation, and Control Conference, Minneapolis, MN, USA, 2008.

[190] I. Skrjanc, G. Klancar, Optimal cooperative collision avoidance between multiple robots based on Bernstein-Bezier curves, Robot. Auton. Syst. 58 (1) (2010) 1-9.

[191] B. Lau, C. Sprunk, W. Burgard, Kinodynamic motion planning for mobile robots using splines, in: Proceedings of the IEEE/RSJ International Conference on Intelligent Robots and Systems, St. Louis, MO, USA, 2009, pp. 2427-2433.

[192] Y. Wang, G.S. Chirikjian, A new potential field method for robot path planning, in: Proceedings of the IEEE International Conference on Robotics and Automation, vol. 2, San Francisco, CA, USA, 2000, pp. 977-982.

[193] S.S. Ge, Y.J. Cui, New potential functions for mobile robot path planning, IEEE Trans. Robot. Autom. 16 (5) (2000) 615-620.

[194] P. Abichandani, G. Ford, H.Y. Benson, M. Kam, Mathematical programming for Multi-Vehicle Motion Planning problems, in: Proceedings of the IEEE International Conference on Robotics and Automation, St Paul, MN, USA, 2012, pp. 3315-3322.

[195] B. Tovar, R. Murrieta-Cid, S.M. LaValle, Distance-optimal navigation in an unknown environment without sensing distances, IEEE Trans. Robot. 23 (3) (2007) 506-518.

[196] C. Zheng, L. Li, F. Xu, F. Sun, M. Ding, Evolutionary route planner for unmanned air vehicles, IEEE Trans. Robot. 21 (4) (2005) 609-620.

[197] E. Besada-Portas, L. de la Torre, J. de la Cruz, B. de Andres-Toro, Evolutionary trajectory planner for multiple UAVs in realistic scenarios, IEEE Trans. Robot. 26 (4) (2010) 619-634.

[198] H. Kurniawati, Y. Du, D. Hsu, W.S. Lee, Motion planning under uncertainty for robotic tasks with long time horizons, Int. J. Robot. Res. 30 (3) (2011) 308-323.

[199] J. De Schutter, T. De Laet, J. Rutgeerts, W. Decra, R. Smits, E. Aertbelian, K. Claes, H. Bruyninckx, Constraint-based task specification and estimation for sensor-based robot systems in the presence of geometric uncertainty, Int. J. Robot. Res. 26 (5) (2007) 433-455.

[200] G.M. Saggiani, B. Teodorani, Rotary wing UAV potential applications: an analytical study through a matrix method, Aircraft Eng. Aerospace Technol. Int. J. 76 (2004) 6-14.

[201] M. Caccia, R. Bono, G. Bruzzone, Variable-configuration UUVs for marine science applications, IEEE Robot. Autom. Mag. 6 (2) (1999) 22-32.

[202] K.B. Lee, M.H. Han, Lane-following method for high speed autonomous vehicles, Int. J. Autom. Technol. 9 (5) (2008) 607-613.

[203] A. Matveev, M. Hoy, A. Savkin, A method for reactive navigation of nonholonomic under-actuated robots in maze-like environments, Automatica 49 (5) (2013) 1268-1274.

[204] J.M. Toibero, F. Roberti, R. Carelli, Stable contour-following control of wheeled mobile robots, Robotica 27 (1) (2009) 1-12.

[205] A. Matveev, M. Hoy, A. Savkin, The problem of boundary following by a unicycle-like robot with rigidly mounted sensors, Robot. Auton. Syst. 61 (3) (2013) 312-327.

[206] J. Kim, F. Zhang, M. Egerstedt, Curve tracking control for autonomous vehicles with rigidly mounted range sensors, J. Intell. Robot. Syst. 56 (2) (2009) 177-197.

[207] T. Yata, L. Kleeman, S. Yuta, Wall following using angle information measured by a single ultrasonic transducer, in: Proceedings of the IEEE International Conference on Robotics and Automation, vol. 2, Leuven, Belgium, 1998, pp. 1590-1596.

[208] A. Bemporad, M.D. Marco, A. Tesi, Sonar-based wall-following control of mobile robots, ASME J. Dyn. Syst. Meas. Control 122 (2000) 226-230.

[209] R. Carelli, E.O. Freire, Corridor navigation and wall-following stable control for sonar-based mobile robots, Robot. Auton. Syst. 45 (12) (2003) 235-247.

[210] L. Huang, Wall-following control of an infrared sensors guided wheeled mobile robot, Int. J. Intell. Syst. Technol. Appl. 7 (1) (2009) 106-117.

[211] J.M. Yang, J.H. Kim, Sliding mode control for trajectory tracking of nonholonomic wheeled mobile robots, IEEE Trans. Robot. Autom. 15 (3) (1999) 578-587.
[212] R. Solea, U. Nunes, Trajectory planning and sliding-mode control based trajectory-tracking for cybercars, Integrated Comput. Aided Eng. 14 (1) (2007) 33-47.
[213] G.N. DeSouza, A.C. Kak, Vision for mobile robot navigation: a survey, IEEE Trans. Pattern Anal. Mach. Intell. 2 (24) (2002) 237-267.
[214] C. Samson, Control of chained systems: application to path-following and time-varying point stabilization of mobile robots, IEEE Trans. Autom. Control 40 (1995) 64-77.
[215] Y. Zhu, T. Zhang, J. Song, An improved wall following method for escaping from local minimum in artificial potential field based path planning, in: Proceedings of the 48th IEEE Conference on Decision and Control and the 28th Chinese Control Conference, Shanghai, China, 2009, pp. 6017-6022.
[216] F. Mastrogiovanni, A. Sgorbissa, R. Zaccaria, Robust navigation in an unknown environment with minimal sensing and representation, IEEE Trans. Syst. Man Cybern. B: Cybern. 39 (1) (2009) 212-229.
[217] S. Fazli, L. Kleeman, Wall following and obstacle avoidance results from a multi-DSP sonar ring on a mobile robot, in: IEEE International Conference Mechatronics and Automation, vol. 1, Niagara Falls, Canada, 2005, pp. 432-437.
[218] F. Zhang, E.W. Justh, P.S. Krishnaprasad, Boundary following using gyroscopic control, in: Proceedings of the 43rd IEEE Conference on Decision and Control, vol. 5, Paradise Island, Bahamas, 2004, pp. 5204-5209.
[219] M. Malisoff, F. Mazenc, F. Zhang, Input-to-state stability for curve tracking control: a constructive approach, in: Proceedings of the American Control Conference, San Francisco, CA, USA, 2011, pp. 1984-1989.
[220] F. Zhang, D.M. Fratantoni, D.A. Paley, J.M. Lund, N.E. Leonard, Control of coordinated patterns for ocean sampling, Int. J. Control 80 (7) (2007) 1186-1199.
[221] A. Matveev, H. Teimoori, A. Savkin, Method for tracking of environmental level sets by a unicycle-like vehicle, Automatica 48 (9) (2012) 2252-2261.
[222] D.W. Casbeer, D.B. Kingston, R.W. Beard, T.W. McLain, S.M. Li, R. Mehra, Cooperative forest fire surveillance using a team of small unmanned air vehicles, Int. J. Syst. Sci. 36 (6) (2006) 351-360.
[223] A. Joshi, T. Ashley, Y.R. Huang, A.L. Bertozzi, Experimental validation of cooperative environmental boundary tracking with on-board sensors, in: Proceedings of the American Control Conference, St. Louis, MO, USA, 2009, pp. 2630-2635.
[224] D. Marthaler, A.L. Bertozzi, Tracking environmental level sets with autonomous vehicles, in: S. Butenko, R. Murphey, P.M. Pardalos (Eds.), Recent Developments in Cooperative Control and Optimization, vol. 3, Kluwer Academic Publishers, Boston, 2003.
[225] S. Srinivasan, K. Ramamritham, P. Kulkarni, ACE, in the hole: adaptive contour estimation using collaborating mobile sensors, in: Proceedings of the International Conference on Information Processing in Sensor Networks, St. Louis, MO, USA, 2008, pp. 147-158.
[226] F. Zhang, N.E. Leonard, Cooperative control and filtering for cooperative exploration, IEEE Trans. Autom. Control 55 (3) (2010) 650-663.
[227] M. Hsieh, S. Loizou, V. Kumar, Stabilization of multiple robots on stable orbits via local sensing, in: Proceedings of the IEEE Conference on Robotics and Automation, Roma, Italy, 2007, pp. 2312-2317.
[228] C. Barat, M.J. Rendas, Benthic boundary tracking using a profiler sonar, in: Proceedings of the IEEE/RSJ International Conference on Intelligent Robots and Systems, vol. 1, Las Vegas, NV, USA, 2003, pp. 830-835.

[229] M. Kemp, A.L. Bertozzi, D. Marthaler, Multi-UUV perimeter surveillance, in: Proceedings of the IEEE/OES Autonomous Underwater Vehicles Conference, Sebasco, ME, USA, 2004, pp. 102-107.

[230] S.B. Andersson, Curve tracking for rapid imaging in AFM, IEEE Trans. Nanobiosci. 6 (4) (2007) 354-361.

[231] A. Bertozzi, M. Kemp, D. Marthaler, Determining environmental boundaries: asynchronous communication and physical scales, in: V. Kumar, N.E. Leonard, A.S. Morse (Eds.), Cooperative Control, Springer Verlag, Berlin, 2004, pp. 25-42.

[232] S. Susca, F. Bullo, S. Martinez, Monitoring environmental boundaries with a robotic sensor network, IEEE Trans. Control Syst. Technol. 16 (2) (2008) 288-296.

[233] J. Zhipu, A.L. Bertozzi, Environmental boundary tracking and estimation using multiple autonomous vehicles, in: Proceedings of the 46th IEEE Conference on Decision and Control, New Orleans, LU, USA, 2007, pp. 4918-4923.

[234] E. Burian, D. Yoeger, A. Bradley, H. Singh, Gradient search with autonomous underwater vehicle using scalar measurements, in: Proceedings of the IEEE Symposium on Underwater Vehicle Technology, Monterey, CA, 1996, pp. 86-98.

[235] C. Zhang, D. Arnold, N. Ghods, A. Siranosian, M. Krstic, Source seeking with nonholonomic unicycle without position measurement and with tuning of forward velocity, Syst. Control Lett. 56 (3) (2007) 245-252.

[236] J. Cochran, M. Krstic, Nonholonomic source seeking with tuning of angular velocity, IEEE Trans. Autom. Control 54 (4) (2009) 717-731.

[237] D.W. Casbeer, S.M. Li, R.W. Beard, T.W. McLain, R.K. Mehra, Forest fire monitoring using multiple small UAVs, in: Proceedings of the 2005 American Control Conference, vol. 5, Minneapolis, MA, USA, 2005, pp. 3530-3535.

[238] D. Baronov, J. Baillieul, Reactive exploration through following isolines in a potential field, in: Proceedings of the American Control Conference, New York, NY, USA, 2007, pp. 2141-2146.

[239] A. Sankaranarayanan, M. Vidyasagar, Path planning for moving a point object amidst unknown obstacles in a plane: a new algorithm and a general theory for algorithm development, in: Proceedings of the IEEE International Conference on Decision and Control, Brighton, UK, 1991, pp. 1111-1119.

[240] V.J. Lumelsky, S. Tiwari, An algorithm for maze searching with azimuth input, in: Proceedings of the IEEE Conference on Robotics and Automation, San Diego, CA, USA, 1991, pp. 111-116.

[241] A. Sankaranarayanan, M. Vidyasagar, A new algorithm for robot curve-following amidst unknown obstacles, and a generalization of maze-searching, in: Proceedings of the IEEE International Conference on Robotics and Automation, Nice, France, 1992, pp. 2487-2494.

[242] H. Noborio, A sufficient condition for designing a family of sensor based deadlock free planning algorithms, Adv. Robot. 7 (5) (1993) 413-433.

[243] H. Noborio, T. Yoshioka, An on-line and deadlock-free path planning algorithm based on world topology, in: Proceedings of the IEEE/RSJ Conference on Intelligent Robots and Systems, Yokohama, Japan, 1993, pp. 1425-1430.

[244] S.L. Laubach, J.W. Burdick, An autonomous sensor-based path-planner for planetary microrovers, in: Proceedings of the IEEE International Conference on Robotics and Automation, Detroit, MI, USA, 1999, pp. 347-354.

[245] S. Kim, J. Russel, K. Koo, Construction robot path-planning for earthwork operations, J. Comput. Civil Eng. 17 (2) (2003) 97-104.

[246] E. Magid, E. Rivlin, CautiousBug: a competitive algorithm for sensor-based robot navigation, in: Proceedings of the IEEE/RSJ International Conference on Intelligent Robots and Systems, Sendai, Japan, 2004, pp. 2757-2762.
[247] R. Langer, L. Coelho, G. Oliveira, K-Bug, a new bug approach for mobile robot's path planning, in: Proceedings of the IEEE International Conference on Control Applications, Singapore, 2007, pp. 403-408.
[248] V.J. Lumelsky, A.A. Stepanov, Dynamic path planning for a mobile automaton with limited information on the environment, IEEE Trans. Autom. Control 31 (11) (1986) 1058-1063.
[249] V.J. Lumelsky, A.A. Stepanov, Path-planning strategies for a point mobile automaton moving amidst unknown obstacles of arbitrary shape, Algorithmica 2 (1) (1987) 403-430.
[250] J. Ng, An analysis of mobile robot navigation algorithms in unknown environments, Ph.D. Thesis, the University of Western Australia, Perth, Australia, 2010.
[251] C. Ordonez, E.G. Collins Jr., M.F. Selekwa, D.D. Dunlap, The virtual wall approach to limit cycle avoidance for unmanned ground vehicles, Robot. Auton. Syst. 56 (8) (2008) 645-657.
[252] A.M. Shkel, V.J. Lumelsky, Incorporating body dynamics into sensor-based motion planning: the maximum turn strategy, IEEE Trans. Robot. Autom. 13 (6) (1997) 873-880.
[253] S.S. Ge, X. Lai, A.A. Mamun, Boundary following and globally convergent path planning using instant goals, IEEE Trans. Syst. Man Cybern B: Cybern. 35 (2) (2005) 240-254.
[254] S.S. Ge, X. Lai, A.A. Mamun, Sensor-based path planning for nonholonomic mobile robots subject to dynamic constraints, Robot. Auton. Syst. 55 (7) (2007) 513-526.
[255] M. Hoy, A method of boundary following by a wheeled mobile robot based on sampled range information, J. Intell. Robot. Syst. 72 (3-4) (2013) 463-482.
[256] J.C. Alvarez, A. Shkel, V.J. Lumelsky, Accounting for mobile robot dynamics in sensor-based motion planning: experimental results, in: Proceedings of the IEEE International Conference on Robotics and Automation, vol. 3, Lueven, Belgium, 1998, pp. 2205-2210.
[257] P. Ogren, N.E. Leonard, A convergent dynamic window approach to obstacle avoidance, IEEE Trans. Robot. 21 (2) (2005) 188-195.
[258] P. Ogren, N.E. Leonard, A tractable convergent dynamic window approach to obstacle avoidance, in: Proceedings of IEEE International Conference on Intelligent Robots and Systems, Lausanne, Switzerland, 2002.
[259] J.L. Fernandez, R. Sanz, J.A. Benayas, A.R. Diaguez, Improving collision avoidance for mobile robots in partially known environments: the beam curvature method, Robot. Auton. Syst. 46 (4) (2004) 205-219.
[260] C. Schlegel, Fast local obstacle avoidance under kinematic and dynamic constraints for a mobile robot, in: Proceedings of the 1998 IEEE/RSJ International Conference on Intelligent Robots and Systems, vol. 1, Victoria, Canada, 1998, pp. 594-599.
[261] C. Stachniss, W. Burgard, An integrated approach to goal-directed obstacle avoidance under dynamic constraints for dynamic environments, in: Proceedings of the 2002 IEEE/RSJ International Conference on Intelligent Robots and Systems, vol. 1, Lausanne, Switzerland, 2002, pp. 508-513.
[262] J.-L. Blanco, J. Gonzalez, J.-A. Fernandez-Madrigal, Extending obstacle avoidance methods through multiple parameter-space transformations, Auton. Robots 24 (1) (2008) 29-48.

[263] N.Y. Ko, R.G. Simmons, The lane-curvature method for local obstacle avoidance, in: Proceedings of the 1998 IEEE/RSJ International Conference on Intelligent Robots and Systems, vol. 3, Victoria, Canada, 1998, pp. 1615-1621.

[264] M. Hoy, A. Matveev, M. Garratt, A. Savkin, Collision free navigation of an autonomous unmanned helicopter in unknown urban environments: sliding mode and MPC approaches, Robotica 30 (4) (2012) 537-550.

[265] S. Horn, K. Janschek, A set-based global dynamic window algorithm for robust and safe mobile robot path planning, Proceedings of the 41st International Symposium on Robotics and the 6th German Conference on Robotics (2010).

[266] P. Krishnamurthy, F. Khorrami, GODZILA: a low-resource algorithm for path planning in unknown environments, J. Intell. Robot. Syst. 48 (3) (2007) 357-373.

[267] A. Brooks, T. Kaupp, A. Makarenko, Randomised MPC-based motion-planning for mobile robot obstacle avoidance, in: Proceedings of the 2009 IEEE International Conference on Robotics and Automation, Kobe, Japan, 2009, pp. 3962-3967.

[268] K. Yang, S. Gan, S. Sukkarieh, An efficient path planning and control algorithm for RUAV's in unknown and cluttered environments, J. Intell. Robot. Syst. 57 (1) (2010) 101-122.

[269] H. Yu, R. Sharma, R.W. Beard, C.N. Taylor, Observability-based local path planning and collision avoidance for Micro Air Vehicles using bearing-only measurements, in: Proceedings of the American Control Conference, San Francisco, CA, USA, 2011, pp. 4649-4654.

[270] J. Minguez, L. Montano, The ego-kinodynamic space: collision avoidance for any shape mobile robots with kinematic and dynamic constraints, in: Proceedings of the 2003 IEEE/RSJ International Conference on Intelligent Robots and Systems, vol. 1, Las Vegas, NV, USA, 2003, pp. 637-643.

[271] J. Minguez, L. Montano, Extending collision avoidance methods to consider the vehicle shape, kinematics, and dynamics of a mobile robot, IEEE Trans. Robot. 25 (2) (2009) 367-381.

[272] G. Manor, E. Rimon, High-speed navigation of a uniformly braking mobile robot using position-velocity configuration space, in: Proceedings of the IEEE International Conference on Robotics and Automation, St Paul, MN, USA, 2012, pp. 193-199.

[273] L. Valbuena, H. Tanner, Hybrid potential field based control of differential drive mobile robots, J. Intell. Robot. Syst. 68 (3-4) (2012) 307-322.

[274] J. Ren, K.A. McIsaac, R.V. Patel, Modified Newton's method applied to potential field-based navigation for nonholonomic robots in dynamic environments, Robotica 26 (1) (2008) 117-127.

[275] A. Masoud, A harmonic potential approach for simultaneous planning and control of a generic UAV platform, J. Intell. Robot. Syst. 65 (1) (2012) 153-173.

[276] A. Masoud, Kinodynamic motion planning, IEEE Robot. Autom. Mag. 17 (1) (2010) 85-99.

[277] D.H. Kim, S. Shin, New repulsive potential functions with angle distributions for local path planning, Adv. Robot. 20 (1) (2006) 25-48.

[278] S. Cifuentes, J.M. Giron-Sierra, J. Jimenez, Robot navigation based on discrimination of artificial fields: application to single robots, Adv. Robot. 26 (5-6) (2012) 605-626.

[279] J. Ren, K.A. McIsaac, R.V. Patel, Modified Newton's method applied to potential field-based navigation for mobile robots, IEEE Trans. Robot. 22 (2) (2006) 384-391.

[280] D.E. Chang, S.C. Shadden, J.E. Marsden, R. Olfati-Saber, Collision avoidance for multiple agent systems, in: Proceedings of the 42nd IEEE Conference on Decision and Control, vol. 1, Maui, HI, USA, 2003, pp. 539-543.

[281] M. Galicki, Collision-free control of an omni-directional vehicle, Robot. Auton. Syst. 57 (9) (2009) 889-900.

[282] S.G. Loizou, K.J. Kyriakopoulos, Navigation of multiple kinematically constrained robots, IEEE Trans. Robot. 24 (1) (2008) 221-231.

[283] L. Lapierre, R. Zapata, A guaranteed obstacle avoidance guidance system, Auton. Robots 32 (3) (2012) 177-187.

[284] I. Ulrich, J. Borenstein, VFH*: local obstacle avoidance with look-ahead verification, in: Proceedings of the IEEE International Conference on Robotics and Automation, vol. 3, San Francisco, CA, USA, 2000, pp. 2505-2511.

[285] R. Sharma, J. Saunders, R. Beard, Reactive path planning for micro air vehicles using bearing-only measurements, J. Intell. Robot. Syst. 65 (1) (2012) 409-416.

[286] A.C. Victorino, P. Rives, J.-J. Borrelly, Safe navigation for indoor mobile robots. Part I: a sensor-based navigation framework, Int. J. Robot. Res. 22 (12) (2003) 1005-1118.

[287] L. Montesano, J. Minguez, L. Montano, Modeling dynamic scenarios for local sensor-based motion planning, Auton. Robots 25 (3) (2008) 231-251.

[288] T. Gecks, D. Henrich, Sensor-based online planning of time-optimized paths in dynamic environments, in: T. Krager, F.M. Wahl (Eds.), Advances in Robotics Research, Springer, Berlin/Heidelberg, 2009, pp. 53-63.

[289] S. Petti, T. Fraichard, Partial motion planning framework for reactive planning within dynamic environments, in: Proceedings of the AAAI International Conference on Advanced Robotics, Barcelona, Spain, 2005.

[290] D. Althoff, J. Kuffner, D. Wollherr, M. Buss, Safety assessment of robot trajectories for navigation in uncertain and dynamic environments, Auton. Robots 32 (3) (2012) 285-302.

[291] E.A. Sisbot, L.F. Marin-Urias, R. Alami, T. Simeon, A human aware mobile robot motion planner, IEEE Trans. Robot. 23 (5) (2007) 874-883.

[292] T. Ohki, K. Nagatani, K. Yoshida, Local path planner for mobile robot in dynamic environment based on distance time transform method, Adv. Robot. 26 (14) (2012) 1623-1647.

[293] A. Foka, P. Trahanias, Probabilistic autonomous robot navigation in dynamic environments with human motion prediction, Int. J. Soc. Robot. 2 (1) (2010) 79-94.

[294] B.D. Ziebart, N. Ratliff, G. Gallagher, C. Mertz, K. Peterson, J.A. Bagnell, M. Hebert, A.K. Dey, S. Srinivasa, Planning-based prediction for pedestrians, in: Proceedings of the IEEE/RSJ International Conference on Intelligent Robots and Systems, St. Louis, MO, USA, 2009, pp. 3931-3936.

[295] Z. Shiller, O. Gal, E. Rimon, Safe navigation in dynamic environments, in: Robot Design, Dynamics and Control, CISM Courses and Lectures, vol. 524, Springer Vienna, Vienna, Austria, 2010, pp. 225-232.

[296] P. Fiorini, Z. Shiller, Time optimal trajectory planning in dynamic environments, in: Proceedings of the IEEE International Conference on Robotics and Automation, Minneapolis, MN, USA, 1996, pp. 1553-1558.

[297] D.H. Shim, S. Sastry, An evasive maneuvering algorithm for UAVs in see-and-avoid situations, in: Proceedings of the American Control Conference, Minneapolis, MN, USA, 2007, pp. 3886-3891.

[298] X. Yang, L. Alvarez, T. Bruggemann, A 3D collision avoidance strategy for UAVs in a non-cooperative environment, J. Intell. Robot. Syst. 70 (1-4) (2012) 315-327.

[299] S. Bouraine, T. Fraichard, H. Salhi, Provably safe navigation for mobile robots with limited field-of-views in dynamic environments, Auton. Robots 32 (3) (2012) 267-283.

[300] F. Belkhouche, Reactive path planning in a dynamic environment, IEEE Trans. Robot. 25 (4) (2009) 902-911.

[301] J. van den Berg, M. Overmars, Planning time-minimal safe paths amidst unpredictably moving obstacles, Int. J. Robot. Res. 27 (11-12) (2008) 1274-1294.
[302] W. Chung, S. Kim, M. Choi, J. Choi, H. Kim, C.-B. Moon, J.-B. Song, Safe navigation of a mobile robot considering visibility of environment, IEEE Trans. Ind. Electron. 56 (10) (2009) 3941-3950.
[303] S.S. Ge, Y.J. Cui, Dynamic motion planning for mobile robots using potential field method, Auton. Robots 13 (3) (2002) 207-222.
[304] C. Wang, A. Matveev, A. Savkin, T. Nguyen, H. Nguyen, A collision avoidance strategy for safe autonomous navigation of an intelligent atomic-powered wheelchair in dynamic uncertain environments with moving obstacles, in: Proceedings of the European Control Conference, Zurich, Switzerland, 2013.
[305] A. Savkin, Coordinated collective motion of groups of autonomous mobile robots: analysis of Vicsek's model, IEEE Trans. Autom. Control 49 (6) (2004) 981-983.
[306] A. Savkin, H. Teimoori, Decentralized navigation of groups of wheeled mobile robots with limited communication, IEEE Trans. Robot. 26 (10) (2010) 1099-1104.
[307] A. Matveev, A. Savkin, The problem of state estimation via asynchronous communication channels with irregular transmission times, IEEE Trans. Autom. Control 48 (4) (2003) 670-676.
[308] A. Savkin, Analysis and synthesis of networked control systems: topological entropy, observability, robustness, and optimal control, Automatica 42 (1) (2006) 51-62.
[309] A. Savkin, T. Cheng, Detectability and output feedback stabilizability of nonlinear networked control systems, IEEE Trans. Autom. Control 52 (4) (2007) 730-735.
[310] A. Matveev, A. Savkin, Estimation and Control Over Communication Networks, Birkhauser, Boston, 2009.
[311] B. Wang, Coverage Control in Sensor Networks, Springer, London, NY, 2010.
[312] A. Nayak, I. Stojmenovic, Wireless Sensor and Actuator Networks: Algorithms and Protocols for Scalable Coordination and Data Communication, John Wiley & Sons, Hoboken, NJ, 2010.
[313] A. Savkin, T. Cheng, Z. Xi, F. Javed, A. Matveev, H. Nguyen, Decentralized Coverage Control Problems for Mobile Robotic Sensor and Actuator Networks, IEEE Press-Wiley, New York, 2015.
[314] A. Savkin, F. Javed, A. Matveev, Optimal distributed blanket coverage self-deployment of mobile wireless sensor networks, IEEE Commun. Lett. 16 (6) (2012) 949-951.
[315] T. Cheng, A. Savkin, Decentralized control of mobile sensor networks for asymptotically optimal blanket coverage between two boundaries, IEEE Trans. Ind. Inform. 9 (1) (2013) 365-376.
[316] T. Cheng, A. Savkin, Decentralized control for mobile robotic sensor network self-deployment: barrier and sweep coverage problems, Robotica 29 (2) (2011) 283-294.
[317] T. Cheng, A. Savkin, Self-deployment of mobile robotic sensor networks for multilevel barrier coverage, Robotica 30 (4) (2012) 661-669.
[318] T. Cheng, A. Savkin, F. Javed, Decentralized control of a group of mobile robots for deployment in sweep coverage, Robot. Auton. Syst. 59 (7-8) (2011) 497-507.
[319] W. Li, C.G. Cassandras, A cooperative receding horizon controller for multivehicle uncertain environments, IEEE Trans. Autom. Control 51 (2) (2006) 242-257.
[320] J. Peng, S. Akella, Coordinating multiple robots with kinodynamic constraints along specified paths, Int. J. Robot. Res. 24 (4) (2005) 295-310.
[321] R. Cui, B. Gao, J. Guo, Pareto-optimal coordination of multiple robots with safety guarantees, Auton. Robots 32 (3) (2012) 189-205.

[322] A. Fujimori, M. Teramoto, P.N. Nikiforuk, M.M. Gupta, Cooperative collision avoidance between multiple mobile robots, J. Robot. Syst. 17 (7) (2000) 347-363.
[323] T. Tarnopolskaya, N. Fulton, H. Maurer, Synthesis of optimal bang-bang control for cooperative collision avoidance for aircraft (ships) with unequal linear speeds, J. Optim. Theory Appl. 155 (1) (2012) 115-144.
[324] J.K. Kuchar, L.C. Yang, A review of conflict detection and resolution modeling methods, IEEE Trans. Intell. Transport. Syst. 1 (4) (2000) 179-189.
[325] S. Mastellone, D.M. Stipanovic, C.R. Graunke, K.A. Intlekofer, M.W. Spong, Formation control and collision avoidance for multi-agent non-holonomic systems: theory and experiments, Int. J. Robot. Res. 27 (1) (2008) 107-126.
[326] E.G. Hernandez-Martinez, E. Aranda-Bricaire, Convergence and collision avoidance in formation control: a survey of the artificial potential functions approach, in: Multi-Agent Systems-Modeling, Control, Programming, Simulations and Applications, InTech, 2011.
[327] D.M. Stipanovic, P.F. Hokayem, M.W. Spong, D.D. Siljak, Cooperative avoidance control for multiagent systems, J. Dyn. Syst. Meas. Control 129 (5) (2007) 699-707.
[328] A. Widyotriatmo, K. Hong, Navigation function-based control of multiple wheeled vehicles, IEEE Trans. Ind. Electron. 58 (5) (2011) 1896-1906.
[329] D.V. Dimarogonas, S.G. Loizou, K.J. Kyriakopoulos, M.M. Zavlanos, A feedback stabilization and collision avoidance scheme for multiple independent non-point agents, Automatica 42 (2) (2006) 229-243.
[330] H.G. Tanner, A. Boddu, Multiagent navigation functions revisited, IEEE Trans. Robot. 28 (6) (2012) 1346-1359.
[331] D.V. Dimarogonas, K.J. Kyriakopoulos, Connectedness preserving distributed swarm aggregation for multiple kinematic robots, IEEE Trans. Robot. 24 (5) (2008) 1213-1223.
[332] D.V. Dimarogonas, K.J. Kyriakopoulos, Decentralized navigation functions for multiple robotic agents with limited sensing capabilities, J. Intell. Robot. Syst. 48 (3) (2007) 411-433.
[333] G.P. Roussos, G. Chaloulos, K.J. Kyriakopoulos, J. Lygeros, Control of multiple non-holonomic air vehicles under wind uncertainty using model predictive control and decentralized navigation functions, in: Proceedings of the 47th IEEE Conference on Decision and Control, Cancun, Mexico, 2008, pp. 1225-1230.
[334] G.P. Roussos, D.V. Dimarogonas, K.J. Kyriakopoulos, 3D navigation and collision avoidance for nonholonomic aircraft-like vehicles, Int. J. Adapt. Control Signal Process. 24 (10) (2010) 900-920.
[335] G.M. Hoffmann, C.J. Tomlin, Decentralized cooperative collision avoidance for acceleration constrained vehicles, in: Proceedings of the 47th IEEE Conference on Decision and Control, Cancun, Mexico, 2008, pp. 4357-4363.
[336] E.J. Rodriguez-Seda, M.W. Spong, Guaranteed safe motion of multiple Lagrangian systems with limited actuation, in: Proceedings of the 51st IEEE Conference on Decision and Control, Maui, HI, USA, 2012, pp. 2773-2780.
[337] S.W. Ekanayake, P.N. Pathirana, Formations of robotic swarm: an artificial force based approach, Int. J. Adv. Robot. Syst. 6 (1) (2009) 7-24.
[338] F. Fahimi, C. Nataraj, H. Ashrafiuon, Real-time obstacle avoidance for multiple mobile robots, Robotica 27 (2) (2009) 189-198.
[339] J. van den Berg, S.J. Guy, M. Lin, D. Manocha, Reciprocal n-body collision avoidance, in: Proceedings of the International Symposium on Robotics Research, Lucerne, Switzerland, 2009.
[340] J. Snape, J. van den Berg, S.J. Guy, D. Manocha, The hybrid reciprocal velocity obstacle, IEEE Trans. Robot. 27 (4) (2011) 696-706.

[341] J. Snape, J. van den Berg, S.J. Guy, D. Manocha, Independent navigation of multiple mobile robots with hybrid reciprocal velocity obstacles, in: Proceedings of the IEEE/RSJ International Conference on Intelligent Robots and Systems, St. Louis, MO, USA, 2009, pp. 5917-5922.

[342] J. van den Berg, J. Snape, S.J. Guy, D. Manocha, Reciprocal collision avoidance with acceleration-velocity obstacles, in: Proceedings of the IEEE International Conference on Robotics and Automation, Shanghai, China, 2011, pp. 3475-3482.

[343] A.T. Rashid, A.A. Ali, M. Frasca, L. Fortuna, Multi-robot collision-free navigation based on reciprocal orientation, Robot. Auton. Syst. 60 (10) (2012) 1221-1230.

[344] J. Alonso-Mora, A. Breitenmoser, P. Beardsley, R. Siegwart, Reciprocal collision avoidance for multiple car-like robots, in: Proceedings of the IEEE International Conference on Robotics and Automation, St Paul, MN, USA, 2012, pp. 360-366.

[345] Y. Abe, M. Yoshiki, Collision avoidance method for multiple autonomous mobile agents by implicit cooperation, in: Proceedings of the IEEE/RSJ International Conference on Intelligent Robots and Systems, vol. 3, Maui, HI, USA, 2001, pp. 1207-1212.

[346] E. Lalish, K.A. Morgansen, T. Tsukamaki, Decentralized reactive collision avoidance for multiple unicycle-type vehicles, in: Proceedings of the American Control Conference, Seattle, WA, USA, 2008, pp. 5055-5061.

[347] E. Lalish, K.A. Morgansen, Distributed reactive collision avoidance, Auton. Robots 32 (3) (2012) 207-226.

[348] C. Belta, A. Bicchi, M. Egerstedt, E. Frazzoli, E. Klavins, G.J. Pappas, Symbolic planning and control of robot motion [grand challenges of robotics], IEEE Robotics Automation Magazine 14 (1) (2007) 61-70.

[349] R.W. Ghrist, D.E. Koditschek, Safe cooperative robot dynamics on graphs, SIAM J. Control Optim. 40 (5) (2002) 1556-1575.

[350] S.A. Reveliotis, E. Roszkowska, Conflict resolution in free-ranging multivehicle systems: a resource allocation paradigm, IEEE Trans. Robot. 27 (2) (2011) 283-296.

[351] T. Nishi, M. Ando, M. Konishi, Distributed route planning for multiple mobile robots using an augmented Lagrangian decomposition and coordination technique, IEEE Trans. Robot. 21 (6) (2005) 1191-1200.

[352] R.V. Cowlagi, P. Tsiotras, Hierarchical motion planning with dynamical feasibility guarantees for mobile robotic vehicles, IEEE Trans. Robot. 28 (2) (2012) 379-395.

[353] A. Krontiris, K.E. Bekris, Using minimal communication to improve decentralized conflict resolution for non-holonomic vehicles, in: Proceedings of the 2011 IEEE/RSJ International Conference on Intelligent Robots and Systems, San Francisco, CA, USA, 2011, pp. 3235-3240.

[354] L. Pallottino, V.G. Scordio, A. Bicchi, E. Frazzoli, Decentralized cooperative policy for conflict resolution in multivehicle systems, IEEE Trans. Robot. 23 (6) (2007) 1170-1183.

[355] M. Farrokhsiar, H. Najjaran, An unscented model predictive control approach to the formation control of nonholonomic mobile robots, in: Proceedings of the IEEE International Conference on Robotics and Automation, St Paul, MN, USA, 2012, pp. 1576-1582.

[356] Y. Kuwata, J.P. How, Cooperative distributed robust trajectory optimization using receding horizon MILP, IEEE Trans. Control Syst. Technol. 19 (2) (2011) 423-431.

[357] M.C. Hoy, Deadlock resolution for navigation of wheeled robots in continuous state-space, in: Proceedings of the International Conference on Automation, Robotics, Control and Vision, Guangzhou, China, 2012.

[358] A. Bemporad, D. Barcelli, Decentralized model predictive control, Lect. Notes Control Inform. Sci. 406 (2010) 149-178.
[359] J. Shin, H.J. Kim, Nonlinear model predictive formation flight, IEEE Trans. Syst. Man Cybern. A Syst. Hum. 39 (5) (2009) 1116-1125.
[360] R.L. Raffard, C.J. Tomlin, S.P. Boyd, Distributed optimization for cooperative agents: application to formation flight, in: Proceedings of the 43rd IEEE Conference on Decision and Control, vol. 3, Paradise Island, Bahamas, 2004, pp. 2453-2459.
[361] Y. Wakasa, M. Arakawa, K. Tanaka, T. Akashi, Decentralized model predictive control via dual decomposition, in: Proceedings of the 47th IEEE Conference on Decision and Control, Cancun, Mexico, 2008, pp. 381-386.
[362] T.H. Summers, J. Lygeros, Distributed model predictive consensus via the alternating direction method of multipliers, in: Proceedings of the Allerton Conference on Communication, Control, and Computing, Monticello, IL, USA, 2012.
[363] V. Desaraju, J. How, Decentralized path planning for multi-agent teams with complex constraints, Auton. Robots 32 (4) (2012) 385-403.
[364] E. Siva, J.M. Maciejowski, Robust multiplexed MPC for distributed multi-agent systems, in: Proceedings of the 18th IFAC World Congress, Milano, Italy, 2011.
[365] S. Adinandra, E. Schreurs, H. Nijmeijer, A practical model predictive control for a group of unicycle mobile robots, in: Proceedings of the 4th IFAC Conference on Nonlinear Model Predictive Control, vol. 4, Leeuwenhorst, Netherlands, 2012, pp. 472-477.
[366] F. Augugliaro, A.P. Schoellig, R. D'Andrea, Generation of collision-free trajectories for a quadrocopter fleet: a sequential convex programming approach, in: Proceedings of the IEEE/RSJ International Conference on Intelligent Robots and Systems, Algarve, Portugal, 2012, pp. 1917-1922.
[367] D. Morgan, S.-J. Chung, F.Y. Hadaegh, Decentralized model predictive control of swarms of spacecraft using sequential convex programming, in: Proceedings of the AAS/AIAA Space Flight Mechanics Conference, Kauai, HI, USA, 2013.
[368] Z. Weihua, T.H. Go, Robust decentralized formation flight control, Int. J. Aerospace Eng. (2011).
[369] K.E. Bekris, K.I. Tsianos, L.E. Kavraki, Safe and distributed kinodynamic replanning for vehicular networks, Mobile Netw. Appl. 14 (3) (2009) 292-308.
[370] K.E. Bekris, D.K. Grady, M. Moll, L.E. Kavraki, Safe distributed motion coordination for second-order systems with different planning cycles, Int. J. Robot. Res. 31 (2) (2012) 129-150.
[371] M. Vaccarini, S. Longhi, Formation control of marine vehicles via real-time networked decentralized MPC, in: Proceedings of the 17th Mediterranean Conference on Control and Automation, Thessaloniki, Greece, 2009, pp. 428-433.
[372] W. Peng, D. Baocang, Z. Tao, Distributed receding horizon control for nonholonomic multi-vehicle system with collision avoidance, in: Proceedings of the 31st Chinese Control Conference, Hefei, China, 2012, pp. 6327-6332.
[373] E. Gratli, T. Johansen, Path planning for UAVs under communication constraints using SPLAT! and MILP, J. Intell. Robot. Syst. 65 (1) (2012) 265-282.
[374] A. Grancharova, E.I. Gratli, T.A. Johansen, Distributed MPC-based path planning for UAVs under radio communication path loss constraints, in: Proceedings of the IFAC Conference on Embedded Systems, Computational Intelligence and Telematics in Control, Wurzburg, Germany, 2012, pp. 254-259.
[375] S. Sundar, Z. Shiller, Optimal obstacle avoidance based on the Hamilton-Jacobi-Bellman equation, IEEE Trans. Robot. Autom. 13 (2) (1997) 305-310.

[376] A. Savkin, M. Hoy, Tangent graph based navigation of a non-holonomic mobile robot in cluttered environments, in: 19th IEEE Mediterranean Conference on Control and Automation, MED 2011, Corfu, Greece, 2011, pp. 309-314.
[377] D.J. Struik, Lectures on Classical Differential Geometry, Courier Dover Publications, Mineola, 1988.
[378] C. Grinstead, J. Snell, Introduction to Probability, American Mathematical Society, New York, NY, 1997.
[379] C. Bolkcom, Homeland Security: Unmanned Aerial Vehicles and Border Surveillance, Congressional Research Service, 2008.
[380] D. Caveney, R. Sengupta, Architecture and application abstractions for multi-agent collaboration projects, in: 44th IEEE Conference on Decision and Control, Seville, Spain, 2005, pp. 3572-3577.
[381] J. Lee, R. Huang, A. Vaughn, X. Xiao, J. Hedrick, Strategies of path-planning for a UAV to track a ground vehicle, in: Proceedings of the AINS Conference, Berkeley, California, USA, 2003.
[382] S. Ge, X. Lai, A. Mamun, Boundary following and globally convergent path planning using instant goals, IEEE Trans. Syst. Man Cybern. 35 (2) (2005) 1-15.
[383] M. Hoy, A. Savkin, A method for border patrolling navigation of a mobile robot, in: Proceedings of the IEEE International Conference on Control and Automation, Santiago, USA, 2011, pp. 130-135.
[384] A. Matveev, M. Hoy, A. Savkin, A method for reactive navigation of nonholonomic robots in the presence of obstacles, in: Proceedings of the 18th IFAC World Congress, Milano, Italy, 2011, pp. 11894-11899.
[385] A. Matveev, M. Hoy, A. Savkin, Boundary tracking by a wheeled robot with rigidly mounted sensors, in: The 12 International Conference on Control, Automation, Robotics and Vision, Guangzhou, China, 2012, pp. 148-153.
[386] S. Sternberg, Lectures on Differential Geometry, AMS Chelsea Publishing, New York, NY, 1999.
[387] Mobile Robot Simulator Mobotsim 1.0, URL http://www.mobotsoft.com/, 2011.
[388] K. Ahnert, M. Abel, Numerical differentiation of experimental data: local versus global methods, Comput. Phys. Commun. 177 (2007) 764-774.
[389] L.K. Vasiljevic, H.K. Khalil, Error bounds in differentiation of noisy signals by high-gain observers, Syst. Control Lett. 57 (2008) 856-862.
[390] V.I. Arnold, The Theory of Singularities and its Applications, second ed., The Press Syndicate of the University of Cambridge, Cambridge, 1993.
[391] C. Wang, A. Matveev, A. Savkin, R. Cloutz, H. Nguyen, A semi-autonomous motorized mobile hospital bed for safe transportation of head injury patients in dynamic hospital environments without bed switching, Robotica (2015), doi:10.1017/S0263574714002641.
[392] H. Choset, K. Lynch, S. Hutchinson, G. Kantor, W.Burgard, L. Kavraki, S. Thrun, Principles of RobotMotion: Theory, Algorithms and Implementations, MIT Press, Englewood Cliffs, NY, 2005.
[393] D.W. Thompson, On Growth and Form, Cambridge University Press, UK, 1966.
[394] B. Fajen, Steering toward a goal by equalizing taus, J. Exp. Psychol. Hum. Percept. Perform. 27 (4) (2001) 953-968.
[395] H. Abelson, A.A. diSessa, Turtle Geometry, MIT Press, Cambridge, 1980.
[396] E. Kreyszig, Differential Geometry, Dover Publications, Inc., NY, 1991.
[397] A. Matveev, M. Hoy, A. Savkin, Mixed nonlinear-sliding mode control of an unmanned farm tractor in the presence of sliding, in: Proceedings of the 11th International

Conference on Control, Automation, Robotics and Vision, Singapore, 2010, pp. 927-932.

[398] R. Millman, G. Parker, Elements of Differential Geometry, Prentice-Hall, Englewood Cliffs, NJ, 1977.

[399] A. Savkin, C. Wang, A reactive algorithm for safe navigation of a wheeled mobile robot among moving obstacles, in: 2012 IEEE International Conference on Control Applications, Dubrovnik, Croatia, 2012, pp. 1567-1571.

[400] C. Wang, A. Savkin, R. Clout, H. Nguyen, An intelligent robotic hospital bed for safe transportation of critical neurosurgery patients along crowded hospital corridors, IEEE Trans. Neural Syst. Rehabil. Eng. (2014), in press.

[401] M. Finlayson, T. van Denend, Experiencing the loss of mobility: perspectives of older adults with MS, Disabil. Rehabil. 25 (20) (2003) 1168-1180.

[402] A. Schultz, Mobility impairment in the elderly: challenges for biomechanics research, J. Biomech. 25 (1992) 519-528.

[403] R. Simpson, Smart wheelchairs: a literature review, J. Rehabil. Res. Dev. 42 (4) (2005) 423-436.

[404] R. Simpson, E. LoPresti, R. Cooper, How many people would benefit from a smart wheelchair? J. Rehabil. Res. Dev. 45 (1) (2008) 53-72.

[405] N.T. Nguyen, H.T. Nguyen, S. Su, Advanced robust tracking control of a powered wheelchair system, in: Proceedings of 29th Annual International Conference, Lyon, France, 2007, pp. 4767-4770.

[406] S. Gulati, B. Kuipers, High performance control for graceful motion of an intelligent wheelchair, in: IEEE International Conference on Robotics and Automation, Pasadena, California, USA, 2008, pp. 3932-3938.

[407] C. Kuo, H. Yeh, C. Wu, Development of autonomous navigation robotic wheelchairs using programmable system-on-chip based distributed computing architecture, in: IEEE International Conference on Systems, Man and Cybernetics, Montreal, Canada, 2007, pp. 2939-2944.

[408] T. Nguyen, S. Su, H. Nguyen, Robust neuro-sliding mode multivariable control strategy for powered wheelchairs, Neural Syst. Rehabil. Eng. 19 (1) (2011) 105-111.

[409] T. Luhandjula, K. Djouani, Y. Hamam, B. van Wyk, Q. Williams, A hand-based visual intent recognition algorithm for wheelchair motion, in: IEEE Conference on Human System Interactions, Rzeszow, Poland, 2010, pp. 749-756.

[410] S. Reahman, B. Raytchev, I. Yoda, L. Liu, Vibrotactile rendering of head gestures for controlling electric wheelchair, in: IEEE International Conference on Systems Man and Cybernetics, New Jersey, USA, 2009, pp. 413-417.

[411] T. Saitoh, N. Takahashi, R. Konishi, Oral motion controlled intelligent wheelchair, in: Annual Conference on Instrumentation, Control, Information Technology and System Integration, Fukuoka, Japan, 2007, pp. 341-346.

[412] L. Fehr, W. Langbein, S. Skaar, Adequacy of power wheelchair control interfaces for persons with severe disabilities: a clinical survey, J. Rehabil. Res. Dev. 37 (3) (2000) 353-360.

[413] L. Gillman, G. Leslie, T. Williams, K. Fawcett, R. Bell, V. McGibbon, Adverse events experienced while transferring the critically ill patient from the emergency department to the intensive care unit, Emerg. Med. J. 23 (11) (2006) 858-861.

[414] U. Beckmann, D. Gillies, S. Berenholtz, A. Wu, P. Pronovost, Incidents relating to the intra-hospital transfer of critically ill patients, Intensive Care Med. 30 (8) (2004) 1579-1585.

[415] B. Fanara, C. Manzon, O. Barbot, T. Desmettre, G. Capellier, Recommendations for the intra-hospital transport of critically ill patients, Crit. Care 14 (3) (2010) 1-10.
[416] J. Papson, K. Russell, D. Taylor, Unexpected events during the intrahospital transport of critically ill patients, Acad. Emerg. Med. 14 (6) (2007) 574-577.
[417] K.E. Bekris, A.A. Argyros, L.E. Kavraki, Angle-based methods for mobile robot navigation: reaching the entire plane, Proceedings of IEEE International Conference on Robotics and Automation (ICRA), New Orleans, USA, 2004, pp. 2373-2378.
[418] S.G. Loizou, V. Kumar, Biologically inspired bearing-only navigation and tracking, Proceedings of the 46th IEEE Conference on Decision and Control, New Orleans, USA, 2007, pp. 1386-1391.
[419] A. Gadre, D.J. Stilwell, Toward underwater navigation based on range measurements from a single location, Proceedings of IEEE International Conference on Robotics and Automation, New Orleans, USA, 2004, pp. 4472-4477.
[420] P. Pathirana, N. Bulusu, A. Savkin, S. Jha, Node localization using mobile robots in delay-tolerant sensor networks, IEEE Trans. Mobile Comput. 4 (4) (2005) 285-296.
[421] A. Matveev, C. Wang, A. Savkin, Reactive navigation of nonholonomic mobile robots in dynamic uncertain environments with moving and deforming obstacles, in: Proceedings of the 31st Chinese Control Conference, Hefei, China, 2012.
[422] E.H. Lockwood, A Book of Curves, Cambridge University Press, UK, 1961.
[423] M. Srinivasan, Honeybees as a model for the study of visually guided flight, navigation, and biologically inspired robotics, Physiol. Rev. 91 (2011) 413-460.
[424] J. Spencer, Continuum Mechanics, Dover Publications, NY, 2004.
[425] J. Thorpe, Elementary Topics in Differential Geometry, Springer, NY, 1979.
[426] J. Leigh, The Essentials of Nonlinear Control Theory, Peter Peregrinus, London, 1983.
[427] P. Raja, S. Pugazhenthi, Path planning for a mobile robot in dynamic environments, Int. J. Phys. Sci. 6 (2011) 4721-4731.
[428] Y. Zhu, T. Zhang, J. Song, X. Li, M. Nakamura, A new method for mobile robots to avoid collision with moving obstacles, Artif. Life Robot. 16 (2012) 507-510.
[429] M. Moussaïd, D. Helbing, G. Theraulaz, How simple rules determine pedestrian behavior and crowd disasters, Proc. Natl. Acad. Sci., 108 (2011) 6884-6888.
[430] A. Savkin, C. Wang, A method for collision free assisted navigation of semi-autonomous vehicles in dynamic environments with moving and static obstacles, in: Proceedings of the 10th Asian Control Conference, Kota Kinabalu, Malaysia, 2015.
[431] A. Savkin, C. Wang, Seeking a path through the crowd: robot navigation in unknown dynamic environments with moving obstacles based on an integrated environment representation, Robot. Auton. Syst. 62 (10) (2014) 1568-1580.
[432] A. Savkin, C. Wang, A simple real-time algorithm for safe navigation of a non-holonomic robot in complex unknown environments with moving obstacles, in: The 13th European Control Conference, Strasbourg, France, 2014, pp. 1875-1880.
[433] A. Matveev, M. Hoy, A. Savkin, A globally converging algorithm for reactive robot navigation among moving and deforming obstacles, Automatica 54 (2015) 292-304.
[434] A. Matveev, M. Hoy, A. Savkin, Real-time kinematic navigation of a mobile robot among moving obstacles with guaranteed global convergence, in: IEEE International Conference on Robotics and Biomimetics, Bali, Indonesia, 2014.
[435] F. Clarke, Optimization and Nonsmooth Analysis, Wiley, NY, 1983.
[436] A. Zakhar'eva, A. Matveev, M. Hoy, A. Savkin, Distributed control of multiple non-holonomic robots with sector vision and range-only measurements for target capturing with collision avoidance, Robotica 33 (2) (2015) 385-412.

[437] S. Reveliotis, E. Roszkowska, Conflict resolution in multi-vehicle systems: a resource allocation paradigm, in: Proceedings of the IEEE International Conference on Automation Science and Engineering, Washington, DC, USA, 2008, pp. 115-121.
[438] Q. Li, A.C. Adriaansen, J.T. Udding, A.Y. Pogromsky, Design and control of automated guided vehicle systems: a case study, in: Proceedings of the IFAC World Congress, Milano, Italy, 2011.
[439] K. Sun, P. Ning, C. Wang, Secure and resilient clock synchronization in wireless sensor networks, IEEE J. Select. Areas in Commun. 24 (2) (2006) 395-408.
[440] D. Zhou, T. Lai, An accurate and scalable clock synchronization protocol for IEEE 802.11-based multihop ad hoc networks, IEEE Trans. Parallel Distrib. Syst. 18 (12) (2007) 1797-1808.
[441] M. Hoy, A. Matveev, A. Savkin, Robust cooperative navigation of multiple wheeled robots in unknown cluttered environments, in: 19th Mediterranean Conference on Control and Automation, Corfu, Greece, 2011, pp. 650-655.
[442] R. Gonzalez, M. Fiacchini, J. Guzman, T. Alamo, Robust tube-based MPC for constrained mobile robots under slip conditions, in: Proceedings of the 48th IEEE Conference on Decision and Control, Shanghai, China, 2009, pp. 5985-5990.
[443] A.L. Wijesinha, Y. Song, M. Krishman, V. Mathur, J. Ahn, V. Shyamasundar, Throughput measurement for UDP traffic in an IEEE 802.11g WLAN, in: Proceedings of the Sixth International Conference on Software Engineering, Artificial Intelligence, Networking and Parallel/Distributed Computing and First ACIS International Workshop on Self-Assembling Wireless Networks, Towson, MD, USA, 2005.
[444] S.M. LaValle, J.J. Kuffner, Randomized kinodynamic planning, Int. J. Robot. Res. 20 (2001) 378-400.
[445] M. Goff, Characterization of Error Growth in Wheel-Odometry Based on Dead Reckoning in a Differential Drive Robot, tech. rep., University of Florida, 2012.

Index

Note: Page numbers followed by *f* indicate figures and *t* indicate tables.

Printed in the United States
By Bookmasters